ROUTLEDGE LIBRARY EDITIONS: ENERGY ECONOMICS

Volume 17

WORLD NUCLEAR POWER

WORLD NUCLEAR POWER

PETER R. MOUNFIELD

LONDON AND NEW YORK

First published in 1991 by Routledge

This edition first published in 2018
by Routledge
2 Park Square, Milton Park, Abingdon, Oxon OX14 4RN

and by Routledge
711 Third Avenue, New York, NY 10017

Routledge is an imprint of the Taylor & Francis Group, an informa business

British Library Cataloguing in Publication Data
A catalogue record for this book is available from the British Library

ISBN: 978-1-138-10476-1 (Set)
ISBN: 978-1-315-14526-6 (Set) (ebk)
ISBN: 978-1-138-30620-2 (Volume 17) (hbk)
ISBN: 978-1-138-30640-0 (Volume 17) (pbk)
ISBN: 978-1-315-14165-7 (Volume 17) (ebk)

Publisher's Note
The publisher has gone to great lengths to ensure the quality of this reprint but points out that some imperfections in the original copies may be apparent.

World Nuclear Power

Peter R. Mounfield

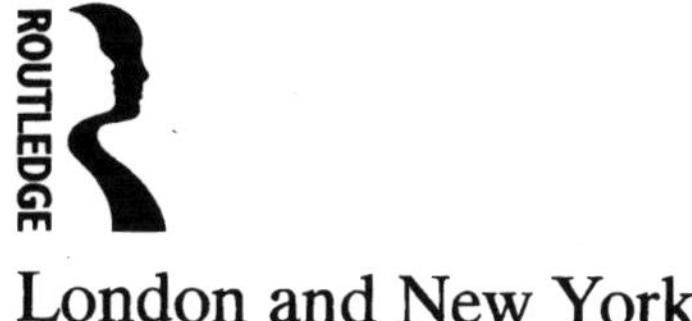

London and New York

First published 1991
by Routledge
11 New Fetter Lane, London EC4P 4EE

Simultaneously published in the USA and Canada
by Routledge
a division of Routledge, Chapman and Hall, Inc.
29 West 35th Street, New York, NY 10001

Typeset by Leaper and Gard Ltd, Bristol
Printed and bound in Great Britain by
Biddles Ltd, Guildford and King's Lynn

British Library Cataloguing in Publication Data
Mounfield, Peter R., *1935–*
World nuclear power.
1. Nuclear power
I. Title
621.48

ISBN 0-415-00463-2

Library of Congress Cataloging-in-Publication Data
Mounfield, Peter R., 1935–
World nuclear power / Peter R. Mounfield.
p. cm.
Includes bibliographical references and index.
ISBN 0-415-00463-2
1. Nuclear power plants. 2. Nuclear power plants—Safety measures. 3. Nuclear fuels. 4. Spent reactor fuels. I. Title.
TK1078.M68 1990 90-45131
333.792′4—dc20 CIP

To Pat, John
and David

Contents

Figures

Tables

Acknowledgements

I am indebted to a host of organizations and individuals who have assisted me over the past twenty-four years in the collection and interpretation of material used in writing this book. Special thanks are due to the people who have read and commented on the manuscript; E.M. Rawstron (Queen Mary College, University of London), J.P. Cole (University of Nottingham), Christine Sharrock (Omega Scientific), who have read it all; E.A. Davis (University of Leicester) and P.M.S. Jones (AEA Technology and University of Surrey) for reading chapters 8 and 9 respectively. All errors of fact and interpretation are, of course, my responsibility, as are the views expressed. Mrs Katherine Moore expertly prepared most of the maps and diagrams and Mrs Ruth Pollington the remainder. The word processing of the manuscript was efficiently undertaken by Miss Susan Haywood, Mrs Barbara Hughes, and Mrs Carole Deacon. All five of these ladies are or have been members of the technical and secretarial staff of the Department of Geography, University of Leicester.

I am grateful to the William Waldorf Astor Foundation whose generosity enabled me to conduct interviews at nuclear power plants across the USA, and to the Research Board of the University of Leicester for contributing to the cost of similar visits in Europe. The University of Cincinnati provided a welcoming base early on for the North American research and Jesus College, Oxford, was a tranquil location for writing the final draft when I was Visiting Senior Research Fellow for the Hilary Term, 1989.

The following have kindly allowed diagrams from their publications to be redrawn and used here: AEA Technology (Figures 0.1, 7.8, 8.7, 9.2, 10.14, 13.1), The American Society of Mechanical Engineers (Figures 10.1 and 10.2), The British Petroleum Co. plc (Figure 2.5), *The Economist* (Figure 8.10), Edinburgh University Press (Figure 2.8), *The Financial Times* (Figures 3.1 and 9.4), J.H. Fremlin (Figure 7.9), The Geographical Association (Figures 7.1 and 7.4), Gower Publishing Co. (Figure 1.1), Guardian News Service Ltd (Figure 12.13), Her Majesty's Stationery Office (Figures 1.2 and 12.5), A.D. Horrill (Institute of Terrestrial Ecology, Figure 8.6b), Institution of Mechanical Engineers (Figure 11.7), International Atomic Energy Agency (Figures 10.5, 10.8, 10.9), International Institute for Strategic Studies (Figure 7.10), MAFF (Figure 12.9), Maxwell Macmillan (Figures 8.5 and 10.13 from *Health Physics*, Figures 9.3 and 10.11 from *Energy*, Figure 10.6 from *Progress in Nuclear Energy*), *Nuclear Safety* (Oak Ridge National Laboratory, Figures 3.4, 8.9, 10.14, 11.2, 11.5), OECD/NEA (Figures 0.2, 2.7, 3.3, 7.6, 7.7, 8.8, 9.1, 12.1, 12.8, 12.10), The Open University (Figure 7.1), Oxford University Press (Figure 8.11), Dr Barry Smith (Meteorological Office, Figure 8.6a), Superintendent of Documents, United States Government Printing Office (Figures 1.9, 1.10, 1.11, 8.3, 11.6a), United Kingdom Nirex Ltd (Figure 12.6).

Last, but certainly not least, my grateful thanks are due to my wife for her patience and for the unstinting loan, when needed, of her own professional academic skills.

Introduction

Feelings run high in the debate about nuclear power. On the one hand there are fervent environmental condemnations, and on the other, stern technocratic justifications. If it were possible to write any book that was truly value free it would hardly be on this topic. Nevertheless, a conscious effort has been made in this volume not to take sides, to be impartial in selecting material and to be restrained in judgement. The objective is not to convert the reader to either an anti-nuclear or a pro-nuclear view. The aim is to inform, and, in so doing, to provide a volume that is more a textbook than a piece of polemic.

The reader must judge whether or not that aim has been achieved, but in attempting to be dispassionate the author has some advantages. He has not worked for a nuclear power company or electrical utility. He has not fallen in love with nuclear power technology and, to the best of his knowledge, does not own shares in companies who profit from it. Neither does he belong to Greenpeace, Friends of the Earth, the Union of Concerned Scientists or any other pressure group committed to an anti-nuclear power position. He has never been on an anti-nuclear power rally and does not display either 'I Like Nukes' or 'Nuclear Power – No Thanks' stickers on his car windscreen! As a professional geographer he has, however, thought and read about, written on, and lectured to many audiences on nuclear power topics since 1958. His view of future world energy supplies accords exactly with that of a paragraph in the Introduction to Edward Teller's *Energy from Heaven and Earth* (1979):

> No single prescription exists for a solution to the energy problem. Energy conservation is not enough. Petroleum is not enough. Coal is not enough. Nuclear energy is not enough. Solar energy and geothermal energy are not enough. New ideas and developments will not be enough. Only the proper combination of all these will suffice.

The organization of the book follows the general sequence of the nuclear fuel cycle. This is a term not entirely precise but widely used in the nuclear industry to cover all aspects of the production of electricity by means of nuclear power, from the mining of uranium to the disposal of radioactive waste. The three main stages of the cycle are illustrated in Figure 0.1. Each contains a particular subset of operations.

(a) The *front-end* of the fuel cycle includes all stages from the mining of uranium ore up to and including the delivery of manufactured fuel element assemblies to reactor sites.

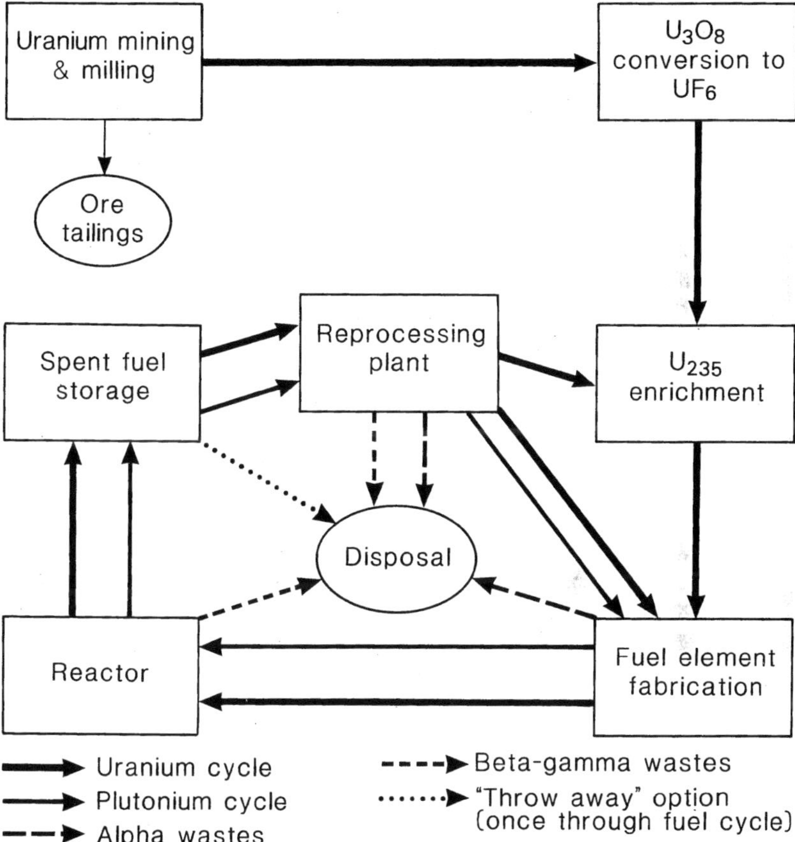

Figure 0.1 The nuclear fuel cycle. The diagram shows the main stages in the nuclear fuel cycle and the main radioactive waste streams requiring disposal. Variants of the cycle are (i) the 'once-through' cycle, in which plutonium is left in the spent fuel elements which are not reprocessed and which become the main radioactive waste stream, (ii) the reprocessing cycle, in which plutonium is stored pending development of fast reactors, (iii) thermal recycling, in which plutonium has some value as replacement fuel burned in thermal reactors, at most 16 per cent of the uranium fuel charge, and (iv) the fast reactor cycle.
Source: UK Atomic Energy Authority

(b) The second major stage is the *in-reactor* phase – the use of nuclear fuel in reactors in which fission energy is used to produce heat. The storage of fresh fuel at reactor sites and the short-term storage of spent fuel after removal from the reactor core are normally also included in this stage.

(c) Third comes the so-called *back-end* of the nuclear fuel cycle, which begins with the transport of spent fuel away from reactor storage ponds and ends with the final disposal of wastes from reprocessing, or of the spent fuel itself.

The detailed sequence of back-end events depends upon national and power company policies. In the *reprocessing cycle* the used fuel assemblies are carried to reprocessing plants where unused uranium and plutonium are separated for further use from the wastes created during the fission process. This cycle is shown in Figure 0.1. Also shown is the *once-through fuel cycle* in which the spent fuel is not reprocessed but disposed of in stores following appropriate treatment. The fissile material is used only once rather than repeatedly as in the reprocessing cycle.

In the reprocessing cycle the uranium recovered from used fuel elements is returned to fuel manufacturing establishments and, after treatment, is sent to the enrichment plant to have its original uranium-235 content restored. On return to the fuel manufacturing establishment the uranium is used to make more fuel elements. Storage of spent fuel or wastes may occur before and after treatment but separate transport costs are not incurred if consecutive phases in the fuel cycle are located at the same site.

The flow over time of nuclear fuel costs for reactors using enriched fuel is shown in Figure 0.2. It is clear that there are high front-end costs, especially for fuel enrichment. It should also be noted that some of the back-end costs indicated by this diagram are notional. Very little of the

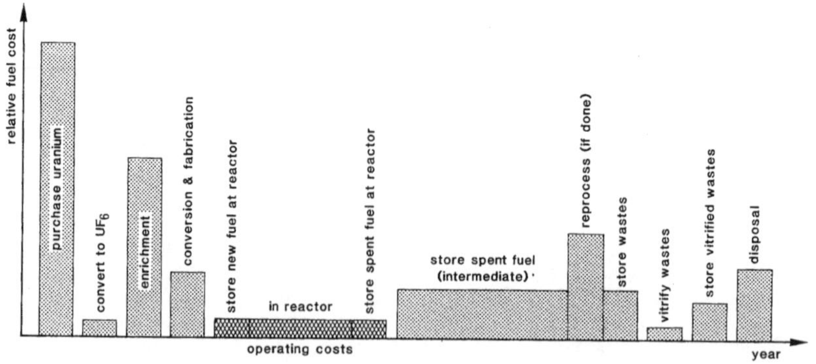

Figure 0.2 Flow of nuclear fuel cycle costs (typical fuel batch for reactors using enriched fuel)
Source: OECD 1983a: 16

high-level waste produced to date has been vitrified and there are no long-term depositories yet in existence. It remains a possibility that these back-end costs may ultimately match or exceed those at present incurred at the front-end.

From this brief account, and from Figures 0.1 and 0.2, it will be appreciated that it is possible to generalize the nuclear fuel cycle as a sequence of closely related operations, as follows:

- uranium mining and milling;
- conversion of ore concentrate to standards required for nuclear fuel;
- fuel enrichment (not present in all variants);
- manufacture of fuel elements;
- fuel burn-up in the reactors;
- interim storage of spent fuel, usually at reactor sites;
- transport of spent fuel for reprocessing or disposal;
- fuel reprocessing (not present in all variants);
- conditioning of radioactive wastes (radwastes);
- interim storage or disposal of radwaste;
- final disposal of radwaste.

This sequence is broadly reflected in the arrangement of chapters in this volume. However, it is a *technical* sequence which, if strictly adhered to, would exclude important areas of geographical concern, political awareness and public debate. It does not provide a specific framework for answering a number of questions. Where are the world's nuclear power stations and what principles have guided their location and siting? Which countries have developed nuclear power as a commercial energy source, and which have not? How competitive is nuclear power with other means of producing electricity in particular energy markets? What environmental issues are involved in the disposal of radwaste? What kinds of terrestrial environments may offer the safest conditions for building long-term radwaste depositories? What are the hazards involved in the use of nuclear power and are the benefits sufficiently large to make the level of risk worthwhile?

A purely technological view of the nuclear fuel cycle therefore requires redefinition. It also has to include environmental, economic, social, political and place-related issues. The redefinition suggested here incorporates the following.

(a) *Technological areas*: the mining and milling of uranium, conversion and enrichment, fuel element manufacture, the transport of used fuel and its reprocessing (Chapter 7) and strategies for interim and final disposal of radwaste (Chapter 12).

(b) *Techno-economic areas*: the technical character and economic potential of different types of reactor, their particular impact in normal operation (Chapter 1) and fuel cycle cost comparison. Each country, and each electrical power utility, has its own particular mix of resources for electricity production offering a variety of spatially variable policy alternatives (Chapter 9).

(c) *Socio-technical areas*: these include attitudes to nuclear power and the susceptibility of the nuclear power industry to public opinion and pressure groups (Chapter 13), safety-related issues in siting and safety policy (Chapters 8 and 10) and the significance of exceptional environmental events on the location of power reactors and radwaste repositories (Chapter 11).

(d) *Politico-technical areas*: these include the impact of international agreements on the spread throughout the world of nuclear power technology (Chapter 2), the possibility of terrorist activity in relation to nuclear materials and reactors and national standards of regulatory control.

These areas of concern are not mutually exclusive, and matters of technology, economics, geography, politics and public policy all interact in each. For example, a decision by any utility to build a nuclear power station may be prompted by economic considerations backed by technical assessments, but in its building and ultimate operation the power station will have a social and environmental impact. The decision whether or not to build it thus becomes a political act in a wider framework of public policy, public opinion and pressure group politics. The text is written in recognition of this fact for, in modern economic geography, political and social issues loom as large and importantly as they do in the real world. In the real world, too, nuclear power is very unevenly distributed. Some countries have nuclear power stations in numbers; others have none at all.

1 Preliminary considerations: reactor types and characteristics

Nature is niggardly in its provision of uranium isotopes capable of use in today's reactors. One particular uranium isotope, uranium-235 (U-235), is the only element found in nature which is fissile. When its nucleus is hit by a subatomic particle, a neutron, it splits into two fragments. In this process of fission a large amount of energy is released and more neutrons are produced; these can be used to split further U-235 nuclei and set up a chain reaction. A nuclear reactor is a means of controlling the chain reaction and converting the energy produced into heat and subsequently into electricity.

The neutrons emitted by the fissioned nuclei travel at speeds which are much too fast for them to be captured by other U-235 nuclei and so they must be slowed down, or moderated. Figure 1.1 illustrates the nature of a chain reaction using U-235. A single neutron, after being slowed down by a moderator, causes a U-235 nucleus to fission, an event which results in the production of two fission products and two fission neutrons. These neutrons, after being moderated, go on to cause the fission of two more U-235 nuclei and the creation of four more fission products and four more neutrons. And so the reaction proceeds.

Uranium ore extracted from the earth normally contains about 0.72 per cent U-235 and 99.27 per cent U-238, which is not fissile. The proportion of fissile isotopes can be increased by fuel enrichment. There are also two man-made fissile elements, plutonium (Pu-239) which is produced from U-238, and U-233 which is produced from thorium (Th-233). The three elements U-235, U-233 and Pu-239 are the basic fuels for nuclear reactors. U-235 and U-233 undergo fission most readily in thermal reactors, i.e. reactors in which the neutrons are slowed down to thermal equilibrium with their surrounding matter. Pu-239 is the most suitable fuel for fast reactors.

THERMAL AND FAST REACTORS

Figure 1.2 is a flow diagram that helps understanding of the difference between thermal and fast reactors. In thermal reactors the neutrons are

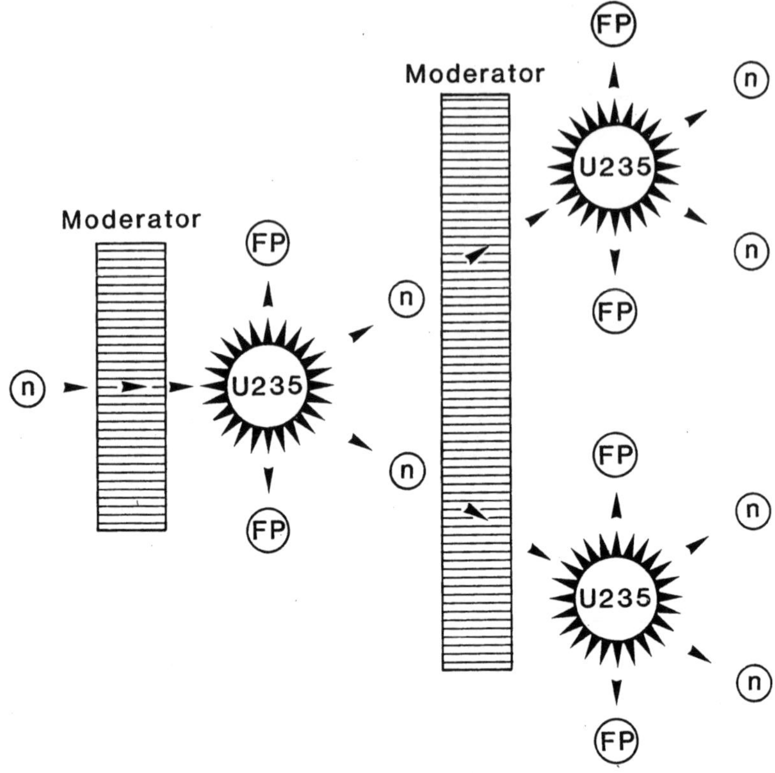

Figure 1.1 Creation of a nuclear chain reaction using U-235
Source: Adapted from Cole 1988: 69

slowed down to a state of equilibrium with their surrounding matter. This is achieved by using a moderator, such as carbon atoms in graphite or hydrogen atoms in water. Fast reactors, whether liquid-cooled or gas-cooled, contain no moderator. This is the essential distinction which gives rise to the broad classification of reactor types into fast and thermal.

In both types, neutrons released by splitting a fissile nucleus start off with a kinetic energy of around one million electron volts (1 MeV), corresponding to a speed of some 4,400 km/s. In the fast reactor, matters are so arranged that these neutrons initiate further fissions before being appreciably slowed down. In the thermal reactor, however, the chain reaction is maintained by neutrons which are slowed down by repeated collisions with the moderator nuclei until they are in thermal equilibrium with the latter; hence the name, thermal reactor.

Only three types of moderator material are in common use for thermal reactors; (i) the light or common isotope of hydrogen in the form of ordinary water (H_2O), (ii) deuterium, the heavy isotope of hydrogen in the

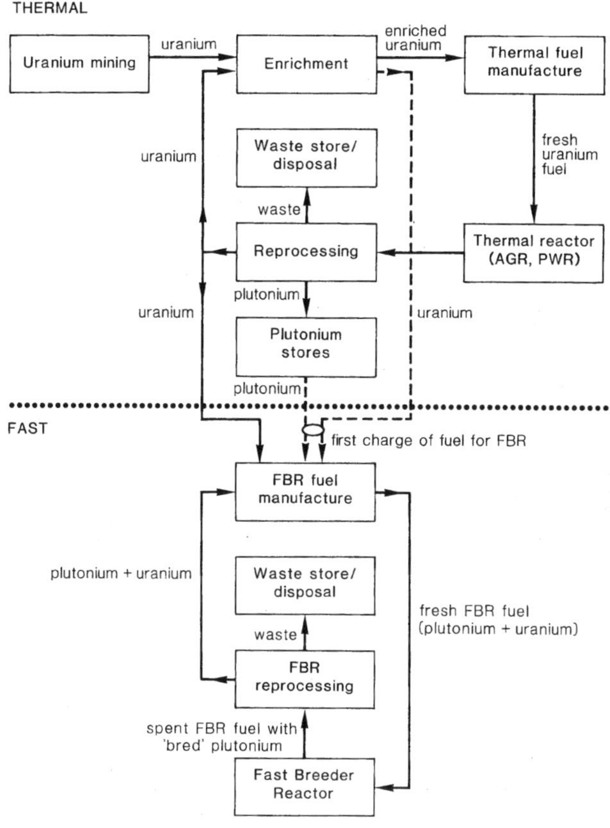

Figure 1.2 Sequence of events in the thermal and fast reactor nuclear fuel cycles
Source: House of Commons 1986: ixxvii

form of heavy water (D_2O), and (iii) carbon, in the form of graphite.

Control rods, made from a neutron-absorbing material such as boron or cadmium, regulate the heat output of a reactor, and they can be used to control an accidental increase in the rate of the fission chain reaction, known as a 'nuclear excursion'. The heat is also controlled by a coolant, either gas or liquid, which passes over the nuclear fuel in the reactor core and then goes to a turbine generator either directly or via a heat exchanger (Figure 1.3).

The coolant is contained within a pressure vessel because the most efficient transfer of heat occurs under pressure. It should be emphasized that the power level at which a reactor can operate safely is limited by the capacity of its cooling system, i.e. the rate at which the primary coolant can carry away the heat generated in the reactor core. It is vitally important for

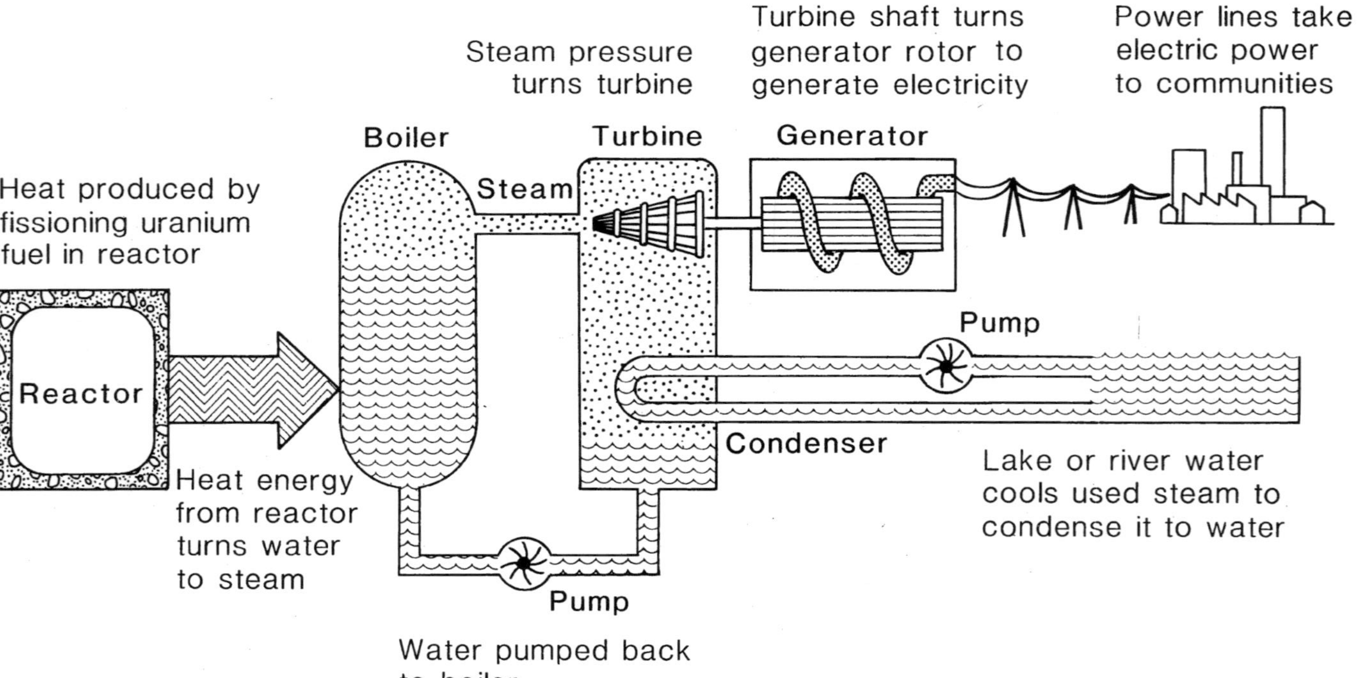

Figure 1.3 Major components involved in the production of electricity at nuclear power stations. In all nuclear power stations the design of the coolant circuit has an important influence on the cost of the plant and the economics of generation

the cooling system to work efficiently and safely. If heat were to be generated at a rate faster than it is carried away by the coolant, the fuel would overheat and might melt or vaporize.

A biological shield, usually made of concrete, is placed around the whole assemblage of fuel, moderator, control rods and coolant as a final engineered defence against any accidental escape of dangerous radiation.

TYPES OF THERMAL REACTORS IN COMMERCIAL USE FOR POWER PRODUCTION

Many reactor types have been developed to the power-producing stage but only a few combinations of coolant and moderator have been used for the commercial production of electricity.

Light water reactors

The world's major concentrations of nuclear power stations are dominated by reactors using water as coolant and moderator (Figures 1.4 and 1.5). There are two types: the pressurized water reactor (PWR), in which water at high pressure is heated within the reactor core and passes to heat exchangers where it boils water in a secondary circuit, and the boiling water reactor (BWR), in which the water in the core is allowed to boil and steam from the coolant drives the turbines directly. Both types use enriched uranium dioxide fuel clad with zirconium alloy (Zircaloy).

PWRs were first developed by the USA and the USSR as a compact reactor for marine propulsion, especially in submarines. Today they dominate the world pattern of nuclear reactors used for electricity generation both numerically and in the proportion of electricity produced. They account for half the completed reactors and 70 per cent of those under construction.

Mainly because of very high power density in the core the pressure vessel of a PWR is relatively small, with thick steel walls (Figure 1.6). The coolant system becomes highly radioactive in use and adequate shielding of personnel is an important requirement. An operating characteristic of the PWR is that it cannot be fuelled on load; before refuelling the reactor has to be shut down and brought to a cold condition, and the vessel head removed. A typical 1000 MWe plant requires a fuel mass loading of about 93,000 kg of uranium, and it discharges about 270 kg of plutonium annually (Marshall 1979). In refuelling, about a third of the spent fuel assemblies in the core are removed and the other assemblies in the core are moved to different positions so that new fuel can be loaded. Maintenance operations take place at the same time and the reactor may be out of use for up to ninety days. Therefore PWRs are designed to operate for at least a year between refuelling operations and the fuel elements must be capable of use over three years before discharge.

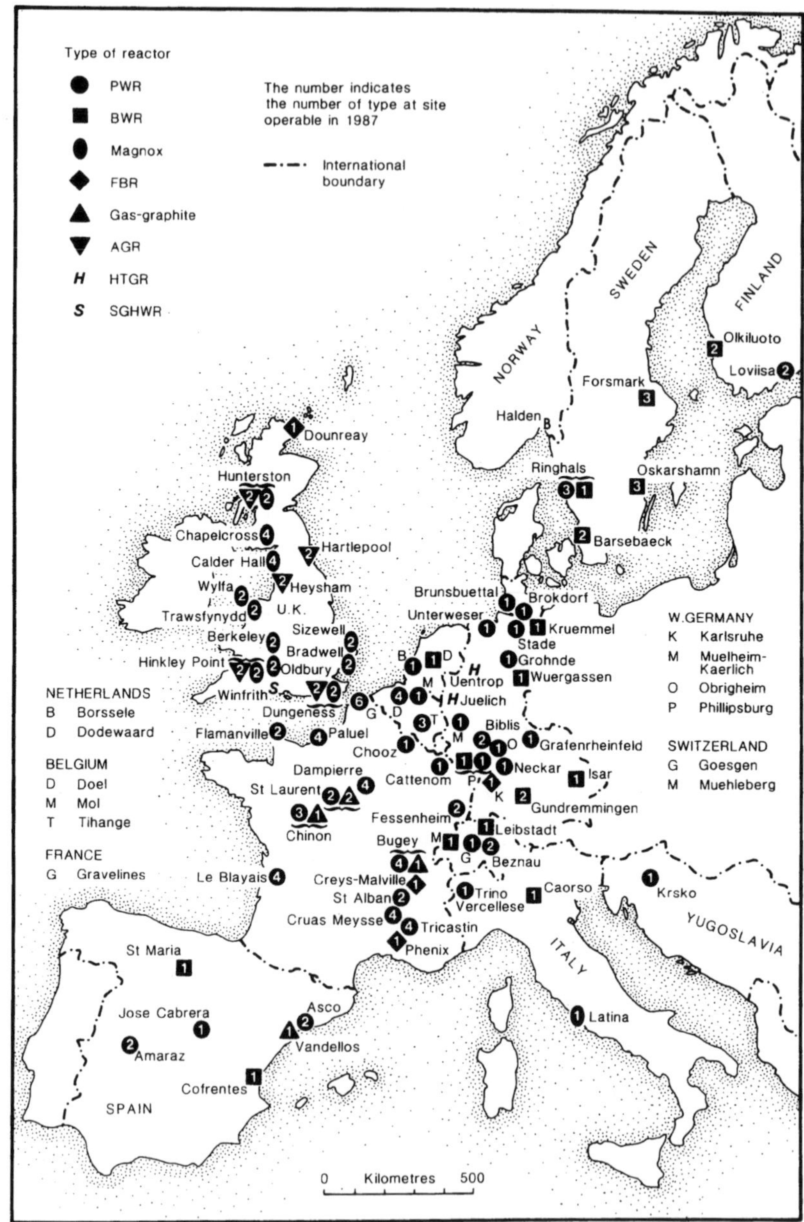

Figure 1.4 Reactor types at nuclear power stations in Western Europe in 1987

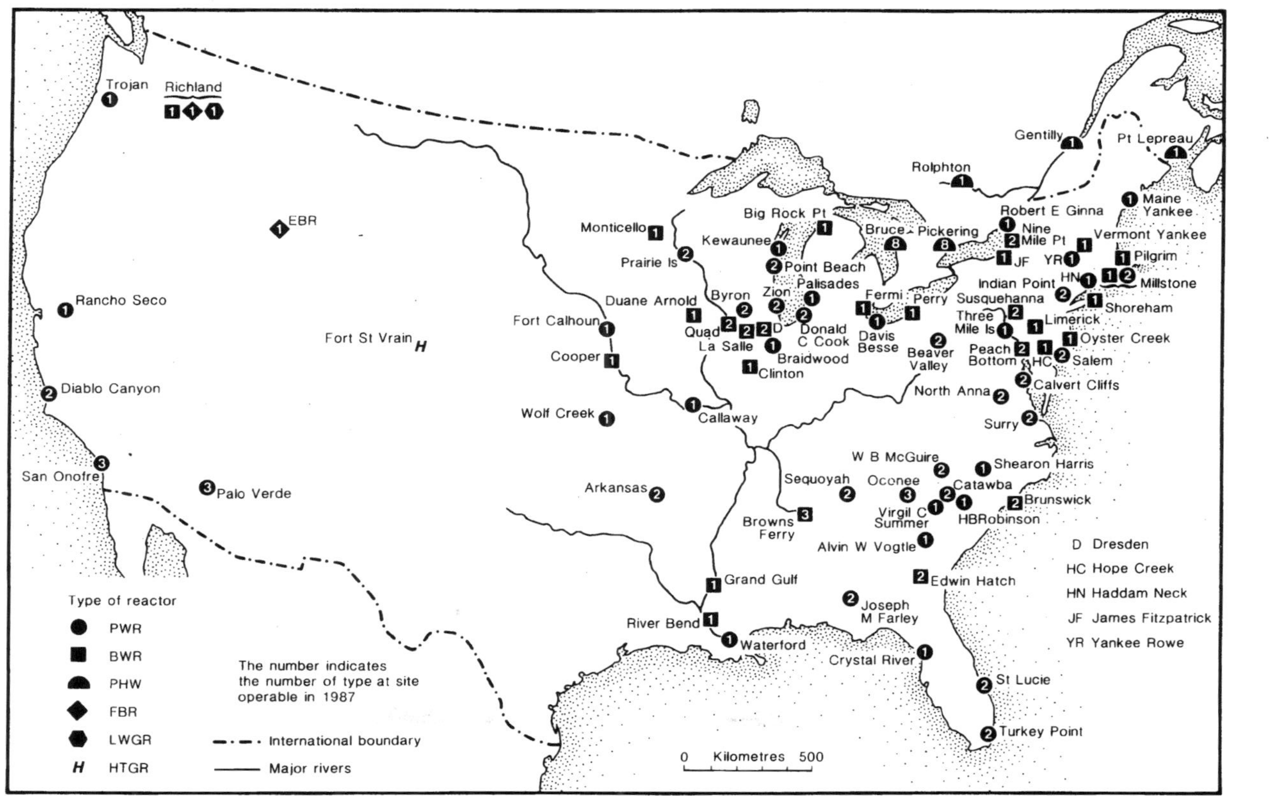

Figure 1.5 Reactor types at nuclear power stations in the USA and Canada in 1987

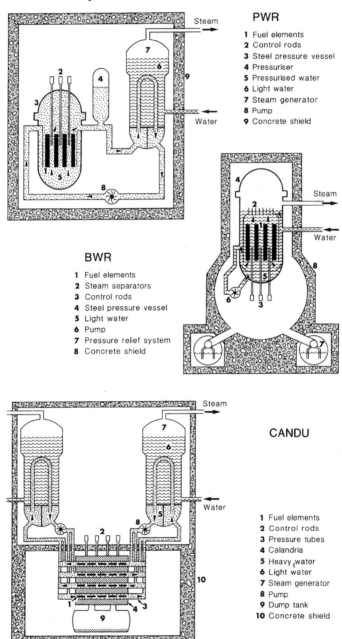

Steam

Water

PWR

1 Fuel elements
2 Control rods
3 Steel pressure vessel
4 Pressuriser
5 Pressurised water
6 Light water
7 Steam generator
8 Pump
9 Concrete shield

Steam

Water

BWR

1 Fuel elements
2 Steam separators
3 Control rods
4 Steel pressure vessel
5 Light water
6 Pump
7 Pressure relief system
8 Concrete shield

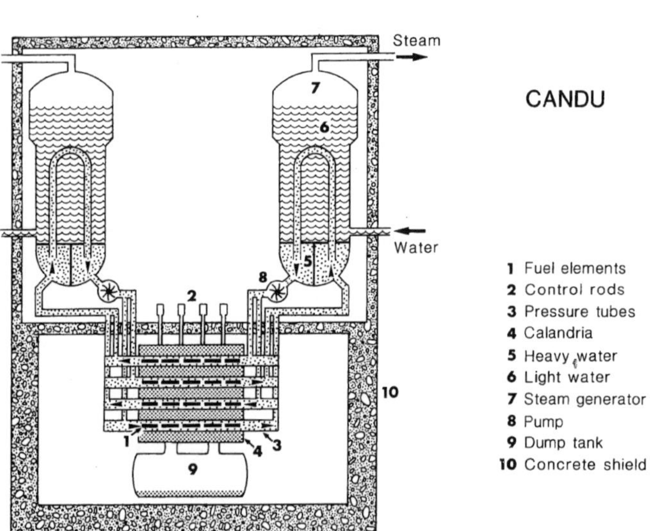

Figure 1.6 Schematic diagrams of the main features of PWRs, BWRs and CANDU reactors

The principal advantages of the PWR are as follows: the considerable operating experience that has been accumulated; high reliability and economic competitiveness with fossil power plants in many energy markets; the use of inexpensive ordinary water as moderator and coolant. The main disadvantages are as follows: the requirement for enriched uranium, calling for either international supply or domestic industrial enrichment capability; heavy investment in engineering facilities to support the technology, e.g. pressure vessel manufacture; the use of saturated steam, demanding a specially designed steam turbine technology; the highly pressurized primary circulation system which is very extensive and presents the possibility of a major rupture; the susceptibility of the heat exchanger to breakdown.

The BWR, which was developed commercially at about the same time, is the PWR's major competitor. There are nearly a hundred units in use totalling around 80,000 MWe. BWRs have been adopted mainly in the USA, Japan, the Federal Republic of Germany (FRG) and Sweden, but not in the USSR or countries belonging to the Council for Mutual Economic Aid (CMEA).

In a BWR the coolant is at much lower pressure than in a PWR so, for a comparable output, the reactor vessel is considerably larger and heavier. Like the PWR, the BWR has to be shut down and depressurized and the vessel opened up before it can be refuelled. Commonly, refuelling takes place at intervals of twelve to eighteen months. For a 1,000 MWe plant the initial core loading is about 140,000 kg of uranium, and about 200 kg of plutonium are produced annually. The water coolant/moderator is allowed to boil in the core. About a fifth of the core is replaced at each refuelling.

The principal advantages of the BWR are as follows: long operating experience; high reliability; ordinary water can be used as moderator and coolant; it is often economically competitive with fossil-fuelled power stations of comparable size in industrialized countries. Disadvantages include the following: the requirement for enriched uranium; the primary coolant system, though smaller than that of a PWR, is still very extensive and poses a potential hazard; heavy industry is needed to support the technology; BWRs require a specially designed steam turbine technology; their economics, like those of the PWR, would be adversely affected if a major delay in breeder reactor technology exerted pressure on plutonium prices.

Gas-cooled reactors (GCRs)

During the early stages of reactor development, while the USA concentrated on light-water-cooled reactors, the UK and France worked on gas-cooled graphite-moderated designs (Figure 1.7). France soon abandoned them in favour of PWRs but their development continued in the UK.

The USSR claims to be the first country to have used nuclear power to generate electricity for commercial purposes, in 1954, and the US Central Intelligence Agency appears to concede this claim (CIA 1985: 52). The

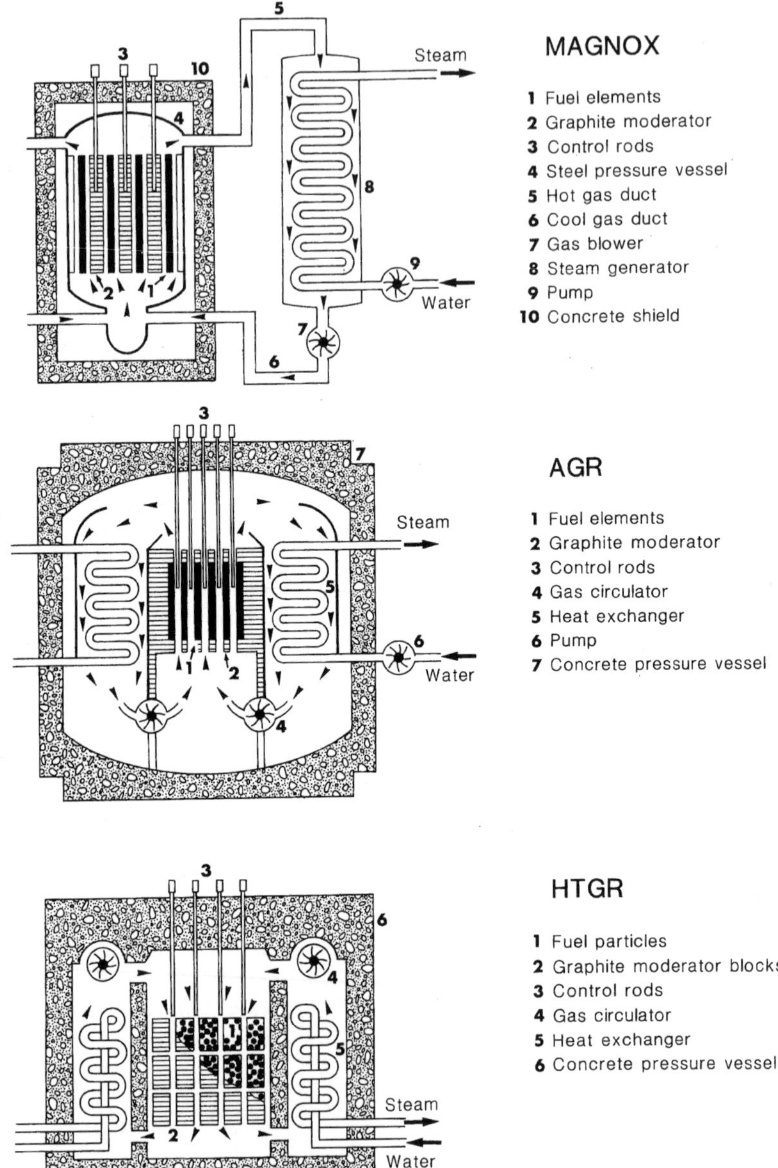

Figure 1.7 Schematic diagrams of the main features of Magnox reactors, AGRs and HTGRs

UK, however, regards the Calder Hall Number 1 reactor as the first in the world to have fed electricity into a national grid (Dancy 1986). This is a Magnox reactor, so named because the natural uranium oxide fuel is clad in a magnesium alloy canister. The Magnox design has been superseded by the advanced gas-cooled reactor (AGR) which, with higher steam pressure and temperature, operates at higher efficiency. These reactors use enriched ceramic uranium dioxide fuel in a stainless steel cladding and have a much more effective fuel burn-up than their Magnox predecessors. The early Magnox reactors have steel pressure vessels but later ones, and the AGRs, have prestressed concrete vessels integrated into the radiation shielding to form a concrete shell over 5 m thick. These are lined with mild steel.

A 140 MWe Magnox plant has an initial core loading of about 230,000 kg of uranium. A 600 MWe AGR has an initial fuel loading of 150,000 kg of uranium. AGRs can be refuelled whilst on load and fuel elements can remain in the core for up to five years. Magnox reactors are not efficient fuel users but AGRs are, with a thermal efficiency of 40 per cent. They use fuel more efficiently than light water reactors. Because they use superheated steam conventional turbine technology can be employed. The inherent safety of reinforced concrete pressure vessels and the large thermal capacity of the graphite core mean that special containment provision is not required to deal with the possible effects of a primary circuit rupture; the AGR inherently is a safe reactor. However, it requires heavy industry to support its technology, it requires enriched uranium and it has not demonstrated sufficient economic advantages to warrant acceptance outside the UK in competition with PWRs and BWRs.

The high temperature gas-cooled reactor (HTGR or HTR) is a development of the gas-cooled concept. The idea is to achieve higher temperatures by using helium as the coolant and graphite-coated spheres of enriched uranium dioxide as fuel. Only experimental and prototype versions have been built so far, in the UK, the USA and the FRG. Dragon, the reactor built in the UK on the OECD project, was closed down in 1975. An attractive feature of the HTGR is that the temperature of the core never reaches the melting point of the fuel and therefore a Chernobyl-type accident would not be possible.

Heavy-water-moderated reactors

To avoid the need for enriched uranium Canada has designed and built heavy-water-moderated reactors fuelled with natural uranium, the Canadian deuterium (CANDU) pressurized heavy water (PHW) reactor. In 1989 thirty-six power reactors of this design were in use, in Canada, India, Pakistan, Argentina, Korea and Romania.

The first commercial CANDU reactor, opened in 1966 at Douglas Point, Ontario, met a variety of operating problems and was closed down a few years later, but it was soon followed by the first four 510 MWe units at

Pickering, Ontario, opened between 1971 and 1973.

In the CANDU design heavy water (deuterium) is used both as moderator and coolant, in separate circuits (see Figure 1.6). The moderator is contained within a tubed tank or *calandria* with fuel bundles in horizontal pressure tubes. The fuel is natural uranium dioxide clad in Zircaloy.

The large core of natural uranium requires the use of on-load fuelling. For a typical CANDU of about 500 Mwe the initial core fuel load is about 93,000 kg of uranium and the annual discharge compared with other reactors is shown in Table 1.1.

Table 1.1 The production of plutonium from different types of reactors expressed as the weight of Pu-239 produced for each gigawatt-year of electricity

Input	0.0	
Creation	+ 710 kg	for PWR
Destruction	− 440 kg	for PWR
Net production	+ 270 kg	for PWR
	+ 617 kg for Magnox	
	+ 493 kg for CANDU	
	+ 173 kg for AGR	

Source: Marshall 1979
Note: All numbers are standardized, e.g. Ekg (Equivalent kilogram) of plutonium from the operation of a 1000 MWe reactor over 365 days (1 MW = 1,000 kilowatts or 1 million watts; e = electric power rating of reactors)

This reactor also has good safety features. A concrete containment building is used, and as an extra safety feature there is a large vacuum building designed to draw steam out of the containment building in the event of excessive pressure.

The CANDU reactor offers a number of advantages: on-load refuelling at full power has enabled the design to lead the world in terms of load factor and reliability; the use of natural uranium results in lower fuel fabrication costs than for PWRs, BWRs and AGRs, which require expensive fuel enrichment facilities. The use of natural uranium makes CANDU reactors virtually inflation proof once they are built. More parts, more processing, more tubing, more pellet-grinding, more hardware and more labour are involved in producing a tonne of LWR fuel than in producing a tonne of CANDU fuel. The use of pressure tubes for fuel in the reactor core allows the coolant to be pressurized without the need for a large steel or reinforced concrete pressure vessel; it is economically competitive with most other proven reactor types for plants of about 500 MWe and over; it is an efficient producer of plutonium; it does not have a high excess of reactivity as do reactors using enriched fuel, and this helps to reduce the likelihood of a major power excursion.

The principal disadvantages are as follows: deuterium is an expensive moderator and the CANDU reactor requires a much larger core than that needed for a PWR or BWR of comparable power output; it requires costly production facilities for deuterium; it uses a saturated steam system requiring specially designed turbines.

The CANDU reactor is not the only heavy-water-moderated design in commercial use. Atucha I (367 MWe) in Argentina was developed in the FRG and came into use in 1974. Its major difference from CANDU is that it has a pressure vessel, not a calandria.

Steam generating heavy water reactors (SGHWRs)

Some countries have developed prototype reactors similar to CANDU in that they use heavy water as the moderator, but light water as their coolant. Examples include Winfrith SGHWR (UK), Fugen (Japan) and Cirene (Italy). The coolant is allowed to boil and the steam produced in the reactor drives the turbine, thereby eliminating the need for a separate steam-raising circuit as in CANDU. The design has not been pursued commercially because financial assessments have demonstrated it to be uneconomic in comparison with alternatives.

Water-cooled graphite-moderated reactors (LWGRs)

The first nuclear power station in the USSR, the 5 MWe plant at Obninsk, was a water-cooled graphite-moderated reactor, and the USSR has twenty-two units of this type in operation. Known as RBMK, this reactor is fuelled with enriched uranium dioxide clad in zirconium alloy and cooled by ordinary water which is allowed to boil. The reactor can be refuelled on load. The design is unique to the USSR and CMEA countries and the ill-fated Chernobyl Reactor Number Four was of this type (Figure 1.8).

THE FAST BREEDER REACTOR (PFR OR FBR)

The fast breeder reactor is not a thermal reactor for it has no moderator. World-wide there are only a few fast reactors in operation (Table 1.2). Variants include the liquid metal fast breeder reactor (LMFBR) using sodium as coolant and the gas-cooled fast breeder reactor (GCFR) cooled with helium.

Figure 1.8 includes a simplified circuit diagram of a sodium-cooled fast reactor. The fission process takes place in the core and produces heat which is carried away by the liquid sodium. This is contained within a large pool and exchanges its heat with a second intermediate circuit of liquid sodium. This intermediate circuit transfers the heat from the sodium pool to steam generators where the heat is transferred to water, thus making steam to drive the turbine.

Because the core contains no moderator the fission neutrons are not

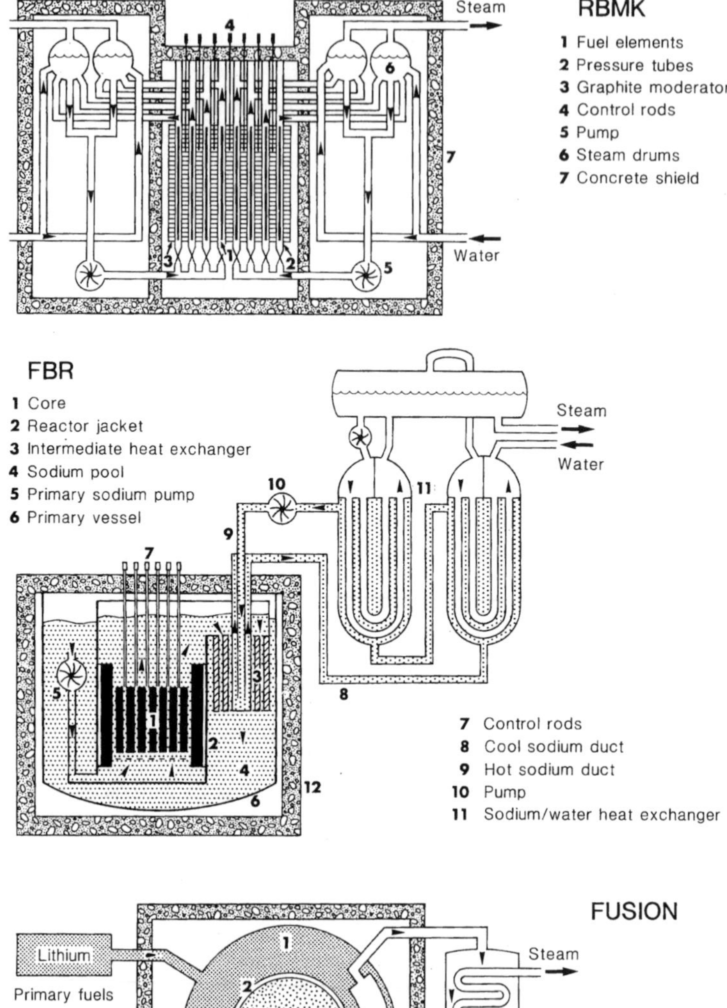

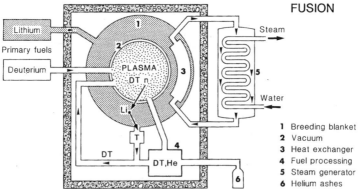

Figure 1.8 Schematic diagrams of the main features of RBMKs (USSR and other CMEA countries), FBRs and fusion reactors

Table 1.2 Fast reactors in operation or under construction in July 1987

Country	*Power rating (MWe)*	*Status*
FRG		
SNR 300 Kalkar	327	Under construction
Karlsruhe KNK 11	20	Operable
France		
Creys-Malville, Super Phénix	1,242	Operable
Phénix	250	Operable
Japan		
Joyo	100	Operable
Monju	280	Under construction
UK		
Dounreay PFR	270	Operable
USA		
EBR 2	20	Operable
Fermi (65 MW) closed down		
Clinch River (375 MW) cancelled		
USSR		
BN350 Shevchenko	150	Operable
BN600	600	Operable
Melekess BOR 60	12	Operable
Obninsk BR5	15	Operable
BN800	800	Under construction

Source: Nuclear Engineering International, *World Nuclear Industry Handbook 1988*
Notes: (a) Super Phénix is the world's first commercial-scale fast reactor, but it produces electricity that is 250 per cent more expensive than that from a standard French PWR. Super Phénix was shut down in May 1987 when a sodium leak was discovered. It had previously achieved full power but had not been commercially accepted. By mid-1989 it had not been reopened. In India the 13 MWe fast reactor at Kalpakkam was put back into operation in 1989 after a 2 year shutdown.

slowed down and retain fast speeds. The central core region consists of a mixed uranium–plutonium oxide fuel (U-235, Pu-239 and U-238) surrounded by a 'blanket' of U-238. The incineration of plutonium in the core is inevitable; the production of plutonium in the blanket is optional. Once launched, however, the fast reactor is self-sufficient and may make excess plutonium, hence the appellation 'breeder'.

Lord Marshall of Goring has provided a vivid description of the fast reactor:

> Imagine a group of castaways on a beach, trying to keep warm. We suppose that they have a small supply of dry wood, together with a large supply of wet driftwood, washed up on the shore. They could simply use the dry wood to build a fire, which would keep them warm while the dry wood lasted, or they could build a blanket of wet wood around the fire.

> Then as the dry wood was burnt up, as well as keeping them warm, it would also dry out some of the wet wood, to give a fresh supply of dry wood. In this way the castaways could keep warm in perpetuity.
>
> We have here a simple but precise analogy to the operation of the fast reactor. The fire of dry wood is analogous to the incinerated plutonium in the core. The drying out of the wet wood is analogous to the production of plutonium from the U-238 blanket. The wet wood is analogous to U-238, both useless waste products unless used in this way. The operation of building up the blanket with wet wood, extracting dry wood from it and throwing dry wood on to the fire are exactly analogous to the fast reactor fuel cycle (Marshall 1979).

Figure 1.2 shows the relationship of the fast fuel cycle to the thermal cycle. The first step in the fast fuel cycle is to accumulate enough plutonium to start it off. At the end of a suitable interval, a year or so, so much plutonium has been incinerated in the core that it is necessary to replenish it with fresh fuel. The blanket fuel elements can be removed and plutonium extracted from them. The blanket and core can be replenished with U-238, which is simply a waste product otherwise.

The fast reactor has held an important position in discussion of the economics of different variants of the thermal fuel cycle. The economics of the reprocessing fuel cycle are often put into the context of a situation where plutonium can be anticipated to have a value not only for weapons manufacture but also as a fuel for fast reactors. The fact that the UK government in the late 1980s has distanced itself from the financial support of fast reactor research and development reflects the exposure of the economics of the reprocessing cycle to reassessment (Sweet 1980, 1983).

Some reactors produce more plutonium than others, but the production of some plutonium is inevitable whatever the type of reactor (see Table 1.1). The question arises, therefore: 'What should be done with it?' If it is left in the fuel element, as in the once-through cycle, potentially useful fuels including plutonium are treated as waste products. In the reprocessing cycle plutonium has to be stored if it is not incinerated as a fuel. Another possibility is to recycle both plutonium and uranium from the reprocessing cycle in thermal reactors, but the most logical step with plutonium produced by thermal reactors is to burn it up in the fast reactors for, in this way, dramatic savings can be made in the consumption of nuclear fuels. In this case a plutonium credit may be allowed in calculating the economics of the thermal reprocessing cycle. However, such a credit makes best sense in the context of building and using fast reactors.

THE FUSION REACTOR: PROSPECT FOR THE FUTURE?

In the process of nuclear fusion two light elements combine to form a heavier one. The mass of the heavier element thus produced is slightly less than the combined mass of the two light elements and the mass that is apparently 'lost' is, in fact, coverted directly into energy.

This is the heat production process used by the sun and in the hydrogen bomb, but controlling it to produce electricity is difficult. The principle is to persuade the nuclei of light elements like deuterium to collide with each other at such enormous velocities that they combine rather than bounce off one another. The principle was shown to work by Rutherford in the 1930s when he used a particle accelerator to propel deuterium nuclei at one another at very high speeds, but in his experiments so much energy was needed that there was a net loss of energy rather then a net gain.

More recent experiments on fusion use a different approach. As the atoms of deuterium are heated, they begin to move about inside their container at increasing speed. At normal temperatures and pressures they move at 3,000 miles/hour and bounce off one another. At 100,000 °C the atoms have broken up into their component nuclei and electrons, which are moving at 170,000 miles/hour but still bouncing off one another. At 100 million degrees Celsius the deuterium nuclei will move at more than 5 million miles/hour, fast enough to fuse together when they collide. The energy produced then will almost equal the energy that has to be used to heat them up. A further increase in temperature, to 380 million degrees Celsius, would make the fusion reaction self-sustaining, producing a net surplus of energy which could be used by man.

Many of the difficulties involved in building operational fusion reactors lie in finding materials capable of withstanding these extremely high temperatures. Others stem from the need to prevent the cloud of deuterium nuclei and electrons from flying apart before anything useful can happen. The solution being attempted is to contain the particles by means of a magnetic field in the shape of a torus. Inside the torus the particles are prevented from hitting the steel walls by the magnetic fields. This is the principle of the Joint European Torus (JET) project at Culham in Oxfordshire.

If it could be made to work, the advantage of fusion power would be energy abundance and not low cost. Deuterium occurs naturally in water, and so the fuel would be plentiful and cheap, but fusion power plants based on magnetic containment are unlikely ever to be cheap to build and run. There are also potential hazards. The magnetic fields in a large fusion reactor would store huge amounts of energy which, if released suddenly, might rupture the system and release into the environment a quantity of hot liquid lithium, which is used as a means of extracting the energy from the torus. Lithium ignites spontaneously in air, and so a very substantial explosion could result. But a fusion reactor would produce lower radwaste releases than a fission reaction in normal operation and would be far easier to dismantle after decommissioning.

ENVIRONMENTAL CONSIDERATIONS

In evaluating sites for nuclear power plants much emphasis is placed on assessments of what might happen under postulated accident conditions (see Chapters 10 and 11). Competent examination of the suitability of a proposed site also requires an assessment of the possible impact of the power station in normal operation on a wide range of environmental systems. These include the hydrological character of the site, both in respect of ground-water movements and sources of cooling and process water, levels of radioactivity in discharged water, gaseous discharges from chimney stacks, and assessment of a variety of ecosystems and food-chains in the surrounding environment which might induce reconcentration of diluted radionuclides or provide pathways back to man (see Chapter 8). In the USA the National Environmental Policy Act 1969 required detailed consideration to be given to the environmental impact of nuclear power plants in the licensing process, and subsequent legislation directed at retaining or improving environmental quality widened the scope of environmental impact statements required of operators by regulatory bodies to include issues such as the quality of the environment measured in terms of amenity and aesthetics, land values and a range of socio-economic concerns not required during the early days of the industry.

Sources of nuclear power plant radioactivity in normal operation

There are two main sources of radio-isotopes in the heat transport system of a nuclear reactor: first, those produced from the uranium dioxide fuel by the fission process (nuclear fission products) and, second, those produced by neutron activation of materials other than fuel which constitute the reactor core. Fission products normally remain within the metallic sheath which encloses the uranium fuel, but during the course of normal operation a proportion of the fuel elements develop holes or cracks in the sheath through which gaseous products can emerge. Some of them have half-lives measured in minutes; longer-lived products are removed from the heat transport system by purification. Neutron activation products are formed by the effect of neutron flux in the reactor core on any substances residing in or passing through the core. Thus the fuel, any pressure tubes and the calandria all become radioactive as their materials absorb neutrons. Water passing through the reactor core becomes activated, as does the circulating water because of traces of impurities from corrosion of the materials used to construct the heat transport system (Neil 1974).

The type of reactor and the technical features of its design affect the amount and composition of the various radioactive products. The use in water-cooled reactors of Zircaloy instead of steel for fuel element cladding, for example, results in a reduction of tritium release by a factor of 100 because the tritium combines with the alloy within the cladding. Thus, in

Western Europe, the nuclear power stations at Chooz and Trino (PWRs) and Hunterston B and Hinkley Point B (AGRs) exhibit relatively high tritium discharges because they use stainless steel clad fuel elements (Luykz and Fraser 1982). Tritium is impossible to remove from wastes and has a long half-life.

Most nuclear power reactors in operation or being built are light water BWRs and PWRs. BWRs release more gaseous activity, by about one order of magnitude, generally in the form of noble gases, whereas PWRs produce more liquid radioactive wastes, especially tritium. Radio-iodine is released in gaseous effluent, particularly by BWRs and GCRs (Lave and Freeburg 1973; Hull 1974).

In normal operation, corrosion products and activated gases comprise the major source of radwaste from BWRs and, depending on the materials used in the primary system, turbine, and feed-water plant, the corrosion will usually contain radioactive isotopes of cobalt, manganese, chromium and iron. Gases taken from the reactor core by steam flow and carried to the vent chimney stack include krypton and xenon. Non-gaseous fission products in BWRs tend to remain in the coolant water, where their concentration is controlled by the reactor coolant clean-up system.

PWR wastes normally consist of corrosion products and minor concentrations of fission gases from uranium, tritium produced by the activation of coolant additives, naturally occurring deuterium in coolant and ternary fission products of U-235. The corrosion products depend on the material in the coolant but usually consist of radioactive isotopes of cobalt, caesium, manganese, iron and chromium. Liquid wastes arise from dilution of reactor coolant water for boron concentration control, from sampling activities, from equipment and process leaks and from decontamination activities (Richardson 1973).

Solid radwaste disposal is dealt with in Chapter 12 as part of the discussion of the back-end of the nuclear fuel cycle, but in this brief summary of reactor radwaste production it is worth noting that, per megawatt of installed capacity, BWRs produce four or five times more solid radwaste for land burial than PWRs.

The methods and equipment used for monitoring the territory around a nuclear power station vary from one country to another, and this can lead to appreciably different results. Despite this, the various methods used reveal amounts of radioactive material released to be typically extremely small, and the quantitative measurements made around the power stations demonstrate the exposure of members of the public to lie well within the variations in exposure that result from natural background radiation (Luykx and Fraser 1982).

Gaseous wastes

It is normal practice for active or potentially active gases extracted from the coolant or waste process systems to be mixed with air discharging from station ventilation systems before release through chimney stacks. Such release may take place on a continuous emission basis or periodically via gas storage tanks provided to permit retention for radioactive decay. Release of the gases from such hold-up tanks can be timed to coincide with weather conditions favourable to dispersion.

Since early work in the UK by Sutton in the 1940s and later studies by Scorer, Barrett and others, a great deal has been written about the behaviour of chimney stack plumes under various meteorological conditions. The plumes from tall power station chimney stacks are usually buoyant and may rise to several hundred metres in the lower atmosphere. Their buoyancy is gradually diluted by entrainment of surrounding air into the plume and at the same time they quickly acquire the horizontal momentum of any wind, causing a characteristic 'bent-over' appearance (Figure 1.9).

The distribution of particulate material and gases within such plumes tends to be fairly uniform for considerable distances downwind, of the order of tens of stack heights (Gifford 1972). Rather less is known about the processes that bring buoyant plumes back to the ground (Figure 1.10). Such events, collectively called fumigations, can occur (i) because the downward growth of the plume base is fast enough relative to its rise to cause it to reach the ground, (ii) because strong vertical convective motions of the atmosphere on a scale approximately equal to the size of the plume bring the plume down to the surface or (iii) because of the influence of some peculiarity of the terrain, such as the presence of a steep ridge–valley system, or of changes in the land surfaces, for instance from water to land or from rural countryside to city. Downward plume spreading by diffusion, sometimes called 'coning', is favoured by a fairly strong wind and an upward mixing that is limited by some kind of inversion. Plumes reach the ground as a rule only at comparatively great downwind distances, of the order of twenty stack heights, but the phenomenon may persist for some periods. Looping action of the plume from the stack (see Figure 1.9), which can occur on a light wind with unstable atmospheric conditions, whilst not usually a concern beyond the plant boundary, may occasionally bring strong concentrations to the ground in the neighbourhood of the stack but the duration is only for a minute or so. Even so, this is an argument in favour of having a large exclusion area around a nuclear power station, and for continuing to pay attention to the density and distribution of population within a radius of 3 km from the reactor building, since Munn and Cole (1967) have indicated that looping plumes need not be considered beyond this distance. Deflection of plume patterns by local winds at valley sites is shown in Figure 1.10, but the terrain-modified and land-, lake- or sea-breeze types of fumigations are complex phenomena.

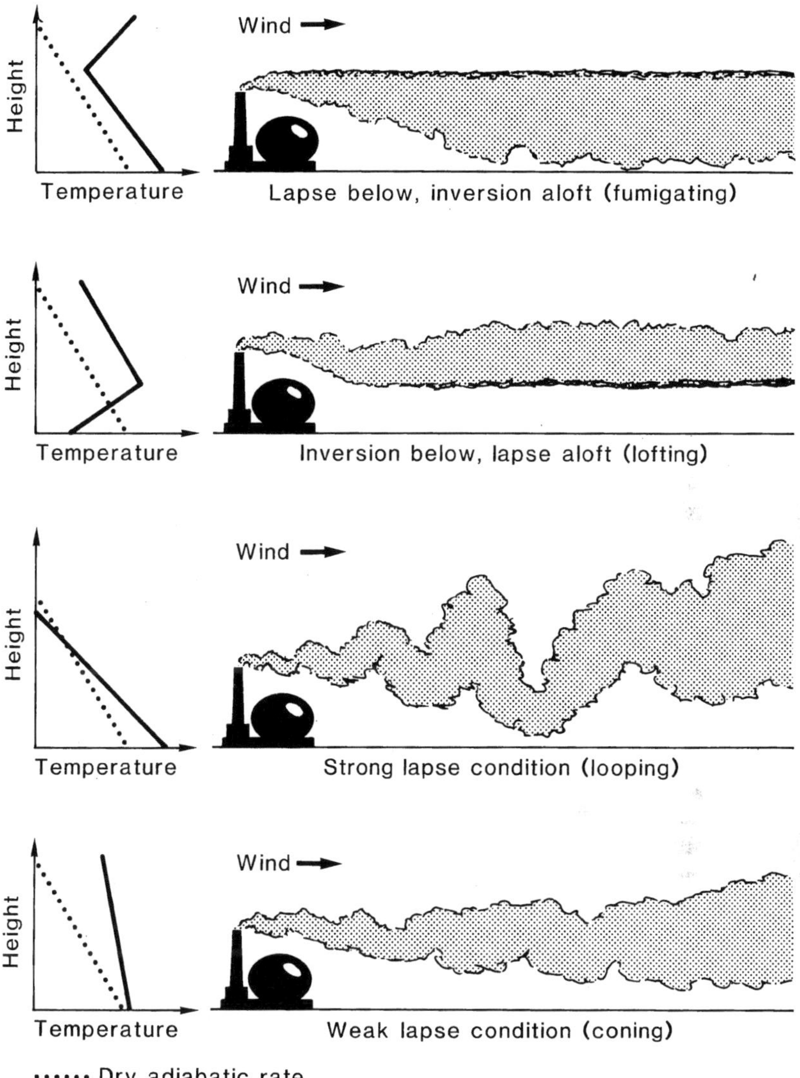

Figure 1.9 Influence of lapse rate conditions on the shape and dispersal of power station chimney stack emissions
Source: Energy Policy Staff 1968: 122

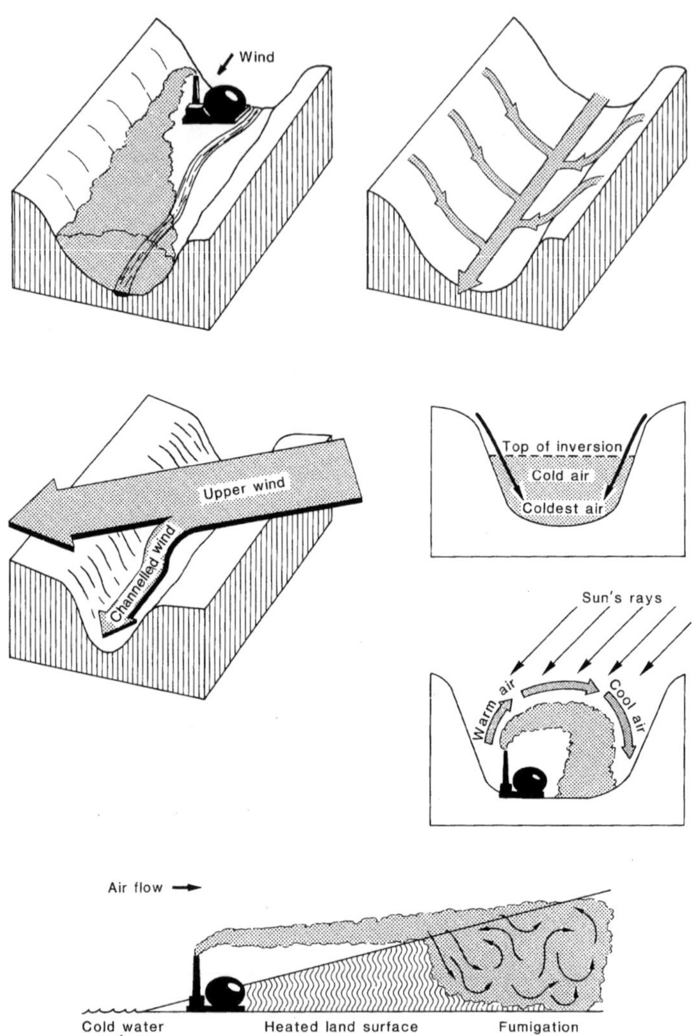

Figure 1.10 Some physiographic influences on the dispersal of chimney stack emissions. Top left, a valley site with wind along the valley can concentrate emissions along the valley downwind from the plant. Centre left, similar but more complex situations may occur if there is a wind across the valley but which may partly become channelled along it by local topography. In the right-hand diagrams, (top) night-time drainage of air from the valley slopes may provide a valley wind, (centre) an inversion of temperature may result from cold air drainage, (lower right) fumigation may be caused by heat from the sun. In the lowest sketch a power station next to a lake or other large cold water surface may experience fumigation from time to time
Source: Energy Policy Staff 1968: 125–6

Cooling-water requirements have prompted the choice of lakeside sites in many countries, and in the early 1980s over thirty nuclear power stations were operating on the edges of such water bodies (Burda *et al.* 1982). The particular meteorological conditions associated with lake and sea shores, including the seasonal and diurnal lake- or sea-breeze phenomenon, have called for some attention, but the details of fumigation, including its location and persistence, are strongly influenced by the circumstances of any particular case. When the Pickering site in Canada was being assessed for Ontario Hydro, for example, it was judged that a continuous release of radioactivity in gaseous form would be sufficiently small for there to be no effect on the surrounding population, although it was also suggested that precautions might have to be taken at the plant during purges of the boiler room (Ontario Hydro 1966).

Short chimneys have their own plume characteristics. The behaviour of waste gases after discharge from a short stack in the vicinity of buildings is dependent on the size and shape of the buildings and the air flow in the neighbourhood of these obstacles. Sometimes a 'cavity' is produced downstream and immediately behind the building which is surrounded by a 'wake' extending a considerable distance downstream. If the stack release becomes entrained in the building wake, high ground-level concentrations will occur. A rule of thumb that is commonly accepted is that building aerodynamic effects can be ignored if the stack height is greater than two and half times the height of adjacent buildings, but the flow over and around a complex of buildings may be quite variable, depending on the particular spacing and orientation of the buildings relative to the wind (Munn and Cole 1967) (see Figure 1.11).

Ground-water hydrology

The hydrological features to be considered in assessing the suitability of a site for a nuclear power station include the capability of adjacent water bodies to provide sufficient cooling water and to further disperse water discharges containing radioactivity from the plant, the ground-water characteristics of the site and the presence of nearby drinking-water intakes and bathing facilities, together with the possibility of reconcentrating radionuclides by fauna or flora.

Ground-water hydrology is important when medium-term storage of spent fuel occurs at burial areas at nuclear power station sites, as happens in Canada and Japan. If the ground-water movement is away from rather than towards such areas, it becomes important to be able to calculate the effect that travel time may have on any contaminants that *might* enter them and be transported to wells or other outlets. The characteristics of the lakes, watercourses or sea-shores by which a station is to be sited and the other uses to which they are put indicate the degree of risk that may be involved and help determine the safety precautions to be observed at the plant in effluent control and management.

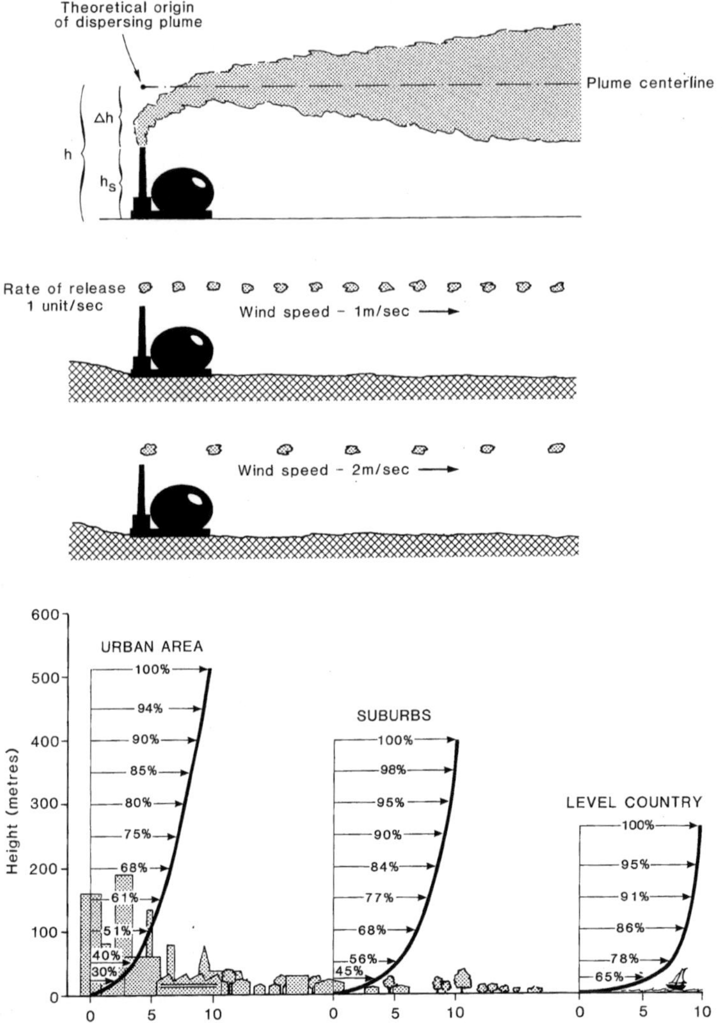

Figure 1.11 Effect of wind speed on plume dispersion and of terrain roughness on wind speed profile. Wind speed can alter the plume, and wind speed in turn may be influenced by the terrain.

The upper diagram shows the relationship between stack height, plume centreline and theoretical origin of a dispersing plume. Since 1950 at least twenty plume rise formulae have been proposed and abandoned, mostly because of lack of definite experimental verification.

In the UK most power reactors have chimney stacks that are essentially at the same height as the surrounding buildings and therefore building entrainment often occurs during operational leakage or reactor blowdowns

Source: Energy Policy Staff 1968: 121

Cooling water

All fossil-fuelled and nuclear power stations are heavy consumers of water for cooling purposes, and access to adequate supplies of cooling water is a very important siting consideration. A 1,000 MWe power station producing base load current may need 2300 million litres a day and in the UK there has been a perceptible trend towards coastal and estuarine sites because most British rivers are relatively small. The UK maximum demand for power station cooling water is equal to half the total normal rainwater run-off and the US electric power industry requires access to about one-sixth of the nation's total available fresh-water run-off.

Radioactivity in the water wastes comes from mixed fission products and from process system corrosion products. Liquid radwastes at nuclear power stations are usually considered in three categories: potentially active liquid wastes, active water wastes and active chemical wastes. Potentially active liquid wastes are those normally expected to be kept as low as reasonably possible and to carry less activity than the maximum permissible concentrations suggested by the International Commission on Radiological Protection (ICRP). These wastes are discharged through a radioactivity monitor and are then diluted in the used cooling water before being returned to the water body from which this water is drawn. Active water wastes are sent to a hold-up tank, where they are periodically monitored. When these tests show that disposal by dilution is possible, these wastes are transferred to a dispersal tank and disposed of in the same way as potentially active wastes. Active chemical wastes are those which, because of their chemical content, cannot be released into lake or river in the plant effluent. They are collected in special tanks, put into drums, mixed with cement and buried at solid waste disposal sites.

The concentration of radioactive wastes in plant liquid effluent is normally very low, often below 15 per cent of the ICRP recommended levels. Given this fact, further dispersal depends upon mixing the discharge water with other water in the lakes or rivers alongside which the power stations are sited.

Nuclear power stations share the condensing steam cycle with other coal-fired and oil-fired power plants. This involves raising steam at high temperature and pressure in a boiler, allowing it to expand and do mechanical work in a turbo-generator and, finally, condensing it close to ambient temperature for re-use as feed-water in the boiler. The lower the condenser temperature, the lower is the pressure at the turbine exhaust and the greater is the mechanical energy that can be extracted from the steam. The laws of thermodynamics dictate that the efficiency of the cycle is fundamentally restricted according to its top and bottom temperatures and therefore, to achieve an optimum electrical output, modern thermal power plants reject upwards of 50 per cent of their combustion energy as heat. The UK Magnox power stations must normally discard more than 70 per cent, because of their lower steam temperatures (Mounfield 1961), although more modern stations can do better. Condensers are cooled either by passing water from the sea, a lake or a large river directly through the condenser back into the river or sea or by using a recirculatory system and cooling this water using cooling towers or some other means. The temperature of the discharged water may be 10 °C higher than the water body to which it is returned.

Intake and outlet points are normally set well apart, or they may be separated by an engineered barrier. This is important since recirculation of the warm discharge would reduce condenser efficiency, leading either to increased heat rejection or to reduced electrical output from the station. When it leaves the outlet, the warm discharge begins to spread over and mix with the receiving body of water. At coastal or estuarine sites, or sites on the shores of inland water bodies approaching the size of the Great Lakes, it rapidly becomes entrained by tidal flow and moves along the shoreline close to which it has been released. When lakes are small, such as Trawsfynydd in North Wales, artificial horizontal circulation patterns have to be engineered.

At coastal or large lake sites, the warm water plume spreads progressively further, ultimately losing its excess heat to the atmosphere. The dispersion of the thermal plume is normally divided into three stages. The initial or 'near-field' stage covers just the first few seconds after the discharge leaves the outfall to form a warm surface layer. The second or 'mid-field' stage lasts for less than a tidal period, while the plume is still clearly distinguishable from the surrounding water with which it subsequently mixes. Finally comes the 'far-field' stage, covering small but generally widespread increases of background water temperature and for which the time-scale extends over many tidal cycles (MacQueen and Howells 1978).

In addition to providing cooling water and an enormous heat sink, water bodies of the size of the Great Lakes, wide estuaries and open coasts make it difficult to demonstrate widespread ecological effects attributable to either the waste heat or the radioactive effluent from nuclear power stations. The middle and lower courses of rivers are not always easy to monitor by normal ground survey techniques because their mud flats, which are exposed to the sun during the day, may have a larger thermal impact than a power station. Until the 1960s the variability of such natural effects made it difficult in practice to isolate temperature changes that could confidently be attributed to a cooling-water discharge. This monitoring problem was partly solved by the availability from the late 1960s of remote-sensing techniques which through thermal imagery have enabled the course and diffusion of warm water discharges to be traced with some accuracy.

SUMMARY

Light water reactors, shared approximately equally between pressurized water and boiling water designs, dominate the world's nuclear power industry. National enclaves of technology diverting from this pattern do exist. The AGR, for instance, is a reactor type restricted to the UK. The CANDU reactor has reached a wider international market through a combination of generous Canadian export loan arrangements and good technical characteristics, but design and construction expertise is limited to Canada. In contrast, the capacity to design and build light water reactors exists not only in the USA but also in France, Sweden, the FRG, the USSR, Japan and elsewhere.

It is tempting to suggest that a nuclear reactor is simply a substitute for the coal-fired or oil-fired furnaces which power conventional electricity generating stations, but things are not that simple. In a nuclear power station steam pressures are lower; the steam is wetter so that turbine blades have to be made to a different design, they corrode quickly unless made with special metals and much water has to be drained away from them. A problem unique to nuclear reactors is that if they are suddenly taken off load when they have been operating at a normal power level, they must be restarted and brought up to two-thirds or more of their previous operating level within about thirty minutes or they cannot be restarted for about forty hours. The reason for this is that neutron-absorbing decay products build up in the core unless the neutron flux is generally restored.

A handful of fast reactors are now in operation and development programmes have attracted the interest of several countries. Some governments, however, including those of the USA and the UK, have made budget reductions in this area, partly because of the potential hazards involved in the use of plutonium as a fuel and partly because of economic assessments. This development weakens the case for a plutonium credit for normal thermal reactors.

It seems generally agreed that fusion power is unlikely to be available for general power station use for at least three decades.

The rest of this volume is concerned, therefore, with the production of electricity as it has been provided commercially by nuclear power stations for the past thirty years or so, i.e. by thermal reactors. In the next five chapters we examine in some detail the geographical distribution of these reactors on a world and continental scale.

FURTHER READING

Alesso, H.P. (1981) 'Proven commercial reactor types: an introduction to their principal advantages and disadvantages', *Energy* 6:543–54.

Cole, H.A. (1988) *Understanding Nuclear Power: A Technical Guide to the Industry and its Processes*, Aldershot: Gower Technical Press.

Patterson, W.C. (1983) *Nuclear Power*, Harmondsworth: Penguin Books, Chapters 1–3.

2 The world pattern of nuclear power production

In 1987 there were in the world 418 nuclear reactors capable of producing commercially useful supplies of electricity. Over two-thirds were in just five countries: the USA, the USSR, France, Japan and the UK (Table 2.1). There were also a further 130 reactors at various stages of construction, many of them being built at sites where other units were already in operation.

These reactor numbers, in conjunction with the data in Figures 2.1 and 2.2 and Table 2.2, make it clear that nuclear power has become an important means of producing electricity in the world today. Nuclear power stations accounted for 9 per cent of the world's total installed electricity generating capacity and produced nearly 13 per cent of the total electricity output in 1984. By the end of the century, they may account for 20 per cent of electricity production. It is also evident that some parts of the world have become much more dependent upon nuclear power than these average figures might suggest, and that other areas have been barely touched by the technology. The latter include Australasia, the whole of Africa, except for South Africa, the Arab States of the Middle East, much of South East Asia and most of Central and South America, except Mexico, Argentina and Brazil and, in the future, perhaps Cuba.

Conversely, Western Europe produces over 20 per cent of its electricity by nuclear means and the amount of nuclear energy produced stands comparison with that provided by the continent's much larger installed hydroelectric power generating capacity. Commonly, nuclear power is used to provide *base load* current, electricity that is needed day in and day out, winter and summer. Much hydroelectric capacity, on the other hand, is limited to supplying *peak load*, and plants dependent upon stream flow rather than water stored in reservoirs may have their usefulness curtailed by seasonality in water flows.

The proportion of total electricity supplies produced is one useful guide to the importance of nuclear power stations within the framework of national and continental energy supplies; another is the amount of nuclear generating capacity installed and in operation. With 110 operable reactors providing 100,323 MWe of installed capacity in 1987, the USA contains

Table 2.1 Nuclear power reactors in operation, under construction or cancelled at the end of 1987

Country	*In operation*		*Under construction*		*Cancelled or indefinitely deferred*		*Shut down*	
	Units	*Rating (MWe)*	*Units*	*Rating (MWe)*	*Units*	*Rating (MWe)*	*Units*	*Rating (MWe)*
Argentina	2	1,005	1	745				
Austria					1	722		
Belgium	8	5,740						
Brazil	1	657	1	1,245				
Bulgaria	4	1,760	3	3,200				
Canada	19	12,553	4	3,740			3	499
China			3	2,172	6	4,300		
Cuba			2	880				
Czechoslovakia	7	3,002	9	6,216			1	143
Finland	4	2,400						
France	49	46,693	14	18,477			7	454
FRG	21	19,911	4	4,325	11	14,704	6	725
GDR	5	1,835	6	2,640				
Hungary	3	1,320	1	440				
India	7	1,243	6	1,410				
Iran					4	4,488		
Italy	3	1,312			2	1,904	1	160
Japan	37	28,146	11	10,068			1	13
Luxembourg					1	1,330		
Mexico			2	1,350				
Netherlands	2	540						
Pakistan	1	137						
Philippines					2	1,302		
Poland			2	930				
Romania			5	3,395				
South Africa	2	1,930						
South Korea	7	5,816	2	1,900				
Spain	8	5,810	2	2,022	11	10,944		
Sweden	12	10,030			1	150	1	12
Switzerland	5	3,065			2	2,000	1	9
Taiwan	6	5,144						
UK	38	12,796	5	3,822			3	51
USA	110	100,323	13	15,809	106	121,555	20	1,984
USSR	57	34,334	33	29,620			2	1,108
Yugoslavia			1	632				
World	418	307,502	130	115,038	147	163,399	46	5,158

Source: compiled from tables in Nuclear Engineering International, World Nuclear Industry Handbook 1988

Table 2.2 The contribution of nuclear power plants to world electricity production, 1984–6

Continents and countries with nuclear electricity generating capacity in operation in 1984	*Net installed electricity generating capacity of all types of generating plants in 1984 (GW)*	*Net installed electricity generating capacity of nuclear power plants in 1984 (GW)*	*Nuclear generating capacity as percentage of total installed electricity generating capacity in 1984*	*Total electricity production in 1984 (GWh, rounded up)*	*Electricity produced by nuclear power plants in 1984 (GWh, rounded up)*	*Nuclear electricity production as percentage of total in 1984 (1986)*
Africa	56.2	0.9	1.7	224.1	3.93	1.8
South Africa	24.7	0.9	3.9	122.4	3.93	3.2 (6.8)
North America	822.2	81.4	9.9	3,048.7	380.3	12.5
Canada	95.2	9.6	10.0	437.8	52.7	12.0 (14.7)
USA	688.4	71.9	10.4	2,472.3	327.6	13.3 (16.6)
South America	86.8	1.0	1.2	330.6	4.6	1.4
Argentina	15.3	1.0	6.6	44.9	4.6	10.3 (11.3)
Brazil	41.7	–	–	175.7	–	– (0.1)
Asia	424.8	26.3	6.2	1,664.0	166.3	10.0
India	47.7	26.3	2.3	165.4	3.8	2.3 (2.7)
Japan	60.7	19.9	32.6	647.4	126.7	19.6 (24.7)
Pakistan	5.0	0.1	2.7	16.0	0.2	0.9 (1.8)
South Korea	15.5	1.9	12.4	58.1	11.8	20.3 (43.6)
Europe	599.3	76.5	12.8	2,362.9	477.7	20.2
Belgium	12.3	3.5	28.2	53.6	27.7	51.7 (67.0)
Bulgaria	13.3	1.8	13.2	44.6	14.0	31.3 (30.0)
Czechoslovakia	18.6	1.1	5.9	78.4	7.2	9.2 (21.0)
Finland	11.3	2.3	20.3	43.3	17.7	41.1 (38.4)
France	84.7	32.9	38.8	306.8	181.8	59.2 (69.8)
FRG	87.1	11.2	12.8	376.6	67.2	17.8 (29.4)
GDR	21.5	1.8	8.5	110.1	11.7	10.7 (11.6)
Hungary	5.8	0.8	13.9	26.3	3.8	14.3 (18.3)
Italy	54.0	1.3	2.4	179.5	6.9	3.8 (4.5)

Netherlands	17.7	0.5	2.8	62.8	3.8	6.0 (6.2)
Spain	31.0	1.9	6.4	115.5	9.0	7.8 ?
Sweden	30.9	7.4	24.0	123.5	50.9	41.2 (50.3)
Switzerland	15.1	2.9	19.1	48.1	17.4	36.1 (39.2)
UK	67.0	6.6	9.8	280.5	53.7	19.2 (18.4)
USSR	304.0	24.1	7.9	1,493.0	142.0	9.5 (10.0)
Yugoslavia	15.3	0.6	4.3	72.3	4.4	6.1 (5.4)
World	2,332.3	210.2	9.0	9,267.4	1,174.9	12.7

Source: UNO 1986: Tables 32, 34; *Nuclear Engineering International*, June 1987

Notes: (a) 1 gigawatt = 1,000 megawatts

(b) Percentages calculated before any rounding of figures

(c) Taiwan is not included in the UNO tables but in 1986 had six reactors totalling 4,918 MW of installed capacity, i.e. 43 percent of the total

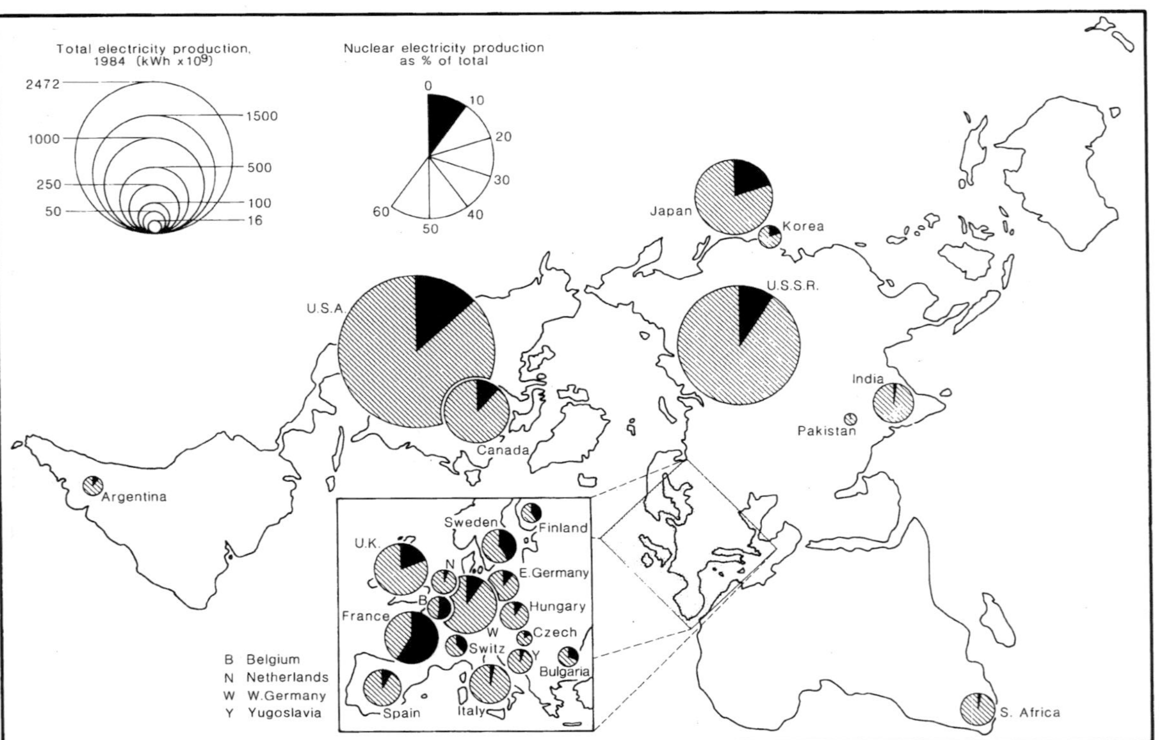

Figure 2.1 Nuclear electricity production as a percentage of total electricity output in 1984
Source: UNO 1986: Tables 32, 34

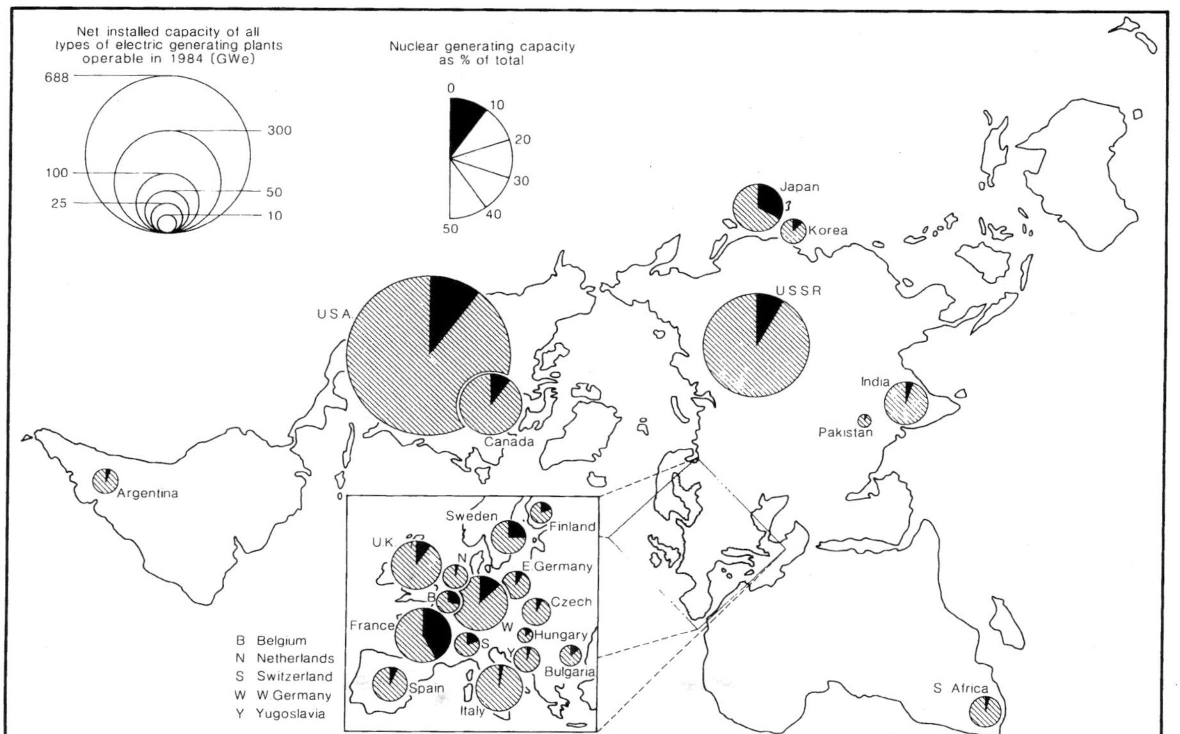

Figure 2.2 Nuclear electricity generating capacity as a percentage of total electricity generating capacity in 1984
Source: UNO 1986: Tables 32, 34

by far the largest national concentration of developed nuclear power. Indeed, the USA possesses *30 per cent of the world's total electricity generating capacity by all means.* The USA and Canada produce around 15 per cent of their electricity through nuclear power. It will be seen from Table 2.2 and Figure 2.2 that by the mid-1980s 39 per cent of France's total electricity generating capacity was nuclear, supplying very nearly 70 per cent of the country's electricity. In 1987 France had 49 units providing 46,693 MWe of capacity. Four other European countries generated a high proportion of their electricity from nuclear power stations: Belgium (67 per cent), Sweden (50 per cent), Switzerland and Finland (each around 40 per cent). However, France had a *total* installed electricity generating capacity in 1984 of 84,700 MWe, well ahead of Sweden (30,863 MWe), Switzerland (15,060 MWe), Belgium (12,309 MWe) and Finland (11,314 MWe), countries with smaller populations than France and lower electricity demands. South Korea obtained 44 per cent of its electricity from nuclear plants and Taiwan obtained 43 per cent.

Other countries generated approximately a quarter to a fifth of their electricity by means of nuclear power in the mid-1980s. These were Bulgaria (30 per cent), the FRG (29 per cent), Japan (25 per cent), Czechoslovakia (21 per cent), the UK and Hungary (each 18 per cent). The GDR (12 per cent), Argentina (11 per cent) and the USSR (10 per cent) hovered close to the world average figure. Well below it were Yugoslavia, the Netherlands, Italy, Pakistan, India, South Africa and Spain, but with the last of these rising quite rapidly as a result of recent nuclear power station completions.

NUCLEAR POWER 'HAVES' AND 'HAVE NOTS'

This factual summary indicates that nuclear power has entered commercial usage for electricity production most markedly in the world's advanced industrial countries. Table 2.1 shows that, in 1987, 86 per cent of the world's total installed nuclear power electricity generating capacity was accounted for by only eight countries: the USA, France, the USSR, Japan, the FRG, the UK, Canada and Sweden. These nations possess the following:

(a) the technology required to obtain, manage and deploy this energy source;
(b) the large urban–industrial markets (load centres) needed to provide scale economies for large capital-intensive power generating units;
(c) grid transmission systems powerful enough to accommodate large power stations working at high load factors, and extensive enough to distribute their current over distance to several load centres.

The geopolitical dimension of world nuclear energy is one of its most striking elements, and it is those parts of the world without large spatially concentrated urban–industrial markets, without powerful grid systems and without a well-developed technology in nuclear physics and nuclear engineering construction where, generally, there is no nuclear power or where it is only lightly represented in the fuel mix used for electricity production. Such countries have found it to be an economic, political or technological necessity to use other means to produce the electricity that they consume.

These generalizations need immediate qualification. In East Asia nuclear power has been part of the recent rapid industrialization of Taiwan and South Korea. India has developed nuclear power to the level of possessing all parts of the fuel cycle, at considerable cost, and in part through scientific expertise that is in many respects 'home-grown'. The Bhabha Institute near Bombay is named in honour of one of the world's most distinguished nuclear physicists, Dr Homi Bhabha, who was the architect of India's nuclear programme. Figure 2.3 shows that Argentina and Brazil possess commercially operable nuclear power plants. However, several power generating utilities in a number of industrialized countries, including the USA, after building large new nuclear power stations, have decided to mothball or dismantle them in response to public opposition expressed in a variety of ways. Sweden has decided to use nuclear power as an interim measure to satisfy electricity requirements while simultaneously trying to develop alternative energy technologies, but France has implemented a policy designed to make nuclear power the mainstay of its electricity production industry for the foreseeable future.

Thus, in both the industrialized West and the Third World, some countries have pursued nuclear power, others have developed modest programmes and some have not adopted the technology at all. Therefore, the difference between the 'haves' and 'have nots' of the world nuclear power economy, made so very clear by Figures 2.1 and 2.2, is not without confusions, contradictions and paradoxes if particular international comparisons are attempted. Underlying all the issues, though, are the facts that massive areas of the world are energy hungry (between different countries of the world the range of energy use per head is at least 1,000 to 1 on a primary energy basis), that nearly 60 per cent of the energy used in the world today is provided by oil and natural gas, still plentiful but almost universally agreed to be finite, and that, despite the clear need to conserve oil as a vital fuel for transportation, a quarter of the world's oil output is being burned each year to produce electricity.

THE EVOLUTION OF THE WORLD PATTERN OF NUCLEAR POWER PRODUCTION

It is well known that the first stages in the development of the nuclear power industry were concerned not with providing the world with a new source of energy, but with the military objective of devising the atomic bomb. Therefore they were dominated by a governmentally perceived need

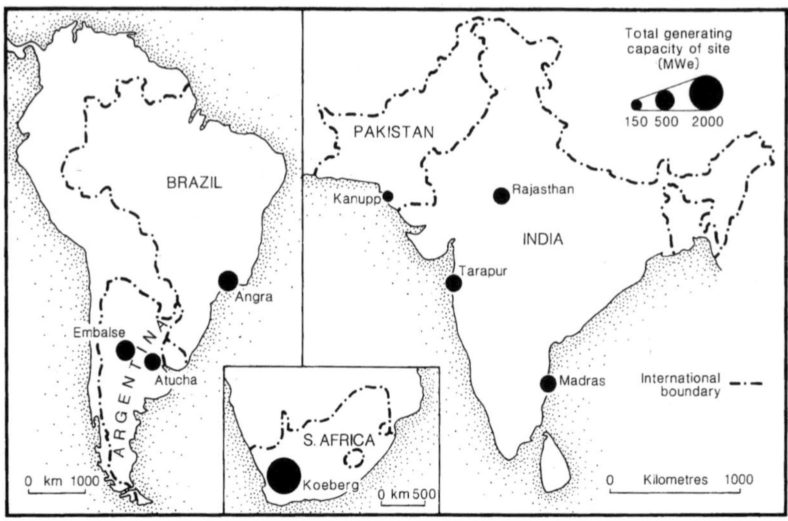

Figure 2.3 Location of nuclear power plants in South America, South Africa, India and Pakistan, 1987
Source: Nuclear Engineering International, *World Nuclear Industry Handbook 1988*
Note: South Africa's two 629 MWe units (Koeberg 1 and 2) at Duinefontain, Cape Province, 27 km north of Capetown, have been in operation since 1985. Only the Capetown region, 1,600 km from South Africa's coalfields, can make a convincing economic case for nuclear power in South Africa.

for secrecy, especially during the Second World War. In definitive work on the history of nuclear power Professor Margaret Gowing has clarified many of the issues involved and has portrayed the detail of decision-making at that time (Gowing, 1974, 1978). In 1932 Rutherford split the atom, in the UK. The results of splitting a uranium atom were discovered by Hahn and Strassman, in Nazi Germany, in December 1938. In September 1939 Germany was at war with the UK, Australia, New Zealand, and France, but two days before the outbreak of war Niels Bohr of Denmark and John Wheeler of Princeton published in the open scientific literature a theoretical explanation of the fission process. In the spring of 1939 Joliot-Curie's team in Paris showed that spare neutrons were also released to make a chain reaction possible, and the French report suggested that atomic bombs might be possible. In February 1940 two Austrian physicists who had become refugees in Sweden, Lise Meitner and Otto Frisch, provided a physical explanation of the process of nuclear fission and wrote a memorandum, 'on the properties of a radioactive super bomb'.

Secrecy successfully guarded the bomb project from Germany until the first weapon was tested, in July 1945, by which time the war in Europe had finished. But secrecy also became a real barrier between the Allies. From mid-1942 to mid-1943 Anglo-American atomic collaboration virtually ceased. It was renewed in a limited way in August 1943, when the UK's military research project was closed down and her scientists joined 'permitted' sections of the US programme. Secrecy also built barriers against the French; the Russians claim to have pioneered nuclear power but probably obtained crucial information through the defection of Klaus Fuchs and by kidnapping German scientists. The 1946 McMahon Act in the USA, designed for the domestic control of atomic energy, virtually prohibited the release of any information about the manufacture or use of atomic weapons, the production of fissile material or its use in electricity production. The Act was aimed equally at friends, enemies and neutrals, and was a disaster for the UK. In January 1947 the UK decided to make its own bomb and the technical consequence was that thereafter the UK's nuclear reactor development diverged very much from that of the USA. The USA maintained the need for secrecy from the early 1950s until late 1970, believing that '... the secret of the bomb was God's sacred trust to the American people and must remain their monopoly' (Gowing 1978).

Figure 2.4 helps to clarify the way in which the geographical pattern of world nuclear power production has evolved. While the USA and USSR produced token amounts of electricity in the early 1950s from small experimental reactors, Calder Hall, in the UK, was the world's first full-sized nuclear power station to produce electricity on a commercial scale, in 1956. France commissioned an experimental reactor at Marcoule, near Avignon, also in 1956, but the French *commercial* nuclear power programme began with Electricité de France's 70 MWe Chinon 1 reactor which went critical in 1962. Like the UK reactors, it was gas cooled and

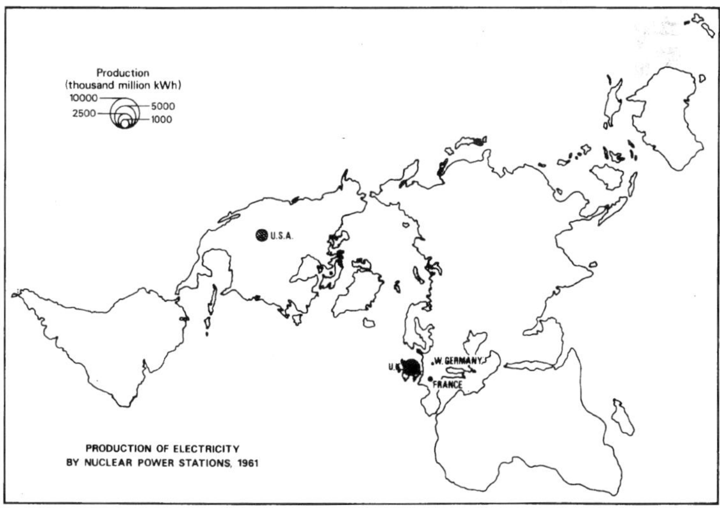

Figure 2.4a

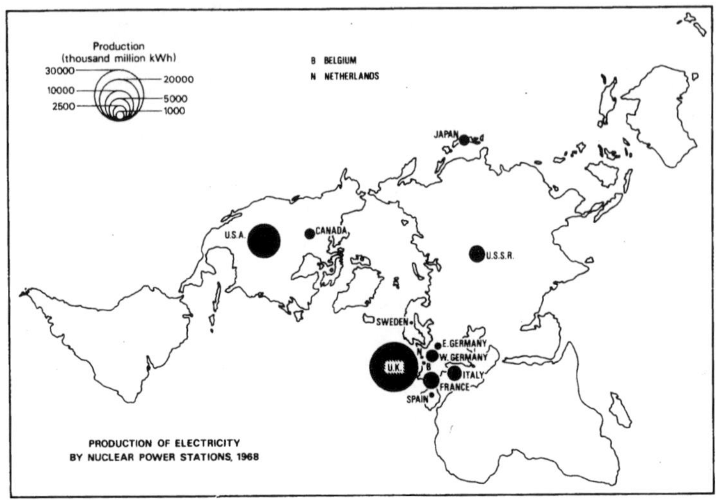

Figure 2.4b

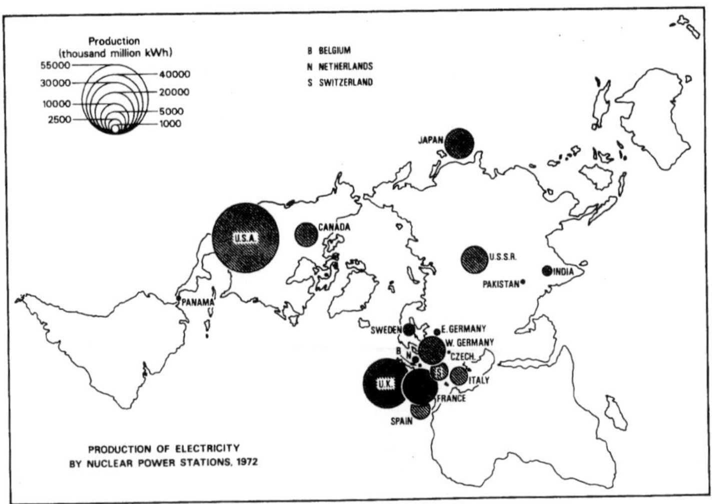

Figure 2.4c

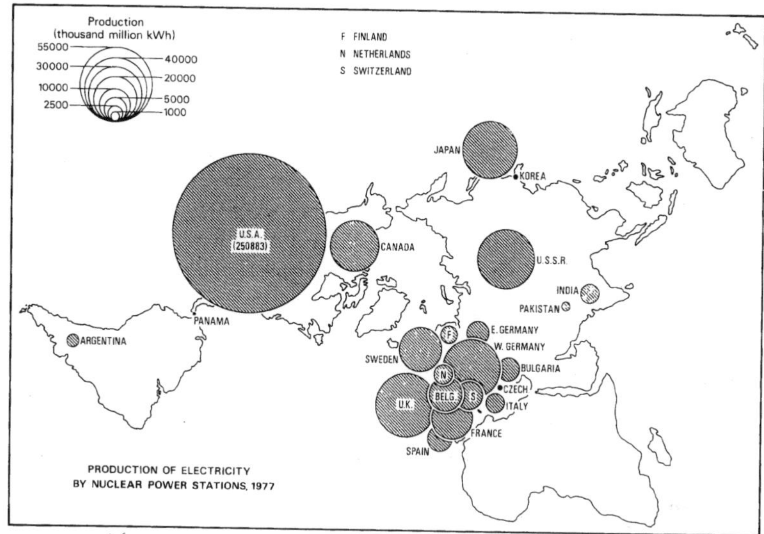

Figure 2.4d

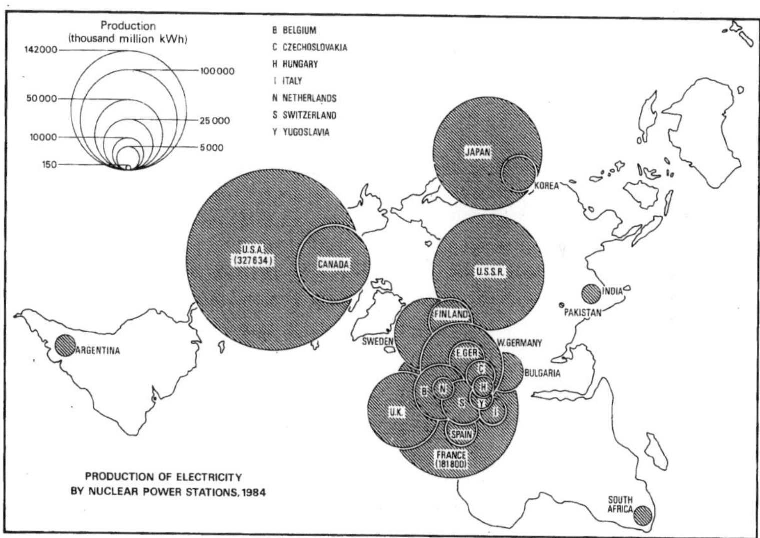

Figure 2.4e

Figure 2.4 The growth and spread of electricity by means of nuclear power
Source: UNO 1986
Note: These maps show the amount of electricity generated by nuclear power stations in the year indicated. The maps make clear (i) the early and continuing dominance by industrialized northern hemisphere countries, (ii) the early lead of the UK during the 1960s, (iii) the very rapid growth of nuclear power to a position of world dominance in the USA during the 1970s and (iv) slower growth in the USA since 1977 but, in Western Europe, the increasingly strong position of France.

graphite moderated in order to use natural uranium fuel. Chapel Cross, on the edge of Solway Firth, was the second nuclear power station opened in the UK.

These early units were not built primarily as commercially viable propositions to produce electricity; the UK reactors, for example, were built to provide weapons-grade plutonium and produced electricity as a by-product simply because the reactors in which this fissile material was produced had to be cooled down to a tolerable temperature; the heat was used to produce steam to drive turbines. Nevertheless, the UK had built up a lead by the mid-1960s, as Figure 2.4 and Table 2.3 show.

Table 2.3 Net electricity output from nuclear power stations in 1965

	Output (kWh)
UK	3,399,300
USA	1,232,400
USSR	895,700
Italy	622,000
France	372,000
Canada	219,300
Japan	168,700
FRG	65,000
Belgium	10,500
Sweden	9,000

Source: Blank 1966

Despite this early start, the UK's nuclear power industry was subsequently overtaken first by the USA, then by France and, most recently, by Japan and the FRG. In world markets the gas-cooled graphite-moderated reactors pioneered by the UK scientists in the 1950s, in a state of technological and scientific isolation from the USA, were effectively pushed aside in the 1960s and 1970s by the pressurized water and boiling water units developed in the USA. These reactors have become world-dominant means of producing nuclear power. Early UK success in selling nuclear power stations to Italy, Japan and India has been overwhelmed by US light water reactors built in many countries under turnkey contracts, as part of a 'nuclear island', or under licence to US firms such as Westinghouse and General Electric. Light water reactors using enriched fuel were a logical choice for commercial electricity production in the USA because they had access to an ample supply of fuel from the three large gaseous diffusion enrichment plants operating at reduced capacity but built originally with US government money to provide weapons-grade materials. It was not until the end of 1970 that the USA began to consider seriously the possibility of sharing fuel enrichment technology with friendly powers.

GAPS ON THE WORLD MAP: THE NON-NUCLEAR POWER NATIONS

To support their growing populations in greater dignity and comfort the Third World countries must build up their material infrastructures and production systems to a more advanced level. They need plentiful supplies of energy to do this, but, in comparison with Western rates of usage, both per capita consumption and availability of energy are pitifully low. The Third World countries, with 70 per cent of the world's population, use only 14–16 per cent of all the manufactured energy (Ott and Spinrad 1985; Hedley 1986). If world population were to double over the next half century, much more of the growth would be in the developing world than in the developed world. If the present per capita energy use in the developed economies were maintained, and per capita energy use in the developing countries increased to a quarter of that in the developed countries, there would be a need to increase world energy supplies by some 125 per cent between 1988 and 2030.

Such estimates of world energy futures are almost irrelevant, however, for the energy famine plaguing the Third World is here *now*, and it dwarfs any and all of the industrial world's periodic energy problems. Traditional 'poor country' energy sources, principally fuelwood, charcoal and forage for draught animals, are growing ever more scarce and expensive. However, the low population density of the African continent, the lack of many large towns, the small and scattered populations, and the shortage of industrial markets all combine to provide a poor economic environment for building large power plants of any kind. Electricity consumption is low except for urban or industrial enclaves such as capital cities or the Copper Belt in Zambia.

Africa is not alone, however, for one of the most noteworthy features of Figures 2.1–2.4 is the dominance by a small number of industrialized countries of the world pattern of nuclear power, and the almost complete absence of this energy form in the Third World. This is not because the nuclear powers have always tried to keep the technology to themselves. It is true that secrecy dominated the early history of the nuclear industry, but on 6 December 1953 President Dwight D. Eisenhower launched the USA's Atoms for Peace programme and later offered the first donation of fissile material, U-235, to the embryonic International Atomic Energy Agency; the USA had abandoned the 9 years long closed-door policy regarding the dissemination of nuclear power information and technology imposed by the McMahon Act. Between 1956 and 1962 the USA went on to spend $8.95 million on programme grants for research reactors to twenty-six countries, including thirteen in the Third World, and provided personnel training and fissile materials, often on a grant basis.

Despite these initiatives, very few of the Third World countries have reached the point of having commercial nuclear power plants connected to their electricity supply networks. In the literature on nuclear power in poorer parts of the world, a number of explanations, many of them seamed with contradictions, have been offered for this situation. Four are of particular importance:

(i) World oil prices fell markedly in real terms for fifteen years prior to 1973 to the point where the economic attractiveness of nuclear power could not always be convincingly demonstrated in many Third World energy markets. There is prima facie evidence to indicate that military as much as civil applications have attracted some of the Third World countries who have adopted the technology. In such circumstances, strict cost comparisons between electricity produced by nuclear and other means become less relevant, but they have been cogent for those without such military aspirations.

(ii) Many Third World countries are too poor, and too much in debt, to buy and sustain such an expensive, complex and demanding technology without sacrificing things perceived as more important for their people. Nuclear power stations are capital intensive. They not only require much money to build but also need scarce foreign exchange. Loaning organizations have a crucial role to play, but indebtedness amongst the Third World countries is endemic. The total medium- and long-term debt of Third World nations stood at US$530 billion in 1983. Just eight countries were responsible for half this debt, and they include those most deeply involved in international trade and Western technology (Clausen 1983). Three of them, Brazil, India and Argentina, have acquired nuclear power technology well beyond the research reactor stage, and Brazil in particular has accumulated a substantial proportion of its indebtedness through this fact. Frequently, loans arranged through the efforts of Western nuclear power station construction consortia on turnkey contracts, sometimes guaranteed by their governments, have been the way in which nuclear power stations have been acquired by Third World countries. Yet to argue in consequence that nuclear power is the wrong technology for power production in the poorer countries may be an oversimplification. Compared with imported oil, nuclear power stations *once in operation* are not readily affected by inflation. Fuel costs account for only 20 per cent of the nuclear fuel cycle costs in a typical light water reactor. The fuel savings are substantial enough for some of the less developed countries to have given serious consideration to the nuclear power option. The aspirations of Third World countries to solve more than a fraction of their energy problems without nuclear power, by using sources such as solar energy or gas and liquids from biogas, are receding very quickly because of the problems of making the alternatives work and of financing their development (12th Congress, Triennial World Energy Conference, New Delhi, September 1983).

By 1980 it was clear that the anticipated strong growth in reactor requirements was not going to materialize as world economic growth had slowed down, energy conservation measures were resulting in more efficient energy use and concern over the environment resulted in opposition to nuclear power by some pressure groups. Nuclear power programmes in most countries were revised downward, in some cases dramatically. But even before this the sometimes desperate efforts of Western nuclear construction consortia to stay solvent in a highly competitive market by selling reactors abroad sometimes had ludicrous results. For example, on 1 May 1969 the UK Atomic Energy Authority announced that long-term credit from the UK government would be used to sell a 450 MWe nuclear power station to Greece, for a site at Lavrion, 30 miles (48 km) from Athens. The deal was linked to a contract involving the purchase of 40,000 tons of Greek tobacco! Ultimately it was abandoned, but intense lobbying from industrial multinationals such as General Electric, Westinghouse, Framatome, Bechtel, Northern Engineering Industries, ASEA-ATM (Sweden), AECL, GEC (UK), Combustion Engineering, Kraftwerk Union and others has made banks and governments more inclined to arrange loans for nuclear power stations than for wind generators or biogas plants. India has 250 million cattle which produce 800 million tons of wet dung during a year. It also has around 600,000 villages. Yet there are only 50,000 biogas plants in use at present, compared with 6 million in China. The World Bank has no specialized energy offshoot. Thus the Third World ends in the same position as the West; if it has to make progress in oil substitution, it has for the medium term at least to use energy provided by coal, gas, hydraulic sources, wood, dung – or nuclear power.

(iii) A generally accepted rule of thumb in the electricity supply industry is that no one power station should account for more than 10–15 per cent of the total system installed generating capacity, otherwise an unexpected shutdown of that station could cause a blackout throughout the system. However, to achieve optimum scale economies in construction and operation, the commonest size of unit now being built in the industrialized countries is in the size range 800–1,200 MWe with 600 MWe as an approximate minimum threshold. Studies have shown that there was little or no market for nuclear power units of 300 MWe or less even in the 1970s, when unit sizes were generally smaller (Falls 1973). If a new nuclear power station were to be contemplated in the 1980s and early 1990s, a network with a minimum installed generating capacity of 6,000 MWe might not be mandatory but would certainly be looked for, with a minimum individual station size of 600 MWe. A large number of Third World countries do not reach this approximate threshold, even though this does not seem to have deterred some of them from acquiring nuclear plants in the past.

The allocation of responsibility for the organization of electricity supply also matters. In India, for example, power supply has been the responsibility of individual states, with central government ministries having powers that are mainly regulatory and advisory with no executive function. Each state is jealous of its own rights and has insisted upon building and operating power stations within its own boundaries even when it would be preferable to join a neighbouring state to set up a larger and more economical power generating unit (Henderson 1975; Hart 1983).

(iv) A particular and continuing problem attached to the dissemination of nuclear power technology has always been that of nuclear weapons proliferation. In 1971 the Fourth International Conference on the Peaceful Uses of Atomic Energy and the General Conference of the International Atomic Energy Agency both produced recommendations that efforts should be intensified to assist developing countries in planning for nuclear power. Subsequently, the IAEA sponsored power reactor survey and siting sessions to less developed countries, conducted feasibility studies, organized technical meetings and awarded training fellowships in the field of nuclear power technology. However, the powers that have developed nuclear technology are aware that it takes only about 10 kg of plutonium to make a powerful nuclear weapon. U-235, U-233 and Pu-239, one at least of which is needed for nuclear power, are all suitable to a greater or lesser degree for atomic explosives and natural, even if not indeed indispensable, for thermonuclear explosives. Thus a nuclear power programme presupposes the existence, in some form, of material that can be used in bombs. Furthermore, if U-238 is present in a power reactor, as it will be if natural uranium or uranium of quite low enrichment is used, the reactor will itself make plutonium. If thorium is present, the reactor will make U-233. The fuel most widely used in power reactors, U-235 of moderately low enrichment, is not directly usable for weapons, but the further enrichment required to make it so is not a formidable undertaking if a country has an isotope separation plant. It has been known for over three decades that an enriched U-235 reactor, embodying normal uranium or thorium, will make material suitable for weapons (Oppenheimer 1957). Thus, once a country has a nuclear power plant, it is well on the way to being able to produce nuclear weapons, and the motives of some Third World countries in wishing to acquire the technology have come under scrutiny from time to time. This has been the case particularly when they have gone to great lengths to acquire not only reactors but the potentially more dangerous enrichment and used-fuel reprocessing facilities as well. Thus a vitally important question as far as the economic and political geography of the nuclear fuel cycle is concerned is: 'What happens to the plutonium?' Controlling the spread of the means by which plutonium is produced, and accounting down to very small quantities for that produced, have become major international preoccupations.

The simplest route available for plutonium production is the reprocessing of the spent nuclear fuel removed from a reactor to separate its

Table 2.4 Parties to the Nuclear Non-Proliferation Treaty

1	Afghanistan	61	Liberia
2	Antigua & Barbuda	62	Libyan Arab Jamahiriya
3	Australia	63	Liechtenstein
4	Austria	64	Luxembourg
5	Bahamas	65	Madagascar
6	Bangladesh	66	Malaysia
7	Barbados	67	Maldives, Republic of
8	Belgium	68	Mali
9	Benin	69	Malta
10	Bolivia	70	Mauritius
11	Botswana	71	Mexico
12	Bulgaria	72	Mongolia
13	Burundi	73	Morocco
14	Cambodia	74	Nauru
15	Canada	75	Nepal
16	Cape Verde	76	Netherlands
17	Central African Republic	77	New Zealand
18	Chad	78	Nicaragua
19	China, Republic of	79	Nigeria
20	Congo	80	Norway
21	Costa Rica	81	Panama
22	Cyprus	82	Papua New Guinea
23	Czechoslovakia	83	Paraguay
24	Democratic Yemen	84	Peru
25	Denmark	85	Philippines
26	Dominican Republic	86	Poland
27	Ecuador	87	Portugal
28	Egypt	88	Romania
29	El Salvador	89	Rwanda
30	Ethiopia	90	St Lucia
31	Fiji	91	Samoa
32	Finland	92	San Marino
33	Gabon	93	Senegal
34	Gambia	94	Sierra Leone
35	German Democratic Republic	95	Singapore
36	Germany, Federal Republic of	96	Solomon Islands
37	Ghana	97	Somalia
38	Greece	98	Sri Lanka
39	Grenada	99	Sudan
40	Guatemala	100	Surinam
41	Guinea-Bissau	101	Swaziland
42	Haiti	102	Sweden
43	Holy See	103	Switzerland
44	Honduras	104	Syrian Arab Republic
45	Hungary	105	Thailand
46	Iceland	106	Togo
47	Indonesia	107	Tongo
48	Iran, Islamic Republic of	108	Tunisia
49	Iraq	109	Turkey
50	Ireland	110	Tuvalu
51	Italy	111	Uganda

52 Ivory Coast	112 UK
53 Jamaica	113 USA
54 Japan	114 United Republic of Cameroon
55 Jordan	115 Upper Volta
56 Kenya	116 USSR
57 Korea, Republic of	117 Venezuela
58 Laos People's Democratic Republic	118 Vietnam
59 Lebanon	119 Yugoslavia
60 Lesotho	120 Zaire

Source: IAEA 1983b

Pu-239 component. This was the means used by Indian scientists at the reprocessing plant at Trombay. The more difficult route is uranium enrichment. Where nuclear fuel requires, on average, a fourfold enrichment of the fissile U-235 from 0.7 per cent to about 2.5 per cent, a nuclear explosive requires nearly fivefold U-235 enrichment. Yet a process capable of achieving low levels of enrichment can be adapted to produce the high levels. What is more, it is easier to engineer a nuclear explosion with U-235 than with Pu-239. Nations known to have enrichment technology capable of making nuclear explosives are the USA, the USSR, the UK, France, China, the FRG, the Netherlands, South Africa, Israel, Argentina and India. Pakistan, Vietnam, North Korea and Colombia are thought to have sufficient nuclear facilities to manufacture nuclear weapons. Every nation with indigenous uranium resources can make a plausible case for possessing enrichment as a legitimate means of doubling the value of its uranium in the world market for nuclear fuel.

The Nuclear Non-Proliferation Treaty (NPT), negotiated in 1968 and in effect since 1970, has been one attempt to establish control and accountability. Generally, if a state has signed and ratified the Treaty, it will find it easier to acquire nuclear power technology than if it has not, but not having signed and ratified the treaty has not prevented some countries from obtaining the technology. The Nuclear Non-Proliferation Treaty consists of a Preamble and eleven Articles. The former contains a commitment to seek an end to the nuclear arms race and a comprehensive weapons test ban treaty. It also affirms the principle that the benefits of the peaceful uses of nuclear technology should be available for peaceful purposes to all parties of the Treaty. Article 1 pledges the nuclear signatories not to transfer nuclear weapons to other states. Article 2 pledges the non-nuclear signatories not to acquire them. Article 3 deals with safeguards, and the remaining substantive articles underline the principles spelled out in the Preamble. For the purposes of the Treaty, a nuclear weapons state is one that had 'manufactured and exploded a nuclear weapon or other nuclear explosive device prior to 1 January 1967'. That means the UK, the USA, the USSR, France and China. The last two have refused to have anything to do with the Treaty, as has India, which developed and exploded a nuclear device on 1 May 1974. Over 120 countries have signed the Treaty but many others, including Argentina, Brazil, China, Chile, Cuba, Pakistan, Israel, South Africa and Spain, as well as the three already mentioned, have not (Table 2.4).

Clearly, therefore, the Non-Proliferation Treaty has not been entirely effective as a means of containing and directing the diffusion of nuclear technology. Various attempts to plug the leaks have been made by individual countries and groups of countries. In 1971 Professor Claude Zangger, a senior Swiss energy official, gathered twenty nuclear nations together in an effort to reach common understanding on the control of nuclear exports. The committee produced a 'trigger list' of items essential to plutonium manufacture, but found it difficult to do this for those parts of the nuclear fuel cycle concerned with uranium enrichment and fuel reprocessing. By 1974 this weakness in the trigger list, and the incomplete acceptance of the Nuclear Non-Proliferation Treaty by the nuclear powers, had become very apparent. The USA in particular became disturbed by India's demonstration of a nuclear explosive in 1974, by negotiations between the FRG and Brazil for the sale of enrichment and reprocessing technologies which the US government had forbidden its own industry to export, and by negotiations by France with Pakistan and Korea for the sale of similar technologies. At the instigation of the USA, representatives of the UK, the USA, the USSR, France, the FRG, Canada and Japan met in London in April 1975 to discuss the sensitive technologies of enrichment, reprocessing and production of heavy water. They agreed not to export nuclear technology unless three assurances could be obtained. The first was that nuclear exports would not be used to make nuclear explosives, peaceful or otherwise. The second was that exports would be adequately protected against the possibility of theft or sabotage. The third was that any re-exporting of nuclear technology by a receptor country would require the same assurances on safeguards, use and physical protection. Subsequently, the Netherlands, Sweden, the GDR, Belgium, Italy and Czechoslovakia joined this 'London Club'. The USA went further in 1978, during the Carter Administration, when Congress passed the Non-Proliferation Act specifying that it could not provide nuclear materials or nuclear technology to countries which refused to accept 'full scope' inspection safeguards of nuclear installations by the IAEA. Safeguards are accounting checks carried out periodically to detect any diversion of nuclear materials to nuclear explosive production, and 'full scope' became a shorthand term to indicate all a country's nuclear facilities. The USA has also argued for the creation of a limited number of multinational regional fuel recycling centres, factories large enough to offer, through economies of scale, attractive prices for services requiring the 'sensitive technologies' to nations with developing nuclear programmes. These would be operated under IAEA safeguards and under high standards of physical security. This has proved to be politically impossible to implement, especially in Latin America and the Middle East.

Many countries who have had no intention of developing nuclear technology in any form have signed the Non-Proliferation Treaty. Some countries have remained aloof from it, either because they feel that it discriminates against poor-world countries or because of their own perceptions of national security. At least seven countries may have developed nuclear weapons capability by 1989 outside the safeguards of the full IAEA inspection arrangements. These are Argentina, India, Pakistan, Israel, South Africa, North Korea and Colombia.

Numerous ramifications have grown out of this complex geopolitical situation and have profoundly affected the channels by which nuclear power technology has or has not spread throughout the Third World. For example, until a few years ago, France, Japan and Belgium used US reactor technology under licence in their own countries and therefore could not tender to build nuclear power stations in countries with which the USA did not want them to deal. Perhaps the largest divergence of view over exports of nuclear technology has occurred between France, which is a nuclear weapons and nuclear power state and not a signatory to the Nuclear Non-Proliferation Treaty, but which endorses nuclear safeguards, and the other Western partners. In the mid-1970s France undertook to sell a nuclear waste treatment plant to Pakistan but, after US and Canadian pressure, decided to abandon the project because of the associated dangers of nuclear weapons proliferation. In 1975 France dropped a contract for a nuclear power plant in South Korea under US pressure, but early in 1983 began talks with Pakistan on supplying a 900 MWe PWR for a site at Chasma, 150 miles (242 km) south of Islamabad on the River Indus. Pakistan was agreeable to safeguards being negotiated with the IAEA for this plant, but refused to place its heavily guarded uranium-enrichment plant at Kahuta, near Islamabad, under international inspection. Most Western sources remain convinced that Pakistan has acquired nuclear weapons capability (Hart 1983). Thanks to the installation by France, without safeguards, of a plant capable of separating out weapons-grade plutonium from spent fuel at Dimona, Israel's goal of making nuclear weapons was achieved by 1967 in the form of several warheads. In 1978 the Dutch threatened to veto the construction by Kraftwerk Union of nuclear power stations in Brazil, under the Anglo-Dutch-German treaty governing the production of enriched uranium at the Almelo in The Netherlands. All three signatories have to agree to uranium exports.

Only four of the barriers that have inhibited the spread of nuclear power stations throughout the Third World have been discussed here. There are many others (Poneman 1982), but cost needs to be emphasized. A 1,200 MWe nuclear power station could easily cost US$3,500 million at 1987 prices. Brazil has one operable reactor (Angra 1 (657 MWe) which is so frequently turned on and off that it has been nicknamed Firefly) and two others under construction in 1989, for completion in 1992 and 1995. By September 1986 the total investment needed to finish the two new plants was US$2,200 million, not including servicing foreign borrowing at US$240 million. Brazil's nuclear power programme has run into trouble financially. Nuclebras, the state company formed to develop the industry, had an operating deficit of US$740 million in 1986 and by the Autumn had failed to pay many of its suppliers' bills. Argentina, however, is self-sufficient in both technology and the fuel cycle for the production, operation and export of heavy water reactors using natural uranium mined within the country. It has a gas-diffusion enrichment plant and a fuel reprocessing facility. Some countries, such as Argentina and India, have no wish to become dependent upon Western or Communist Bloc supplies for reactor fuel supplies and fuel reprocessing; some do not have sufficient qualified personnel to handle the technology; some do not have the transport facilities to move plant construction items to possible sites. Some parts of the developing world are in seismically sensitive areas where it is difficult to guarantee earthquake-free sites; others have water supply problems.

However, there does not seem to be a consistently positive relationship between apparent need and actual acquisition of nuclear power plants. Indigenous energy resources vary from one Third World country to the next, and it might be expected that a reasonable sufficiency of fuel and energy resources for a foreseeable future might reduce the wish to acquire nuclear electricity generating capacity. This has been the case in Nigeria, which is a significant oil producer. Yet so is Mexico, a country which in 1989 had an active programme to install 1,350 MWe of nuclear capacity by 1993. The reason for this apparent inconsistency is that in virtually every case the economic assessments made by individual countries are tempered by and sometimes led by political calculation. One example is the Pakistan programme, which is interpreted by some Western observers as an attempt to keep up with India (Hart 1983). In Latin America, the nuclear power programmes in Brazil and Argentina may not be unconnected with the competition between these two states for ascendancy in Latin American affairs. Argentina established its Atomic Energy Commission (CNEA) in 1950, but refuses to sign the Non-Proliferation Treaty.

Getting past, or under, the barriers may be a demanding task for a Third World country wishing to acquire nuclear power stations, but, by various routes, some have done it and their number promises to increase (Table 2.5). Argentina, Brazil, India, Pakistan and South Africa have made considerable financial investment in their own research and development efforts to preserve as much independence from the nuclear powers as possible in acquiring not just reactor technology but also enrichment and reprocessing plant. The Philippines, South Korea and Taiwan have bought nuclear power stations 'off the peg' from Western consortia, often on turnkey contracts, with the objective of obtaining a large tranche of

Table 2.5 Nuclear power reactors in operation in developing countries on 31 July 1987

Power station	*Location*	*Power rating (MWe)*	*Year when first operable*	*Financial/contract arrangements*	*Benchmarks*
Argentina					
PHW (CANDU)					
Embalse	Cordoba Province	684	1984	Embalse contract was with AECL (Canada) and Italiampianti (Italy); large loan from Canada's Export Development Corporation with ancillary technology transfer agreement	Argentina received Atoms for Peace programme grant for a research reactor in 1962 and with India and Pakistan was one of the three Third World adopters of nuclear power for electricity production; National Atomic Energy Commission (CNEA) formed in 1950; not an NPT signatory but reactors are under IAEA safeguard
PHWR					
Atucha 1	Lima, on the Parana River 110 km northwest of Buenos Aires	357	1974		
Total		1,041 (2 units)			
Brazil					
Angra dos Reis 1	Itaorna, RdJ, on the coast, on Angra dos Rios Bay, half-way between Rio de Janeiro and Sao Paulo	657	1982	Half cost met by US banks led by Export-Import Bank; expensive programme with not much to show for it; Kraftwerk Union major contractor for Brazil's reactors; has large uranium reserves	Went on line in 1982 but frequent operating problems since (the Firefly); Angra units 2 and 3 unfinished 1988 despite planned 1984 completion date; National Nuclear Energy Commission created 1956 when Atoms
Total		657 (1 unit)			

Power station	Location	Power rating (MWe)	Year when first operable	Financial/contract arrangements	Benchmarks
					for Peace grant allocated for 5 MWe research reactor; not an NPT signatory
Pakistan					
PHW					
Kanupp	CANDU reactor, near Karachi	137	1972	Turnkey project with Canadian General Electric Co.; has very little uranium; may have underground enrichment plant at Multan 320 km south of Rawalpindi (Hart 1983)	Recipient of Atoms for Peace research reactor 1960; Atomic Energy Commission set up 1955; not an NPT signatory
India					
BWR					
Tarapur 1	Maharashtra, 100 km north of Bombay	160	1969	General Electric main contractor for Taraput; AECL for Rajasthan; Canada's Export Development Corporation supplied a loan for RAPS 1 and 2; slender uranium reserve but plenty of	Firewood the single most important energy source; regional grid systems only partly interconnected; cities main markets for electricity; first Third World country to operate a nuclear power station; not
Tarapur 2		160	1969		
FBR					
FBTR	Kalpakkam, Tamil Nadu	13	1988		
PHW (CANDU)					
Madras MAPS 1	Kalpakkam, Tamil Nadu,	235	1984		

Table 2.5 continued

Madras MAPS 2	near Madras	235	1986	thorium (Karela); MAPS 1 and 2 have minimal foreign involvement	an NPT signatory; Atomic Energy Commission set up 1948 but technological kernel lay in Tata Institute of Fundamental Research set up in Bombay in 1945, at instigation of H.J. Bhabha, based on the West's nuclear power programmes; Indian civil nuclear power programme began in earnest in 1954 with the establishment of a Department of Atomic Energy
Rajasthan RAPS 1	Kota Rajasthan	220	1975		
Rajasthan RAPS 2		220	1983		
Total		1,243	(7 units)		

Sources: Nuclear Engineering International, *World Nuclear Industry Handbook 1988*; World List of Nuclear Power Plants, *Nuclear News* 30(2), 1987; IAEA, *Nuclear Power Reactors in the World* (The 'Vienna Index'); Poneman 1982; Hart 1983

Note: Power reactors *in operation* are only part of the Third World nuclear energy story. In 1972 Mexico began to build Laguna Verde 1 and 2 (660 MWe each) at Cardel, Vera Cruz, with US General Electric and financial backing from the US Export-Import Bank. A change in government policy resulted in suspension of work for 4 years; the nominal completion date is 1990. Iguape 1 and 2 (each 1,245 MWe) in Brazil are years behind their construction schedule. Iran, under the Shah, signed a turnkey project with Kraftwerk Union for two 1,200 MWe reactors at Bushehr, on the Persian Gulf, but the contract was terminated in 1979 and whilst a preliminary agreement was made between Kraftwerk Union and the Khomeni regime to complete one of the reactors in 1983, the completion date is indefinite. In order to ensure a supply of enriched uranium for its reactors Iran lent the French Atomic Energy Commission US$1,000 million over a 15 year period for a 10 per cent share in France's Tricastin gaseous diffusion fuel enrichment plant. In Iraq, a 70 MWe reactor under construction at Osarik with French assistance was destroyed by an Israeli air attack in June 1981 as it was suspected that the reactor would have been used to produce weapons-grade materials. Israel has a 26 Mwe 'research reactor' at Dimona built with French assistance in the early 1960s, and was a recipient of an Atoms for Peace grant for this reactor in 1958. There is a high probability that Israel has nuclear weapons. The Philippines carried on with building large (2 × 620 MWe) PWR units at Napot Point, Morong, on the Bataan Peninsula 70 km west of Manila to a late stage and then cancelled the project in 1987. Cuba has nominated Cienfuegos, on the southern coast, for a 440 MWe Soviet PWR, with a completion date in 1991.

generating capacity as quickly as possible. When they have shown signs of wishing to obtain other parts of the fuel cycle, they have generally acceded fairly quickly to Western pressure not to do so. China, with a total installed electricity generating capacity in 1984 of 83,000 MWe, slightly below that of France, is on the brink of obtaining a nuclear power plant. Israel has nuclear weapons capability and has announced plans to build nuclear power plants but has done very little to follow them through. Early in 1987 the French and FRG nuclear plant manufacturers Framatome and Kraftwerke Union announced their co-operation on a contract for a 600 MWe station for Indonesia. Turkey and Greece have made efforts in the past to acquire nuclear power stations. They proved abortive, and neither country currently seems very committed to a renewed effort.

In the longer term, it is possible that some of the oil-rich Middle East states may provide a market for nuclear power. In the medium term oil may prove to be too valuable a source of export earnings to be burned locally as power station fuel. The Gulf States have the capital and, with the rising living standards created by oil revenues plus the large quantities used in extracting oil, electricity demand is rising very rapidly in many states (Kuwait's per capita electricity consumption is exceeded only by that of the USA and Sweden, but the five Middle Eastern states with the fastest-growing capacities are not likely to reach 10,000 MWe before the year 2000). Natural gas is more valuable to reinject for enhanced oil recovery than to burn. For the Gulf States nuclear power would have particular attractions for water desalination combined with electricity production. Almost certainly, however, Egypt will be the first Arab state to generate its own nuclear power. The rise of its relatively industrialized economy gives the necessary base load and scale economies, the Aswan Dam has not left much more hydroelectric power potential undeveloped and relations with the USA are good. Egypt ratified the Non-Proliferation Treaty in 1981 and since then an agreement has been made in principle with France on the construction by Framatome of two 1,000 MWe reactors, the USA has undertaken to supply enriched uranium, bilateral co-operation accords have been agreed with the FRG, the USA and Canada, and a site has been chosen near Alexandria (El Dabaa).

FUTURE PROSPECTS

Past projections of the future growth of nuclear power over the medium and longer term have not been very accurate and have had to be frequently revised, usually downwards. For example, in a report published in 1987 the Nuclear Energy Agency of the Organization for Economic Co-operation and Development (OECD/NEA) lowered projections for the year 2025 by half from those it had made in 1982. The reasons given were the world economic recession, reduced power station ordering rates and lower growth rates in electricity demand (OECD/NEA 1987). It is not unknown for computer-based energy production models from prestigious sources, which include nuclear power centrally in their scenarios, to seem less convincing under detailed independent scrutiny. A well-known example is the two-volume world energy model entitled *Energy in a Finite World*, published in 1981 by the International Institute for Applied Systems Analysis (IIASA), an influential East–West resource research organization based in Laxenburg, Austria. For a time this model, produced by an international team headed by Dr Wolf Hefele of the FRG, held a central place in the energy planning of many western governments (Tucker 1984). However, in November 1984 the journal *Policy Sciences* devoted an entire issue to deconstructing the Hefele model (Keepin *et al.* 1984). The critical analysis revealed the apparently sophisticated, complex and supposedly 'desensitized' model developed by the IIASA team to be highly sensitive in virtually all aspects. In this context 'sensitivity' means a large change in predicted patterns if a small change is made to a single assumption. The elaborate IIASA model had been assumed to be the most sensitive, the most robust and the best available, but the *Policy Sciences* review demonstrated that many of the compartments of the highly prized model comprised the interaction of merely two sets of information and that, in its various sections, it could be mimicked by very simple models which turned out similar, and insensitive, results.

Even worse, it was shown that the Hefele scenarios stemmed from shifts of assumptions which Keepin *et al.* argued to be without foundation. In the context of nuclear power needs, this proved particularly important. The model, even on a low rate of energy growth, indicated that large amounts of nuclear power would be required by the end of this century and a large fast breeder reactor programme would be essential thereafter. In the model uranium prices were given an upward shift around the turn of the century, a sudden rise nowhere justified in the analysis. Yet it was the shift which made the world seem to need large numbers of costly fast breeder reactors.

Thus the IIASA model, now largely abandoned, has cast a cloud over predictive modelling as a basis for long-term energy policy. Yet the future position of nuclear power requires some grasp of the future world energy scene.

Decisions to build many of the world's existing nuclear power stations were taken during a period of very rapid growth in world primary energy requirements, which lasted from 1950 to the first major oil price rise by the Organization of Petroleum Exporting Countries (OPEC) in 1973 (Figure 2.5), and when forecasts from authoritative sources were predicting an enormous demand for energy by the end of the century. In July 1971, for example, the Resources and Transport Division of the United Nations Department of Economic Affairs produced almost unbelievable estimates of total world energy consumption in the year 2000 of 30,216 million tons of coal equivalent (mtce) for an exponential trend and 28,086 mtce for a logistic trend (UNO 1971). The forecasts seem to have given little

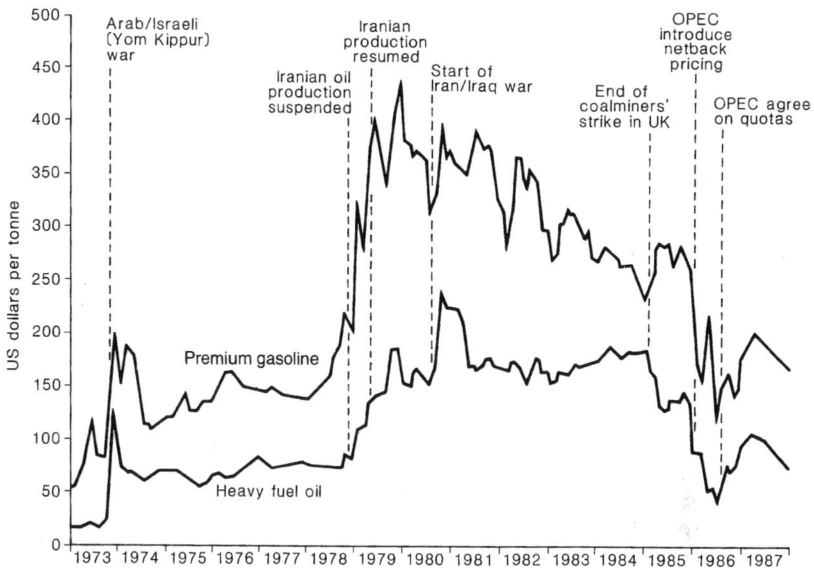

Figure 2.5 Trends in Rotterdam oil prices 1973–87. It is clear that the 1978–9 oil price rise was larger than that of 1973–4 and that since 1980 (before Iraq's 1990 invasion of Kuwait) there has been a marked downward trend in prices
Source: *Platt's Oilgram, Price Report and Statistical Review of World Energy*, June 1988

recognition to the fact that Third World countries could not realistically expect to make a substantial increase in their consumption. Nevertheless, had the growth rate of the 1960s continued to the mid-1980s, the world's demand for primary energy in 1984 would not have been the figure of 8,981 mtce actually required, but somewhere between 12,000 and 15,000 mtce on a simple arithmetic extrapolation. The option of being able to postpone or cancel nuclear power stations would not have been so readily available to power utilities. Thus demand and trends in the energy market-place have an essential part to play in the decision-making environment of the nuclear fuel cycle.

The Armstead–Tester graph

Armstead and Tester (1987) published a graph showing both the world's historic primary energy demand from 1900 to 1982 and the postulated requirements to the year 2000. Figure 2.6 reproduces this diagram, updated by using the same sources for historic trends but retaining the forecasts for the period 1984–2000. Numerous organizations have engaged in energy forecasting. The shaded area on the graph incorporates forecasts from nine such sources.

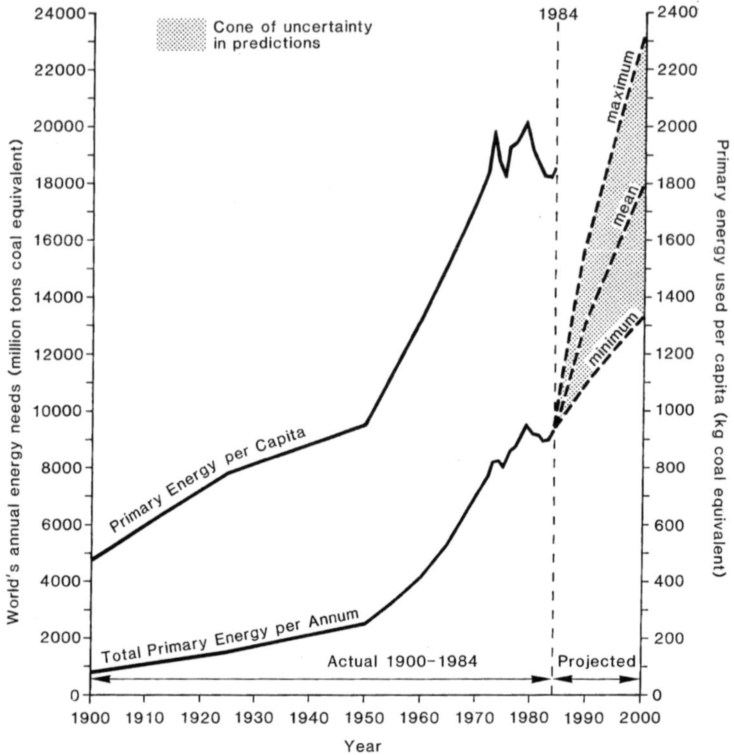

Figure 2.6 The Armstead–Tester graph of world primary energy consumption from 1900 to 2000 AD
Source: Updated from a diagram in Armstead and Tester 1987: 6

The sharp check in the energy demand growth rate over the decade 1973–83 is very obvious, but since 1983 growth has resumed and the minimum extrapolation on this graph, the lowest of the lines defining the shaded cone of uncertainty, indicates a world primary energy demand approaching 14,000 mtce by the year 2000. This figure is based on a demand curve falling away from the extremely steep trend line of 1950–73, but it is still a daunting target. The rest of the twentieth century seems unlikely to be disturbed by insurmountable problems in world energy supply. There is no fear of a sustained energy gap for the industrialized economies of the northern hemisphere. However, over 90 per cent of the world's present energy usage is supplied by fossil fuels and the broad implication of Figure 2.6 is that, world-wide, fossil fuel consumption will rise inexorably toward the end of the century and beyond. Only rapid and concerted international action to help counteract the acceleration of the greenhouse effect is likely to alter this substantially.

A slackening of the demand rate for primary energy gives governments and energy supply utilities a breathing space to reflect on their preferred medium- and longer-term fuel mix for electricity supplies. They must surely consider very carefully the need for a mix sensibly balanced between fuels, not only to safeguard security of electricity supplies but also to provide some degree of insurance against increased prices of any particular fuel. It is still possible to conceive a world soon after the year 2000 with dwindling oil supplies again dominated by the politically capricious and cartel-minded Gulf States and with the greenhouse effect becoming a vital international policy issue. Viewed in these terms it is difficult to see how the world's energy supplies can meet the demand in prospect without a substantial tranche of nuclear power additional to that in existence now, even though the nuclear fuel cycle has introduced in the public mind an element of fear not yet attached to any other energy source.

The OECD/NEA forecasts

In 1987 the OECD/NEA published a substantial report entitled *Nuclear Energy and its Fuel Cycle: Prospects to 2025* (OECD/NEA 1987). The report accepted a depressed market for nuclear power because both electricity growth rates and nuclear power programme expectations had declined over the decade 1977–87. It described how the nuclear reactor manufacturers and equipment and service suppliers had taken measures to reduce surplus capacity and skilled manpower. Reasonably enough, however, the report's authors also pointed out that construction work in progress on schemes already authorized guaranteed a substantial rise in nuclear generating capacity to the year 1995 and a more modest rise from 1995 to 2000. The cone of uncertainty widens rapidly, however, by the year 2025 (Table 2.6). For the period 1985–95 average annual nuclear growth rates are anticipated to be 4.4 per cent for the OECD, 6.1 per cent for the developing world outside the centrally planned countries (WOCA), and 8.9 per cent for the centrally planned economies (CPEs).

Table 2.6 Projected range of world nuclear power capacity

Year	*Capacity (GWe)*		*Low–high rounded net (GWe)*	
	1985	*1995*	*2000*	*2025*
(a) OECD*	207	307	240–429	555–1,150
(b) Developing WOCA	12	22	36–71	120–405
(c) WOCA	219	329	376–500	675–1,555
(d) CPEs	35	85	96–146	200–605
World (a + b + d)	253	414	472–646	875–2,160

Source: OECD/NEA 1987; this table incorporates later estimates for the OECD

The problem of projections

Very few medium- and long-term projections of nuclear power capacities made in the past have come anywhere near to the installed capacities actually achieved. The range of projections for nuclear power in OECD countries is made clear by Figure 2.7. It is also obvious that, whilst installed

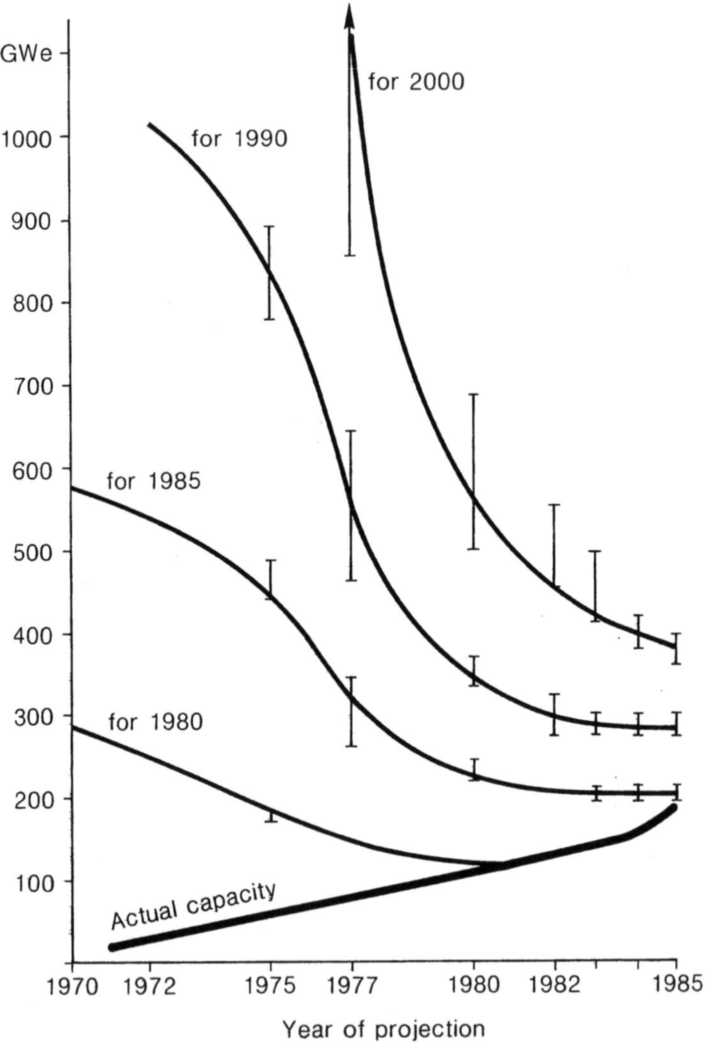

Figure 2.7 Projections of installed nuclear generating capacity for OECD countries, 1972–85
Source: OECD/NEA 1985a: 25

capacity has increased steadily since 1972, accelerating from 1984, *the forward projections invariably had to be forced downwards.* Clearly, the optimistic character of some projections could result from the fact that many of the data-collecting agencies with good time series from which to project, such as the IAEA and the NEA, may have some interest in an expanding market for nuclear power. It should be noted, therefore, that Figure 2.7 is taken directly from an OECD/NEA source.

SUMMARY

History shows changes in the dominant fuel used in the world. Some time ago in developed countries coal was substituted for the fuelwood still in use in the poorer countries today. During the earlier twentieth century petroleum and natural gas tended to substitute for coal in many energy markets. Today, although coal and oil remain dominant, nuclear and renewable energy sources may be substituting for oil and natural gas (Figure 2.8). However, the use of nuclear power to produce electricity in the world today is firmly in the hands of the world's industrially and technologically advanced nations. Eight countries dominate the world pattern of installed nuclear generating capacity; four of them were deeply involved in the scientific work that produced the world's atomic weapons. Hence, from the beginning, the development of nuclear power for civilian uses has inherited a legacy of fear in the public mind, even though coal is the riskiest of all the energy sources (Ott and Spinrad 1985).

A set of interlinked economic, technical and geopolitical considerations have inhibited the spread of nuclear power technology to Third World

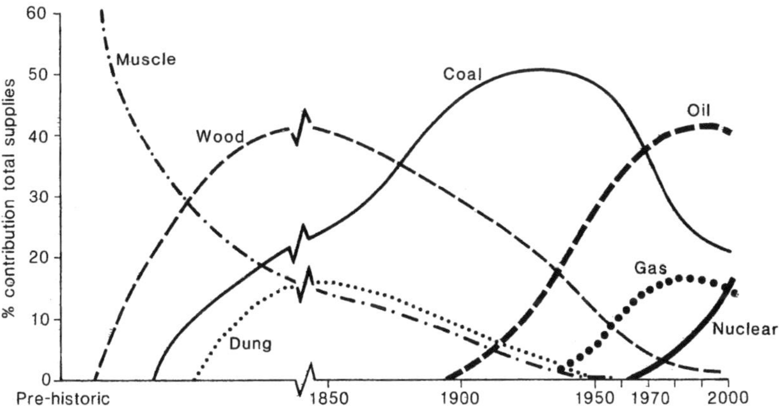

Figure 2.8 Generalized pattern of world energy supplies from prehistoric times to the year 2000
Source: Harvey and Newland 1969

countries, and it has been argued by some commentators that the adoption of nuclear power to produce electricity, even when technically and economically possible, to provide the energy needs of urban and industrial enclaves would strengthen rather than diminish the technological and commercial dependence of poor countries upon rich ones (Hayes 1976). Technology is a commercial asset; if it is sold and bought, it is a commodity of commerce. As such it is produced, distributed, financed, stored, imported and imitated, and the transference of the technology is linked to its commercial value. Therefore transfer of nuclear power technology may be argued to be the transfer of something valuable. Yet it is a demanding technology, one that requires many of the poorer countries to improve radically the management and organization of their power industries. They would need to adopt suitable safety standards and licensing requirements so that nuclear power stations can be built in an orderly manner. They must also train adequate manpower to plan and build the plants and operate them safely. It may be reasonably argued that a demanding technology is best handled by technologically advanced countries. If those countries were to accept the responsibility of further exploiting nuclear power to satisfy their energy needs they could, perhaps, ease the burden that increasingly will be imposed on the world's fossil fuel resources pending the large-scale commercial development of the renewable energy producing systems involving solar power, wind energy and wave energy (Mounfield 1986).

Some form of restraint by the industrialized nations in the fossil fuel market could provide more than a marginal easing of demand and prices, for the USA alone consumes around a quarter of the world's manufactured energy each year. Known reserves of oil are sufficient to satisfy the world's demand level of the late 1980s for perhaps another thirty-five to forty years. Proven reserves of natural gas are more plentiful and, at present consumption rates, would last until at least the middle of the next century. The consensus view is that economically exploitable coal reserves are sufficiently generous to last at least three times as long as oil, and internationally traded coal supplies are currently available at low prices. Oil has been cheap; in real terms the price in 1988 was as low as in the early 1970s (Figure 2.5) although rising temporarily after Iraq's invasion of Kuwait in 1990.

A cheap energy decade gives a more relaxed climate in which to tackle the particular environmental and social issues attached to nuclear power, to develop and implement more carefully considered siting strategies and to devise ways of involving the public fully in siting issues. It gives society and governments a breathing space – a window of time to examine calmly the future course of the nuclear option. However, the debate about nuclear power is only one element in a wider dialogue about future world energy needs. The demand for and supply of energy is a world issue, and most of the questions involved in the commercial development of the nuclear fuel cycle are international in scale. A decision by the world's developed industrial countries *not* to build more nuclear power plants might quieten anti-nuclear power pressure groups and reduce public anxiety about this energy form. It would exacerbate the energy situation in poor countries by doing nothing to reduce the level of world demand for traditional fossil fuels such as oil and coal. It would not help to reduce the output of sulphur dioxide and nitric oxide from the chimneys of conventional power stations burning coal and oil, which are implicated in acid deposition (acid rain), and it would do nothing to reduce the greenhouse effect.

The fossil fuels burned in conventional power plants have been claimed to help cause the largest immediate global risk to human beings from uncontrollable chemical wastes in the form of air pollution (Fremlin 1986). The burning of fossil fuels throughout the world is contributing to a marked increase in atmospheric carbon dioxide and other gases which, by absorbing infra-red radiation, are claimed to be warming the earth's surface and threatening to produce major and irreversible climatic changes through the so-called greenhouse effect. The greenhouse gases let sunlight through but prevent the heat from escaping back into space. Conventional power stations burning coal and oil are sources of such gases contributing world-wide about 12 per cent of the total. For each kilowatt-hour of electricity produced, nuclear power plants contribute less than 4 per cent of the carbon dioxide levels emitted by coal-fired power stations (House of Commons 1989).

The greenhouse effect is not a new idea. A form of it was described by the British scientist John Tyndall in the 1860s. A possible link between global warming and increasing levels of atmospheric carbon dioxide was made in the 1890s. Since then, public concern has fluctuated more or less in line with climatic trends. In the 1930s, following the spread of the US dustbowl and three decades of global warming, it increased. It decreased between the 1940s and 1970s, when the global temperature fell. Recent media attention to the theme began after the Ethiopian famines and subsequently the US drought heightened public awareness. In June 1988 Dr James Hansen, director of the USA's NASA Institute for Space Studies, told a Senate Committee: 'It's time to stop waffling so much and say the evidence is pretty strong that the greenhouse effect is here' (*Guardian*, June 1988).

The four warmest years in the past hundred have been in the 1980s. The first five months of 1988 were globally a record 0.17 °C warmer than the 1950–80 baseline. The most conservative estimate is that world temperatures will rise in the next fifty years by 1 °C; the most pessimistic estimate is another 4 °C. An estimate of a non-cyclical rise of 1–2 °C by the year 2030 would imply a large change. A warming of 1 °C would probably make the world warmer than at any time in the last 120,000 years. A difference of 4 °C is the difference between the depths of the last Ice Age and now. It would cause shifts in world climate that would have quite cataclysmic effects on world agriculture and human settlement. The whole greenhouse effect may be the major environmental issue to be faced in the next century, and the cause relates to the whole world's way of life, including energy production and use.

It is difficult to quantify this risk properly. At the moment, all that the world can do about it is to invest in energy conservation and burn as little fossil fuel as possible. Governments could contemplate a 'carbon tax' on fossil fuels to encourage the building of more nuclear power stations but, as the Chernobyl accident in 1986 showed, nuclear power stations, like other large industrial plants, can kill people when subject to catastrophic failure. Moreover, to make a significant impact on the burning of fossil fuels, nuclear power stations would have to be built in large numbers in many countries at a rapid rate. Therefore improvements in energy conservation and efficiency in use may offer a better opportunity for the world to reduce carbon dioxide emissions (Keepin and Kats 1988). Nevertheless, if nuclear power were to continue to displace fossil fuels in electricity generation, and it it were substituted for other primary energy sources now used for heating and in industry, it could make a worthwhile contribution. In evidence to the UK's House of Commons Energy Committee (House of Commons 1989: vol. 1, p. xxxix) the UK Atomic Energy Authority has asserted that, without the world's existing nuclear generating capacity, global warming would be 3 per cent higher and the energy component 7 per cent higher than they are now.

FURTHER READING

Hart, D. (1983) *Nuclear Power in India: a Comparative Analysis,* London: Allen and Unwin.

Poneman, D. (1982) *Nuclear Power in the Developing World,* London: Allen and Unwin.

Ramberg, B. (1986) *Global Nuclear Energy Risks: the Search for Preventive Medicine,* Boulder, CO, and London: Westview Press.

Surrey, J. (1988), 'Nuclear power: an option for the Third World', *Energy Policy,* 16 (5): 461–79.

3 The USA and Canada

THE USA

1942–56: First steps and initial applications

At precisely 3.25 p.m. Chicago time, on 2 December 1942, in what had been a squash court beneath the West Stands of Stagg Field, Chicago, a group of scientists made history. Led by Enrico Fermi, they began and then at 3.53 p.m. stopped a man-made self-sustaining nuclear reaction. The USA had entered the Nuclear Age.

At the time of this demonstration, the USA and its allies were in the grip of the Second World War, and immediately following the successful operation of the Chicago pile the US government embarked on a large-scale high priority programme to build plutonium-producing reactors. The sense of urgency was reinforced by the knowledge that German laboratories had been trying to develop a nuclear weapon. Thus, the first stage of nuclear power in the USA was devoted to military ends and the programme was managed by the Manhattan Engineer District of the Army Corps of Engineers, under the command of Lieutenant General Leslie R. Groves. During 1943, a pilot plant for plutonium production was built on federal government property at Oak Ridge, Tennessee, and construction of three full-size plutonium reactors followed at the Hanford Engineer Works, on a 162,000 hectare (600 square mile) desert reservation at the side of the Columbia river near Pasco, Washington State. In selecting this site prime consideration was given to its relative isolation, the small number of residents to be displaced, the initial availability of power from the Bonneville Power Administration and ample water for once-through cooling from the river (Edinger 1964).

For some years following Fermi's successful project, the design and construction of nuclear reactors was directed to the production of plutonium for nuclear weapons, an effort which came to a head with the explosion of the first nuclear device in New Mexico on 16 July 1945. In the field of nuclear technology, power and weapons programmes intersect at many points, but they are not one and the same thing. All nuclear power plants incorporate nuclear reactors, but not all nuclear reactors are nuclear power plants. Heat is one of the two main products of a controlled nuclear reaction; the other is radiation. It is the heat that is used in the production of electricity; when the heat they produce is held at a low level, reactors become useful for the production of plutonium but less capable of raising steam to run turbines. Consequently, when the US Atomic Energy Commission (USAEC) came into being on 1 January 1947, as successor to the Corps of Engineers, it inherited a mixed assortment of low temperature reactors built for military purposes and not designed to produce electricity (Zinn *et al.* 1964).

During the final months of its existence, in the immediate post-war period, the Manhattan organization initiated a programme intended to lead to the production of electricity by nuclear power stations, but the first serious planning to use nuclear technology for this purpose in the USA did not begin until two years after the Atomic Energy Act (the McMahon Act) was passed in 1946. This Act brought the USAEC into being and, through it, reserved to the federal government the right to own both special nuclear materials and the installations capable of their production.

The first task facing the USAEC was to assess the programme for civilian nuclear power that had been prepared by the Manhattan District and to prepare long-term guidelines and objectives for a civilian nuclear power programme. Most of the projects suggested or begun by the Manhattan District were cancelled or curtailed pending this exercise. The assessment was ready by the end of 1948 and, following its publication, the USAEC moved into the first phase of a positive policy aimed at the eventual production of commercially competitive electricity from nuclear power plants.

The decision was made to advance over a broad front, trying out different types of reactors. During the period 1948–51 the USAEC authorized the construction of a materials testing reactor (MTR) and an experimental fast breeder reactor (EBR1), but the most important decision was to build a prototype pressurized water reactor (STR) for submarine propulsion. All three of these projects were located at a site at Idaho Falls, Idaho, which became the National Reactor Testing Station (NTS). A major step forward was taken in 1951 when the USAEC invited electrical utility companies and other industrial interests to participate with it in studying the feasibility of designing and building dual-purpose reactors to produce both plutonium and electrical power. This was the first time that US industrial firms were given access to the information on nuclear power that was a necessary prerequisite for investment in the field. This cooperative effort between the USAEC and private industry was to become one of the hallmarks of the US nuclear power programme. In 1951 electricity was produced, in token amounts it is true, but for the first time, from a nuclear reactor (EBR1) on the NTS site.

In 1953, the USAEC stressed to Congress the importance of the early development of economic nuclear power as a national objective and, as a result, the idea of a dual-purpose reactor was pushed aside in favour of all-out concentration of effort on central station reactors for electricity alone. In the autumn of 1953, the USAEC declared its intention to build what was then described as a large-scale (60 MWe) PWR power plant, and invited industrial participation in the project. The purpose was twofold. First, it would demonstrate the feasibility of nuclear power in a utility system and, second, it was envisaged as a facility in which PWR technology could be tested. This was particularly important because it was PWR reactors that provided the design basis for nuclear propulsion units in submarines. Considerable industrial interest was roused by the proposal, and construction of the power station began in 1954 at Shippingport, Pennsylvania, on the south bank of the Ohio River. The Shippingport site was chosen because the Duquesne Light Company, who owned it, produced the most advantageous proposal in response to the USAEC's general invitation for industrial participation. Under its contract for participation in the project, Duquesne contributed US$5 million to the cost of the reactor and built all the accompanying conventional facilities at its own cost. In addition, the company furnished the site and agreed to pay for the steam produced by the reactor at a rate equivalent to 8 mills/kWh of electricity generated (1 mill = 0.1 cents) (Duquesne Power Company, personal interview, 24 March 1967). The company also agreed to operate the plant. Although scheduled for completion in June 1957, Shippingport did not come into operation until December of that year, and the original cost estimate of US$37.5 million was well below the eventual cost of US$55 million, a cost overrun which was to look quite modest against those incurred by later US nuclear power plants.

1954–64: Progress towards the privatization of nuclear power

The development of nuclear power for central station use began in earnest in 1954 when the Atomic Energy Act became federal law on 30 August. This Act, with subsequent amendments, was designed to make the way clear for full industrial participation in nuclear power development by permitting private ownership of nuclear reactors and private use of nuclear fuels under lease arrangements. It also established procedures by which the nascent nuclear power industry could obtain the hitherto classified data needed for nuclear power development and, in contrast with the restrictive terms of the MaMahon Act, it opened up the possibility of sales of US power reactors abroad through clauses designed to encourage international co-operation in the central station field and other peaceful uses of atomic energy.

The Power Demonstration Reactor Programmes

In recognition that investor-owned and other electricity generating utilities could not readily justify to their shareholders the level of investment required, the USAEC and Congress devised the Power Demonstration Reactor Programme (PDRP) the first round of which was announced early in 1955. The prime aim was to encourage the construction of several nuclear power stations. The USAEC offered financial incentives to co-operating utilities, including research and development assistance, and a 'waiver' of normal fuel charges during the first five years of plant operation. Three projects were undertaken in response to the first round of the PDRP: Yankee (175 MWe PWR, Rowe, Massachusetts); Hallam (75 MWe SGR, Nebraska); Vallecitos (5 MWe BWR, Pleasanton, California). Two projects, Dresden No. 1 (180 MWe BWR, Morris, Illinois) and Indian Point (255 MWe PWR, Indian Point, New York) were undertaken by utilities without direct government financial involvement. In the autumn of 1955, the USAEC announced the second round of the PDRP, inviting proposals for small plants in the 5–40 MWe range. Financial incentives were the same as for the first round, with the significant addition that this time the USAEC offered to finance the reactor system. Four projects were tabled on this basis, two of which were subsequently cancelled. The two completed were a small (11 MWe) organically cooled plant at Piqua, Ohio, and a 20 MWe BWR with external super-heating (Elk River, Minnesota). A little later, an additional second-round project was undertaken at Genoa, Wisconsin (La Crosse BWR, 50 MWe). In December 1956 the USA gained its first operating experience with a reactor system designed expressly for electricity production, the Experimental Boiling Water Reactor (EBWR) at the Argonne National Laboratory (Zinn *et al.* 1964).

Early in 1957, the USAEC announced the third round of the PDRP which was aimed at encouraging electrical utilities to build prototype plants in support of its development objectives and offering financial incentives comparable with those made available under the first round. Under this effort, four projects were undertaken up to the end of 1961. Two were BWRs of advanced design (Big Rock Point, Michigan, 48 MWe, and Pathfinder, Sioux Falls, South Dakota, 59 MWe), one was a 17 MWe heavy-water-moderated pressure-tube plant (Carolinas-Virginia, Parr, South Carolina) and one was a 40 MWe HTGR at Peach Bottom, Pennsylvania. Late in 1957, Shippingport began operation and was soon followed by Dresden No. 1 and Yankee. The total capacity of these three plants was a mere 350 MWe, and this accounted for virtually all the nuclear power station capacity in operation in the USA at the end of 1961. However, as far as the development of the programme as a whole was concerned, the most important event was that, after a comprehensive review of progress, in 1960 the USAEC issued a report setting out both short- and longer-term objectives and outlining the steps that were proposed to achieve them. The short-range objective was defined as *making nuclear power economically competitive by 1968 in those parts of the USA dependent on high cost fossil fuels.* The long-range objective was the development of breeder reactors.

The first indication of the onset of a vigorous period of expansion in the development of nuclear power occurred in 1960 when a West Coast utility, SoCal Edison, announced its intention of building a 375 MWe PWR, San Onofre 1, near San Clemente, California, as an additional PDRP third-round project. A further indication came in 1961 when another West Coast company, Pacific Gas and Electric, issued a declaration of intent to build a 315 MWe BWR at Bodega Bay, California, without federal assistance. This announcement was accompanied by cost estimates showing the proposed nuclear plant to be economically competitive with an oil-fired power station, the only practicable alternative under the company's particular circumstances. Bodega Bay was subsequently cancelled, but both these power stations were much larger than any nuclear power plant previously attempted.

In the spring of 1962 the US President called for a study of the role of nuclear power in the economy of the US electricity generating industry, taking into account prospective energy needs and resources and anticipated advances in alternative means of power generation. In response, the USAEC submitted its findings in the autumn and these amounted to a blueprint for a strategy of central station nuclear power development. It reaffirmed the long-range objective of developing breeder reactors and called for increased efficiency in nuclear fuel utilization. In the summer of 1962 the USAEC announced a modified third round of the PRDP. Seeking to encourage the building of large plants, the USAEC offered, in addition to the original third-round incentives, partial support of design costs. In 1963 two projects were put forward on this basis – Malibu, at Corral Canyon, California, where the City of Los Angeles Power Company proposed a 460 MWe PWR (the project was subsequently cancelled) and Connecticut Yankee, at Haddam Neck, Connecticut, where the Connecticut Yankee Power Company took the lead in suggesting a 460 MWe PWR. Soon afterwards, the long-delayed San Onofre I proposal advanced to the point of requesting a construction permit. The state of the US Power Development Reactor Programme in 1963 is shown in Table 3.1.

1964–73: a decade of development

In 1964 two 640 MWe BWRs were ordered, as commercial decisions independent of PDRP, at Nine Mile Point (Scriba, New York) and Oyster Creek (Forked River, New Jersey). A detailed analysis of the economic rationale for the Oyster Creek plant hinted that it might be the breakthrough that the USAEC had been waiting for (Jersey Power and Light Company 1964).

Table 3.1 Electricity generating nuclear power reactors in the USA in 1963

Name or other identification	*Location*	*Owner (1964 designation)*	*Electrical power rating in 1963 (MWe)*	*Year of initial operation*	*Chief nuclear contractor, actual or proposed in 1963*	*Operational status July 1987*[a]
Light water reactors (PWR and BWR)						
Operable						
EBWR	Argonne	AEC	4	1956	Argonne	*
Vallecitos VWBR (shut down in 1964)	Pleasanton Almeda Co., CA	GE and Pacific Gas	5	1957	GE	* M
Shippingport	Shippingport, PA	AEC and Duquesne	60[b]	1957	West	*
Dresden I	Morris, IL	Commonwealth Edison	208	1960	GE	*
Yankee	Rowe, MA	Yankee Atomic Elect. Co.	175	1960	West	+
Saxton	Saxton, PA	NUC. Experimental Corp.	5	1962	West	* M
Big Rock Point	Big Rock Point, MI	Consumers Power Co.	48	1962	GE	+
Elk River	Elk River, MN	AEC and Rural Co-op.	20	1964	AC	* D
Indian Point	Indian Point, NY	Com. Edison	255	1963	B&W	*
Humboldt Bay 3	Humboldt Bay, CA	Consumers Power Co.	49	1963	GE	*
Under construction or planned						
La Crosse BWR	Genoa, WI	AEC & Dairyland Power Co.	50	1966	AC	*
Bodega Bay	Bodega Bay, CA	Pacific Gas & Electric	313	–	GE	c
San Onofre I	San Clemente, CA	South Cal. Ed.	375	1966	West	+
Malibu	Corral Canyon, CA	City of Los Angeles	463	–	West	c
Connecticut Yankee	Haddam Neck, CT	Conn. Yankee Power Co.	463	1967	West	+
Nine Mile Point I	Oswego, NY	Niagara Mohawk	500	1969	GE	+
Oyster Creek I	Oyster Creek, NJ	Jersey Central Power Co.	500	1969	GE	+

Operable heavy water reactor						
Carolinas–Virginia tube reactor	Parr, SC	Carolinas-Virginia Nuclear Power Associates	17	1963	West	* M
Organic reactor operable						
Piqua	Piqua, OH	AEC and City of Piqua	11	1963	AI	* E
Nuclear superheat reactors under construction						
Pathfinder	Sioux Falls, SD	Northern States Power Co.	59	1964	AC	* M
Bonus	Puerto Rico	AEC & Puerto Rico Water Authority	16	1964	GNEC	* E
Operable sodium graphite reactors						
SRE	Santa Susana, CA	AEC	6	1957	AI	*
Hallam	Hallam, NE	AEC & Consumers Power Co.	75	1963	AI	* E
Gas-cooled under construction						
HTGR (Peach Bottom 1)	Peach Bottom, PA	Philadelphia Electric	40	1967	GA	* M
EGCR	ORNL	AEC	20	1969	GE/AC	*
Fast Breeders						
EBR-2	NRTS	AEC	16	1963	Argonne	?

Table 3.1 continued

Name or other identification	*Location*	*Owner (1964 designation)*	*Electrical power rating in 1963 (MWe)*	*Year of initial operation*	*Chief nuclear contractor, actual or proposed in 1963*	*Operational status July 1987*[a]
Enrico Fermi I	Lagoona Beach, Monroe County, MI	Power Reactor Development Co.	65	1966	Owner	* M

Source: *Nuclear Engineering*, 9(96): 161, 1964; Nuclear Engineering International, *World Nuclear Industry Handbook 1988*; *Radioactive Waste Management*, Proceedings of a Conference held in London, 5–6 March 1985, Oyez Scientific and Technical Services, 1985, p. 112

Notes: (a) Abbreviations: AC, Allis-Chalmers; AI, Atomics International; B&W, Babcock and Wilcox; GA, General Atomic, Division of General Dynamics; GE, General Electric; GK, Gesellschaft für Kernforschung (FRG); GNEC, General Nuclear Engineering (Combustion Engineering); West, Westinghouse Electric Corporation.

(b) Other reactors operable in 1964 were experimental facilities contributing so little electrical current that they have been excluded from this table. They were two heavy water reactors on the Hanford Reservation, designated PRTR and HWCTR, two Super-heat reactors, one at the NRTS in Idaho (BORAX 5) and the other at Pleasanton, CA, and a fast breeder (EBR1) at NRTS. Vallecitos EVSR (5 MWe) was opened in 1965 and closed by 1987, but the largest commercial casualty was Three Mile Island 2, a 906 MWe facility commissioned in 1978 but inoperable in 1987. After 1963 other reactors were opened, and then closed down by 1987: SEFOR (Stickler, AZ), WTR (Waltz Mill, PA), NASA Plumbrook (Sandusky, OH) and GE EVESR (Almeda County, CA). All these have been mothballed. The B&W reactor at Lynchburg, VA, has been dismantled.

(c) It is possible from this table to identify among the larger reactors those likely to be the first canditates for decommissioning, having served for 20 years or more in operation.

[a] *, Closed; +, operating; c, cancelled; M, mothballed; D, dismantled; E, entombment.

[b] Increased to 150 MWe by 1964.

Jersey Power and Light Company produced an analysis of the comparative costs of power generation for three alternatives: (i) building the nuclear plant, (ii) building a coal-fired plant of the same capacity at the same site and (iii) long-distance transmission of electricity from a coal-fired plant at a mine-mouth location. The nuclear cost estimates were based on a fixed-price turnkey bid submitted by the General Electric Company; those for the coal alternatives reflected the best coal price that the utility had been able to obtain from its suppliers. In the months preceding the Oyster Creek evaluation, the utility had been contracting for coal at a delivered price of 30–31 cents/million BTU of energy content. Under the pressure of nuclear competition, coal was offered for the Oyster Creek site at 26 cents, the average price paid for coal by US utilities in 1964 (only those operating in or close to coal-mining areas enjoyed coal prices as low as 20 cents). However, the Oyster Creek analysis showed that coal would have to be available at less than 26 cents to be competitive (Jersey Power and Light Company, personal interview, 23 March 1967).

There was much discussion of the Oyster Creek figures (Atomic Industrial Forum 1964). Jersey Power and Light Company emphasized that its analysis was specific to its own circumstances and might not apply elsewhere. Niagara Mohawk Power Corporation, the company sponsoring the Nine Mile Point project, which involved the same reactor manufacturer and virtually the same reactor design, did not publish a comparable cost analysis but released some general cost estimates that were more conservative. Independent judgement was that there was insufficient profit margin in the turnkey project. But the feeling grew that for nuclear power the break-even coal price in an investor-owned system was 25–29 cents/million BTU. Essentially, this meant immediate market opportunities for nuclear power in the areas of the USA with higher cost fuel (see Chapter 9, Figure 9.6). However, important changes were taking place in coal costs, notably in the costs of transportation. In the mid-1960s the expense of transport in many parts of the country accounted for a third to half of the utilities' fuel costs, or for roughly a fifth of the total cost of power generation. Most power station coal in the USA had traditionally been moved by rail, and substantial reductions in transport costs were being achieved as a result of three innovations in rail practice. The first came in the mid-1950s when the railways received permission from the Interstate Commerce Commission to offer low 'incentive rates' to utilities. The second came in 1959 when the first trainload rates became effective (prior to that date, freight rates had been based on car load (waggon load) quantities). The third and most important innovation came with the introduction in 1962 of the unit train, i.e. a train like the UK 'merry-go-round' which shuttles constantly back and forth between mines and a power station. By 1964 the use of unit trains was spreading rapidly in the USA, coal slurry pipelines were coming into use and substantial reductions in tariffs were being made.

There were other factors pertinent to the situation in the mid-1960s.

Increased interconnection between utilities and the emergence of regional power pools was increasing the capability of utilities to build larger power plants. There were also indications of an increased public awareness of the environmental impact of the power industry, with some quarters resisting nuclear power because of reservations about its safety and others favouring it because of its apparent cleanliness compared with coal burning. Another important event in 1964 was the enactment by Congress of legislation allowing private ownership of nuclear fuel supplies by 1969.

The coal industry managed to hold off its nuclear competitor for eighteen months. In situations where utilities were known to be seriously considering a nuclear option, substantial cuts were made in coal prices, mainly through the granting of favourable unit-train rates for coal delivery. In a number of cases the price concessions extended to other coal-burning plants on the same system, thereby benefiting the utility's overall power generating economy. Capital costs of building a nuclear plant at about US$132 per kilowatt were 20 per cent higher than for a coal-fired plant of the same capacity (Zinn *et al.* 1964). But the utilities held off ordering nuclear power plants for only a little while. Nine nuclear power plants were ordered during 1965, twenty-two in 1966 and twenty-four in 1967, with the orders in the two latter years accounting for nearly half of all the new power generating capacity ordered by US utilities. In the process, unit sizes leap-frogged from 500 MWe to 800 MWe and then to 1,100 MWe. With only one exception, the orders were for either PWR or BWR plants, divided almost equally between Westinghouse and General Electric. Later, the Babcock and Wilcox Company and Combustion Engineering established themselves as suppliers. It was not without justification that in March 1967 *Fortune* magazine announced that 'US nuclear power suddenly and dramatically came of age in 1966'.

Reactor contracts began to take several forms. In 'nuclear island' contracts, the reactor manufacturer supplied all the essential components of the steam generating system. In a pressurized water plant, for example, these components would include the reactor vessel and some supports, the fuel handling apparatus, the reactor control system complete with control rods and drives, the primary coolant pumps and valves, the pressurizer and the primary heat exchangers. The contract might also cover supply of the first fuel charge and fuel element performance warranties. It could be written as a prime contract with the utility, or as a sub contract to (or joint contract with) an engineering organization retained by the utility to engineer and construct the plant. Another type of contract that emerged was the turnkey contract in which a single firm took complete responsibility for all the plant.

In the years 1965–7 two main factors were working in favour of nuclear power. First, the coal interests hardened their pricing policies in the winter of 1965. With a record volume of business in hand and a record number of coal-burning power plants under construction, with nuclear power at that stage threatening only the higher cost coal areas and with the national electricity demand doubling every ten years, the coal industry felt that it could concede those areas and still anticipate excellent growth prospects. Nuclear interests, however, kept their marketing drive going long after the initial breakthrough, partly because of competition amongst reactor manufacturers and partly to achieve deeper penetration into coal territory. It was not until the latter part of 1966 that nuclear equipment prices began to seek higher ground and not until 1967 that they increased substantially. By then it was too late for coal to stem the flow of nuclear power plant sales (Hogerton 1968).

The second element in the nuclear breakthrough was a growing conviction among utilities that nuclear power would be the way of producing electricity in the future. This was particularly true of the utilities that operated coal-fired or oil-fired power stations in urban or highly industrialized areas. The Federal Clean Air Act was strengthened by the Air Quality Act 1967 (Public Law 90–148) as a response to public concern over air pollution, and the utilities were being confronted almost daily by new local ordinances setting limits on the emission of sulphur dioxide and other stack gases. Low sulphur fuels quickly commanded high prices, but removal of sulphur from run-of-mine coal, either at the mine or in the operation of the power plant, promised to be expensive. As a general practice, long-distance transmission of electricity from power stations located at some distance from the urban and industrial markets would also be costly. In addition, coal mining was itself encountering increasing public criticism on environmental issues such as river pollution and land despoliation. In short, coal's days seemed to be numbered.

During 1968 and 1969 the orders for nuclear plants fell away a little as the industry adjusted its profit margins to the new market circumstances, but in 1970 (seventeen orders) and 1971 (twenty-eight orders) the growth rate accelerated again to an all-time peak of forty-two orders in 1973, the year of the first massive round of OPEC oil price increases, of a temporary embargo by Arab states on oil exports to the USA and of crisis planning by federal government agencies in energy matters. The following year, 1974, saw twenty-four units ordered, but then there was a sudden fall in 1975, and a dearth of orders thereafter. Completion rates on previously ordered plants became very erratic. The exponential character of the ordering curve from 1964 to 1973 should have been matched by a similar curve in the commissioning of the nuclear power plant, but this occurred only from 1967 to 1974. Since then, the start-up curve of the nuclear power industry in the USA has become very ragged. All the nuclear power plants ordered since 1974 have been either cancelled or placed on an indefinite construction schedule.

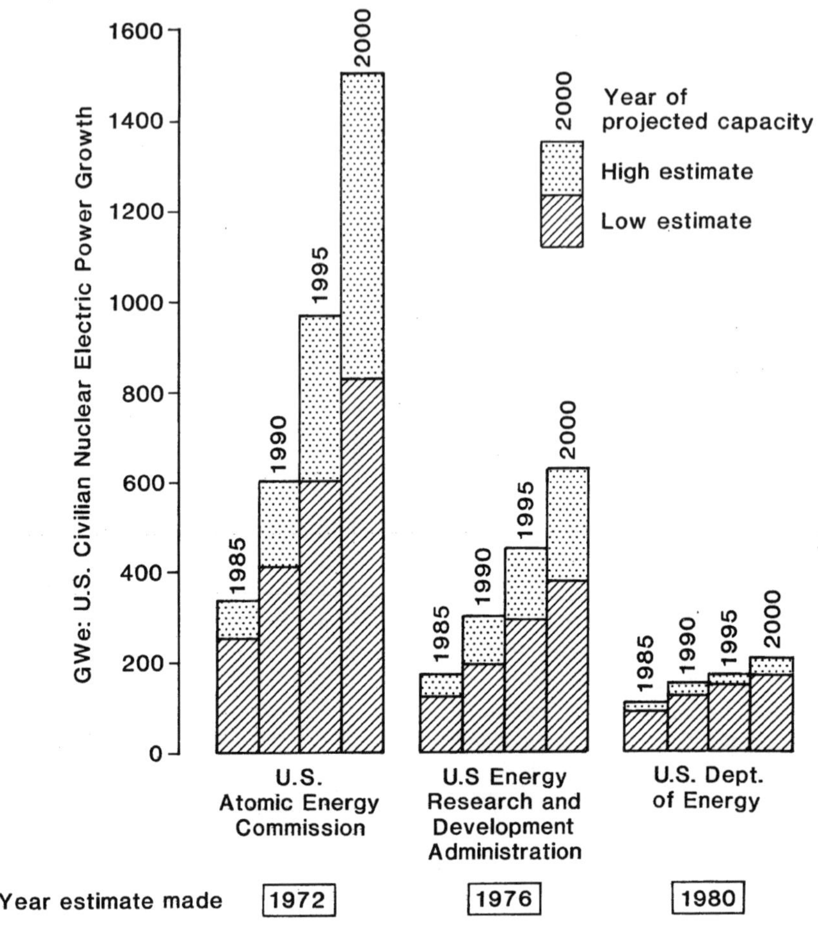

Figure 3.1 Trends in USA nuclear power projections. The diagram shows clearly the dramatic reversal in the fortunes of nuclear power in the USA in the late 1970s
Source: US General Accounts Office, adapted from *Financial Times*, 18 August 1982

1974–?: Disarray, disillusion and retrenchment

Figure 3.1 shows how authoritative projections of US nuclear electrical generating capacity were revised downwards after 1972, and there is now a real possibility that no more nuclear power plants will be ordered before the mid-1990s. More than 100 reactors, representing half the total capacity of nuclear power stations ever ordered in the USA, have been cancelled since that year, some at an advanced state of construction (Lambert 1983; Dodsworth 1984). Outstanding examples include the Washington Public

Power Supply Service, which in 1983 abandoned two nuclear power reactor projects and mothballed one other in order to save two at an advanced state of construction, the Tennessee Valley Authority, which cancelled eight plants by 1983, and Cincinnati Gas and Electric Company, which spent US$1,700 million building the W.H. Zimmer nuclear power plant at Moscow, on the Ohio River, only to decide in 1984 to convert it to coal before it was commissioned. In January 1984 the Public Service Company of Indiana abandoned a reactor on which it had spent US$2,400 million. In May 1988 Long Island Lighting Company finally bowed to public political pressure and decided to write off a 900 MWe nuclear plant at Shoreham, Long Island, completed in 1983 but never used because evacuation facilities were considered inadequate. It had cost US$5,300 million.

A number of reasons underlie the domestic policy reversal. The most important are public opposition and environmental concern, and also issues of a directly economic nature.

Environmental impact

Environmental issues were not ignored in the building of the early commercial nuclear power plants in the USA. On 30 June 1966 the USAEC issued a sixty-four page *Guide for the Organization and Contents of Safety Analysis Reports.* Section 2 specified site details, including evaluation of population distribution, meteorological characteristics, hydrology, geology and seismology. Section 2 is only five pages long, however, and the other fifty-nine pages are concerned mainly with the design and technical details of a proposed plant. The Safety Analysis Reports prepared for nuclear power stations during the 1960s were balanced accordingly. At many sites, moreover, sufficiently detailed meteorological and hydrological data over adequate time periods were often hard to come by when the reports were being prepared. In general, the environmental details of a site were not much more than a gloss on the technical details of the plant which were written by engineers for other engineers to assess.

Things soon changed. The widespread concern for environmental quality which developed in the USA in the late 1960s and early 1970s led to the enactment of legislative and administrative measures intended to decrease the quantities of pollutants released to the air and to water bodies. This concern extended to the thermal effect of all power plants on lakes, rivers and shore-waters and to the radioactivity released to the environment in nuclear power plant effluent streams under normal operating conditions.

Three major environmental federal statutes were enacted during 1969–70; the National Environmental Policy Act (NEPA) in 1969, the Water Quality Improvement Act (WQIA) of 1970, followed by the federal Water Pollution Control Act Amendments of 1972, and the Clean Air Amendments in 1970. All had implications for the environmental impact assessment procedures required for nuclear power stations. State legislatures also began to act on environmental issues as they became increasingly politically sensitive.

The purpose of NEPA was to establish a national policy for protecting the environment under the Act; a Council on Environmental Quality was set up to assess any major project that might affect the environment. WQIA established a national policy for the control of water pollution. The Act enabled limits to be set and legally enforced to prevent undue temperature rises in rivers, lakes and streams as a result of hot water discharges by power stations. Different limits were set by different states. For example, discharges into Lake Michigan should not exceed the lake temperature by more than 1 °F whereas in Tennessee discharges of up to 10 °F above lake and river temperatures were permitted. For power stations, WQIA required a certification by an appropriate state agency that plant discharges met applicable water quality standards, and certificates were required as a condition for the issue of a construction permit. The restrictions on the discharge of heated water have had a direct influence on the design and operation of nuclear power plants (Bryan *et al.* 1972), but by requiring minimal environment impact from any major new industrial plant NEPA became even more important in its effects on the nuclear power industry.

During the first year under NEPA, the USAEC relied on assessment by other federal agencies and state-level authorities for meeting its responsibilities, but the Calvert Cliffs court decision on 23 July 1971 brought the nuclear power industry into the spotlight of environmental impact concern. The Calvert Cliffs nuclear power station is run by the Baltimore Gas and Electric Company and the site is on the shores of Chesapeake Bay, in Maryland (Figure 3.2). The power plant proposal was originally announced in 1967 and the USAEC granted a construction permit for two 800 MWe PWRs at the site in 1969, when construction began. A number of groups of objectors considered that the power station violated the terms of NEPA and formed the Calvert Cliffs Co-ordinating Committee. Through their lawyers the objectors petitioned the US Court of Appeal for the District of Columbia Circuit for a review of USAEC rules governing consideration of non-radiological environmental issues. On 23 July 1971 the court ruled that the USAEC was remiss in its policies and regulatory responsibilities in the licensing of nuclear power plants under NEPA. The court imposed stringent requirements on the USAEC to conduct full environmental reviews of all nuclear plants that did not have full operating licences by 1 January 1970, the date NEPA became effective. By 4 August the USAEC had issued interim revised rules to applicants for construction permits and operating licences, but the utilities reacted with dismay. A total of 88 nuclear site licences were affected by the Calvert Cliffs decision.

After the Calvert Cliffs decision, and a similar court ruling in 1971 regarding the Quad Cities nuclear power station on the Mississippi River,

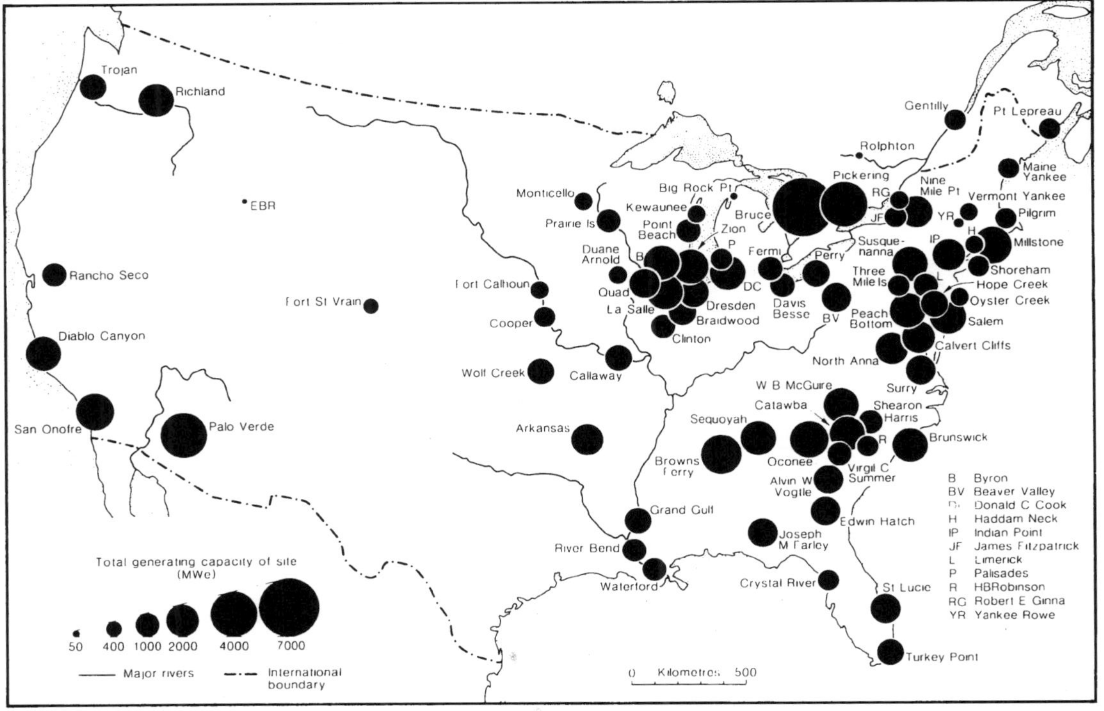

Figure 3.2 Location of nuclear power stations in operation in the USA and Canada 1987. Nuclear power stations are most strongly concentrated in the heavily populated and industrialized areas of the eastern USA and southern Canada. Much of the western USA is sparsely populated and there are large supplies of cheap coal available for electricity generation in the northwestern states such as Dakota and Montana. In the southwest, Texas has natural gas for power station use

Source: Nuclear Engineering International, *World Nuclear Industry Handbook 1988*; World List of Nuclear Power Plants, *Nuclear News* 30 (2), 1987; IAEA, *Nuclear Power Reactors in the World*

the task of balancing environmental costs and economic benefits became the responsibility of the USAEC and its successor the Nuclear Regulatory Commission (NRC), and a detailed environmental impact statement had to be filed. The revised USAEC Regulatory Guide 4.2 (2 March 1973) entitled *Preparation of Environmental Reports for Nuclear Power Plants* required thirteen chapters of environmental impact assessment from an applicant, extending but not restricted to items such as the construction and operation of water intake and discharge structures, operation of radwaste systems, construction and operation of transmission and switching facilities and transportation of fuel elements to and from the plant, as well as the construction and operation of the reactor and associated facilities. In making a decision on a project it required 'unquantified amenities and values [to] be given appropriate consideration in the decision making along with economic and technical considerations'.

This wealth of regulatory control became territory in which, during following years, one anti-nuclear power group after another found fertile ground for delaying and halting the construction of nuclear power stations, with cost increases due to time overruns arising as a consequence.

Economic issues

Predictions of electricity demand, on the basis of which many power stations, including coal- and oil-fired as well as nuclear, were ordered, proved much too optimistic. From 1966 to 1972 growth in summer peak loads was projected at rates of 7 per cent per annum or more, and on that basis the utilities set out to double their generating capacity over a period of ten years. In reality, between 1974 and 1983 the growth in demand averaged only 2.1 per cent and in 1982 US peak load demand actually fell for the first time since the Second World War. The annual growth rate from 1983 to 1991 is not expected to be much more than 3 per cent. In parenthesis, it should be pointed out that rates of growth in the demand for electricity show considerable regional variations in the USA, suggesting that forecasts at national level should be used cautiously as the basis for energy policies at regional or state levels (Chern and Just 1980).

As the demand for electricity fell away, the cost of building nuclear power plants escalated. In 1971, the estimated cost of building a 1,000 MWe nuclear plant was US$345 million; by 1980 the figure had climbed to US$3,200 million. San Onofre 1 (450 MWe), ordered by Southern California Edison in 1963, went onto full power in 1968 and cost US$90 million to build; units 2 and 3, each of 1,100 MWe, took eleven years to build. They went into operation in 1982 and 1983 respectively, but cost US$4,500 million compared with the original cost estimate of US$1,600 million. It was not just planned plants that were affected by such cost overruns. Commonwealth Edison completed its twin-reactor 1,500 MWe Quad Cities nuclear power station at Cordova, Illinois, in

1972 at a cost of US$262 million. Subsequently it was obliged to spend another US$180 million on modification required by the NRC to meet the new safety and environmental regulations, and such 'backfitting' has affected a large number of plants.

As electricity demand forecasts were revised downwards, all types of power plant schemes, oil-fired, coal-fired and hydroelectric, as well as nuclear, were cancelled or shelved. However, the capital cost characteristics of nuclear power plants, with actual construction charges between 30 and 100 per cent higher than those of coal- or oil-fired plants of equivalent capacity, have made nuclear plants particularly vulnerable as financing costs have become an increasingly difficult problem for the utilities. Generally, utilities provide only a third of construction expenditures from internal funds; the rest has to come from large issues of both debt and equity. From 1973 to 1983 total common equity in the industry jumped from US$40.9 billion to US$100.4 billion, while preference stock rose from US$14.7 billion to US$28.6 billion and long-term debt from US$63.6 billion to US$123.7 billion. The lead time for designing and building nuclear units increased from 78 months in 1978 to 150 months in 1983, largely as a result of increasing regulatory intervention, growing public opposition to the rapid growth of nuclear power and construction shortcomings. The cost of paying interest on these funds as they were held for a much longer period than originally planned has been financially crippling for many utilities. Electricity price increases can only by charged to customers after state and federal approval, which have been subject to delays. Several parts of the country allow the power generating companies to recover only a portion of construction costs on new current tariffs; customers can only be charged for the rest when the plant starts up. Thus many utilities have been unable to get cash flowing in and, as electricity demand has not risen as forecast, they have been forced to write off projects with book values far in excess of their own net worth.

In a detailed study Komanoff (1981b) provided evidence that the average excess of nuclear capital costs over those of coal-fired plants increased from 6 per cent for 1971 completions to 52 per cent for 1978 completions. In 1985 the Federal Energy Information Administration indicated that new nuclear power plants would have an economic advantage over new coal-fired plants only in New England and the South Atlantic regions. Coal-fired plants would have a distinct edge in the Southwest and North Central regions. Elsewhere there was not much to choose between the two in terms of the cost of the electricity they produced.

The future for nuclear power in the USA

Over the past decade there has been a sharp reduction in expectations for the future of commercial nuclear power in the USA. In 1973 the nuclear capacity projected for 1990 was 250 GWe. In July 1987 100 GWe of nuclear power plant was in operation and another 15 GWe was in various stages of construction. The immediate problems in the 1990s are economic and financial, but the underlying long-term problems are political and very much related to public acceptance and the attitude of government at federal and state level.

Competing fuels for electricity generation

Oil and natural gas account for three-quarters of the total energy consumption of the USA; the remainder comes from coal, hydroelectricity and nuclear power. In November 1974 the US Government launched Project Independence, a programme designed to achieve a high degree of independence from imported energy supplies. In 1973 the USA imported 29 per cent of its energy needs; in 1984 it imported 45 per cent, mostly oil. The oil, once supplied by Canada and non-Arab members of OPEC, is coming increasingly from the Middle East (25 per cent in 1984 compared with 16 per cent in 1974). In 1971 the Federal Power Commission estimated US oil reserves at 2,380 thousand million barrels but in 1975 the US Geological Survey cut its estimates of US oil reserves by a factor of four, to somewhere between 47 and 130 thousand million barrels. Domestic oil production has declined since 1973, but US oil consumption now accounts for well over a third of the West's demand for oil.

Even the most optimistic forecasters see only a minor role for energy sources such as solar power, wind, geothermal energy and tidal power. At the very best, they are likely to provide no more than 20 per cent of the USA's energy demand by the year 2000. Therefore if oil is running out and soft energy paths are not ready to take the lead, coal and nuclear fission are still likely to be at centre stage for the next fifteen to twenty years at least.

The USA still has huge coal reserves. A third of the world's proved coal resources lie within her borders and more than half of that, a fifth of the global total, lies west of the Mississippi River. From the Great Plains to Appalachia 217,000 million tonnes of recoverable coal lie waiting, almost two centuries' worth at double the present consumption rate – the energy equivalent of some 5 million barrels of oil and natural gas a day. But coal carries with it three severe problems which influence its competitive position *vis-à-vis* nuclear power:

(a) The sulphur content of much of the coal is high enough to cause serious aggravation of air pollution problems unless the electrical utilities, who are the coal industry's chief customers, install chimney stack 'scrubbers' and other costly clean air equipment. Western coals, from areas such as eastern Montana, have lower heating qualities than Virginia or Ohio coals, and are removed from the major markets in the East, but their content of sulphur and other polluting chemicals is low.
(b) Strip mining is the quickest, cheapest and most profitable way of

extracting about a third of the coal remaining in the USA, particularly the environmentally desirable low sulphur coal in the West, which is often close to the surface. However, uncontrolled strip mining in the past has resulted in savage destruction of land and watersheds in eastern Montana, while other western coal districts may yet see the kind of landscape degradation that has devastated eastern Kentucky (Atwood 1975).

(c) Coal mining operations in the USA kill one miner every third working day, and injure thousands more each year, despite mining safety requirements enacted by Congress.

Thus the critical issue is not so much whether the USA should use more coal, but how and when the accompanying social and environmental costs should be dealt with, and by how much such costs will increase the price of coal delivered to electrical utilities. Each of the fuel cycles has an environmental impact and an environmental cost.

The role of the federal government

A further imponderable is the role and attitude of the federal government. In the past, the federal government has played a key role in subsidizing the development of all US energy resources, including nuclear power. It has been calculated that by 1980 US$18,000 million had been spent by the federal government in assisting the commercial development of nuclear power (Bezdek and Cone 1980). Under PDRP the USAEC offered financial incentives to co-operating utilities to help build nuclear plants, funded research and development activities, waived fuel use charges, and provided for fuel fabrication and the training of personnel. The portion of costs assumed by the USAEC for the demonstration projects was about 20 per cent. Currently, all steps in the fuel cycle except enrichment and radwaste management are handled by private industry, but government involvement has been an important element in launching nuclear power. However, the US Government has now developed an ambivalent attitude to the nuclear option, as decisions have become increasingly politically charged. Reports by Congress's Office of Technology Assessment and by the NRC have blamed poor management and shoddy workmanship by inexperienced power companies for the delays and cost overruns. There is no doubt that too many power companies embraced nuclear technology too quickly, without fully understanding its complexities. The utilities and some construction companies say that NRC safety codes have become so bureaucratic and divorced from the actual technology of the nuclear power stations that licensing procedures are now described in terms of 'a hopeless muddle' and 'a mass of paperwork'. Yet Congress has refused to streamline the labyrinthine procedures which the industry has blamed for many of its financial troubles.

The size of electrical utilities and the need for interconnection

One of the constraints on the development of nuclear power in the USA may have been the relatively small size of many power utilities. Even by 1985, the largest system was less than half the size of the UK's Central Electricity Generating Board's (CEGB) network (Table 3.2), and the smaller systems on their own would have had insufficient load to operate a large-base-load nuclear plant.

Table 3.2 The ten largest electricity utilities in the USA 1985

	Total capacity (MWe)	*No. of plant sites*	*No. of generating units*
Tennessee Valley Authority	22,003	19	59
Commonwealth Edison	18,189	15	35
US Corps of Engineers	15,617	39	196
US Dept of Interior	14,724	26	71
Duke Power Co.	12,361	11	32
Southern California Edison	12,152	20	52
Georgia Power Co.	11,730	12	30
Florida Power and Light	11,705	11	26
Texas Power and Light	10,239	11	20
Houston Lighting and Power	9,833	8	25
Other systems (more than 500 MW) – 145	359,890	561	1,309

Source: IEA/OECD 1985, 375

Some areas, therefore, have deliberately enlarged the effective size of their operational area and market partly through co-operative effort in developing nuclear power. New England depends on nuclear power stations for 35 per cent of its electricity supplies, and the Northeast has witnessed considerable co-operation between utilities in the development of nuclear power stations.

Both the creation of large regional markets and power pools will need to continue to provide the right environment for nuclear power development. The North American bulk transmission systems have developed through the linkage of systems of a variety of smaller-sized electricity utilities. In 1968 the industry formed the North American Electric Reliability Council (NERC) to promote the reliability and adequacy of bulk power supply in the electric utility systems of North America. NERC consists of nine regional reliability councils encompassing virtually all the power systems in the USA and major areas of Canada (Figure 3.3).

In the USA and parts of Canada there are now three large highly interconnected and synchronized systems:

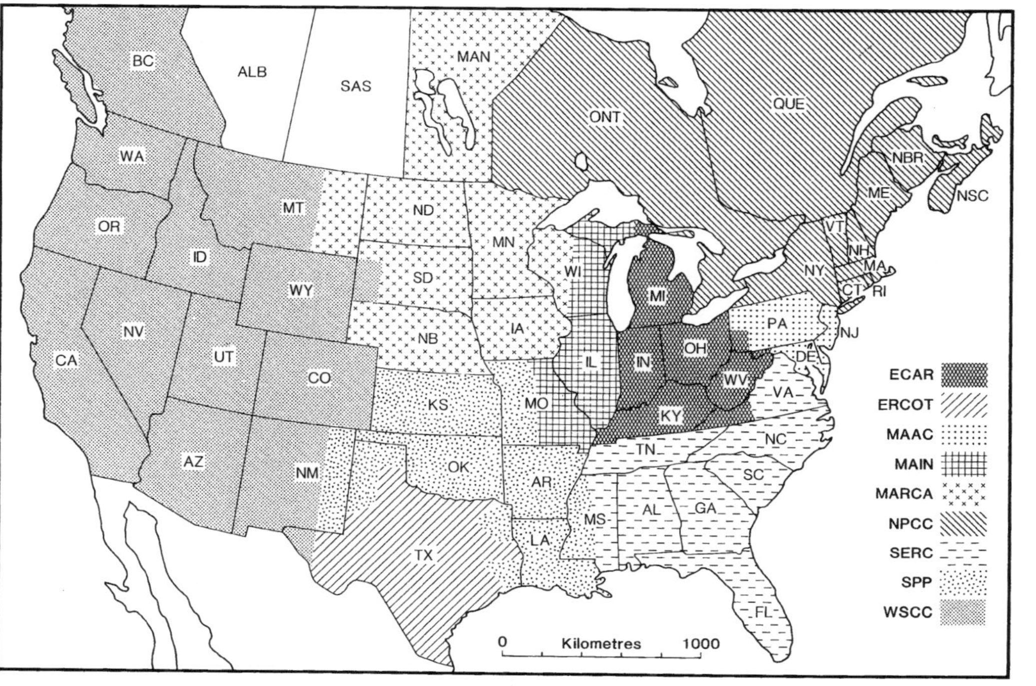

Figure 3.3 Regional power blocks in the USA and Canada: ECAR, East Central Area Reliability Co-ordination Agreement; ERCOT, Electric Reliability Council of Texas; MAAC, Mid-Atlantic Area Council; MAIN, Mid-American Interpool Network; MARCA, Mid-Continent Area Reliability Co-ordination Agreement; NPCC, Northeast Power Co-ordinating Council; SERC, Southeastern Electric Reliability Council; SPP, Southwest Power Pool; WSCC, Western Systems Co-ordinating Council
Source: IEA/OECD 1985: 95

(a) *The Western Interconnected System*, which includes the Rocky Mountain States and the West Coast as well as the Canadian Provinces of British Columbia and Alberta;
(b) *The Texas Interconnected System*, a comparatively small independently operated grid; Texas is highly dependent on the use of natural gas for electricity generation;
(c) *The Eastern Interconnected System*, which covers the rest of the lower forty-eight states as well as the Canadian provinces of Manitoba, Ontario and New Brunswick.

Energy transfer between the three networks requires separate direct current tielines or transformer facilities to synchronize purchased power to the buyer's system. In recent years the bulk transmission systems have been used extensively to transfer electricity from low cost areas where generation is based on hydraulic sources, coal and nuclear energy to areas like New York, New England and Florida where the generating system depends heavily on oil. On the basis of the argument that there is still underdeveloped potential to expand bulk transfer of current from low to high cost electricity producing areas, the US Government has encouraged the industry to increase inter-regional co-operation, but more powerful interties are required in the Pacific Northwest and between the Midwest and New England.

Bulk transmission often requires wheeling of power, i.e. selling power to another utility by using a third utility's transmission system. There are difficulties in this process. The request to a third utility to wheel current may conflict with the third utility's own selling interests. Also, some utilities view the wheeling of power from another utility as competition that erodes the utilities' monopoly franchises.

A future locational and siting strategy for US power reactors

It has been suggested that the future development of the US nuclear energy system might be made more publicly acceptable if power reactors were confined to fewer sites, perhaps by expanding existing sites rather than looking for new ones. It has been argued that this would reduce the environmental impact and lower the industry's profile. In this way, the extent to which the public is affected by a nuclear site will be defined by sites already chosen. Such an 'existing site' policy might create a distribution pattern for nuclear power in the USA akin to that shown in Figure 3.4 and devised by the Institute for Energy Analysis, Oak Ridge. This policy, it is argued by its authors, would lead to a nuclear energy system comprising perhaps 100 large and slowly growing nuclear enclaves operated by a small number of large and experienced organizations dedicated to safe nuclear power generation (Burwell 1981).

Excluding existing nuclear sites with insufficient water supplies for more

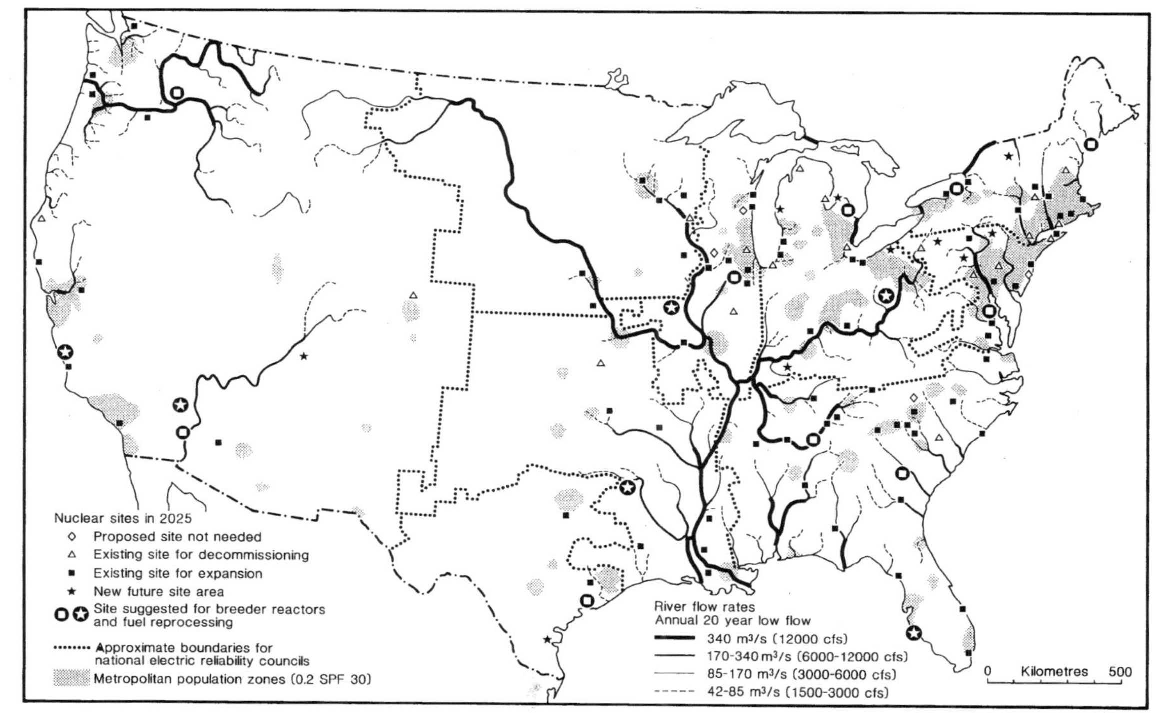

Figure 3.4 Postulated nuclear power sites in the USA in the year 2025
Source: Burwell (1981) 'An existing site policy for the US nuclear energy system', *Nuclear Safety* 22 (2): 159, after Institute for Energy Analysis

reactors, those constrained by a lack of land and space for expansion, those with high close-in population densities and those with little expectation for growth in power demand in a reasonably sized service area, seventy-seven out of the nearly 100 existing sites are identified as being well suited to expansion. These sites, it has been suggested, could accommodate anticipated growth in the nuclear power generating system until past the year 2000. Less than 20 per cent as much land would be needed for power transmission corridors if existing sites were expanded rather than new ones added, because existing sites already have spare grid linkages to ensure reliability and many existing transmission lines can be expanded by adding circuits or increasing voltages. Safety considerations would also reinforce the desirability of such a locational strategy. Some of the early nuclear power stations have already been closed down and wholly or partially dismantled (see Table 3.1). However, they were small plants, and by the year 2025 a nuclear capacity of 130 GWe will have completed the anticipated 40 year life of useful service. Maintained intact on the site, the irradiated decommissioned reactor poses much less of a hazard to the public than if it is dismantled and transported. On-site management of low level waste, spent fuel and decommissioned reactors becomes an integral part of nuclear power operations, and the handling and transport of radioactive materials is minimized.

The argument is persuasive, and the adoption of such a strategy would leave very few consumers without access to power generated by nuclear reactors. By 1991, all but 8 per cent of the US population will be living in areas served to a greater or lesser degree by nuclear power. The present dilemmas should not be allowed to hide a considerable achievement. In July 1987 nuclear reactors in the USA provided over 100 GWe of generating capacity and produced 15 per cent of the country's electricity (Table 3.3).

CANADA

Canada has large reserves of oil, natural gas, coal and tar sands, together with a substantial number of undeveloped hydroelectric sites, but since the 1960s, particularly through construction programmes in Ontario, a large tranche of nuclear generating capacity has been developed. This is due to the huge size of the country and the fact that major population centres are a long way from fossil fuel deposits or potential hydroelectric sites. The fossil fuels are found principally in the west and far north, and the remaining hydroelectric sites are also in the north, but industry and population are concentrated along a strip adjacent to the USA in southern Ontario and Quebec, two provinces which account for approximately a third of Canada's total electricity output and demand. Thus nuclear power had become economically competitive especially east of Manitoba, and by 1976 was claimed to have a massive fuel cost advantage in Ontario

Table 3.3 Nuclear power stations in operation in Canada and the USA in July 1987

Type and name of power station	*Location*	*Reactor power (MWe gross)*	*Year commissioned*
Canada			
CANDU			
Rolphton NPD[a]	Rolphton, Ont.	25	1962
Pickering 1	Pickering, Ont.	542	1971
Pickering 2		542	1971
Pickering 3		542	1972
Pickering 4		542	1973
Pickering 5		540	1983
Pickering 6		540	1984
Pickering 7		540	1985
Pickering 8		540	1986
Bruce 1	Tiverton, Ont.	815	1977
Bruce 2		825	1977
Bruce 3		815	1978
Bruce 4		825	1979
Bruce 5		885	1985
Bruce 6		890	1984
Bruce 7		890	1986
Bruce 8		890	1987
Gentilly 2	Becancour, Quebec	685	1983
Point Lepreau 1	Point Lepreau, New Brunswick	680	1983
Total (19 units)		12,553	
Under construction			
Darlington 1	Darlington, Ont.	881	
Darlington 2		881	
Darlington 3		881	
Darlington 4		881	
USA			
FBR			
EBR2	Idaho Falls, ID	20	1964
LWGR			
Hanford N16	Richland, WA	860	1966
HTGR			
Fort St. Vrain	Platteville, CO	342	1973
PWR			
Alvin W. Vogtle 1	Waynesboro, GA	1,210	1987
Arkansas 1-1	London, AR	883	1974
Arkansas 1-2		897	1980
Beaver Valley 1	Shippingport, PA	860	1987
Beaver Valley 2		860	1987
Braidwood 1	Braidwood, IL	1,175	1987
Byron 1	Byron, IL	1,175	1985
Byron 2		1,175	1987

Table 3.3 continued

Type and name of power station	*Location*	*Reactor power (MWe gross)*	*Year commissioned*
Callaway 1	Fulton, MO	1,174	1984
Calvert Cliffs 1	Lusby, MD	900	1975
Calvert Cliffs 2		900	1977
Catawba 1	Clover, SC	1,205	1986
Catawba 2		1,205	1986
Crystal River 3	Crystal River, FL	860	1977
Davis Besse 1	Oak Harbor, OH	960	1977
Diablo Canyon 1	San Luis Obispo, CA	1,125	1985
Diablo Canyon 2		1,130	1986
Donald C. Cook 1	Bridgman, MI	1,056	1975
Donald C. Cook 2		1,100	1978
Fort Calhoun 1	Fort Calhoun, NB	510	1973
H.B. Robinson 2	Hartsville, SC	769	1971
Haddam Neck	Haddam, CT	610	1968
Indian Point 2	Buchanan, NY	900	1974
Indian Point 3		1,000	1976
Joseph M. Farley 1	Dothan, AL	860	1977
Joseph M. Farley 2		860	1981
Kewaunee	Kewaunee, WI	563	1974
Maine Yankee	Wiscasset, ME	864	1972
Millstone 2	Waterford, CT	888	1975
Millstone 3		1,209	1986
North Anna 1	Mineral, VA	982	1978
North Anna 2		982	1980
Oconee 1	Seneca, SC	934	1973
Oconee 2		934	1974
Oconee 3		934	1974
Palisades	South Haven, MI	845	1971
Palo Verde 1	Wintersburg, AZ	1,303	1986
Palo Verde 2		1,303	1986
Palo Verde 3		1,303	1987
Point Beach 1	Two Rivers, WI	524	1970
Point Beach 2		524	1972
Prairie Island 1	Red Wing, MN	534	1974
Prairie Island 2		531	1975
Rancho Seco	Sacramento, CA	966	1975
Robert E. Ginna	Ontario, NY	498	1970
Salem 1	Salem, NJ	1,170	1977
Salem 2		1,170	1981
San Onofre 1	San Clemente, CA	456	1968
San Onofre 2		1,127	1983
San Onofre 3		1,127	1984
Sequoyah 1	Chattanooga, TN	1,183	1981
Sequoyah 2		1,183	1982
Shearon Harris 1	New Hill, NC	960	1987
St. Lucie 1	Fort Pierce, FL	872	1976
St. Lucie 2		892	1983

Table 3.3 continued

Type and name of power station	*Location*	*Reactor power (MWe gross)*	*Year commissioned*
Surry 1	Surry, VA	856	1972
Surry 2		856	1973
Three Mile Island 1	Middleton, PA	824	1974
Trojan	Prescott, OR	1,178	1976
Turkey Point 3	Miami, FL	699	1972
Turkey Point 4		699	1973
Virgil C. Summer 1	Jenkinsville, SC	937	1984
W.B. McGuire 1	Cornelius, NC	1,220	1981
W.B. McGuire 2		1,220	1984
Waterford 3	Taft, LA	1,153	1985
Wolf Creek 1	Burlington, KS	1,192	1985
Yankee Rowe	Rowe, MA	185	1961
Zion 1	Zion, IL	1,080	1973
Zion 2		1,080	1974
Total (69 units)		65,299	
BWR			
Big Rock Point	Charlevoix, MI	73	1965
Browns Ferry 1	Decatur, AL	1,098	1974
Browns Ferry 2		1,098	1974
Browns Ferry 3		1,098	1977
Brunswick 1	Southport, NC	849	1977
Brunswick 2		849	1975
Clinton 1	Clinton, IL	985	1987
Cooper	Brownville, NB	801	1974
Dresden 2	Morris, IL	832	1970
Dresden 3		824	1971
Duane Arnold	Palo, IA	565	1975
Edwin Hatch 1	Baxley, GA	813	1975
Edwin Hatch 2		820	1979
Fermi 2	Monroe, MI	1,154	1986
Grand Gulf 1	Port Gibson, MS	1,036	1985
Hope Creek 1	Salem, NJ	1,170	1986
James A. Fitzpatrick	Oswego, NY	849	1975
La Salle 1	Seneca, IL	1,122	1982
La Salle 2		1,122	1984
Limerick 1	Pottstown, PA	1,092	1986
Millstone 1	Waterford, CT	684	1970
Monticello	Monticello, MN	576	1971
Nine Mile Point 1	Seriba, NY	620	1969
Nine Mile Point 2		1,162	1988
Oyster Creek	Forked River, NJ	670	1969
Peach Bottom 2	York Co., PA	1,098	1974
Peach Bottom 3		1,098	1974
Perry 1	Perry, OH	1,252	1987
Pilgrim 1	Plymouth, MA	691	1972

Quad Cities 1	Cordova, IL	833	1972
Quad Cities 2		833	1972
River Bend 1	St Francisville, LA	991	1986
Susquehanna 1	Berwick, PA	1,100	1983
Susquehanna 2		1,100	1985
Vermont Yankee	Vermont, VT	540	1972
WNP2	Richland, WA	1,154	1984
Total (36 units)		32,912	

Sources: Nuclear Engineering International, *World Nuclear Industry Handbook 1988*; IAEA, *Nuclear Power Reactors in the World*
Note: [a] Shut down by 1989.

compared with gas, oil, and coal for electricity generation (Maine 1976).

Canada has a history of scientific research and development in the field of nuclear energy as long as that of the UK and the USA. This has provided Canada with a complete and largely autonomous fuel cycle depending on Canadian uranium resources and Canadian technology. The result in July 1987 was 12,553 MWe of capacity installed at five sites (see Figure 3.2), with four more units under construction at a fifth location (see Table 3.3).

Canadian experience preceded wartime co-operation with the USA and the UK, but the Allied effort to produce nuclear weapons during the Second World War played a crucial role. The Canadian National Research Council set up an atomic energy research programme in 1942 and in 1943 Canadian, UK and French scientists and engineers began to design and construct a heavy water research reactor at Chalk River, Ontario, Canada's oldest and most important atomic research establishment (Eggleston 1965). A characteristic feature of the Canadian nuclear power programme, and one that has considerable economic importance, is the use of heavy water (D_2O) as a moderator for the reactor core and coolant for the reactor fuel. This stems in part from a historical accident. In 1943 the two most practical moderators for nuclear reactors appeared to be graphite and heavy water. With the fall of France in 1941, researchers who had been conducting experimental work fled to the UK with about 200 kg of heavy water which earlier had been smuggled from Norway. Thus, in the early 1940s, only a very limited amount of heavy water was available and graphite for reactor use was already being produced and used in the USA. It was agreed between the Allies that the necessary heavy water should be allocated to Anglo-Canadian research, whilst work proceeded on graphite-moderated reactors in the USA.

After 1945, the Canadian nuclear programme lost its military urgency and in the USA the McMahon Act of 1946 restricted the international exchange of information, thus impelling the UK into an independent nuclear programme. Canada had to decide whether to stay in the field or to turn away from the commercial opportunities offered by several years of research and development. With a nucleus of excellent research facilities, especially at Chalk River, with a degree of national pride at stake, and with enough trained scientists and engineers to launch and sustain a worthwhile programme, the Canadian government decided to underwrite further work. In 1946 the Atomic Energy Control Act was passed, establishing the Atomic Energy Control Board to supervise the development and use of atomic energy. In 1952 a Crown Company, Atomic Energy of Canada Ltd (AECL), was formed under the Act and was given operational responsibility for nuclear research and development. Also in 1952 another Crown company, Eldorado Mining and Refining Ltd, was set up to act as the purchasing and marketing agency for Canada's considerable uranium resources.

During this time, experimental design work continued at Chalk River, in conjunction with isotope production, and in 1954 the Canadian civilian nuclear power programme began in earnest when a group of engineers and scientists from electricity utilities, industry and the universities assembled at Chalk River to study the heavy water reactor as a means of raising steam in an electricity generating station. The group recommended the construction of a nuclear power demonstration plant (NPD) for actual electricity production, and the result was a 25 MWe plant commissioned in 1962 at a site near Rolphton, Ontario. In 1959 the decision was taken to build a 218 MWe station at Douglas Point, Tiverton, Ontario. This reactor, now closed down, went critical on 15 November 1966 and first produced power on 7 January 1967. Orders for nuclear plant and equipment rose rapidly in 1964–5 and toward the end of 1965 construction work began on the first two 542 MWe units at Pickering, on the shore of Lake Ontario to the east of Toronto. They were commissioned in 1971, on time and within budget. In March 1967 the Atomic Energy Control Board issued a permit to Ontario Hydro for two more units of the same size at this site. These were ready in 1973. In September 1966 a permit was granted to AECL in conjunction with the Quebec Hydro-Electric Power Commission for the building of a 266 MWe station at Gentilly, Quebec, 10 miles (16 km) downstream from Trois-Rivières on the St Lawrence River. This CANDU variant was completed in 1971 but proved unsatisfactory and has never been commercially operated. Near to it is the only working nuclear plant in Quebec, Gentilly 2, a 685 MWe unit which has been in operation since 1983.

Although Canada is a major oil producer, by the early 1970s it had become a net oil importer to the tune of 10–12 per cent of annual oil demand. The Yom Kippur War of October 1973 awakened Canadians to new realities in energy supply and stimulated a wish for reduced dependence on imported oil (see Figure 2.5). However, electricity and water power were no longer synonymous terms in some areas. In Ontario in particular most of the suitable hydroelectric sites within reasonable transmission distance of the main load centres had been exploited and the use of remote sites was increasing capital and running costs through the use of long transmission lines. In 1965 it was estimated that the largest additional power increment available to Ontario from hydroelectric sources by 1975 was only 1,500 MWe, with a possible 2,000 MWe of pumped storage capacity at a site within 100 miles of the main load centre (H.A. Smith 1965). To an increasing extent, therefore, electricity supplies from hydroelectric sources were having to be supplemented by building coal-fired power stations. However, most of the coal had to be imported from the USA, and it was argued that without nuclear power the total value of the coal imports would reach unacceptably high levels by the 1980s (Haywood 1967). Ontario also contained 340 million tons of economically exploitable uranium ore. Developed in the 1950s to satisfy demand from the USA and the UK, the presence of this uranium was a stimulus to Canadian involvement in nuclear power. With a system using natural uranium on a once-through fuel cycle there was no need to commit foreign exchange to the purchase of supplies from the USA.

Thus investment in nuclear power accelerated. Construction work on Pickering Units 5 and 6 began in 1975 and they were completed in 1983–4; Units 7 and 8 came into use in 1985–6. The eight-reactor Pickering site now contains enough nuclear power generating capacity (4,328 MWe) to supply 3 million people with electricity, a massive concentration on one site by any standard. It is more than matched, however, by the Bruce nuclear park near Kincardine, on the shore of Lake Huron, where the retired Douglas Point reactor is dwarfed by eight units commissioned at various dates between 1977 and 1987 to provide 6,835 MWe of installed generating capacity. Under construction in 1988 at Darlington, near Newcastle, Ontario, and scheduled for completion by 1992 are four units, each of 935 MWe. In New Brunswick, Point Lepreau I, a 680 MWe plant, went into service in 1983.

The heavy Canadian concentration of nuclear powered electricity generating capacity in Ontario is clearly shown in Figure 3.2 and it represents an important change in Ontario Hydro's fuel mix. In 1962, 83 per cent of the province's total electricity output came from hydroelectric plants, the balance being mainly coal fired. By 1983, 31 per cent was hydroelectric, 31 per cent nuclear and 31 per cent fossil fuel based, with 7 per cent purchased from other utilities. Moreover, the arrival in strength of nuclear power has initiated changes in the way that power stations fuelled by other means are used in the system. In the 1960s the mode of operation was to use coal-fired plants to make up the difference between the load demand and the hydroelectricity output. Because of their high fuel costs, the coal-fired plants were used mainly for peak load, and the low operating costs, plus high capital costs, of the hydroelectric stations favoured their use for base load. The low fuel costs and high capital costs of the nuclear units has made it imperative for them to be used for base load, and the older hydroelectric and coal-fired plants are now used to meet daily and seasonal peaks.

Economic aspects of the CANDU fuel cycle and deuterium production

The CANDU reactors that form the backbone of the Canadian nuclear power effort differ substantially from most other nuclear power reactors in use elsewhere in the world, including the USA. They have a high burnup natural uranium system depending on a once-through fuel cycle and they use deuterium (heavy water) as the moderator and coolant (see Chapter 1).

A high burnup system is one that achieves high neutron economy, and it is here that heavy water has a critical role in the Canadian scheme of things. In the fission process, for every neutron lost or wasted, an extra atom of fissile uranium must be used or, in effect, purchased. Neutrons thus have a real monetary value that is directly reflected in the cost of the fuel cycle. The most neutron economical moderator is heavy water. By using heavy water, very few neutrons are wasted (i.e. a high burnup is achieved) and the uranium is left in the reactor for so long that most of the fissionable U-235 is consumed.

Thus Canadian technology has provided the Canadian electrical utilities with a reactor design that offers high burnup of uranium giving low fuelling costs, without enrichment, without fuel reprocessing, without any plutonium credit and without costly waste disposal problems. These advantages have led to nuclear power stations for which the overall fuelling costs are low (Williams and Page 1951). On the debit side is the fact that heavy water is expensive and its use adds considerably to the capital cost of the reactor. At 1982 prices the initial charge for 500 tonnes of heavy water for a 600 MWe CANDU reactor accounted for Can$150 million of the front-end capital cost.

Heavy water is produced in large-scale capital-intensive plants by continuous chemical exchange process with hydrogen sulphide, followed by distillation, which uses large quantities of process steam and electricity. In the beginning the supply of heavy water for the Canadian nuclear power programme was a private sector affair, but the federal government supported Deuterium of Canada Ltd (Glace Bay, Nova Scotia) and Canadian General Electric (Port Hawkesbury, Nova Scotia) in their initiatives to construct and operate heavy water plants. Both experienced technical and financial problems, however, and in the end were bought by AECL. In 1968, with forecasts of substantial growth in Ontario's nuclear power programme, AECL built the Bruce A heavy water plant, selling it to Ontario Hydro in 1973. In 1974 AECL began to build another heavy water plant at La Prade, Quebec. However, in the late 1970s rates of growth in electricity demand and reactor orders were well below the anticipated levels and overcapacity in heavy water output was threatened. In 1978 work was suspended at La Prade and Bruce D was mothballed; Bruce

C had been cancelled previously. By the 1980s heavy water was produced by four plants, Bruce A and B, Glace Bay and Port Hawkesbury. These units have a total nominal capacity of 2,400 tonnes/year.

Government involvement

The Canadian federal and provincial governments have been deeply involved from the beginning in the development of nuclear power, acting through the Crown companies and in other ways. Ontario Hydro itself is a publicly owned utility, a power commission appointed by the provincial government. Gentilly I is owned by AECL. Rolphton was built to establish the general technical feasibility of electricity production using heavy water technology. It was connected to the Hydro's grid system but the reactor and steam raising was AECL's responsibility, with the Hydro providing other parts of the station. Douglas Point was built by AECL on a site provided by Ontario Hydro, who also provided the interconnection. Ontario Hydro's share of the total capital cost was 7 per cent, against AECL's 93 per cent. The first two-unit phase at Pickering was financed under a co-operation agreement whereby the Hydro contributed 40 per cent of the capital cost, the federal government 33 per cent and the provincial government 27 per cent.

Important indirect subsidies have also been given. The bulk of nuclear research and development had been undertaken by AECL and private industry, and not the electrical utilities. Moreover, the Canadian nuclear fuel industry is a co-operative effort between the federal government and private industry. Eldorado Mining and Refining look after the mining and refining end but private companies manufacture the fuel elements.

In 1964 AECL contracted to underwrite the sale of the first five years output from Deuterium of Canada Ltd, and in 1966 offered to do the same for the first ten years of heavy water production from General Electric's Glace Bay and Port Hawkesbury plants in Nova Scotia.

During the 1980s Ontario Hydro, like many other North American power utilities, found itself in a new and difficult business environment characterized by slow growth, higher costs and major uncertainties about the future. The decades of 7 per cent growth have ended and any policy to sustain the Canadian nuclear power industry will require continued financial support from the federal government.

Outlook

Most current energy forecasts and electricity demand trends since the early 1980s suggest a relatively slow growth of future energy demand in Canada, and this is confirmed by forecasts relating to Ontario. Depending upon which forecast is used, projected demand in Ontario, the historic and geographic hub of Canada's nuclear power effort, may require the building of two to four reactor units after Darlington, to be in operation by 2000, but the argument depends mainly on the need to replace plants nearing the end of their useful lives rather than to meet fresh increments of demand. Low electricity demand growth and Ontario Hydro's large recent construction programme have led to surplus capacity, financial constraints and upward pressure on the rates charged to the consumer. The Canadian nuclear construction industry, like its US counterpart, has been feeling the pinch. Canadian reactors have been exported to Romania, Argentina and Korea, and in 1982 Canada tried very hard to win contracts to build a 2,300 MWe plant at Luguna Verde in Mexico, but now it is electricity that is being exported rather than reactors. The export of power and any other energy product requires the approval of the Canadian National Energy Board and the federal government as surplus to Canadian needs but in effect a 'common market' in electricity exists between the USA and Canada. There are thirty Canada–USA interconnections at 69 kV or over with a total international power transmission capacity of over 12,000 MWe. Electricity, mainly from Ontario Hydro, accounts for 12 per cent by value of Canada's export trade in merchandise (i.e. excluding invisibles) to the USA. Some of these exchanges arise from the fact that Canada's peak demand is in winter, and is therefore complementary to the USA, which has a summer peak. Some are exports used in the USA to displace high cost oil-fired and other fossil-fuelled generation. Increasingly, however, export contracts are beginning to be long term over twenty years or more, as Canada and the USA are tied together in large regional electricity reliability co-ordination agreements (see Figure 3.3).

SUMMARY

Nuclear power in Canada and the USA has developed with different technologies and under the ethos of contrasting national philosophies. Scientists and engineers from both countries were involved in the early stages of weapons-focused research, but by a set of almost fortuitous historical events Canada's trajectory was into heavy-water-moderated reactors while the USA emerged with light water PWRs and BWRs. Thus Canada was able to use for commercial purposes a fuel cycle depending on natural uranium without fuel reprocessing and without recourse to the plutonium cycle. This fitted comfortably with the country's growing aversion to nuclear weaponry and its proliferation. In the USA private utilities dominate the electricity supply industry but in Canada provincial government organizations hold a virtual monopoly.

There are, nevertheless, many similarities. In both countries there has been heavy government involvement and financial assistance. In each, sheer size has underpinned marked regional variations in the natural resources available for electricity production and major regional contrasts in the distribution of industry and population and, thus, in the demand for current. Both nations have had the advantage of substantial uranium resources within their territories and a uranium mining lobby prepared to push for nuclear power plants. Both have experienced a flattening of the rate of increase in the demand for electricity through economic slowdown since the mid-1970s. In both, construction companies and consortia have experienced retrenchment as their sales in the nuclear power area have decreased. Problems of the acceptability of nuclear power to the public have been common to both but opposition has been stiffer in the USA than in Canada.

While two or three more nuclear power reactors may be ordered in Ontario before the year 2000, to replace ageing coal-fired capacity, it is difficult to anticipate any date by which the ordering and building of nuclear power plants will begin again in the USA. In April 1983 the US Supreme Court allowed individual state legislatures powers to prevent the construction of reactors, and the accidents at Chernobyl and Three Mile Island and the activities of environmental lobbies have made too strong an impression for a decision by any USA utility to order a nuclear power station to be anything other than politically and economically hazardous. Not one has been ordered since 1978 and sixty-six previous orders have been cancelled. In January 1988 the Public Service Company of New Hampshire applied for the first bankruptcy petition filed by a USA public utility since the Depression. The firm's US$5.2 billion (£2.9 billion) new nuclear station at Seabrook on the New Hampshire coast has been ready since 1986 but never opened because of objections over its emergency evacuation procedures. In 1988 the delay was reported to be costing the company US$50 million (£28 million) a month. To say that things have become difficult for the US nuclear industry would be a gross understatement.

FURTHER READING

Komanoff, C. (1981) *Power Plant Cost Escalation: Nuclear and Coal Capital Costs, Regulation and Economics*, New York: Van Nostrand Reinhold.

Okrent, D. (1981) *Nuclear Reactor Safety: On the History of the Regulatory Process*, Madison, Wi; University of Wisconsin Press.

Openshaw, S. (1986) *Nuclear Power Siting and Safety*, London: Routledge and Kegan Paul, Chapter 6, pp. 189–242.

Zinn, W.H., Pittman, F.K. and Hogerton, J.F. (1964) *Nuclear Power, USA*, New York: McGraw-Hill.

4 Western Europe

THE DEMAND FOR NUCLEAR POWER IN WESTERN EUROPE

Western Europe is the second-largest energy-consuming region in the world after the USA, but its heavy dependence on *imported* energy, mainly oil but also natural gas and coal, is comparatively recent. Before the Second World War 90 per cent of Western Europe's energy needs were met from its coal resources. In the post-war years the abundance of cheap oil from the Middle East oilfields resulted in the conversion to an oil-based economy. Over the period 1956–73 the demand for energy grew by an average of 50 million tons of oil equivalent (mtoe) per annum, an average annual growth rate of more than 5 per cent. This rapid growth came to an end in 1973 as a result of the shock to the continent's economy of the OPEC oil price rises that year (see Chapter 2, Figure 2.5). Average annual growth dropped to less than 1 per cent (8 mtoe) over the period 1973–8 and was negative in 1974 and 1975. The recession was the main cause for this reduction in demand but the efficiency with which energy was used also improved after 1973.

During the 1980s oil and oil prices underpinned most discussions of West European energy issues, but the heavy ordering of nuclear power plants that had occurred in the 1960s and 1970s slowed dramatically. Public confidence in nuclear power began to dwindle, high interest rates created a poor environment for investment in capital-intensive projects, there was much slower growth in the demand for energy following the 1973–4 and 1978–9 oil crises, and oil prices themselves were slipping downwards. Except for Greece, Spain, Turkey and Switzerland the *rate of increase* of energy consumption in Western Europe declined between 1970 and 1986, even though primary energy consumption increased at an average of around 1 per cent per annum. Between 1973 and 1987 the industrialized countries of Western Europe learned to use 15 per cent less energy for each unit of output (Economist, 14 November 1987), but the substitution of nuclear power for oil was less successful.

The first Geneva Conference on the peaceful uses of nuclear energy in 1955 focused the attention of governments on the commercial possibilities of nuclear power but did little to develop a European development strategy. The European Community (EC) set up Euratom in 1958 but that organization's economic *raison d'être* soon vanished with a coal surplus in the early 1960s and the increasing flow of cheap oil. Euratom's first President remarked that Western Europe had 'no need for commercial nuclear power now or for many years in the future' (Kingshott 1960). There have been times since its foundation when Euratom has seemed a frail vessel; it has suffered from the 'fair return' syndrome under which no EC government has been prepared to contribute to common programmes unless it is sure that its own nuclear industry or national economy benefits sufficiently. The EC has had an Energy Committee since 1970 and there exist several quasi-official organizations who research, report and advise on energy matters.[1] However, the Treaty of Rome contained nothing on energy and there has never been a common energy policy. Thus nuclear power has remained fundamentally the responsibility of national governments and has developed piecemeal within the context of nationally based energy programmes pursued with greater or lesser vigour according to the political colour of the party or parties in power.

In so far as the EC had any energy policy at all in the early 1960s it was essentially one of minimizing costs through unimpeded oil imports, using world oil prices to determine the prices of other fuels and providing temporary aid to the EC coal producers pending a rise in oil prices sufficient to correct the balance between coal and oil (Harris and Davies 1980). These objectives were embodied in the *First Guidelines on Community Energy Policy* at the end of 1968. At that time nuclear power was seen as the best long-term option for providing cheap and abundant energy, but conventional fossil fuels were cheap and energy policy remained a matter not seeming to deserve urgent attention either in Europe or internationally. The 1968 guidelines were updated in 1972 to include acceptance of the need for increasing EC capacity for uranium enrichment but in 1971 the nine countries then making up the EC were in a position where 58 per cent of the energy used was oil.

At the end of 1972, prior to the OPEC oil price increases of the following year, seven West European countries had a total installed nuclear power generating capacity of only 12,300 MWe. These were the UK (4,500 MWe), France (2,600 MWe), the FRG (2,100 MWe), Spain (1,100 MWe), Switzerland (1,000 MWe), Italy (600 MWe) and Sweden (400 MWe). It is important to recognize, however, that before the oil crisis nuclear power plants were being ordered in large numbers. By 1974 France had 3,200 MWe, the FRG had 4,900 MWe, Sweden had 2,600 MWe and the UK had 7,600 MWe, and many of the nuclear plants in service in Western Europe now were ordered *before* 1973. However, for more than a decade until 1973 Europe had access to abundant supplies of cheap energy in the world market. The 1973–4 oil price rises gave a substantial impetus to nuclear power programmes in many, but not all, West European countries

Table 4.1 Nuclear power generating capacity and production in West European countries in operation in 1988 and planned for 1995

Country	*Total primary 1986 energy consumption (mtoe)*	*Consumption of nuclear energy in 1987 (mtoe)*	*Installed nuclear generating capacity in 1973 (GWe)*	*Coal lignite reserves in 1967 (tonnes)*	*Installed electricity generating capacity (GWe)*						*Electricity generation (net TWh[a])*					
					Actual 1985			*Estimated 1995**			*Actual 1987*			*Estimated 1995*		
					Total	*Nuclear*	*% Nuclear*	*Total*	*Nuclear*	*% Nuclear*	*Total*	*Nuclear*	*% Nuclear*	*Total*	*Nuclear*	*% Nuclear*
Belgium	49.4	9.0	0.4	5,000	14.2	5.5	38.7	13.3	5.5	41.4	59.9	39.6	66.1	69.3	39.6	57.1
Finland	22.9	4.6	0.0	Neg.	11.4	2.3	20.2	13.0	3.3	25.4	50.7	18.5	36.5	60.9	16.5	27.1
France	192.9	53.4	2.8	4,050	88.1	35.7	40.5	118.6	66.2	55.8	359.9	251.3	69.8	442.7	348.2	78.7
FRG	269.2	29.0	2.1	44,000	92.0	16.4	17.8	92.9	22.7	24.5	391.7	122.6	31.3	431.0	140.0	32.5
Italy	141.5	<0.05	0.6	650	58.1	1.3	2.2	75.0	6.1	8.1	190.8	0.2	0.1	261.0	6.3	2.4
Netherlands	72.0	1.0	0.5	2,400	13.5	0.5	3.7	13.6	1.5	11.0	66.9	3.5	5.2	71.7	3.5	4.9
Spain	79.8	8.6	1.1	2,000	38.9	5.5	14.1	46.5	9.2	19.8	134.9	39.2	29.1	160.2	49.8	31.1
Sweden	53.0	16.2	0.4	60	33.1	9.4	28.4	33.65	9.4	27.9	142.0	64.5	45.4	137.5	63.0	45.8
Switzerland	28.8	5.6	1.0	Neg.	15.2	2.9	19.1	15.3	2.9	19.0	56.7	21.7	38.3	59.4	21.2	35.7
UK	207.3	11.7	7.0	48.700	64.0	10.0	15.6	66.3	11.2	16.9	282.5	48.5	17.2	320.6	61.0	19.0
Total																
Ten countries	1,116.8	139.15[b]	15.9		428.5	89.5	20.9	488.2	138.0	1,736.0	1,736.0	609.6	35.1	2,128.5	749.1	35.2

Sources: NEA, *Electricity, Nuclear Power and Fuel Cycle in OECD Countries*, Paris, 1988 and previous years (annual). Estimates for 1995 are those of NEA, installed capacity (GWe) made in 1985–6, generation figures (TWh) made in 1987–8. The April 1973 nuclear installed capacity figures predate by only a month or two the oil price rises later that year and are from OECD/IAEA. Total primary energy consumption data from Union Bank of Switzerland *Business Facts and Figures*, December 1987: 9. Consumption of nuclear energy expressed in mtoe (the amount of oil required to fuel an oil-fired plant to generate the same amount of electricity) from British Petroleum PLC, *Statistical Review of World Energy*, 1988: 28. The 1967 figures are provided because they *precede* the substantial nuclear power programmes of the 1970s; coal and lignite figures are added together for proved or measured reserves. From UKAEA Lo Report No. 4, *Prospects for Nuclear Power in Overseas Countries*, 1968

Notes: Only these West European countries with commercial nuclear power stations operating in 1988 are included in this table. Austria virtually completed a 700 MWe nuclear power station at Zwentendorf in 1978 but a national referendum held in November of that year rejected nuclear power and the plant has remained idle.

[a] 1 terawatt = 1 trillion (10^{12}) watts.

[b] 12.1 per cent of total primary energy consumption.

as they tried to reduce their heavy dependence on imported oil. In 1975 Italy imported 80 per cent of its fuel requirements, three-quarters in the form of oil, Sweden and Switzerland each imported 75 per cent of their energy needs, France bought in 72 per cent, Spain 69 per cent, Austria 65 per cent and the FRG 51 per cent. Of the leading industrial countries of Western Europe, only the Netherlands, with the Groningen gas field, and the UK, with plentiful coal supplies and North Sea oil and gas could view their medium-term energy situation with equanimity in the mid-1970s. Their fortunate position underlined the apparent shortage of indigenous alternative fuel resources elsewhere in the continent, and the need for an alternative energy source was emphasized by many governments.

In retrospect, it seems probable that insufficient attention was given to natural gas in the mid-1970s assessments (Odell 1985), but the answer then seemed to be in the 'technological fix' offered by further development of nuclear power. Even in coal-rich countries comparative cost estimates were often in its favour, and the growth of nuclear power after the mid-1970s was as much demand led as it was politically inspired by the nuclear industry lobby. The first oil price shock served to concentrate attention within the EC on the need to put effective conservation measures in place to reduce the growth in energy demand, to increase oil exploration in the North Sea and to maintain coal output. In May 1974 the first thoughts of the European Commission on the rate at which Western Europe could switch from oil to nuclear were that the Nine would need to have installed 200 reactors of 1,000 MWe each by 1985. Later in the year the target was slashed when the International Union for the Production and Distribution of Electricity (UNIPEDE) argued it to be 25 per cent too high. In 1974, also, the IEA/OECD identified a need for 114,000 MWe of nuclear capacity by 1985. In the event, the total amount installed by that year was 89,500 MWe. Many of the national programmes begun in the 1970s were substantially cut back well before Chernobyl, and only France showed sufficient determination to press on with a really large programme (Figure 4.1 and Table 4.1). By July 1987 there were 150 power reactors in operation in Western Europe providing a total installed generating capacity of 108,297 MWe.

By mid-1986 the EC in general was putting less stress on nuclear power in its statements regarding energy needs and electricity planning. There was a retreat from the previous commitment to produce 40 per cent of the EC's electricity from nuclear power by 1995. Moreover, there seemed to be no likelihood of a comprehensive energy strategy emerging in the near future. Nor is there now. The EC is dominated by a politically conservative outlook in which national governments can summon little enthusiasm for funding supra-national schemes despite the efforts of energy commissioners to make Brussels the centre of an integrated EC network of electricity and gas supply systems, involved in identifying waste storage sites and decommissioning projects, and aiding in the search for uranium. Yet

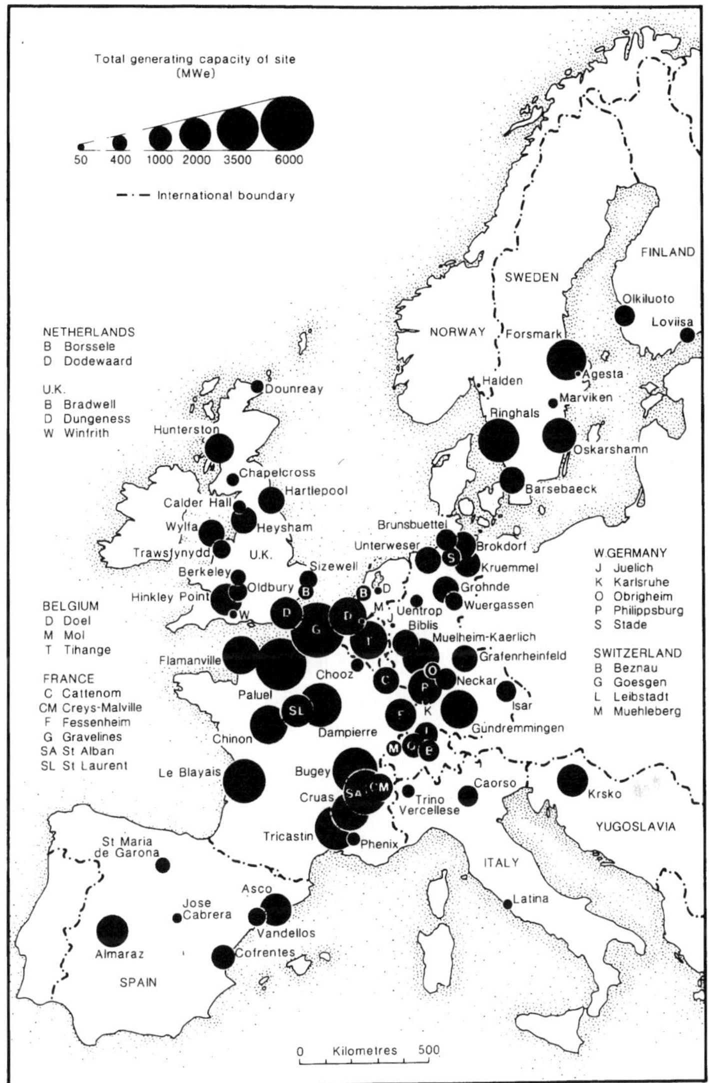

Figure 4.1 Location of nuclear power stations in operation in Western Europe 1987. The leading position of France and the FRG is clear, with the UK and Sweden also firmly in the picture. The most obvious laggard is Italy which in terms of its energy balance might seem to have the strongest case of all for going nuclear

Source: Nuclear Engineering International, *World Nuclear Industry Handbook 1988*; Foratom, *Nuclear Power in Western Europe: Status Report*, London, 1987; IAEA, *Nuclear Power Reactors in the World*

the oil price rises of the 1970s ought to be taken as a salutory warning because, generally, the 'quick-fix' for an energy supply gap is not available in modern industrial societies. At least a decade is needed to provide a substantial tranche of new generating capacity for a large utility from the time that the decision is made to build the power stations to the time that they come fully into operation.

The slackened rate of economic growth in Western Europe after 1973 led to downward revisions in electricity demand forecasts. The average annual rate of growth in demand halved from 6 per cent in 1960–73 to 3 per cent in 1973–80, and changes in the scale and timing of both conventional and nuclear power programmes became inevitable. However, it is also clear that the attempt to solve the first-level problem of securing adequate energy supplies by means of nuclear power has revealed a number of second-level problems, in many of which policy issues and geographical circumstances have been closely intertwined. They include acceptability of nuclear power by the general public, the problem of finding nuclear power station sites which satisfy technical, economic, safety, environmental and amenity criteria, the patchy record of power station construction programmes and operations, the problem of disposing of liquid, solid and gaseous radwastes, particularly at the later stages or back-end of the fuel cycle, and the delicate and often 'hidden' negotiations in the international uranium market (Moss 1981; Mounfield 1981, 1985a).

Each of these issues has a policy inequality dimension. Most obviously, the countries of Western Europe vary considerably in size and population density and, in general, the smaller and more densely populated the country, the narrower becomes the range of acceptable sites. Similarly, 'benign' alternatives to nuclear power are not equally available to all; tidal power, for example, cannot be considered by Italy, even though it has 7,500 km of coastline, because the Mediterranean is a tideless sea. In Sweden, public opinion, through referenda and the ballot box, has exercised a considerable degree of control over decision-making. In contrast, in France even senior political figures have sometimes found it difficult to exercise effective control over nuclear power planning as the country has careered along a *tout-nucléaire* pathway.

FRANCE

France was the first country to patent a nuclear power station design, in 1939, and the first Five Year Plan, which came into force in 1952, included positive plans for nuclear energy. Yet in 1960 nuclear power accounted for only 0.2 per cent of electricity output. By 1987 the proportion had risen to 70 per cent of a total of 360 TWh (see Table 4.1). This remarkable achievement reflects the fact that even in the 1950s the French programme held a will and urgency not found in many other West European states and it retained the support of a national consensus in favour of the military and civil uses of the atom unparalleled in any other leading western country.

The establishment of France as an independent nuclear *military* force meant that in the 1960s nuclear power policy with a strong military tinge was being pursued vigorously, and reactors were being built for the dual purpose of producing electricity and plutonium. Reactors G1, G2 and G3 at Marcoule were built primarily to produce plutonium and the first of these operated in 1959. In the Loire Valley Chinon became the focus of joint ventures between Electricité de France (EdF) and the Commissariat à l'Energie Atomique (CEA) to provide demonstration plants for electricity production.[2] EdF 1 was a 65 MWe unit completed in September 1962 using gas-cooled graphite-moderated technology similar to that developed in the UK, with natural uranium as a fuel. EdF 2 (175 MWe) and EdF 3 (375 MWe) followed at the same site, but even in 1963 EdF was rather sceptical about the commercial prospects of nuclear power and referred to the Chinon reactors as 'experiments on a commercial scale' (*Times Review of Industry*, July 1963: 38). However, at the same time a factory was being built at Cap de la Hague on the Cherbourg peninsula for reprocessing fuel, and the Pierrelatte uranium enrichment factory had already been built on an island in the River Rhône north of Marcoule. However, it is true that, by 1969, many French power interests were disenchanted with the high cost of the electricity being produced by the country's nuclear power stations.

Many of the early French nuclear power plants were located near the load centres and on major rivers,not on the coast as in the UK. The key to this difference in locational pattern lay in the fact that the UK had quite a powerful grid transmission system compared with that in France. The main French power lines ran north to south, with only two major lines to the west.

After the OPEC oil price rises of 1973 the Messmer–Pompidou government announced that it was embarking on a programme to build fifty-five nuclear reactors of 1,000 MWe each by 1985. The French government insisted on a single reactor design and one consortium for building the power stations, and thus avoided the engineering chaos of the UK where several consortia were used to build somewhat varying designs. The objective was to supply by means of nuclear power 70 per cent of the electricity and 30 per cent of the total energy estimated to be needed by France in that year. At the time that this ambitious programme was announced, nuclear was considered to be the cheapest alternative and EdF assumed almost dictatorial powers to push the programme through. The French economist and philosopher Louis Puiseux has argued that a nuclear power programme on the French scale has had clear totalitarian implications (Puiseux 1981). Certainly an almost fanatical drive sometimes seemed to be present in the programme. Between 1973 and 1976 the cost of a 1,000 MWe nuclear power station rose from 1,200 million to 2,000 million French francs and in December of the same year a parliamentary committee expressed grave concern that in the previous four years the cost of nuclear electricity had gone up from 3.83 to 7.00 centimes/kWh, overtaking the cost of power from coal-fired plants. President Giscard d'Estaing reviewed the programme when he came into office, only to conclude that the first phase might be scaled down to between forty-five and fifty-five reactors.

Tacit recognition that nuclear power might not after all be the cheapest option came in December 1977 when the government changed the emphasis in its justification of the programme to grounds of 'national independence' in power supplies. The target of increasing nuclear power capacity by 6,000 MWe per year over the period 1974–85 remained substantially intact, however, and added impetus was given by the breakdown of national electricity supply on 19 December 1978 when almost the whole of France lost electricity for almost 4 hours. By mid-1979, before the oil price increases later that year, nuclear reactors had again begun to look the cheapest means of generating electricity (8.68 centimes/kWh from PWRs compared with 12.68 centimes/kWh for coal and 15.33 centimes/kWh for oil) and in 1988 nuclear electricity was costing 22.7 centimes/kWh compared with 32.9 centimes/kWh for that from coal-fired plants and 80 centimes/kWh for that from those fuelled by oil (*Datafile France* 1988). By 1980, though, the justification for the programme had become much more sophisticated than a simple cost comparison. It was argued by EdF and the Government that a continued rapid switch to nuclear power held four economic advantages for France.

(a) It would help the balance of payments by substituting a large percentage of home-produced uranium for imported oil (French uranium is mined mainly near Nantes in the Vendée, in the Forez and Lyonnais Mountains near Vichy and Autun, and in the district of La Crouzille near Limoges. Workable deposits have been found in the Vosges Mountains and in the Hérault Department).
(b) It would reduce the real cost of electricity because of the growing price differential between oil and nuclear energy.
(c) It would shift the balance of energy supplies from politically unstable areas such as the Gulf States to the more stable regions of the world, including France itself, Canada, Australia, the USA, Niger and Gabon.
(d) An added bonus was the development of a nuclear power station construction industry with export potential.

Even though the nuclear power station construction programme represented an enormous investment effort, which stretched resources to the full, there was little official challenge to the programme until the election of the Mitterand Government in 1981. M. Giscard d'Estaing instituted a policy which brooked virtually no opposition. The entire programme was tightly controlled by the executive, huge resources were poured into the industry, the need for nuclear power was endlessly emphasized in the media by the government's communication machinery and nuclear power station construction came to be regarded as a locomotive for the French economy. In 1980 EdF was the largest borrower on the Paris Bourse, mainly to fund its nuclear power programme. In 1982 investments by EdF amounted to 30 billion French francs (£2.5 billion), of which 21 billion French francs (£1.7 billion) was for nuclear power. By the end of 1983 EdF had amassed debts of 189 billion French francs (£15.7 billion), 44 per cent in foreign currencies, making it one of the world's largest borrowers on international capital markets.

Thus, opposition to nuclear power has tended to be local and regional, and has focused on particular sites rather than upon the total programme. In theory, French local government is strong enough for local communities to refuse to have nuclear power stations imposed upon them; Erdeven, near Quiberon in Brittany, attempted to do this in the mid-1970s, but at a certain point the national interest has regularly been used to override regional objections. A highly effective weapon of persuasion in the hands of EdF is the *patente*, the licence money that a power station must pay each year to the commune in which it is located. At Brennilis, for example, a tiny lake-water-cooled plant of only 70 MWe, a *patente* of around £50,000 per annum enabled the mayor of the village to modernize his offices and to build a football stadium, a schoolmaster's lodging and a recreation room for the elderly, before closure of the plant in 1986. To make them even more locally palatable, in 1979 the Giscard Government authorized a 10 per cent discount in electricity tariffs for those living or working within 10 km of a nuclear electricity generating station. Each power station provided work for several thousand people at the peak of construction, and in some cases up to 75 per cent of these were recruited locally.

With these inducements it is not surprising that public willingness to accede to the nuclear power programme has been characteristic of the French situation. Nevertheless, there has been opposition, some of it genuinely local and some coming from travelling groups of international anti-nuclear protesters. And international questioning of some sort is not unjustified, for it is clear from Figure 4.1 that in terms of the sites chosen so far for many of its nuclear power stations, France is no island. The Gravelines nuclear power station is only 32 km from Belgium and 64 km from Dover, Fassenheim is close to the German border and Bugey is only 11 km from Geneva. Flamanville, on the Cotentin Peninsula in Normandy, is barely 43 km from St Helier in Jersey and France's large nuclear fuel reprocessing plant, belonging to Compagnie Génerale des Matières Nucléaires (COMEGA) and located near Cherbourg, in Normandy, is even closer. In such circumstances, France has a clear safety responsibility to its neighbours.

The original EdF master plan of 1975 envisaged 200 nuclear reactors by the year 2000, grouped in 'parks' along the sea coast and on the banks of major rivers, particularly the Rhine, Rhône and Loire. Such sites could readily provide cooling water. Thirty-four reactors on nine nuclear parks were projected for the 500 km stretch of the River Rhône between the

Swiss frontier and Marseille. The fact that depressed industrial regions were chosen for some of the power stations (e.g. Gravelines in the north and Cattenom in Lorraine) helped to keep opposition muted. Moreover, EdF decided to build 1,300 MWe reactors, grouped together on shared sites or 'parks', not only for scale economies but also because fewer sites meant less frequent local opposition.

Public interest in the nuclear power programme began to grow from mid-1975. In August of that year the filtration plant and chimney stack of the experimental reactor at Brennilis were damaged by a small bomb planted by the Breton Liberation Front, three months after a similar attack by a German-based group at Fassenheim, then under construction. In 1976 a nuclear power plant was announced for Nogent-sur-Seine, very close to the centre of the town and barely 100 km from Paris. The Seine river basin in this area provides a quarter of Paris's water supplies, and one of the last protests made by the then Minister for the Quality of Life, M. André Jarrot, before he was dismissed in 1976, was a letter to his Cabinet colleagues expressing bewilderment at the Nogent project, which was also condemned by the Seine–Normandy River Basin authority. By the end of 1976 EdF was having to send out 24,000 documents a month to defend its policy and was finding it increasingly necessary to answer public doubts on safety and the environmental impact of nuclear power plants.

The first of the modern generation of large reactors in France, the 880 MWe Fassenheim I, went into service in 1977 despite massive anti-nuclear power demonstrations on both sides of the border. These were followed by similar demonstrations, in which a man was killed and a hundred protestors injured, against the 1,200 MWe Creys-Malville fast breeder reactor then being built beside the Rhône at Morestel (Isère), between Lyon and Geneva. In April and May 1978 a series of demonstrations took place against the construction of the nuclear power station at Flamanville, near Cherbourg.

Throughout 1979 and 1980 the opposition continued to focus on specific sites, particularly those at Golfech and Plogoff. The Golfech (Agen) site is on an island formed by the Garonne river 80 km northwest of Toulouse and was initially planned to be one of the largest nuclear 'parks' in Europe, with four 1,300 MWe reactors. In 1975, a referendum in the area produced an 80 per cent 'No' vote, and the regional council rejected the plan as did the Tarn-et-Garonne departmental council. However, EdF went ahead with a 'public utility inquest' and a building permit was granted on 5 May 1981, less than a week before the end of M. Giscard d'Estaing's Presidency. In 1980, France's nuclear programme suffered its largest public setback at the small Breton community of Plogoff, one of Brittany's beauty spots. The Plogoff site was important because most of the opposition was local, and because in the summer of 1980 fierce clashes between demonstrators and police made headlines throughout Europe. A public inquiry found in favour of the site and the project was given approval on 2 December 1980 only to be 'frozen' and ultimately cancelled when the Socialist Government came into power in 1981. Creys-Malville was subject to another violent incident in 1982, this time involving the use of an anti-tank rocket launcher.

Before his election, M. Mitterand had promised to review the nuclear power programme. Immediately on taking office his government froze the existing plans to start building nine new nuclear reactors in 1982–3, pending its own review and assessment. Then, on 25 November 1981, it gave permission for a six-reactor programme, unfreezing the sites at Chinon on the Loire (900 MWe), at Cattenom near Thionville in Lorraine (1,300 MWe), at Chooz (Charleville-Mézières) on the Franco-Belgian border (1,300 MWe), where the local community had been in favour, at Nogent (1,300 MWe), at Penly, in Normandy near Dieppe (an additional 1,300 MWe unit) and at the much-disputed Golfech. An attempt was made to make Golfech palatable to the local community by specifying one reactor, not two; by promising local businesses construction work worth 1,100 million French francs (£100 million), by guaranteeing local labour at least half of the peak figure of 2,000 construction jobs and by EdF's paying the region 10 million French francs (£900,000) a year whilst building proceeded as an incentive for industrial development and a further 6 million French francs a year under the plant's *patente.* In 1981, also, local representatives approved sites at Civaux, near Poitiers, and Le Pellerin, Nantes; the latter was regarded by EdF as necessary to secure the electricity supplies of western France following the cancellation of Plogoff

By 1984, about a decade after it had launched its massive nuclear power programme, France had commissioned thirty-two reactors and had twenty-eight more under construction including fifteen of 1,300 MWe design. In 1984 it obtained nearly half its electricity from nuclear power stations, three times the UK proportion, and by 1987 the figure was 70 per cent (Table 4.1). There can be no question but that France has taken the lead in Europe in exploiting nuclear energy and in developing the advanced technologies needed to sustain its progress. A major branch of industry has been created in the process, with 160,000 jobs spread over 600 companies. The cost, however, has been enormous. EdF is heavily in debt and France now has an excess of electricity generating capacity. The nuclear power construction programme was based on forecasts of energy demand which even by 1983 were clearly optimistic and by 1991 there are likely to be three to five reactors more than the economic optimum. Falling oil prices in the late 1980s and the decline in the rate of increase in energy consumption because of slower economic growth, together with conservation efforts, have altered most of the planning assumptions on which French energy policy was based. There has been a marked slowdown in nuclear power station building since President Francois Mitterand came to power in 1981. New reactor projects, then running at six a year, fell to two in 1984 and one in 1985 as the new administration adopted a tapered construction programme. The construction industry found itself with excess production capacity in the mid-1980s and was severely shaken in 1984 by the collapse of one of the leading participant firms, Creusot-Loire. CEA owns 50 per cent of Framatome, the nuclear power station builder; Creusot-Loire, when it became bankrupt, owned the other half. Framatome had made a profit of 276 million French francs in 1982 and 200 million French francs in 1983, but whereas two French reactor orders a year could support one factory, one was not enough. Local protests about nuclear power turned to protests about the cuts in building programmes, for example around Civaux, near Poitiers, after the government decided in 1986 to cut the number of reactors to be built there from four to two.

The reduced programme has provided some relief for EdF but the utility now has an excess of electricity generating capacity and has reacted by advertising the 'advantages' of all-electric homes and by boosting exports. The objective is to increase exports from the 23.4 billion kWh achieved in 1985 to 40 billion kWh in 1990. In 1986 current worth 5.3 billion French francs was sold abroad, mainly to Switzerland and Italy, bringing in profits of 3.5 billion French francs and EdF is urging its customers to become financial partners in its power station building programme. In terms of location, this could lead to an increase in the number of nuclear power stations at sites near the French frontiers. Following the installation of the direct current cross-Channel link in 1983, the UK imported 4.47 thousand kWh. In 1986 Creys-Malville was connected to the grid and this fast breeder reactor is designed to export 49 per cent of its output to Italy and the FRG, who have been partners with France in the project.

The French public became accustomed to large and expensive power projects with the building of the Rhône hydroelectric schemes in the 1960s, and public opposition to nuclear power has been less than in the FRG or Sweden. The need for the programme was not challenged at a high political level until 1980 when the General Commission for the National Plan argued that overproduction of nuclear energy seemed inevitable in view of the country's slow economic growth. In 1979 the total electricity requirement for 1990 was estimated to be 460 billion kWh. This figure was subsequently progressively reduced, first to 415 billion kWh and then in 1983 to 350 billion kWh. Future growth is forecast at 2–3 per cent per year compared with the 5 per cent recorded from 1976 to 1986.

Resistance at local and regional levels has emphasized safety, environmental and ecological matters. From the safety point of view concern has been expressed that many French reactors have been located too close to cities and it has also been suggested that the ability of France to defend itself may have been diminished by the close proximity of many nuclear power stations to such high density population centres. This siting strategy, it is argued, has provided potential targets for an enemy and hence a 'nuclear yoke' for France. It has also been argued that the nuclear power programme has been pressed home with such vigour that it has been destructive of some traditional French democratic institutions. In consequence, the Mitterand government devised a three-tier system of appeal against reactor siting, at local, regional and national levels. Implementation of this scheme has been facilitated by the fact that more sites are available than are actually needed.

INTERNATIONAL ELECTRICITY FLOWS IN WESTERN EUROPE

Figure 4.2 shows international flows of electricity in 1983, and in its efforts to sell surplus current France has been helped by the existence of an electricity pool. The pool began in the 1950s as a means of conserving energy by allowing surplus hydroelectric power to move across frontiers, thus preventing it from running to waste when the reservoirs and rivers were full. By the same token nations heavily dependent upon hydroelectric power could import thermal power in times of low rainfall. The exchange of electricity generated from hydroelectric power in the Alps, especially Austria and Switzerland, and thermal power produced in other countries is still one of the major flows, but the existence of large and well-interconnected national grids has been significant for the development of nuclear power in western Europe. It has allowed some countries to install nuclear electricity capacity in the form of larger units than might be justifiable in terms of national demand forecasts or risks to supplies arising from an unscheduled shutdown. It has provided France with the opportunity to increase trans-boundary trade by exporting surplus nuclear generated current. It has enabled countries which keep deferring decisions to install nuclear power stations to import nuclear electricity, as Denmark does from Sweden (see Figure 4.2). Austria, which by referendum has voted not to operate its first completed nuclear power station, at Tullnerfeld (Zwentendorf), has a number of inter-ties with the transmission networks of the FRG, Switzerland and Czechoslovakia, all of which are becoming increasingly dependent on nuclear electricity.

It is possible that in the longer term stronger interconnections between the networks of eastern and western Europe may be built. The USSR has proposed measures which have not been greeted with much enthusiasm in the west, partly because the existing eastern and western grids are not of the same frequency and tension but also because the existing size of UCPTE already permits substantial scale economies without the rather frequent supply disruptions that tend to occur at peak periods in the CMEA. There are some 'back-to-back' converter stations (connecting alternating current grids in eastern and western Europe by high voltage direct current links), but Austria's special circumstances have led it to co-operate with East European countries on a larger scale. Austria lacks primary fuel resources and needs electricity imports in winter because of its inadequate base load generating capacity. Therefore it has entered into agreements to import electricity from Czechoslovakia, Hungary, Poland

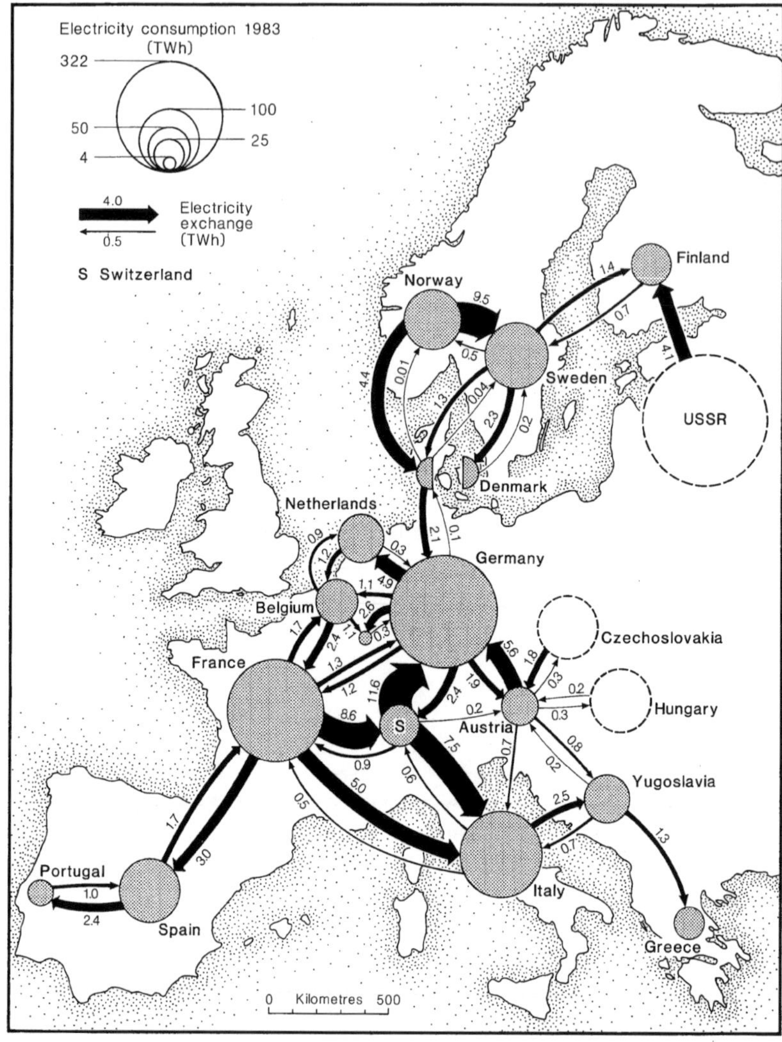

Figure 4.2 International electricity transfers in Western Europe 1983. The bulk of the electricity generated in western Europe is consumed in the producing country but the national grids of most continental European countries are interconnected across national boundaries maintained between these blocks. One block is the Union for the Co-ordination of Production and Transport of Electricity (UCTPE)) which includes Luxembourg, the Netherlands and Switzerland, with four associated countries, Greece, Portugal, Spain and Yugoslavia. The other is NORDEL consisting of the utilities of Denmark, Norway, Sweden, Finland and Iceland (which cannot participate in the actual interchange of electricity). The French and mainland UK systems are connected by a 2,000 mW HVDC cross-Channel cable opened in 1986 between Folkestone in Kent and Sangette in France

Source: IEA/OECD 1985

Note: 1 terawatt hour (TWh) = 1 billion kWh.

and the USSR in winter, exporting it in summer from its seasonal surplus hydroelectric capacity (see Figure 4.2).

FEDERAL REPUBLIC OF GERMANY

The FRG re-entered the nuclear power arena rather late in the post-war period; in 1954 it made a solemn renunciation of nuclear weapons production, but not until that year was it able legally even to research in the field. The embryonic nuclear industry initially had to rely upon foreign licences but, supported by federal subsidies, the first experimental facilities, prototype plants and nuclear power demonstration plants were constructed. The main atomic research centre was built at Karlsruhe, not far from the French border, in 1956. This was the home of the FRG's first working home-produced reactors (Otto Hahn or FR2; FR1 was never built) which started up in 1962. This was a heavy-water-moderated pile and continued the technology of the German war effort, which did not actually succeed in producing a nuclear chain reaction. Once the FRG regained its sovereignty in 1955, the Otto Hahn reactor became a symbol of the country's legitimized nuclear research efforts, redirected to peaceful purposes, but in 1968 attention turned away from heavy water technology to US light water reactor designs. The Karlsruhe complex has played host to four different German nuclear power reactor prototypes including a 20 MWe plutonium-burning fast reactor (KNK) which went on stream in 1974, but although one Karlsruhe-designed reactor, a heavy water model built by Siemens, was sold to Argentina in the 1970s and agreement was reached in 1988 to supply a 200 MWe high temperature reactor to the USSR, no Karlsruhe design has been developed commercially in the FRG. Of the country's twenty-one reactors, eighteen are based on PWR and BWR technology originally acquired from Westinghouse and US General Electric. The first commercial 600 MWe units, at Stade and Würgassen, were ordered in 1967 and began production in 1972.

The FRG does not at present reprocess spent fuel, but a pilot reprocessing plant has been in operation since 1971 and this provided the technological basis for the 350 tonnes per annum reprocessing plant initially scheduled for 1996 at Wackersdorf in East Bavaria, a project which has now been cancelled. The drastic fall in the value of plutonium since the mid-1980s challenged the value of this DM 10 billion project. The country has a uranium-enrichment plant at Gronau in North Rhine Westphalia with a current annual capacity of 400 tonnes. Hanau in Frissen is the centre of fuel fabrication.

The FRG was the first West European country to publish a co-ordinated energy plan, early in 1973, and even before the oil price increases of that year the Bonn government had decided that nuclear power must meet much of the growth in energy demand then being forecast for 1985. Initially the attraction of nuclear power to the energy planners was that it promised cheaper electricity than that generated by coal-fuelled power stations, but the sharply increased oil bill arising from OPEC price increases soon became a key element in reducing the visible trade surplus and increasing its current account deficit. As a proportion of primary energy consumption, nuclear power was expected to grow from 1 per cent in 1973 to 15 per cent by 1985, when it would have accounted for 45 per cent of installed electricity generating capacity. This meant that 45,000–50,000 MWe of nuclear capacity would have needed to be built by 1985. However, nuclear power's anticipated contribution for 1985 was subsequently reduced, first to 30,000 MWe and then to 24,000 MWe, the latter figure representing some 29 per cent of the total electricity generating capacity projected for 1985. By mid-1982 there were fifteen power reactors in existence, giving a total capacity of 10,358 MWe, and another twelve with a capacity of 13,155 MWe under construction. Twenty-one units were in operation in mid-1987, providing a total installed capacity of 19,911 MWe; their names and locations are shown in Figure 4.1 and Table 4.2.

With a larger population than France, the FRG has less than half the land area, and the search for nuclear power station sites has been correspondingly more difficult. Moreover, the programme has met stiff and well-organized resistance on many fronts, including high level political activity, and this opposition has been helped both by the way in which the electricity supply industry is organized and by the complex structures which have existed for granting or refusing approval for particular sites. Unlike France, where the industry is in the hands of one nationalized organization, there are nearly 700 utilities varying widely in size, function and funding. The nine main 'first-level' utilities (the Nine) shown in Figure 4.3 own and operate most of the generating capacity, including all of the nuclear plants, and nearly all the high-voltage national grid. The overall public network has a generating capacity of 85,000 MWe.

Most of the seventy-four second-level regional utilities have some generating capacity of their own, complemented by energy received from one of the Nine. Except for the larger cities that operate their own power stations, the 500 third-level municipal utilities mainly distribute supplies via low voltage networks. There are also about sixty companies owning and operating power plants and supplying electricity exclusively to other utilities. Outside the public supply system, industrial companies and the federal railways have generating capacity totalling 15,000 MWe.

This structure influences the commitment to nuclear power and the location of nuclear plants. For example, Rheinisch Westfälisches Elektrikactswerk (RWE) accounts for 40 per cent of the public electricity supply of the FRG, distributing current to large parts of North Rhine–Westphalia, which is heavily populated and industrialized, Rhineland Palatinate and Lower Saxony. It is the largest of the Nine, with an installed capacity of 28,000 MWe. It owns the Biblis A and B nuclear plants on the Rhine,

Table 4.2 Nuclear power stations in operation in Western Europe in July 1987

Type and name of power station	*Location*	*Reactor power (MWe gross)*	*Year of commercial operation*
Belgium[a]			
PWR			
Doel 1	Doel-Beveren (between Antwerp and Ghent)	412	1975
Doel 2		412	1975
Doel 3		936	1982
Doel 4		1,059	1985
Mol BR3	Mol (closed down 1987)	12	1962
Tihange 1	Huy, Liège (between Liège and Namur)	920	1975
Tihange 2		941	1983
Tihange 3		1,048	1985
Total operable (8 units)		5,740	
Finland			
BWR			
Olkiluoto 1	Eurajoki (NW of Rauma)	735	1979
Olkiluoto 2		735	1982
PWR			
Loviisa 1	Loviisa (between Kotka and Helsinki)	465	1977
Loviisa 2		465	1981
Total operable (4 units)		2,400	
France[b]			
FBR			
Creys-Malville, Super-Phenix 1	Morestel, Isère	1,242	1986
Phenix	Avignon	250	1974
Gas graphite			
Bugey 1	Lyon	555	1972
Chinon A3	Avoine	375	1967
St Laurent A1	Orléans	405	1969
St Laurent A2		465	1971
PWR			
Bugey 2	Lyon	955	1979
Bugey 3		955	1979
Bugey 4		937	1979
Bugey 5		937	1980
Cattenom 1	Thionville	1,330	1987
Chinon B1	Avoine	919	1984
Chinon B2		919	1984
Chinon B3		919	1987
Chooz, SENA	Chooz	320	1967
Cruas Meysse 1	Montélimar	921	1984
Cruas Meysse 2		921	1985
Cruas Meysse 3		921	1984
Cruas Meysse 4		921	1985

Dampiere 1	Gien	937	1980
Dampiere 2	Gien	937	1981
Dampiere 3	Gien	937	1981
Dampiere 4	Gien	937	1981
Fessenheim 1	Mulhouse	920	1978
Fessenheim 2	Mulhouse	920	1978
Flamanville 1	Cherbourg	1,345	1986
Flamanville 2	Cherbourg	1,345	1986
Gravelines 1	Dunkerque	951	1980
Gravelines 2	Dunkerque	951	1980
Gravelines 3	Dunkerque	951	1981
Gravelines 4	Dunkerque	951	1981
Gravelines 5	Dunkerque	951	1985
Gravelines 6	Dunkerque	951	1986
Le Blayais 1	Bordeaux	951	1981
Le Blayais 2	Bordeaux	951	1983
Le Blayais 3	Bordeaux	951	1983
Le Blayais 4	Bordeaux	951	1983
Paluel 1	Fécamp	1,345	1985
Paluel 2	Fécamp	1,345	1985
Paluel 3	Fécamp	1,345	1986
Paluel 4	Fécamp	1,345	1986
Saint Alban 1		1,348	1985
Saint Alban 2		1,348	1986
St Laurent B1		921	1983
St Laurent B2		921	1983
Tricastin 1		955	1980
Tricastin 2		955	1980
Tricastin 3		955	1981
Tricastin 4		955	1981
Total operable (49 units)		46,693	
FRG			
GHTR			
AVR Jülich[c]	Jülich	13	1968
FBR			
KNK Karlsruhe[c]	Leopoldshaven	18	1978
PWR			
KWO Obrigheim	Mosbach, Baden	357	1960
KKS Stade	Stade	662	1972
Biblis A	Biblis	1,204	1975
Biblis B	Biblis	1,300	1977
GKN1 Neckar	Neckarwestheim	855	1976
KKU Unterweser	Esenshamm	1,300	1979
KKG Grafenrheinfeld	Grafenrheinfeld	1,300	1982
KKG Grohnde	Hamein	1,361	1984
KKP2 Philippsburg	Philippsburg, Rhein	1,349	1985
KBR Brokdorf	Brokdorf, Elbe	1,365	1986
BWR			
KKB Brunsbüttel	Brunsbüttel	806	1977
KK1 Ohu/Isar	Ohu, Landshut	907	1979
KKP1 Philippsburg	Philippsburg, Rhein	900	1980

Table 4.2 continued

Type and name of power station	*Location*	*Reactor power (MWe gross)*	*Year of commercial operation*
KKK Krummel	Krummel, Geesthacht	1,316	1983
KRBB Gundremmingen	Gundremmingen	1,300	1984
KRBC Gundremmingen	Gundremmingen	1,308	1984
KKKW Wurgassen	Beverungen	670	1972
Total operable (21 units)		19,911	
Italy			
BWR			
Caorso	Caorso, Piacenza	882	1981
Magnox			
Latina[d]	Borgo Sabotino	160	1964
PWR			
Trino Vercellese	Trino, Piedmont	270	1965
Total operable (3 units)		1,312	
Netherlands			
BWR			
Dodewaard	Dodewaard	59	1969
PWR			
Borssele	Vissingen	481	1973
Total operable (2 units)		540	
Spain[e]			
BWR			
Cofrentes	Cofrentes	974	1985
Santa Maria de Garona	Burgos	460	1971
Gas graphite			
Vandellos 1	Hifrensa, Hospitalet de l'Infant	496	1972
PWR			
Almaraz 1	Almaraz, Cáceres	930	1981
Almaraz 2	Almaraz, Cáceres	930	1984
Asco 1	Tarragona	930	1984
Asco 2	Tarragona	930	1986
Jose Cabrera	Guadalajara	160	1969
Total operable (8 units)		5,810	
Sweden[f]			
BWR			
Barsebäck 1	North of Mälmo	615	1975
Barsebäck 2	North of Mälmo	600	1977

Forsmark 1	North of Uppsala	1,004	1980
Forsmark 2		1,004	1981
Forsmark 3		1,090	1985
Oskarshamn 1	Oskarshamn	460	1972
Oskarshamn 2		617	1974
Oskarshamn 3		1,100	1985
Ringhels 1	Varberg	780	1986
PWR			
Ringhels 2	Varberg, south of Gothenburg	840	1975
Ringhels 3		960	1981
Ringhels 4		960	1983
Total operable (12 units)		10,030	
Switzerland			
BWR			
Muehleberg	West of Berne	336	1972
Leibstadt	Leibstadt	1,045	1984
PWR			
Beznau 1[g]	Doettingen	364	1970
Beznau 2[g]		350	1972
Goesgen	Daeniken	970	1979
Total operable (5 units)		3,065	
UK			
AGR			
Dungeness B1	Kent	660	1985
Dungeness B2		660	1988
Hartlepool 1	Co. Durham	666	1985
Hartlepool 2		666	1984
Heysham 1-1	Lancashire	666	1984
Heysham 1-2		666	1985
Hinkley Point B1	Somerset	660	1976
Hinkley Point B2		660	1978
Hunterston B1	Ayshire	660	1976
Hunterston B2		660	1977
FBR			
Dounreay PFR		270	1977
Magnox			
Berkeley 1[h]	Gloucestershire	167	1962
Berkeley 2[h]		167	1962
Bradwell 1	Essex	173	1962
Bradwell 2		173	1962
Calder Hall 1	Cumbria	61	1956
Calder Hall 2		61	1957
Calder Hall 3		61	1958
Calder Hall 4		61	1959
Chapelcross 1	Dumfriesshire	60	1959
Chapelcross 2		60	1959
Chapelcross 3		60	1960
Chapelcross 4		60	1960

Table 4.2 continued

Type and name of power station	*Location*	*Reactor power (MWe gross)*	*Year of commercial operation*
Dungeness A1	Kent	285	1965
Dungeness A2		285	1965
Hinkley Point A1	Somerset	282	1965
Hinkley Point A2		282	1965
Hunterston A1	Ayrshire	169	1964
Hunterston A2		169	1964
Oldbury 1	Gloucestershire	313	1968
Oldbury 2		313	1968
Sizewell A1	Suffolk	325	1966
Sizewell A2		325	1966
Trawsfynydd 1	N. Wales	290	1965
Trawsfynydd 2		290	1965
Wylfa 1	Anglesey, N. Wales	655	1972
Wylfa 2		655	1972
SGHWR			
Winfrith	Dorset	100	1968
Total operable[i] (38 units)		12,796	

Sources: Nuclear Engineering International, *World Nuclear Industry Handbook 1988*; Foratom, *Nuclear Power in Western Europe: Status Report*, 1987, London; IAEA, *Nuclear Power Reactors in the World*; author's files

Notes: [a] Belgium has co-operated with France on joint reactor projects on French and Belgian soil since the 1960s. The 320 MWe SENA PWR at Chooz near the Franco-Belgian border was commissioned in 1967 as a joint venture of the Belgian utilities and EdF, as was Tihange 1. In 1984 Belgium and France signed an agreement allowing for 25 per cent Belgian participation in two further 1,400 MWe plants being built in 1988 at Chooz. Mol BR3 was closed down in late 1987 and was built as a non-commercial pilot plan.

[b] France's earliest reactors producing electricity were the gas-cooled graphite-moderated natural-uranium-fuelled units used for Chinon 1 (70 MWe) and Chinon 2 (170 MW), now closed down, Chinon 3, St Laurent A1, St Laurent A2 and Bugey 1. These followed reactors G1, G2 and G3 built in the 1950s at Marcoule primarily to produce plutonium. Reactor EL4 at Brennilis was a heavy-water-moderated gas-cooled enriched-uranium-fuelled prototype. The first PWR, Chooz SENA, using enriched uranium, was built under an agreement between the USA and EURATOM.

[c] These two are pilot plants not in commercial operation.

[d] Shut down 1989.

[e] Under the 1984 National Energy Plan five reactors (four of them at an advanced stage of construction) were indefinitely deferred. Two are at Lemoniz, near Bilbao, paralysed by terrorist action of the Basque separatist ETA movement and now finally written off by the government, two are at Valdecabelleros in Badajoz province near the Portuguese border and one is the second reactor at Trillo, northeast of Madrid. According to the Plan nuclear energy is due to provide 29.5 per cent of Spain's electricity in 1992.

[f] Sweden's first nuclear power station was the 65 MWe Agesta unit operated from 1964 to 1973. This was followed by work on a 200 MWe boiling heavy water reactor, Marviken, cancelled in 1970. Oskarshamn 1 was the first commercial BWR; Ringhels 2 was the first PWR. There are no plans to build more nuclear units.

[g] Used for combined heat and power on the Refuna heating network.

[h] Closed in 1989.

[i] Under construction at 31 July 1987: 5 units totalling 3,822 MWe.

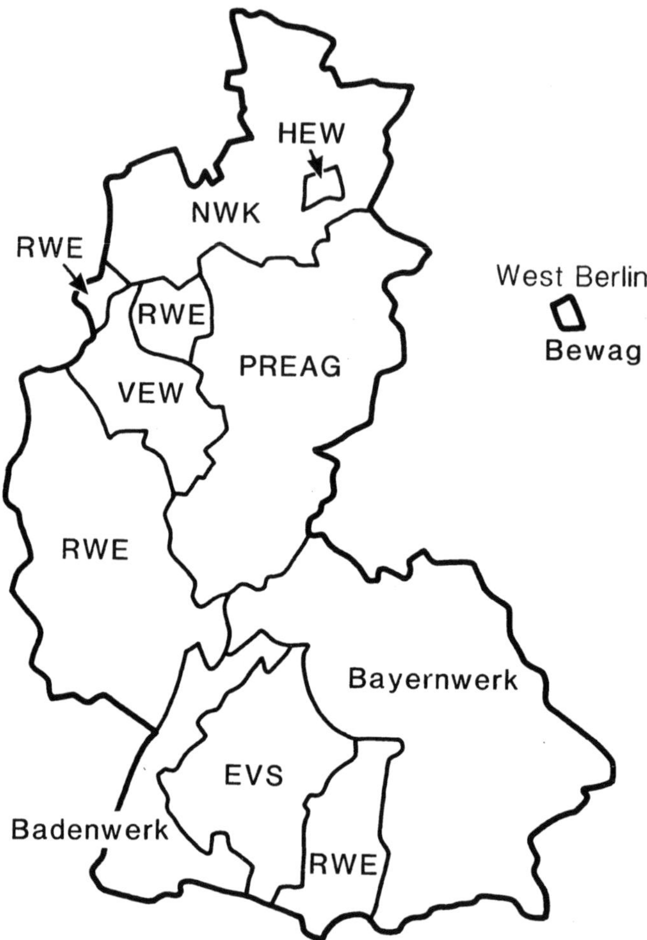

Figure 4.3 The public electricity supply system in the FRG, 1987. Badenwerk is mixed ownership, over 75 per cent by the State of Baden-Württemberg and the rest private; Bayernwerk is publicly owned, 60 per cent by the State of Bavaria and 40 per cent by the Federal Republic; Bewag is mixed ownership, 58.3 per cent State of Berlin, 8.6 per cent Elektrowerke AG, 8.6 per cent PREAG and 24.5 per cent other private; EVS is publicly owned, 10.4 per cent by the State of Baden-Württemberg and the balance by various regional and municipal public bodies; HEW is mixed ownership, 73 per cent by Hamburg and 27 per cent private; NWK is a subsidiary of PREAG; PREAG is mixed ownership, 6.8 per cent Frankfurt, 2.7 per cent State of Hesse, 86.5 per cent VEBA and 4.0 per cent others; RWE is mixed ownership with a private majority of nominal share capital and a public majority of voting capital with cities and municipalities; VEW is mixed ownership, 52.56 per cent various cities and regional public bodies, 25.32 per cent private holding and 22.12 per cent other private

north of Mannheim, and one plant at Mülheim Kärlich on the Rhine upstream from Bonn. The political need to be seen to be supporting North Rhine–Westphalia's indigenous coal and lignite resources, as well as the SPD's anti-nuclear stance, has been reflected in the location of these plants, to the extent that none of them is in North Rhine–Westphalia, even though this state contains the dominant load centres.

The FRG has an extremely complex licensing system for nuclear power stations involving many agencies at state and federal level. The crucial licence for the erection and operation of a nuclear power plant is granted by the licensing authority of the state containing the site. Thus there are eleven licensing authorities, with considerable power, to which the federal Ministry of the Interior must refer and give instruction in its capacity as supreme federal authority. The Ministry also has to approve the basic decision with regard to the site and technical characteristics of the plant. Citizens' groups and environmentalists have so successfully used the courts to resist the choice of particular sites that they effectively imposed a five year ban on nuclear power station construction from 1977 to 1982.

As in France, much of the public resistance has focused on particular sites. In February 1975 inhabitants of the town of Whyl, in Baden–Würtenberg occupied the site of a planned nuclear power station as the bulldozers moved in to clear it. For a time the conservative state government tried to defend the site with police, barbed wire and water cannon, but it soon realized that it was dealing not just with a few extremists but with a popular political force of some magnitude. The courts imposed delays on construction of the power station and the site was occupied by protestors for nearly two years, until November 1976.

During 1975–6 there was a growing feeling that it was irresponsible to continue the nuclear power programme without solving the problem of what to do with nuclear waste. It also became clear, however, that Lower Saxony, the state singled out as being geologically the most suitable for a radwaste storage facility, was not keen to accept it. The most favoured site was Gorleben, where the federal government planned to construct facilities for the storage and reprocessing of spent fuel, waste treatment, the refabrication of reclaimed uranium into new fuel elements and the ultimate storage of high level waste. Gorleben occupies a border salient into the GDR, but the thick salt strata in this part of Lower Saxony are considered to provide good conditions for the burial of highly active radwastes (see Chapter 12). On 12 February 1977 Herr Helmut Schmidt, then Federal Chancellor, confronted Herr Ernest Albrecht, the Prime Minister of the state of Lower Saxony, with the fact that the nuclear programme hung on his decision on whether or not to allow the development at Gorleben. In the following month there were savage clashes between police and radical groups at the site of the nuclear power station at Grohndé (Hamien) in Lower Saxony, in which 300 people were injured. The protest groups came equipped with wire- and bolt-cutters, ropes, grappling irons, aluminium foil kits to block police radios and flame-cutting equipment. The forces of law and order had dogs, gas grenades, water cannon and barbed wire. Grohnde was soon a battlefield, like Whyl and Brokdorf before it. Nevertheless, Gorleben took a step forward in December 1978 when Herr Schmidt and Herr Albrecht agreed on finance for the project. Herr Albrecht maintained that, whilst his state was reluctantly prepared to take on the political burden of Gorleben for the good of the country, it could not be expected to bear the financial costs as well. Therefore the federal government agreed to provide DM 20 million (£50 million) over four years to help cover 'extraordinary expenses' including, amongst other things, securing the site against violent demonstrations. At the end of April 1984 more than 2,000 anti-nuclear demonstrators blockaded access roads at Gorleben. Lower Saxony's lack of enthusiasm for its role as the country's nuclear dustbin may not be unrelated to the fact that its gas fields provide a third of the country's natural gas and its coal mines produce around 85 million tons of lignite a year.

In November 1976 a site for a 1,300 MWe nuclear power station at Brokdorf, a small village on the Elbe estuary north of Hamburg, became the focus of opposition as violent clashes took place between 2,000 militant demonstrators and police using water cannon and tear gas grenades to prevent the protestors from occupying the site. Twenty-six police and a hundred demonstrators were injured.

The pitched battle at Brokdorf received major coverage by the media and had a considerable impact on the attitude of politicians. Following these ugly confrontations, the Schleswig–Holstein administrative court ordered an indefinite ban on construction work at the site until the country could decide what to do with the radwaste resulting from the nuclear power programme, and in January 1977 the state government of North Rhine–Westphalia also decided not to allow any new nuclear power plants before the problem of waste disposal was solved.

During the 1970s nuclear power became an issue on which many German politicians became increasingly uncertain as to the strength and direction of public opinion. Yet policy to counter the dependence on imported oil, and the hesitancy with which the nuclear power programme was being pursued, were also exposed as major structural weaknesses threatening the country's plans for economic growth. Until the election of the Christian Democrat Government under Chancellor Helmut Kohl, most of the failure to push through a nuclear power policy was blamed on the ruling Social Democrat coalition, which was hampered by a vocal wing opposed to nuclear power. In November 1975 the party congress of the Free Democrats, the Liberal partners in the Bonn coalition government, passed a resolution asking Liberal ministers to make sure that no further permits for nuclear power stations should be granted until the waste disposal problem had been solved. The then Chancellor, Helmut Schmidt, a supporter of nuclear power, adopted the demand of the liberal rank and file in his year-end Government Declaration and added that the price of nuclear electricity must in future cover all aspects of the fuel cycle, including waste disposal. He also said that nuclear safety must have precedence 'over all economic considerations'. Thus the nuclear power plans of the FRG were downgraded to the status of a necessary evil and, although the coalition government plodded on doggedly with its reduced nuclear power programme through the latter part of the 1970s, making progress when and where it could, it made a major priority of boosting domestic coal output. A real change in prospects for the nuclear industry came only with the election of the Christian Democrat Government, which declared itself to be in favour of an expansion of the country's nuclear power capacity. The log-jam in construction was broken in August 1982 with the granting of the first two building licences for five years, the Isar II reactor in Bavaria and a plant at Emslend-Lingen in Lower Saxony. These power stations marked a change in licensing procedure to a 'convoy system' under which standardized approval began to be given to a series of reactors rather than each plant having to be approved individually. In May 1983 Herr Albrecht gave permission for the building of a nuclear fuel reprocessing plant at Dragahn, near the GDR border. Under the revived programme it was suggested in 1984 that the country should aim for an installed nuclear generating capacity of 37,000 MWe by 1995, a target which, if achieved, might then provide 17 per cent of the country's total primary energy requirements (Mounfield 1985a).

The impact in the FRG of the Chernobyl disaster has been sufficient to throw into disarray the consensus gradually emerging in the mid-1980s in favour of expanding the nuclear power programme. On 25 May 1986 police used water cannon to repel protestors who tried to dig under a fence at the Wacksdorf construction site, 20,000 people gathered to demonstrate at the Biblis nuclear power station near Frankfurt and several thousand blocked the motorway near Saarbrücken in protest at Cattenom. An opinion poll taken on the same day showed 83 per cent of West Germans to be opposed to expansion of the nuclear programme. Politicians in touch with the public mood have argued for a moratorium on new reactors and for phasing out those already in operation, the so-called 'transitional solution' to the energy crisis. The SDF, in power from 1969 to 1982 and then generally supportive of nuclear power, has switched its stance to one of opposition despite a cost advantage in its favour in 1987 of 0.04 to 0.05 pfennigs/kWh, and this has been interpreted by many observers to indicate the beginning of a long slow march away from the nuclear power option. The result is likely to be increased reliance on coal from the already heavily subsidized German mines and from South Africa, and increased use of gas. There is time to work out a new direction for electricity production because by 1987 the utilities had accumulated substantial excess generating capacity, especially fossil fuelled, but electricity charges are already high compared with those of France. In 1987 the cost of electricity from some of the newer German nuclear power stations was 14 pfennigs/kWh whereas EdF was offering potential German industrial users prices as low as 5 pfennigs/kWh.

THE UNITED KINGDOM

The first programme for the commercial development of nuclear power in the UK, as distinct from reactors built primarily to produce military plutonium, was announced in 1955. At first, twelve nuclear power stations were envisaged, to generate between them 2,000 MWe of electricity by 1965. In 1957 a revised programme was published under which the total was increased to 5,000–6,000 MWe. In June 1960 the proposals were revised again, and the 1965 target was rescheduled for 1966–8, largely reflecting easier circumstances of fuel supply. Eventually, nine relatively small power stations were built under the first programme, providing a total installed capacity of 4,786 MWe. Except for the one at Trawsfynydd, in North Wales, which makes use of a lake for its cooling water, all the sites chosen under this first and later programmes are coastal (see Figure 4.3). This fact reflects the continuous need for large quantities of water for cooling purposes (Mounfield 1961, 1967). Most of the rivers in the UK have summer flow rates that necessitate the use of cooling towers at many of the inland sites used for fossil-fuelled units (Rawstron 1951, 1955), but the siting strategy used has not gone uncriticized (Openshaw 1982).

Before the last of the power stations built under the first programme was completed, the government announced the second programme, in April 1964, adopting a figure for planning purposes of 5,000 MWe by 1970–5. After much debate, a UK designed AGR was used for this programme and units were built at Dungeness, Torness, Hinkley Point, Heysham, Hunterston and Hartlepool. Heysham and Hartlepool were much nearer to major population concentrations than any units previously built.

The AGR programme ran into massive cost overruns, delays, construction difficulties and design changes, and in 1979 the government announced that units built under any third programme would be PWRs. Sizewell was chosen as the first PWR site in the UK, but work began there in the mid-1980s only after a long and expensive public inquiry. However, it does now seem, that the UK will at last have joined the mainstream of nuclear power reactor technology by the end of the 1990s. Having concentrated upon what ultimately turned out to be the wrong reactor, the UK saw an early lead in the commercial nuclear power field lost to the USA, France, the FRG and Japan, and it is clear from Figure 4.1 and Table 4.1 that the UK now ranks with Sweden as a European nuclear power, well behind the two European leaders. It seems unlikely that there will be additional PWRs to follow Sizewell for some time to come even though in 1990 Ministerial approval was given in principle for PWRs at Hinkley Point.

SWEDEN

Nuclear power in northern Europe shows contrasting patterns of development. Denmark and Norway have not ventured into commercial nuclear power. Finland, like Yugoslavia, shows its geographical position between East and West by mixing technologies in what are wryly called 'Eastinghouse' reactors. Sweden has stopped building power reactors under the influence of a politically powerful anti-nuclear lobby but was one of the first European states to develop nuclear technology and to begin a civilian nuclear programme.

In 1959 Sweden imported around two-thirds of its primary fuel requirements: its only indigenous energy resource was hydroelectricity. By 1977 it had five reactors in operation, rising to twelve in 1987, giving a total installed generating capacity in the latter year of 10,030 MWe.

There was much well-organized resistance to nuclear power in Sweden in the 1970s, focusing particularly on the controversial Barsebäck power station near Malmo, almost directly opposite the Danish capital city. A national referendum in 1980 produced a substantial pro-nuclear result (58 per cent for, 37 per cent against). In December 1982, however, the Swedish Parliament took a decision unique in the history of nuclear power in Europe. It voted to continue to expand the country's nuclear capacity to a point where by 1990 it would provide around half the country's electricity but simultaneously it would prepare for an end to nuclear power generation. The problem is that hydroelectric stations provide the only indigenous alternative to nuclear power and there is long-standing environmental opposition to harnessing the unused hydroelectric capacity of the northern rivers. Coal imports may have to increase in consequence.

POWER REACTOR SITING AND THE BOUNDARY PROBLEM

Fifty-three of the 150 West European power reactors in operation or under construction in 1988 were within 30 miles (48 km) of a national frontier, many of them on sites chosen without benefit of effective consultation with the neighbouring country. The only consultations required under Euratom regulations relate to the amount of radioactivity in gaseous and liquid emissions. Belgium has four reactors at Doel on the Scheldt estuary within 2 miles (3 km) of the Dutch frontier. The proximity of the Swedish Barsebeck nuclear power station to the capital of Denmark, with only a narrow stretch of water in between, has been widely publicized. The containments of the Barsebaeck reactors in November 1985 were connected to filtering and venting equipment designed to limit release of radionuclides in the event of a severe core accident. France's fuel reprocessing plant on the 750 acre (300 ha) site at Cap de la Hague on the tip of Normandy's Cotentin peninsula is only 15 miles (24 km) from the Channel Islands. In 1977 Mr Edward Collas, then Guernsey's Chancellor, said 'we are 120,000 people on an archipelago threatened by a major political power of 55 million people'. Diluted liquid radwastes are discharged into the sea from Cap de la Hague through a pipeline 3 miles (5 km) long. By 1991 the Cattenom nuclear power station will have four 1,300 MWe reactors located on the French side of the river Moselle, 5 miles (8 km) from the Luxembourg border, and appears a real environmental threat to many Luxembourgers. Construction work began in 1978, only a little after Luxembourg had decided not to build a nuclear power station at Remershem, also on the Moselle. Luxembourg imports 95 per cent of its electricity, mostly from the FRG, but rejected nuclear power after a vigorous campaign by the environmental pressure group *Mouvement Ecologique.* Two-thirds of Luxembourg's population live within a 25 mile (40 km) radius of the Cattenom site, but Luxembourg has lacked any kind of control over the siting decision. French limits on the release of radioactivity in gaseous and liquid discharges are set five times higher than those in the FRG. A Strasbourg court has rejected attempts by Luxembourg and the FRG to block the commissioning of Cattenom. A Franco-German convention signed in 1986 gives Luxembourg a limited say in how the plant is run, but many groups in Luxembourg regard the terms of the convention as inadequate.

In the post-Chernobyl era there has been a widespread call in the FRG for a nuclear *Ausstieg* (exit), but an *Ausstieg* limited to German territory would still leave the country exposed to the consequences of a nuclear accident in a foreign plant close to its borders. Indeed, the FRG's first commercial fast breeder reactor, the 300 MWe plant being built at Kalkar on the lower Rhine, is close to the border with the Netherlands, but this is a facility jointly funded with Belgium and the Netherlands. In fact, the presence and configuration of international frontiers sometimes provide a context and opportunity for international co-operation. At a point just south of Dinant the French border extends a long salient into Belgium, 15 km long and 3 km wide. At the tip of this salient is Chooz, where the French are building two 1,400 MWe reactors for completion in 1991–3. A quarter of the power produced will be taken by Belgium.

A resolution backing the European Commission's proposal for an EC-wide consultation system in siting nuclear power plants was adopted at a plenary session of the European Parliament in 1981, and in 1986 a senior official of the Commission's Department of Environmental Protection said: 'The real issue is not a disaster like Chernobyl, but the routine discharge of radioactivity from normally functioning reactors day after day'. Standards of control vary between countries. In the FRG four reactors of the type at Cattenom should not be allowed to emit more than 3 curies each of liquid effluent each year. The French authorities have set the Cattenom limit at 15 curies, all of which will discharge into the River Moselle. When the French chose the Cattenom site they did not deign to consult their neighbours, and there was no agreement on radwaste disposal levels and no co-ordination of emergency procedures. All these ingredients can be found at other power stations occupying frontier sites in Western Europe.

SUMMARY

Although the economic justification for the programme has been questioned from time to time, the groundswell of feeling against nuclear power in the FRG since the mid-1970s has had little to do with the financial cost of electricity produced by this compared with other means, or with the locational and system economics of the electrical utilities. It has been prompted much more by questions such as: What will a nuclear power station do to the environment and ecology of its surrounding area? What are the risks of irradiated fuel transport accidents, reactor accidents and terrorist activity? How can highly active radwaste be safely stored and ultimately disposed of? What is the long-term genetic effect on human populations of the addition of low level radioactivity to that already present in the environment? Anti-nuclear power feeling became mingled with nuclear weapons issues and overwhelmed for a time the process of patient logical persuasion that has been a feature of modern German political life. It introduced the citizens' action group, the *Bürger Initiativen*, as a political federation of environmentally and ecologically conscious groups led most conspicuously by Petra Kelly and her *Grune Liste Umweltschutz* (GLU). It is, perhaps, paradoxical that the most violent resistance has taken place in the FRG, rather than France where such attitudes are less habitual. In both countries, however, the heart of the problem has been the question of how far citizens should be allowed to participate in decisions regarding nuclear power programmes and nuclear power plant siting (Touraine *et al.* 1983).

Exactly the same fundamental question underlies the nuclear power programme of every other West European country that has adopted the technology to produce electricity. Austria has gone so far as to complete a nuclear power station, at Zwentendorf, only to decide by referendum not to commission it. In Spain, as in the UK, no national movement has yet emerged to challenge nuclear power, but there is evidence that the Spanish private electricity companies, in conjunction with the most important financial interests in the country, have had a very free hand in choosing sites. There are, for example, two reactors at Asco, on the River Ebro, less than a mile (1.5 km) from the town centre and only half a mile above the point from where much of the town's drinking water is drawn. Indeed, the bleak countryside of the lower Ebro is the most outstanding of some half dozen poverty stricken and depopulated districts chosen by the Spanish electricity companies for nuclear development, regions which by their very nature are unlikely to offer much local resistance. In Sweden, governments have fallen and risen on the nuclear power issue. Sweden does have a number of commercially viable undeveloped hydroelectric sites, but the major obstacle to their development is that, because of environmental considerations, the four major rivers offering these sites are to be left untouched. The possibility of introducing natural gas as a significant energy source in Sweden depends upon Norway's exploitation of its Haltenbaaken field near

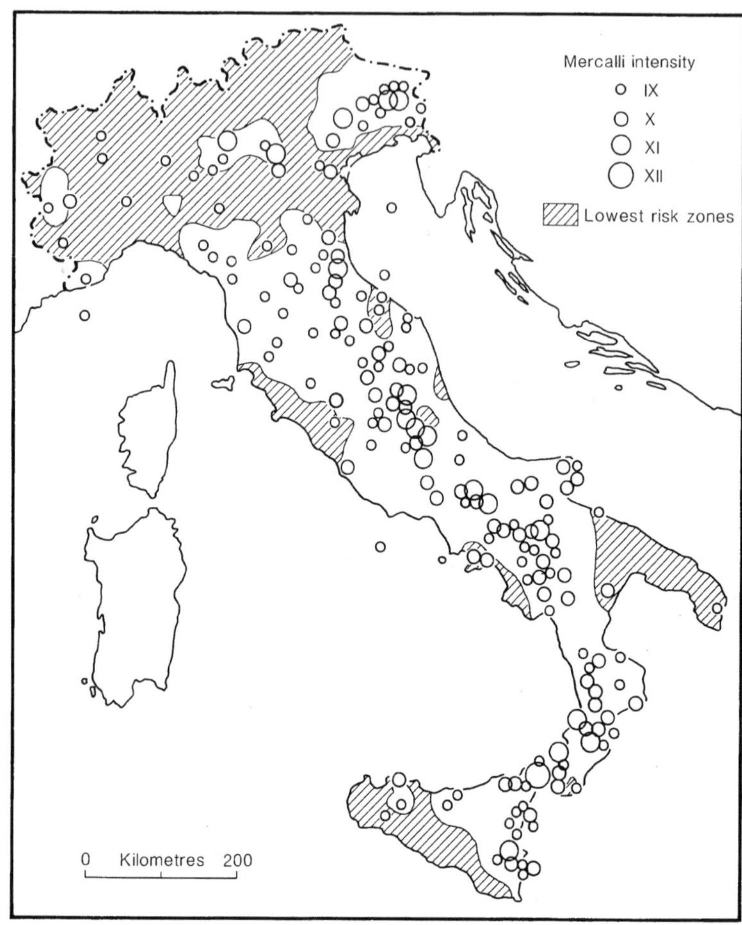

Figure 4.4 The location of known epicentres of earthquakes in Italy over the last 2,000 years. Northern Italy, where there is the heaviest demand for electricity, is a zone of relatively low risk of destruction or severe damage from earthquakes, but there are no zones of absolute safety
Source: Italian National Research Council (CNR) 1980

Trondheim, which is unlikely to begin before 1995. Italy, with the strongest objective case for going nuclear of all the major West European industrial powers, has been the slowest to implement its nuclear energy plans. The reasons for the delay have been part environmental, part political and part bureaucratic. There are many geologically stable parts of Italy where medieval cathedrals still stand in sound condition, but Figure 4.4 shows that over much of the country some risk of seismic activity must be taken into account in reactor siting. Water for cooling is scarce in summer at inland sites. Political opposition has tended to concentrate on issues of compensation rather than principle, and Italian regional and local authorities have considerable delaying powers. The fundamental problem, however, has been that short-lived national governments have been unable to push through the necessary decisions against any well-established local politicians inclined to oppose them, and not until March 1983 did the Italian Parliament pass a law enabling the central government to impose its siting decisions on local authorities. Plans were drawn up for modest expansion of the nuclear power programme and consensus support of all political parties was gained for these plans in 1985, but in May 1986, following the Chernobyl accident, a public opinion poll revealed 71 per cent of the polled population to be against nuclear power, and the Communists and Socialists immediately broke ranks to oppose the programme.

Belgium typifies the way in which most European countries moved from a domestic coal-based energy economy in the 1950s to one of increasing dependence on cheap imported oil. In the early 1950s Belgian-mined coal met more than 90 per cent of the country's total energy needs and some was exported. By 1974 oil dependence had reached 88 per cent. All Belgium's nuclear power stations are co-operative ventures built with foreign help, and they have served to reduce oil's share in electricity generation from 50 per cent in 1973 to 6 per cent in 1987. Most (66 per cent) of the electricity generated now comes from the Westinghouse-type PWRs at Doel (four reactors) and Tihange (three reactors).

Belgium has been relatively free from the worst confrontations, but elsewhere in Europe controversy over safety and environmental issues in nuclear power station location has moved in waves from one country to another and has been sharper in some than in others. It is clear that nuclear energy has aroused very powerful fears and has posed questions about the whole direction of modern industrial development which are not easily answered within familiar geographical frameworks or known social and political categories. It has been argued that public safety is strictly a national responsibility which each government must shoulder for itself, a stance clearly made untenable by the geographical locations, including border sites, of many existing nuclear power plants.

It is worth noting, however, that the West European dependence on imports is developing in natural gas as well as oil. A third of the gas consumed in the EC is imported, and while the burning of gas in power stations has been discouraged in the UK, elsewhere in Europe it has not. Gas burned in power stations accounts for 30 per cent of gas consumption in the FRG, 20 per cent in the Netherlands and 18 per cent in Belgium, with the most important import streams originating in Algeria, Norway and the USSR. If nuclear programmes develop only slowly in the post-Chernobyl era, natural gas may be the fuel to which Europe increasingly turns. Gas pipelines from the USSR are already pumping energy into Western Europe and for the short and medium term the USSR is the most easily available source of additional gas. It may provide a third of the expected growth in gas demand to the end of the century. A scenario in which gas dependence on the USSR is added to oil dependence on the Middle East, with both coupled to reduced nuclear power programmes, seems entirely feasible for Western Europe. In the wider field of world geopolitics, it may not be to everybody's liking, but it must be remembered that dependence on international supplies is not unique. The USA imported half its oil in 1987, and in the post-war period Japan has created a powerful industrial economy with extremely limited indigenous energy resources.

NOTES

1. For example, the International Energy Agency (IEA) which consists of twenty countries covering Western Europe (except France), the USA, Canada, Australia, New Zealand, Japan, Sweden, Greece and Turkey. Ironically, its headquarters are in Paris.

2. CEA was set up by the French government in 1945 to promote the development of the nuclear industry through research and organization of material supplies including uranium and participation in industrial programmes.

FURTHER READING

Bunyard, P. (1988) 'The myth of France's cheap nuclear electricity', *Ecologist*, 18 (1): 4–8.

Nuclear Engineering International (1988) 'Datafile: France', *Nuclear Engineering International* 33 (413): 60–6.

Puiseaux, L. (1981) *Le Babel Nucleaire: Energie et Developpement*, Paris: Galilee.

Touraine, A., Hegedus, Z., Dubet, F. and Wieviorka, M. (1983) *Anti-nuclear Protest: The Opposition to Nuclear Energy in France*, Cambridge: Cambridge University Press.

5 The USSR and CMEA countries of Eastern Europe

In 1984 the USSR's nuclear power plants generated 142,000 million kWh of electricity, 9.5 per cent of the country's total, and the proportion is planned to exceed 15 per cent in 1991. In 1987 the total installed electricity generating capacity of the fifty-seven nuclear generating units in the USSR was 34,334 MWe, with a further 29,620 MWe under construction, out of a total electricity generating capacity of 338,000 MWe. Over 28,000 MWe of nuclear capacity came into use between 1976 and 1987 and the rate of introduction of nuclear power plants since 1976 has sometimes been twice that of fossil-fuelled units. Some projections indicate that by the year 2000 there might be 100,000 MWe of nuclear power capacity in the USSR and Eastern Europe, and the USSR plans to obtain half its electrical energy from nuclear power stations by then. However, a number of problems need to be solved if these targets are even to be approached.

Soviet scientists began research into atomic fission during the Second World War, especially after 1942, and succeeded in bringing about a controlled chain reaction on the night of 24 December 1946. After the end of the War, the consequences of neglecting nuclear power technology could not be ignored by one of the world's super-powers. Following the betrayal of US nuclear weapons secrets by Pontecorvo and Fuchs, and after several years of design and construction work, the first Soviet nuclear research reactor facility was opened in 1954 at Obninsk, southwest of Moscow. Other small research reactors were constructed subsequently, and a major military nuclear installation consisting of six 100 MW reactors, commissioned at various dates between 1958 and 1963, was built at Novotroitsk to produce weapons-grade plutonium. This research and military phase was followed by commercial nuclear power using graphite-moderated BWRs at Beloyarsk (108 MWe) in the Urals (opened in 1968 but now shut down) and a PWR at Novovoronezh on the River Don (278 MWe, 1964). The latter has become the USSR's major site for PWR experimental and design work. Derivatives of these two early reactors, designated RBMK (uranium graphite channel type) and WWER (pressurized water type), have formed the backbone of the nuclear power construction programme in the USSR and most other CMEA countries. Serial reactors

Table 5.1 Nuclear power stations in operation in the USSR and East European CMEA countries in July 1987

Type and name	*Location*	*Reactor power (MWe gross)*	*Year of commercial operation*
USSR			
LWGR (RBMK)			
Beloyarsk 2	Sverdlovsk	194	1969
Bilibino 1	Bilibino, Chukotka	12	1974
Bilibino 2		12	1975
Bilibino 3		12	1976
Bilibino 4		12	1977
Chernobyl 1	Kiev, Ukraine	1,000	1978
Chernobyl 2		1,000	1979
Chernobyl 3		1,000	1982
Ignalina 1	Ignalinski, Lithuania	1,500	1984
Ignalina 2		1,500	1987
Kursk 1	Kursk	1,000	1977
Kursk 2		1,000	1979
Kursk 3		1,000	1983
Kursk 4		1,000	1986
Leningrad 1	Leningrad	1,000	1974
Leningrad 2		1,000	1976
Leningrad 3		1,000	1981
Leningrad 4		1,000	1981
Melekess VK50	Dimitrograd	62	1966
Obninsk APS	Kaluga	5	1954
Smolensk 1	Smolensk	1,000	1983
Smolensk 2		1,000	1985
Novotroitsk 1	Urals, Orenbung oblast	100	1958
Novotroitsk 2		100	1959
Novotroitsk 3		100	1960
Novotroitsk 4		100	1960
Novotroitsk 5		100	1961
Novotroitsk 6		100	1963
Subtotal (28 units)		16,909	
PWR (WWER)			
Armenia 1	Iktembryan	408	1977
Armenia 2			
Balakovo 1		1,000	1986
Kalinin 1	Kalinin, Volga	1,000	1984
Kalinin 2		1,000	1987
Khmelnitskiy 1		1,000	1985
Kola 1	Murmansk	470	1973
Kola 2		470	1975
Kola 3		440	1982
Kola 4		440	1984
Nikolayev, South Ukraine 1	Nikolayev	1,000	1983
Nikolayev, South Ukraine 2		1,000	1985

Table 5.1 Nuclear power stations in operation in the USSR and East European CMEA countries in July 1987

Type and name	*Location*	*Reactor power (MWe gross)*	*Year of commercial operation*
Nikolayev, South Ukraine 3	Nikolayev	1,000	1987
Novovoronezh 1		278	1964
Novovoronezh 2		365	1970
Novovoronezh 3	Voronezh	44	1972
Novovoronezh 4		440	1973
Novovoronezh 5		1,000	1981
Rovno, West Ukraine 1		440	1981
Rovno, West Ukraine 2	Rovno, Ukraine	440	1982
Rovno, West Ukraine 3		1,000	1987
Zaporozhe 1		1,000	1984
Zaporozhe 2	Energodar, Ukraine	1,000	1986
Zaporozhe 3		1,000	1987
(Reactor no. 4 at Chernobyl entombed)			
Subtotal (24 units)		16,643	
OMR			
Melekess Arbus	Dimitrograd	5	1963
Subtotal (1 unit)		5	
FBR			
BN350, Shevchenko	Mangyshlak Peninsula	150	1973
BN600, Beloyarsk	Sverdlovsk	600	1980
Melekess BOR 60	Dimitrograd	12	1970
Obninsk BR5	Kaluga	15	1959
Subtotal (4 units)		777	
Total (57 units)		34,344	
Reactors closed down			
LWGR			
Beloyarsk 1	Sverdlovsk	108	1964
Chernobyl 4	Kiev, Ukraine	1,000	1984
Total (2 units)		1,108	
Czechoslovakia			
PWR (WWER)			
Bohunice VI-1		413	1979
Bohunice VI-2	Jaslovske Bohunice	413	1981
Bohunice V2-1		440	1985
Bohunice V2-2		440	1986
Dukovany 1		440	1985
Dukovany 2	Dukovany	440	1986

Dukovany 3	Dukovany	440	1987
Subtotal (7 units)		3,026	
Reactors closed down			
GCHWR			
Bohunice A1	Jaslovske Bohunice	104	1979
GDR			
PWR (WWER)			
Nord 1	Lubmin	440	1974
Nord 2	Lubmin	440	1975
Nord 3	Lubmin	440	1978
Nord 4	Lubmin	440	1979
Rheinsberg AKWKI	Greitswald	75	1966
Total (5 units)		1,835	
Bulgaria			
PWR (WWER)			
Kozloduy 1	Kozloduy	440	1974
Kozloduy 2	Kozloduy	440	1975
Kozloduy 3	Kozloduy	440	1981
Kozloduy 4	Kozloduy	440	1982
Total (4 units)		1,760	
Hungary			
PWR (WWER)			
Paks 1	Paks	440	1983
Paks 2	Paks	440	1984
Paks 3	Paks	440	1986
Total (3 units)		1,320	

Sources: Nuclear Engineering International, *World Nuclear Industry Handbook 1988*; World List of Nuclear Power Plants, *Nuclear News* 30(2), 1987; IAEA, *Nuclear Power Reactors in the World*

Notes: (a) There were no units in operation in Romania or Poland, but nuclear power stations were under construction in July 1987 in both countries.

(b) In this table and for calculating figures used in the text the East European CMEA countries are taken to include USSR, Czechoslovakia, Bulgaria, GDR, Hungary, Poland and Romania.

(c) Reactors designated RBMK are graphite moderated, water cooled; those designated WWER are PWRs.

of both types are now in operation at nineteen locations and others are under construction (Figure 5.1 and Table 5.1).

The development of sodium-cooled FBRs has been a major part of Soviet reactor development because, from the beginning, the view was taken that a large-scale long-term nuclear power programme could not be realized without them. Small experimental and prototype FBRs such as

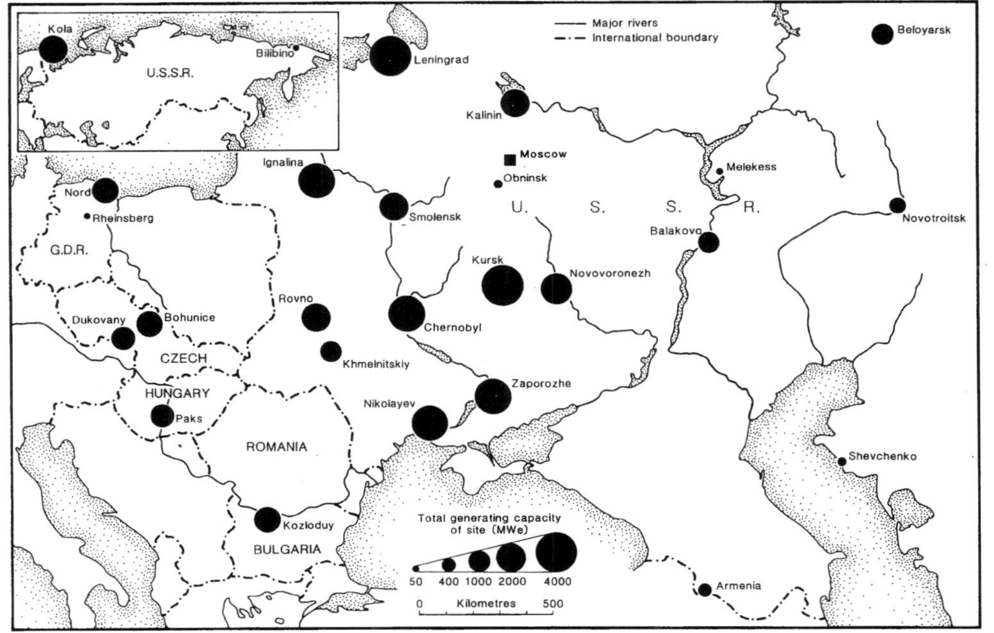

Figure 5.1 Nuclear power stations in the USSR and other CMEA countries of Eastern Europe 1987

Sources: Nuclear Engineering International, *World Nuclear Industry Handbook 1988*; World List of Nuclear Power Plants, *Nuclear News* 30 (2), 1987; IAEA, *Nuclear Power Reactors in the World*

Notes: (a) Except for Bilibino there is no commercial nuclear power station in the USSR east of the Urals.

(b) Yugoslavia, a socialist state but not a CMEA member, has a 664 MWe nuclear power plant at Krsko, using Westinghouse reactor technology, which has been in operation since 1983.

Obninsk BR5 were built from the late 1950s, but the first industrial scale installation (150 MWe) was a multipurpose unit commissioned in 1973 near Shevchenko on the Caspian Sea. The third unit at Beloyarsk (BN600) is also an FBR (600 MWe) dedicated entirely to electricity production.

Ultimately the operational experience of the Beloyarsk and Shevchenko power stations will do much to determine the future direction and commercial development of fast breeder technology in the USSR and its CMEA partners. Some Western press sources have reported that in the mid-1970s the steam generators at Shevchenko suffered serious sodium leaks, accompanied on one occasion by an explosion and fire, but Soviet authorities have consistently denied the seriousness of these incidents (Gillette 1978). However, construction difficulties have put the FBR at Beloyarsk several years behind schedule.

In 1955 it was reported that the cost of electricity produced by the first commercial Soviet nuclear power plants considerably exceeded the average cost of electricity produced by comparable coal-burning power stations (Blokhintsev and Nikolayev 1955). It was argued, therefore, that it was best to build nuclear power plants in peripheral areas remote from coal deposits and this indeed seems to have been the locational policy followed during the early stages of the programme (Voskoboinik 1959), just as it was in the USA. By 1958, however, it was being argued that nuclear fuel costs were lower than for comparable conventional fossil-fuelled power stations and that this, coupled with reductions in construction costs as building experience was gained, would make nuclear power plants competitive elsewhere in the country (Skvortsov and Siderenko 1958). But Donbass coal is costly, and that from Kuzbass is much cheaper at the pit (Cole 1984). At Novovoronezh the production cost of electricity supplied to consumers in 1975 was 0.641 kopek/kWh, while at large modern fossil fuel plants in the European USSR costs ranged from 0.898 to 0.712 kopek/kWh. Vasiliev (1982) has suggested that these figures were typical for the European USSR but Semenov (1983) has stated that the average cost of nuclear-generated electricity in 1979 was 0.793 kopek/kWh, with an average cost from conventional power plants of 0.753 kopek/kWh.

The perception of a political need to develop nuclear power technology has not been the only incentive in the USSR. The sheer size of the country and the discrepancy between the geographical distribution of energy demand and fuel resources have provided real economic incentives to build nuclear power stations in the west. The main problem is that over 70 per cent of the USSR's total proven energy resources are located to the east of the Urals, where only 20 per cent of the people live. Demand for current and the most important load centres are concentrated in the west. Coal-fired power stations to serve the western load centres require either the transport of coal or the transmission of electricity over very long distances, whereas new nuclear power stations, because of their locational flexibility with regard to fuel supplies, can be built close to the markets they are intended to serve. The transport of fuel from Siberia and Kazakhstan to the western regions accounts for 40 per cent of the country's rail freight turnover (Semenov 1983) and the long-distance bulk transmission of electricity is very expensive. Doubts have been expressed by observers in the West about the economic viability of the 1,500 kV DC transmission line that carries power from the Yenisei–Angara hydroelectric power stations to the Urals and Centre industrial regions, and about the Soviet–East European electrical system (Mathieson 1980). Thus the low fuel transport costs and the ability to build close to load centres have given nuclear power plants a clear advantage in western USSR. In 1981 the Twenty-sixth Party Congress agreed that almost all growth in electricity production in the European USSR should be achieved by the construction of nuclear and hydroelectric generating plants, a decision reinforced by the steady depletion of oilfields in the European USSR.

However, even if the vigorous development of recent years continues unabated, despite Chernobyl, it seems unlikely that the contribution of nuclear power to the total energy budget of the USSR will exceed 15 per cent because only 25 per cent of the fossil fuel resources consumed in the USSR are used for electricity production. Twenty per cent is used for central heating, and an extension of nuclear power to include district heating as well as electricity is therefore a prime target of Soviet energy planners. Bilibino, a mining town in the northeastern USSR, has been provided with heat as well as electricity by its nuclear power station since 1973. The BN350 fast reactor on the Mangyshlak Peninsula was built partly to supply the 100,000 inhabitants of the town of Shevchenko with fresh water as well as electricity. Surplus heat from the Beloyarsk, Leningrad, Kursk and Chernobyl plants has been put to use. Nuclear boiler plants have been built at Gorka and Voronezh, and a large combined heat and power (CHP) plant to produce both heat and electricity has been built near Odessa. Before the Chernobyl accident, the USSR took the view that nuclear boiler plants are sufficiently safe sources of heat supply to be located near densely populated areas, thus eliminating expensive long-distance district heating pipelines. Since Chernobyl, however, a number of nuclear power plants, some of them CHP units, planned for Armenia, Azerbaijan, Georgia, Odessa, Minsk and Krasnadov have been cancelled (Soviet Weekly, 24 December 1988).

It is in the question of safety that the siting strategy adopted in the USSR and its CMEA partners may need to be reviewed. The Soviet State Committee for the Peaceful Uses of Atomic Energy has stated that, if even the slightest danger for the population hád existed, not one single nuclear power plant would have been built (Appleyard 1978). It is a philosophy that has permitted the new large nuclear power stations to be built quite close to large urban centres, a location policy for which a high price was paid when the Chernobyl accident occurred in April 1986.

The relationships between the USSR and the other CMEA countries in energy matters are not always simple and straightforward. The USSR prefers to sell its oil and gas for hard currency rather than increase its oil exports to Eastern Europe, and so the need for a new power source is clearly evident in some of the CMEA countries. Czechoslovakia is a heavy user of non-nuclear energy sources (6.7 tonnes of coal equivalent per head, second only to the USA) but relies upon the USSR for 93 per cent of its oil and has had to struggle to make up its natural gas supplies. Hard coal is becoming more difficult to mine as shafts are having to be driven deeper and while brown coal is available, it gives off so much sulphur dioxide in the rather inefficient boilers and furnaces of the elderly industrial districts in which it is burned that its use has blighted many of the towns and cities in northern Bohemia. This has inclined Czech environmentalists to regard nuclear power as the lesser of two evils (Buchan 1982). Czechoslovakia does have uranium deposits, accounting for 14 per cent of the uranium ore output in the CMEA countries, but the target of producing 30 per cent of the country's electricity from nuclear power stations by the mid-1990s seems to be prompted by a degree of desperation about the non-availability of other energy sources. Hungary has one uranium mine, heavily subsidized, in the Mecsek hills. All of the output has to be delivered to the USSR which supplies fuel elements for the Paks power station.

The GDR exhausted the last of its hard coal reserves in 1978. Brown coal reserves exist in sufficient quantity for about half a century at present extraction rates but they are becoming increasingly expensive to mine. Since the early 1970s the East European CMEA countries have been experiencing a rapid increase in industrial and domestic demand to per capita levels approaching those of the industrialized Western market economy countries. In 1987 the CMEA (Europe) members had under construction 55,241 MWe of nuclear generating capacity, of which 29,620 MWe was in the USSR, targeted for completion by the mid-1990s.

The USSR's domination of nuclear power production in the CMEA is made clear by the fact that the total nuclear capacity in operation in mid-1987 in the six East European countries was only 7,317 MWe, but 25,621 MWe was under construction. Five PWR (WWER) power reactors were operating in the GDR, four at one site (Nord 1–4), with six under construction. Four units were in use in Bulgaria (Kozlodvy 1–4) with four more under construction. Seven units were operable in Czechoslovakia, split between two sites (Bohunice and Dukovany), and nine were under construction. Three were in operation in Hungary (Paks 1–3) with one being built. There were no nuclear power stations in operation in Poland or Romania, but two units were being built in Poland at Zarnowiec and Romania had five units under construction at the Cernavoda site.

The USSR, endowed as it is with large deposits of all the major fossil fuel sources, a developed hydroelectric capacity and an established nuclear industry, has been able to provide most of the requirements of energy-deficient countries within the European CMEA (*OECD World Energy*

Outlook 1982). Indeed, these centrally planned economies have generally maintained a net energy export position *vis-à-vis* the rest of the world, with oil, coal and natural gas being exported in some quantities by the USSR to Western industrialized countries. The USSR is particularly rich in natural gas, with proven reserves approaching 40 per cent of the world total.

Traditionally electricity and electrification play an important role in the energy policies of the centrally planned economies as a means of providing energy supplies. The USSR controls all aspects of the nuclear fuel cycle in its CMEA partners including uranium enrichment, fuel fabrication and the reprocessing of spent nuclear fuel, and is committed to assisting CMEA countries with nuclear technology. Ostensibly this control is to guard against weapons proliferation; in reality it has been partly a matter of political control and partly to ensure a large market for the Soviet nuclear industry. The development and use of energy resources in the USSR have absorbed three-tenths of all production investment in industry since the beginning of the Five Year Plans in 1928. The USSR, the East European Six and Yugoslavia have signed an Agreement on Multilateral International Specialization and Co-production on Mutual Delivery of Equipment for Nuclear Power Plants (Interatomenergo), and nearly all the East European countries have relied heavily on Soviet reactor technology, initially the 440 MWe PWR (WEAR) and now the 1,000 MWe (WEAR) units, which are the standard serial reactors for CMEA countries (Vasiliev 1982). Only Romania and Yugoslavia have tried to escape complete Soviet domination of their nuclear power programmes. Although Romania had no operable units in July 1987 it was building the Cernavoda power station in co-operation with AECL of Canada, and Yugoslavia had one Westinghouse PWR in operation (Kisko). The fact that CANDU reactors use natural uranium fuel provides Romania with the opportunity to use its own sizeable uranium supplies without having to send them to the USSR for enrichment, as Czechoslovakia does.

The standardized 440 MWe reactors that have been used to equip nuclear power stations in the USSR and CMEA countries have been manufactured at a large works south of Leningrad. The large new 1,000 MWe units are made at Atommash, a new engineering complex at Volgodonsk, near the Volga-Don canal, built specifically to make standardized reactors and components at high production rates. These are key plants in the Soviet–CMEA nuclear power programme. Czechoslovakia, which has the largest commitment to nuclear power in Eastern Europe outside the USSR, is the only other country with important nuclear power construction capacity, mainly through the activities of the Pilsen-based Skoda company (heavy duty reactor equipment) and firms such as Vitkovice (special steels) and Sigma (pipes and pumps). The Czech domestic reactor programme was allocated nearly a third of total national investment in the 1981–5 plan. Although the USSR has tried to persuade Skoda of the desirability of concentrating on 440 MWe WWER units, leaving

1,000 MWe and larger reactors to Atommash, at least two 1,000 MWe reactors as well as twelve 440 MWe units are due to be supplied by the firm to the Czechoslovakian electricity industry in the early 1990s, and Skoda is working with the USSR to supply nuclear plant to the rest of the CMEA as far afield as Cuba (Buchan 1982).

On *prima facie* grounds, the economic position of nuclear power in the CMEA countries generally should have been strengthened by two sets of circumstances, one political, the other organizational:

(i) A felt need for particular caution in siting power reactors close to population centres has been less evident in the USSR and the other CMEA countries than in the West, and construction delays resulting from the activities of opposition groups have not been a problem for plant builders. There was public debate about the safety issues related to nuclear power in Yugoslavia and Romania after the Three Mile Island incident at Harrisburg, Pennsylvania, but until Chernobyl there was not much public discussion of hazards in any part of the nuclear fuel cycle in the GDR, where the safety of nuclear power has been infrequently questioned. Nuclear power stations in the GDR and in some other East European countries have not always been provided with a thickness of concrete and steel containment comparable with that provided in the West and may be less able to withstand a direct hit by a crashing aircraft or to contain the results of a serious reactor accident (Brayne 1978). The GDR provided extremely limited and low-key media coverage of the anti-nuclear demonstrations of the 1970s in the FRG; the general government line was that under socialism misuse of nuclear power was inconceivable. By interpreting the West's concern over nuclear power only in political terms, the East Germans were able to ignore questions regarding operational accidents, reprocessing hazards, and ecological monitoring, at least until the Chernobyl accident in 1986.

The lack of a safety issue in public consciousness and the absence of an effective anti-nuclear power lobby has meant that the only brake applied to a location policy concentrating the new large nuclear power stations within the industrial heartlands of CMEA countries has been a military–strategic one. Not until the Chernobyl accident were the disadvantages inherent in this siting strategy clearly revealed, and because of the long lead times involved in building large power plants it is likely that, even if it is made, any change in locational strategy towards a remote-siting policy will take at least a decade to have effect. The nuclear power stations intended to provide current from the mid-1990s onwards were being built before Chernobyl and are continuing towards completion. They are the preferred solution to the provision of energy, rather than long-distance energy transmission, although the economic opportunities offered by transmission and international trade in electricity are not being ignored.

(ii) The standardization of Soviet reactor designs in a limited number of sizes, modular buildings and ancillary equipment should have allowed the central control and planning operations of individual CMEA countries, with co-ordination for economies of scale, to provide opportunities for savings not always achieved by the nuclear power industry of the West. There are, however, indications that expansion of nuclear capacity, frequently emphasized by Soviet leaders, has not progressed as rapidly as planned. These were apparent well before the Chernobyl accident, the major policy result of which so far seems to have been a shift away from graphite-moderated reactors to light water reactors rather than a move away from nuclear power. Ineffective co-ordination of the production of the various reactor and power station components, some of them supplied by the USSR and some by other CMEA countries, seems to have caused delays. Atommash has been one major bottleneck. Planned and built for a manufacturing capacity of 8,000 MWe per annum, it produced only seven 1,000 MWe reactors over the period 1981–5, and backlogs in the production of reactors and other parts of new nuclear power plants have been frequently mentioned in *Izvestiya* and other parts of the Soviet press. It seems unlikely that the full manufacturing capacity of Atommash will be achieved before the mid-1990s. As a result, the plans virtually to double the installed commercial nuclear capacity in the USSR and European CMEA from 45,646 MWe in 1987 to 89,932 MWe in the mid-1990s may prove optimistic by up to 20 per cent. The possibility of error in such forecasting is high, and much also depends on the extent to which the USSR succeeds in its efforts to introduce nuclear boiler houses for district heating schemes in urban areas.

Energy-deficient countries in Eastern Europe, of which Czechoslovakia is the most notable, will continue to make major efforts to increase their energy supplies by building nuclear power stations, but so much of the new and projected nuclear capacity lies within the USSR that growing interconnection of electricity grid transmission systems within European CMEA will be needed to allow increased Soviet exports of electricity to parts of Eastern Europe. Time differences between Eastern Europe and the various time zones of the USSR offer scope for covering peak loads on either side by trading electricity. Early in 1979 Czechoslovakia, Hungary and Poland signed an agreement with the USSR to finance jointly one of the two new large plants then being built in the Ukraine to supply the electricity grid in the European part of CMEA and Yugoslavia. A major route for long-distance transmission of current is the high voltage transmission line linking the Ukraine and Hungary between Vinnitsa and Albertirsa. Soviet exports of electricity are expected to double once the Ukrainian Khmelnitskiy nuclear power station is completed. The first 1,000 MWe unit at this site came into service in 1985; the second and third units are due to be commissioned in 1990. Completion of a further high voltage inter-tie at Rzeszow has enabled electricity from Khmelnitskiy to be supplied to Poland. Interconnection between the Konstantinovka nuclear power station in the South Ukraine and Romania will also facilitate exports in that direction. Thus, in the short and medium term, additions to Soviet nuclear generating capacity may be readily absorbed by domestic demand and exports to East European neighbours.

SUMMARY

The USSR dominates the nuclear power picture in the contiguous CMEA countries, not only in terms of installed nuclear power generating capacity, but also through its key role in research and development and in the manufacture of the components needed for nuclear power stations, especially reactors. Only Romania has taken an independent line.

Because of the maldistribution of fuel resources in relation to population distribution and demand for electricity, nuclear power stations have been and are being built mainly in western USSR. A close-in siting policy has been adopted; it is not easy to assess the extent to which the Chernobyl accident may change this policy, but the probability is that it will influence the type of stations built rather than their location.

Within the six other CMEA (Europe) countries, Czechoslovakia in particular has turned to nuclear power to help to solve some of its energy supply problems. In this it has been helped by the presence of an indigenous industrial nuclear manufacturing capacity that is larger than those in its five East European neighbours. The USSR plays a major role in covering the energy requirements of those of its neighbours in CMEA who suffer an energy deficit and while CMEA as a whole earns hard currency by being a net exporter of energy, electricity flows along new transmission system inter-ties, especially from the large new Soviet nuclear power stations in the Ukraine, will contribute to this pattern of interdependence.

FURTHER READING

CIA (Central Intelligence Agency) (1985) *USSR Energy Atlas*, Washington, DC: CIA.

Marples, D.R. (1987) *Chernobyl and Nuclear Power in the USSR.* London: Macmillan in conjunction with the Canadian Institute of Ukrainian Studies, University of Alberta.

Medvedev, Z. (1990) *The Legacy of Chernobyl*, New York: W.W. Norton.

Wilson, D. (1983) *The Demand for Energy in the Soviet Union*, Totowa, NJ: Rowman and Allenheld; Beckenham: Croom Helm.

6 East Asia

The geographical pattern of nuclear power plants in East Asia is very uneven and is dominated by Japan, with a smaller number of units in use in South Korea and Taiwan (Figure 6.1). China has two nuclear power plants under construction (1988) and Indonesia has been conducting feasibility studies. The Philippines, like Austria, has abandoned for political reasons a reactor that was approaching completion. The nuclear power programme in Taiwan has been halted since 1985 when the government ordered the Taiwan Power Company to stop site preparation work at Yenliao for the country's fourth nuclear power stations (units 7 and 8; see Table 6.1). This action was prompted partly by doubts regarding projections of electricity demand growth (in 1986 the electricity generating capacity was twice the level of demand) and partly by increasing public unease about safety and waste disposal. Taiwan is astride a major geological fault zone running southward from Japan through the Ryuku Islands and out into the Pacific north of the Philippines. South Korea has an ambitious nuclear power programme and the main contractors for two units ordered in September 1986 (KNU 11 and 12, each 950 MWe PWRs) were, for the first time, Korean rather than American or Canadian. Until very recently, General Electric, Westinghouse and AECL have been the leading exponents of nuclear power plant technology transfer, channelling scientific and engineering expertise from the USA and Canada to national programmes in East Asia. Things are changing, however. As Japanese firms have gained experience, they have become increasingly responsible for the design, equipment supply and construction of the country's nuclear power stations, to the point where the nuclear industry is now virtually self-sufficient.

JAPAN

A remarkable feature of Japan's economic geography in the post-war period has been the country's emergence as a world-ranking industrial power without the benefit of substantial indigenous fossil fuel supplies. Modest coal resources, in the northern island of Hokkaido and the southern island of Kyushu, both some distance from the great industrial

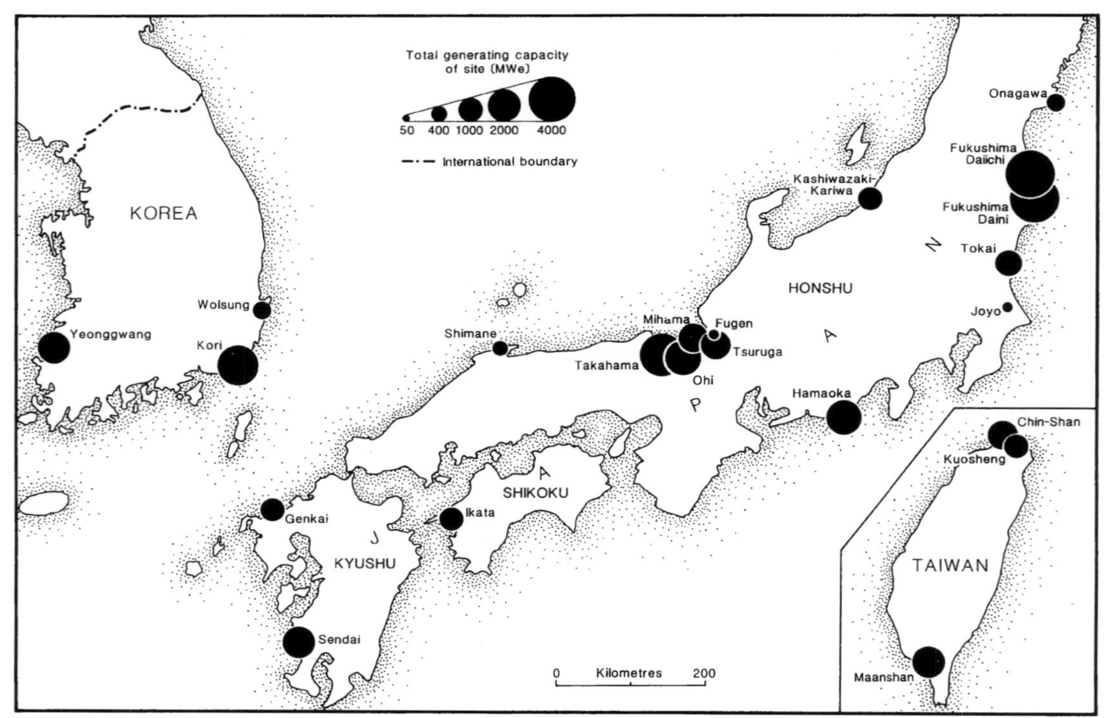

Figure 6.1 Nuclear power stations in operation in East Asia 1987
Source: Nuclear Engineering International, *World Nuclear Industry Handbook 1987*

Table 6.1 East Asian nuclear power plants in operation in July 1987

Power station	*Location*	*Reactor power (MWe gross)*	*Year of commercial operation*
Japan			
BWR			
Fukushima Daichi 1	Ohkuma-machi, central Honshu	460	1971
Fukushima Daichi 2		784	1974
Fukushima Daichi 3		784	1976
Fukushima Daichi 4		784	1978
Fukushima Daichi 5		784	1978
Fukushima Daichi 6		1,100	1979
Fukushima Daini 1	Naraha-machi, central Honshu	1,100	1982
Fukushima Daini 2		1,100	1984
Fukushima Daini 3		1,100	1985
Fukushima Daini 4		1,100	1987
Hamaoka 1	Hamaoka-cho, southern Honshu	540	1976
Hamaoka 2		840	1978
Hamaoka 3		1,100	1987
Kashiwazaki-Kariwa 1	nr Niigata, northern Honshu	1,100	1985
Onagawa 1	Onagawa Miyagi, nr Sendai, Honshu	524	1984
Shimane 1	Kashima-cho, nr Matsue	460	1974
Tokai 2	Tokai-Mura	1,100	1978
Tsuruga 1	Tsuruga	357	1970
Subtotal (18 units)		15,117	
Magnox			
Tokai Japco	Tokai-Mura	166	1966
Subtotal (1 unit)		166	
FBR			
Joyo	Oarailbarakiken	100	1979
Subtotal (1 unit)		100	
LWCHWR			
Fugen ATR	Tsuruga	165	1979
Subtotal (1 unit)		165	
PWR			
Genkai 1	Genkai, Saga	559	1975
Genkai 2		559	1981
Ikata 1	Ikata-cho, southwest of Matsuyama	566	1978
Ikata 2		566	1982
Mihama 1	Mihama-cho	340	1970
Mihama 2		500	1972

Mihama 3	Milhama-cho	826	1976
Ohi 1	Chi-cho	1,175	1979
Ohi 2		1,175	1979
Sendai 1	Sendai, nr Kagoshima	890	1984
Sendai 2		890	1985
Takahama 1	Takahama-cho, southwest of Fukui, southern Honshu	826	1974
Takahama 2		826	1975
Takahama 3		870	1985
Takahama 4		870	1985
Tsuruga 2	Tsuruga, nr Fukui, southern Honshu	1,160	1987
Subtotal (16 units)		12,589	
Total operable (37 units)		28,146	
Republic of Korea			
PHWU (CANDU)			
Wolsung 1, KNU3		679	1983
Subtotal (1 unit)		679	
PWR			
Kori 1, KNU 1	nr Pusan City	595	1978
Kori 2, KNU 2		650	1983
Kori 3, KNU 5		950	1985
Kori 4, KNU 6		950	1986
Yeonggwang 1, KNU 7	Gyaema, nr Kwang Ju	996	1986
Yeonggwang 2, KNU 8		996	1987
Subtotal (6 units)		5,137	
Total operable (7 units)		5,816	
Taiwan			
BWR			
Chin-shan 1	Shihmin Hsain	636	1978
Chin-shan 2		636	1979
Kuosheng 1	Wanli Hsain	985	1981
Kuosheng 2		985	1983
Subtotal (4 units)		3,242	
PWR			
Maanshan 1	Maanshan	951	1984
Maanshan 2		951	1985
Subtotal (2 units)		1,902	
Total operable (6 units)		5,144	

Sources: Nuclear Engineering International, *World Nuclear Industry Handbook 1988*; World List of Nuclear Power Plants, *Nuclear News* 30 (2), 1987

areas of Honshu, produce sufficient coal to meet only 15 per cent on average of annual home demand, output is running down and by the mid-1990s Japan is expected to be importing 85 per cent (100–110 million tonnes) of its anticipated coal requirements. Between 40 and 45 per cent of the imports may be steam coal destined for power stations and factories. South Africa and Australia are expected to be the major suppliers, but the Japanese are also investing in the coalfields of western Canada. Hydroelectric power resources have been and still are being developed, but little room is left for their further exploitation.

Imported oil has been the energy mainstay of Japan's industrial growth and, in 1982–3, 88 per cent of *all* the energy used in Japan was imported. Of the total consumption 62 per cent was provided by oil, 99 per cent of which was imported. Hence the threats to Japanese prosperity posed by the OPEC price increases of 1973 and 1979 and by the Iran–Iraq war have been of serious concern, one that has prompted an energetic search for both conservation measures and diversification of energy use.

The concern is recent, however. During the 1960s heavy import dependence seemed to carry no undue risks for Japan for it was assumed that world oil supplies would remain plentiful and cheap and that export expansion would provide sufficient foreign exchange to cover import requirements. An assumed elastic world supply of oil and rapid export expansion appeared to justify a largely non-interventionist energy policy (Surrey 1974). The increase in demand for electricity averaged 11.3 per cent per annum from 1960 to 1972 (Ipponmatsu 1973), but the country's nine private electricity generating utilities, who had built hydroelectric power stations to meet the growth in demand during the 1950s, simply turned to oil-burning power stations in the 1960s. There seemed to be little incentive to embark on a major nuclear power programme.

Nevertheless, the preconditions for the emergence of a nuclear power industry, in the form of trained scientific manpower, some limited manufacturing experience and embryonic organizational structures, had been established well before the first oil price crisis. The Atomic Energy Basic Law, enacted in December 1955, laid down the legislative framework for the development of nuclear energy in Japan. The Japan Atomic Energy Commission (JAEC) was set up in the Prime Minister's Office to supervise the promotion of nuclear power. Six experimental and materials-testing reactors, designated JRR or JMTR and ranging from 10 to 50 MWe, were started up between 1957 and 1963, and the decision to build the first commercial nuclear power station was taken in 1956. It was built at Tokai Mure, 65 miles (105 km) northeast of Tokyo, and went into operation in 1966 (Imahori 1981). A UK-designed gas-cooled reactor of the Magnox type, it is fuelled with natural uranium, has an installed generating capacity of 166 MWe and contains a number of specially engineered safeguards designed to deal with earthquake conditions (Burkett 1959). For subsequent reactors, however, the Japanese electricity utilities turned to US designed enriched-uranium LWRs, divided equally between BWRs and PWRs built under licence from General Electric and Westinghouse. The three main Japanese firms involved in this phase were Mitsubishi, Toshiba and Hitachi. Japan now has its own standardized LWRs of mainly Japanese design.

Since 1966 the Japanese Government's investment in nuclear power has steadily increased, and in March of that year the nuclear industry and Government jointly adopted a planning target of 30,000–40,000 MWe of light water nuclear plant by 1985 (Narita 1973). The target was not achieved; the installed capacity in July 1987 was 28,146 MWe. However, in 1985 the proportion of electric power generated by nuclear reactors exceeded that from oil-fired power stations for the first time and in 1986 accounted for 29 per cent of the total.

After importing the first of each design from the Americans, the Japanese nuclear plant suppliers purchased the necessary design and production licences from the USA to enable them to build subsequent plants. The first of the BWR series was selected in 1965 by Japco for its site at Tsuruga, and this began operation in 1970 (Figure 6.1). This was followed by Mihama I, a 340 MWe PWR for the Kansai Electric Power Company which was commissioned in 1970, and Fukushima I, a 460 MWe BWR built for the Tokyo Electric Power Company between 1967 and 1971. In March 1971, well before the first oil price upheaval, the planning target was raised to 60,000 MWe, so that by 1985 nuclear power stations would provide 25 per cent of total generating capacity, oil-fired (mainly) and coal-fired plants 55 per cent and hydroelectric plants 20 per cent (Inouye 1973; Owen 1974). This decision was based on calculations indicating that nuclear power would gradually become competitive with that from fossil-fuelled power stations. In 1972 the electricity produced by the nuclear power plants then in operation was 16 per cent more expensive per kilowatt hour sent out than that generated by oil-fired plants, but the oil price increases of 1973 and 1979 shifted costs in favour of nuclear power. Since 1973 the aim has been to reduce the dependence on oil, and the building of nuclear power stations, combined with the conversion of some oil-burning plants to coal burning has received added impetus, despite the fact that recently construction costs have tended to rise faster than inflation. Japan accepts that there will be severe competition for fossil fuels in the next century and that it needs an energy source providing a realistic cost-stable alternative to oil. This role is to be filled by nuclear power.

By the middle of 1987, eighteen BWRs and sixteen PWRs were in operation, with an installed total generating capacity of 28,146 MWe, over 20 per cent of the installed electricity generating capacity. They provided 30 per cent of the country's total electricity supplies. For power plants starting operation in 1983 the cost per kilowatt hour of nuclear power was 12 per cent lower than that for plants fuelled with imported coal and 36 per cent below that from oil-fired plants (Brown 1984). In July 1987 eleven units were under construction, and when the last of these comes into operation in 1995, a further 10,068 MWe will have been added to give a total of 38,214 MWe. This puts Japan into fourth position internationally, behind the USA, France and the USSR, but well ahead of the UK and the FRG. However, it is substantially below the target figure set in 1972. The nuclear power programme kept very close to the target until 1976, after which it fell below target. The problem has not been the UK one of enormously long construction times; in general the Japanese have preferred to solve engineering problems on the drawing board and in the factory rather than on the site, and their power stations have been built with impressive speed. Fukushima II-2, an 1100 MWe BWR, took only 61 months to build, and the average building period from construction permit to operation of the units under construction is expected to be 65 months. Great emphasis is placed on standardization of components and construction sites are organized with immense thoroughness. Nor have Japanese nuclear power stations produced a poor operational record. On the contrary, with an average capacity factor of 76 per cent in 1986 they have an enviable operating record. Some technical problems have arisen in the operation of both reactor types, but more than anything it has been general problems of siting and public resistance that have combined to show up the execution of the 1972 programme to the point where by the late 1980s it could be clearly seen to have been over-ambitious. Only 1,823 MWe of nuclear power capacity was in operation by the end of 1972. The revised planning target of 60,000 MWe required twenty to thirty new sites to accommodate sixty reactors and the installation of 4,500 MWe a year up to 1980, and 6,000 MWe a year thereafter to 1985, a compound target so formidable that it was probably unrealistic even in a Japanese context.

After 1971 the whole generating plant programme was seriously delayed by environmental opposition groups. In 1972 only a third of new plant proposals were approved (3,800 MWe of 12,000 MWe proposed) and in 1973 the proportion was only one-eighth (2,000 MWe of the 16,000 MWe proposed). Until the early 1970s no arrangements existed for public hearings for power reactors, but the Japanese nuclear industry adopted a remote siting policy, concentrating on coastal sites for reasons of water supply and to enable rapid disposal of warm water discharges. During the 1970s rural communities became increasingly inclined to claim that they were paying the costs of nuclear power development while urban populations reaped the benefits (Imahori 1981). Farming and fishing communities in particular have objected to the environmental disturbance and possible loss of income caused by power plants, conventional as well as nuclear, but opposition has been particularly strong in the case of nuclear plants (Ipponmatsu 1973; Owen 1974). The chief worries have been the familiar litany of radioactive product releases, emergency core cooling, thermal pollution effects on marine life and the vulnerability of nuclear plants to earth tremors, a particular hazard in Japan (see Chapter 11).

Offshore sites have been considered, and some of the electrical utilities have responded by adopting an 'existing site' policy, building up selected sites into large nuclear parks. Tokyo Electric Power Company's Fukushima site, for example, has become one of the world's largest. It has ten reactors, giving a total installed capacity of 9,096 MWe.

In order to reduce the level of opposition, Japan has emulated France in providing a system of financial recompense for local communities who find themselves in the vicinity of nuclear power stations. The money compensations include a mix of property tax payments, special grants to local authorities and direct compensation payments.

Understandably, in view of the bombings of Hiroshima and Nagasaki in the Second World War, successive Japanese governments have been reluctant to adopt nuclear weapons and until recently this self-imposed embargo has extended to the building of fuel enrichment plants. Early stages of the nuclear power programme included fuel supply arrangements with the UK and the USA, but the USA announced steep price rises in 1973 for contracts into the 1980s and since then uranium purchases and UF_6 conversion have been sought from a diversity of world-wide suppliers. Now the objective has become independence in fuel element production by the mid-1990s. Japanese uranium resources are very limited. There is a mine at Ningyo Toge, but in 1970 an Overseas Uranium Resources Development Company was set up to work with the French on exploration in Niger. The world's first factory for the extraction of uranium from seawater has been built at Nio, southwest of Takamatsu.

A small fuel enrichment centrifuge pilot plant began operation at Ningyo Toge in central Japan in 1979 and reached full capacity in 1982. A larger centrifuge demonstration plant (10,000–200,000 separative work units (SWU)) was nearing completion at the same site in 1988. These two plants are expected to pave the way for a full-scale commercial enrichment factory, capable of supplying a third of national nuclear fuel requirements, which is to be built by the late 1990s as part of a large nuclear fuel cycle complex also comprising spent fuel reprocessing and waste storage facilities. This complex will be in Aomori prefecture in northern Honshu.

In 1988 fuel reprocessing facilities were limited to one very small facility using French technology at PNC's Tokai works just north of Tokyo. The commercial-sized reprocessing plant to be built as part of the Aomori complex will not be in operation before 1995–7. In the meantime, spent fuel arising from Japanese commercial reactors will continue to be sent to France and the UK for reprocessing, as it has been in the past.

The back-end of the nuclear fuel cycle in Japan is beginning to require careful consideration. High level liquid waste is being stored in tanks, and in 1987 there were 400,000 200 litre drums of low and medium level solid waste stored under surveillance at reactor sites pending the development of a coherent strategy of land or sea disposal. By 1990 this number will have increased to 500,000 drums and by 2000 to nearly a million drums. The government has attempted to accelerate the development of a very high temperature reactor (VHTR) partly in order to solve some of these problems. A high temperature reactor offers higher thermal efficiency, more flexibility in operation and reduced potential radwaste pollution compared with current LWRs. Ultimately, if an outlet temperature of 1000 °C can be achieved, such a reactor could also be used for direct reduction steelmaking and heavy chemicals manufacture. The Japanese steel industry consumes 20 per cent of the country's annual energy supplies, ranking first as a consumer of coal, gas and electricity, and third in oil consumption. Thus an application of the VHTR in this industrial sector could reduce the size of annual incremental increases in energy demands in the industrial sector.

Like many other countries which have developed nuclear power on a scale where it makes a significant contribution to electricity output, Japan regards all fission reactors as an intermediate step on the way to fusion power. Meanwhile, nuclear power targets based on established LWR technology have been officially rescheduled from 28 GWe (1987) to 46 GWe by 1995 and 105 GWe in 2030. If these targets are achieved, 35 per cent of Japan's electricity would come from nuclear power stations in 1995, rising to 60 per cent by 2030. The Japanese Institute of Energy Economics has independently produced more conservative estimates, but considers that the probable installed capacity by 1995 will be 36,000 MWe.

The objective in Japanese energy planning is to retain a steady oil contribution but to reduce its proportion in the energy budget to around 35 per cent by 1995. The government believes that increases in energy demand can be met by substituting imported coal and natural gas for oil, by increasing hydroelectric generating capacity and by continuing with a vigorous nuclear power programme. Energy conservation has been very successful in containing the rate of energy demand increases. Over the six year period 1973–9 a 24.5 per cent increase in gross national product (GNP) was obtained with only an 8.25 per cent rise in total energy consumption and a 1 per cent increase in oil consumption. From 1979 to 1983 total energy demand fell by 3 per cent per annum even though the country's GNP continued to grow at around the same rate. Among the industrialized nations Japan's per capita energy consumption is relatively modest.

CHINA

The GNP per inhabitant of roughly US$500 per year puts China indisputably in the class of Third World countries. Nevertheless its very large population (1,057 million in 1986) and considerable natural resource endowment have enabled China's planners to undertake enterprises in some regions of the country which in certain respects resemble those in industrial countries. One feature is an interest in nuclear power, a source of electricity likely to be attractive in regions distant from the main reserves of fossil fuel.

With estimated recoverable reserves of about 100,000 million tonnes (UNO *Energy Statistics Yearbook* 1982) China has enough coal to last well into the twenty-second century at present rates of output. However, its 2,600 million tonnes of oil reserves would last only 25 years at present rates of production while its natural gas reserves are very small. There is a substantial hydroelectric potential, and large new hydroelectric power plants have been and are being built at Gezhouba (2,700 MWe), Wujiangxi (1,700 MWe) and Longyanxia (1,700 MWe), but many of the undeveloped sites are in the far southwest of the country, away from the main centres of population.

In 1985 about 165 million tonnes of coal were consumed by the power industry, 20 per cent of total production (State Statistical Bureau, People's Republic of China, *Statistical Yearbook of China* 1987: 252, 306). In contrast, only about 10 million tonnes of oil were consumed by the power industry, 10 per cent of total production. Oil has been one of the main items of export despite the small total amount produced. Therefore the production of electricity depends mainly on coal.

Almost all the provinces of China produce some coal and much of that required to generate electricity can be obtained locally. The industrialized provinces of northeast China (Liaoning and Jilin) are reasonably well provided for, as is Hebei province with the two large cities of Beijing and Tianjin. However, Shanghai and the many other industrial cities on the lower Chiang Jiang river are in an area with virtually no coal. Various other zones of industrial development along the coast are also deficient. Such areas depend on coal from elsewhere, especially Shanxi province, which has the largest coal reserves in China and produces a quarter of the country's total output. However, the rail distance from Taiyuan in Shanxi to Shanghai is 1,200 km. China's rail network is only the size of that of France but serves a country twenty times bigger and with twenty times the population. In consequence there is much pressure on the very limited carrying capacity of the rail network.

While exploration for coal and gas is under way in various coastal areas of China, and offshore, it is unlikely that these fuels would be available in large quantities for the generation of electricity.

Many of China's coal mines are small, inefficiently run and produce expensive coal. The sites for large hydroelectric stations are too far inland for power to be transmitted economically to coastal areas by the electricity grid system. There are five major regional grid systems, but transmission technology is antiquated and power losses are high.

On account, therefore, of the difficulty of providing fossil fuels to coastal districts the development of nuclear power in these areas is an attractive prospect. What is more, it could be very difficult to transport large items of equipment, either home produced or imported, anywhere inland because of the complete dependence on rail and gauge constraints.

China has had access to nuclear power technology for some time. The first research reactor was opened in 1956, an atomic device was exploded in 1964 and a hydrogen bomb was exploded in 1967. The country contains perhaps 100,000 tonnes of exploitable uranium but until 1978, with Israel, it was one of only two countries having nuclear weapons to be without an active civilian nuclear power programme. It announced that it was embarking on such a nuclear power programme in February 1978 and a 300 MWe PWR of indigenous design is due for completion at Qinshan in 1991. The plant was approved in 1976 but delayed by the chaos of the Cultural Revolution. Work began at the site in 1984. Daya Bay, 50 km northeast of Hong Kong in Guandong province, is the location chosen for two 936 MWe PWRs in a joint venture using Western technology. Under contracts signed in 1986, Framatome (France) is supplying the 'nuclear islands', GEC (UK) the turbine generators and Electricité de France the architectural and consultancy services. China Light and Power Company will share the output with Hong Kong.

Official government pronouncements in the early 1980s envisaged an installed nuclear capacity of 10,000 MWe by the year 2000 (Qin Tun-Luo 1981). At the start of 1986 Kraftwerk Union were engaged in discussions regarding a 2000 MWe station at Sunan, 37 miles (59 km) west of Shanghai. Proposals were tabled for a 900 MWe reactor at Jinshan, also near Shanghai, and for a 1800 MWe station in the northeastern province of Lianoning. By late 1987, however, the provisional programme had been cut back, and China has decided to rely more on indigenous technology with only limited imports of key components (*Economist*, 12 December 1987). The policy reversal came with the transfer of responsibility for the nuclear power programme from the Ministry of Water Resources and Electric Power, which had favoured imports, to the Ministry of Nuclear Industry, which prefers indigenous technology based on its experience of building the PWR at Qinshan. The principal reasons for the change of policy were the foreign exchange costs of imported plants, the realization that they might be uncompetitive with coal-fired stations in some regions, doubts over the lack of management experience, the emergence of public anti-nuclear sentiment and an emphasis away from large power plants to smaller ones located nearer to population centres. Foreign connections have not been severed, however. In 1987 it was agreed that Kraftwerk Union would help with the 'planning and construction' of two further 600 MWe PWRs on the Qinshan site as part of planned long-term co-operation between the two countries. The agreement included a comprehensive technology transfer stream and foresaw 'supplies and services' from the FRG being paid for in natural uranium for German utilities (Masters 1987).

China's plan is to increase total electricity generating capacity by 30,000 MWe in the four years to 1990 and to sustain a growth rate of plant capacity of 8–9 per cent a year. By the year 2000, it expects to need 240,000 MWe of capacity compared with 100,000 MWe available at the end of 1987. Under the Five Year Plan up to 1990, nuclear power, hydroelectric schemes and coal-fired power stations were expected to contribute to the national need at the maximum feasible rate. Power shortages in 1987 were still acute, with blackouts and industrial shutdowns common in most parts of the country. There is, however, increasing scepticism about the economic benefits of nuclear power. It may be that any expansion of the nuclear programme beyond the Qinshan and Daya Bay projects will be shelved for many years. However, to fulfil China's Four Modernizations Programme (1980–2000) *far more* electricity is needed as soon as possible from all available sources. In 1986 only 450,000 million kWh were produced, only 4.0 per cent of world output, and the output per head was only 365 kWh compared with the world average of 1916 kWh.

SUMMARY

The lack of indigenous fossil fuel resources, and the fact that many accessible hydroelectric power station sites have already been used, have prompted Japan to develop a vigorous nuclear power programme. This is despite the unpromising seismic history of the islands and growing public concern over safety.The commitment to a high level of industrial development in Japan in the post-war period geographically concentrated industrial–urban markets for power, and the presence of a well-developed high voltage grid transmission system have encouraged the construction of large nuclear power stations at coastal sites, especially in Honshu (see Figure 6.1). Particularly important concentrations of nuclear generating capacity have grown up near Niigata and Fukui on the western coast of Honshu and near Fukushima on the east coast.

The economic rationale for nuclear power in Japan is clear: it is regarded as the main replacement for imported oil. Neither the impact of the Chernobyl accident nor falling oil prices have deflected Japanese energy planners from this view.

Nevertheless, as in many of the other leading nuclear power countries, the role of central government in developing the energy form is obvious. Most of the technological development required to support the nuclear power programme has been financed by government either as a direct allocation of government funds or as a special tax account derived from sales of electricity by the utilities. By 1987 government funds had provided 70 per cent of the total nuclear development budget (Huggard 1987). Since the first oil crises of the mid-1970s the Japanese Development Bank has extended low interest loans to the nine major electrical utilities for the purchase of nuclear equipment made in Japan. This has spurred both research and development and production by nuclear engineering equipment producers such as Hitachi, Toshiba and Mitsubishi.

There has been public opposition to nuclear power in Japan, and a sample public opinion poll carried out by Asahi Shimbun shortly after Chernobyl indicated that more Japanese opposed nuclear power (41 per cent) than supported it (34 per cent). This was the first time, however, that the majority had not been in favour. Environmental and other pressure groups have not yet managed to block a nuclear power project to the point of its cancellation, but opposition had led some power utilities to adopt a centralized nuclear power siting philosophy, with many reactors grouped at one site or nuclear park.

South Korea has had no domestic anti-nuclear lobby to hamper the vigorous development of nuclear power and the Korean Electric Power Company (Kepco), a state-owned utility, is likely to continue to build nuclear power stations. China briefly decided upon a moderately large nuclear power programme but has now significantly reduced it. For different reasons, Taiwan has also decided to limit its programme in the post-Chernobyl era. The Taiwan Power Company's projections of increases in electricity demand and the seismic suitability of the Taiwanese environment for nuclear power plants have come under close scrutiny since 1985.

In the technologies that have been adopted for their nuclear power stations Japan, South Korea and Taiwan have all been very much within the US and Canadian spheres of influence, but Korea is now emulating Japan in rapidly increasing the indigenous commercial and technological input that goes into its new nuclear power plants. Technological transfer has become a dominant keynote. China's dilemma over nuclear power is that it makes no sense to develop such a difficult technology in a piecemeal way. But the imported reactors and turbines for the Daya Bay plant will cost China at least US$1,700 million.

There have been virtually no co-operative ventures into nuclear power within East Asia. National programmes have grown without much contact despite the fact that the countries involved have often faced similar problems. However, the next stage of nuclear power development could involve three important new elements.

(a) Japan is poised to become an exporter of nuclear technology and is ready to lead a new form of technological transfer, both to other countries in the region and in the form of a reverse transfer back to the USA (Stinson 1987). Japanese firms are conducting major research programmes on advanced LWR designs, plant automation and the application of artificial intelligence to nuclear plants. With fewer nuclear power stations under construction, and no immediate likelihood of new USA orders, fewer and fewer US firms have had the incentive or revenues to support nuclear research and development fully. Japanese competitors see this as a business opportunity.

(b) There are activities in which regional co-operation would be mutually beneficial to participants: for example, in radiological protection, long-term storage of radwaste and shared reactor operating experience. In these activities Japan is well placed to take the lead in East Asia.

(c) Over the medium term there is the possibility of China becoming a moderately important market for nuclear power. The country's large population, aggregate resources, total GNP and growing industrial capacity put the Chinese economy among the ten largest in the world, despite its Third World status. Coal provides 70 per cent of the country's electricity, and electricity generation is heavily dependent on the northeastern coalfields. A number of large coal-fired power plants were completed in the mid-1980s (e.g. Shenton (1,300 MWe), Datong (1,200 MWe) and Jinshou (1,200 MWe)), and new hydroelectric power stations have been built, but transmission systems and the rail network are both inadequate. Two World Bank reports in the 1980s identified energy shortages as one of the major problems facing the managers of China's economy and this remains the case.

FURTHER READING

China Nuclear Industry Corp. (1989) 'Peaceful use of nuclear energy', *Beijing Review* 32 (42): 17–24.

'Datafile Japan', *Nuclear Engineering International* 340 (420): 53–8, July 1989.

Huggard A.J. (1987) 'The Japanese nuclear power programme', *Atom* 372: 2–6.

Qin Tun-Luo (1981) 'Some aspects of energy policy in China', *Energy* 6 (8): 745–7.

7 From uranium ore to fuel element: the front-end of the nuclear fuel cycle

The hillside of ore that comprised the Mary Kathleen uranium deposit in Queensland, Australia, was discovered in 1954 during weekend prospecting by a taxi-driver who sold off his rights for what now seems a derisory sum (£A250,000). If he had made the discovery a decade earlier, however, he would have obtained even less because, until the invention of the atom bomb, uranium, discovered by H.M. Klaproth in 1789, was little more than a laboratory curiosity, a material of very limited commercial value. Before the Second World War it had a tiny market for colouring yellow paint, to produce yellow glaze on ceramics and in medical research. Madame Curie, who in 1902 extracted radium from uranium, at first used pitchblende from St Joachimsthal in Bohemia (Curie 1938). The Jachymov uranium factory in Czechoslovakia was the first place at which uranium was processed industrially; the production of radium there lasted until 1939, with 2.5–8.5 kg of radium being produced each year. This was a third of world production at the time.

The Bohemian ores were a government monopoly and the need for alternative radium sources led to the discovery of alternative supplies: autunite in Portugal, pitchblende in Cornwall and in Schneeberg and Johanngeorgenstadt in Saxony, and pitchblende and carnotite in the USA, especially in southwestern Colorado and Utah. From 1911 to 1923 the southwestern USA was the world's principal source of the mineral, but in 1913 high quality pitchblende deposits were discovered in Zaire (then the Belgian Congo) and from 1923 these dominated the world market, for a while eclipsing the USA.

In 1930 extensive high grade deposits were discovered at Great Bear Lake in Canada, and from 1930 to 1940 the world's still very small uranium industry, devoted almost entirely to the recovery of radium from the ores, was dominated by Canada, Belgium and Czechoslovakia. When Germany invaded Belgium at the beginning of the Second World War, the Zairean mines were flooded in an attempt to keep them out of German hands, and it was by a stroke of good fortune that the USA found itself the custodian of 1,200 tonnes of uranium oxide which had been stored on Staten Island by Belgian mining interests.

Until the outbreak of the Second World War there was only fragmentary knowledge among mining companies and geologists about the location and size of the world's uranium deposits. There was no pressing commercial reason for them to be well informed; the total US demand was only about 200 tonnes per annum, an amount easily met by the African mines. Uranium is now used in small quantities for the manufacture of inertial guidance devices and gyro compasses, as a counter-weight for missile re-entry vehicles, as shielding material and as X-ray targets. Uranium compounds are used in photographic toners, in coloured glass, in ceramic glazes, in tiles and as catalysts. This list does not disguise the fact that there are only two significant markets: first, the military one, which has been saturated for some time (once refined as an explosive material, the material can be recycled and refashioned into new types of weapons and only a nuclear war would alter this pattern) and, second, electricity generation. Moreover, the owners of nuclear power stations have no option but to buy nuclear fuel, for there is no way of converting a reactor to burn another fuel. Consequently, there is a high degree of interdependence between customer and supplier.

THE NATURE AND OCCURRENCE OF URANIUM: THE SUPPLY SIDE

In nature uranium is a dense hard white element which occurs on average in 4 parts per million in the earth's crust; it is the world's eighth most plentiful mineral. Uranium's physical and chemical properties are such that it

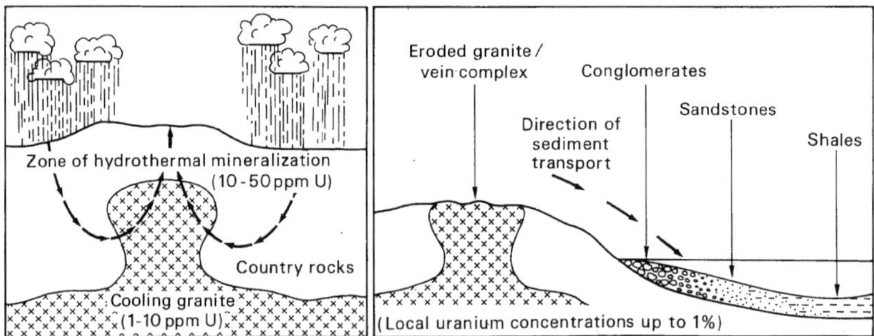

Figure 7.1 Geological formation of uranium deposits. This diagram shows the way in which some commercially exploitable uranium deposits may have been formed. Dominantly meteoric water probably circulated in a hydrothermal convection system which attacked a sub surface mass of cooling granite. As the granite broke down, uranium and other elements were redeposited in a near-surface system of veins and lodes. The right-hand diagram suggests a way in which such primary deposits and even part of an exposed granite mass may have been eroded, leading to the transportation downslope of insoluble uranium minerals which become redeposited, with accompanying sediments, in a water body
Source: Adapted from Brown *et al.* 1975: 16; Simpson *et al.* 1979

Table 7.1 Location of reasonably assured and estimated additional uranium resources in 1985 in specific cost ranges

Country (ranked according to reasonably assured resources in both cost ranges)	*Reasonably assured cost ranges*		*Estimated additional resources (category 1) cost ranges*		*Production 1984 (tonne U)*	*Total production to end of 1984 (tonne U)*
	Up to US$80/ kg U	*US$80–130/ kg U*	*Up to US$80/ kg U*	*US$80–130/ kg U*		
Australia	463.0	63.0	251.0	126.0	4,390	26,242
USA	131.3	266.8	–	–	5,722	305,613
South Africa	256.6	102.1	97.5	27.1	5,732	117,335
Canada	155.0	59.0	105.0	92.0	11,170	172,750
Niger	180.0	2.2	283.6	16.7	3,276	32,849
Brazil	163.3	–	92.4	–	117	552
Namibia	104.0	16.0	30.0	23.0	3,700	28,679
France	56.0	11.1	26.8	18.3	3,168	44,259
India	35.1	11.0	2.1	14.5		
Sweden	2.0	37.0	2.0	44.0	0	200
Spain	26.7	6.2	9.0	–	196	1,918
Denmark	–	27.0	–	16.0		
Algeria	26.0	–	–	–		
Gabon	16.7	4.7	1.3	8.3	918	16,442
Argentina	15.4	3.6	7.7	0.2	129	1,453
Central African Republic	8.0	8.0				
Republic of Korea	–	10.0				
Portugal	6.8	1.4	1.3	–	115	2,755
Mexico	4.5	3.2	–	2.98	0	40
Japan	7.7	–	–	–	4	66
Somalia	–	6.6	–	3.4	31	408
Italy	4.8	–	–	1.3		
FRG	0.9	3.8	1.6	5.7		
Turkey	2.1	1.8	–	3.2		
Zaire	1.8	–	1.7	–	0	25,600

Finland	–	1.5			0	30
Peru	0.5	–	5.0	–		
Greece	0.4	–				
Belgium[a]	–	–			45	195
Total (rounded)	1,669	646	925	407	35,278	777,386

Source: OECD/NEA/IAEA 1986: 24, 46

Notes: Resource estimates are expressed here in terms of metric tons (tonnes) of uranium (U). *Reasonably assured resources* (RARs) refers to uranium occurring in known mineral deposits of such size and grade that it could be recovered within the particular cost ranges indicated, using current mining and processing technology. *Estimated additional resources* category 1 (EAR 1) refers to uranium in addition to RARs that is expected to occur mostly on the basis of direct geological evidence in extensions of well-explored deposits and in deposits in which geological continuity has been established but where knowledge is insufficient to justify RAR classification. EAR category 2 refers to additional uranium that is expected to occur in deposits believed to exist in well-defined areas of mineralization with known deposits. Each of these major categories is subdivided into cost categories which include the direct costs of mining and processing and the cost of capital required to provide and maintain a production centre (one or more are processing plants or mines). In this table the costs are expressed in terms of US dollars as at 1 January 1985. EAR 2 deposits are not included but since 1983–4 the USA has included all its EAR resources (893,000 tonnes up to $130/kg U) in EAR 2. Previously 11 per cent was in EAR1.

[a] Uranium extracted from imported phosphates.

behaves rather differently in the natural environment from other metallic minerals, in a way that affects exploration. Under the conditions prevailing at or near the earth's surface uranium will readily change from a relatively insoluble state to a water-soluble state, and vice versa. Thus, during a geological history spanning billions of years, each uranium atom has repeatedly gone through the process of solution and precipitation and may have been transported hundreds of miles from its original position. The problems that this mobility poses for the exploration geologist are partly offset by the element's radioactive properties which make it possible to use airborne gamma ray spectrometer surveys to find ore bodies.

The average quality of the ore fed to the production mills is 0.15 per cent; ore giving 0.4 per cent value would constitute a very rich source (Bowie 1979). The element appears to have been concentrated in the upper crustal rocks of the earth's mantle in early Precambrian times, and 90 per cent of known resources occur either in shield areas or in the sediments immediately overlying Precambrian rocks, as in the case of the Colorado–Wyoming uranium province in the USA.

The consensus view among geologists is that uranium was introduced to the earth's crust in granite magma generated over subduction zones. Because they contain only 0.001 per cent uranium such granites are not economically exploitable, but they contain uraniferous zircon crystals which, in some places, have been eroded and dissolved by hydrothermal fluids. These carry the uranium out into the country rocks and economically exploitable deposits may be redeposited in veins in those rocks (Figure 7.1). These constitute the *primary* uranium occurrences.

A further set of processes can occur when primary deposits are weathered. The sediments can be transported downhill and redeposited, along with insoluble uranium minerals, in a lake environment to produce *secondary* deposits; many of the Precambrian sedimentary occurrences originated in this way (Figure 7.1). Variations on the secondary theme are *roll-front deposits* which occur when the uranium is dissolved under oxidizing conditions and enters river or ground-water systems. The mineral can be reconcentrated when the water filters through porous sandstones. *Black shale deposits*, like those in the Stensele area of Sweden, occur if there is organic matter present during the filtering process. Uranium mineralization in calcrete sediments within drainage channels dating from the Tertiary period is widespread over 400,000 km^2 of southwestern Australia.

Uranium reserves and production

World reserves of uranium are not known with a high degree of accuracy, and the data presented in Table 7.1, on which Figure 7.2 is based, should be regarded as recent best estimates.[1]

It should also be noted that Table 7.1 and the accompanying maps refer to the world outside Communist areas (WOCA) and thus do not include

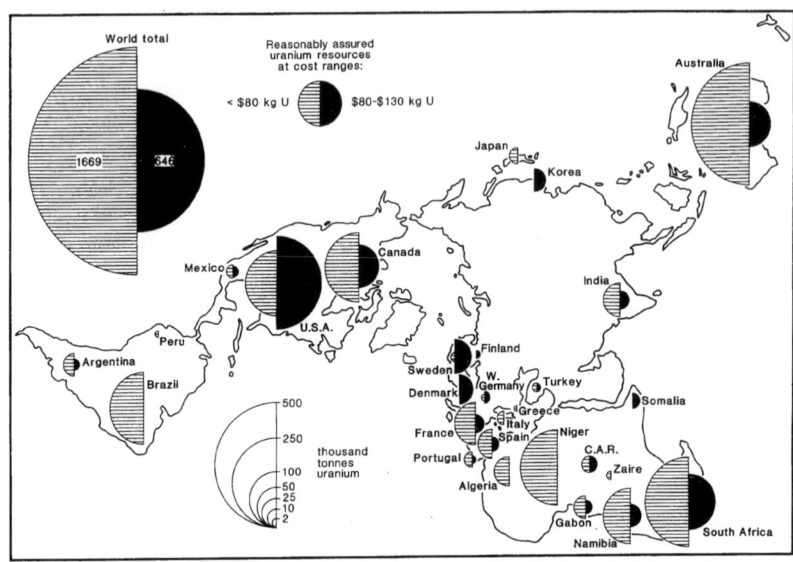

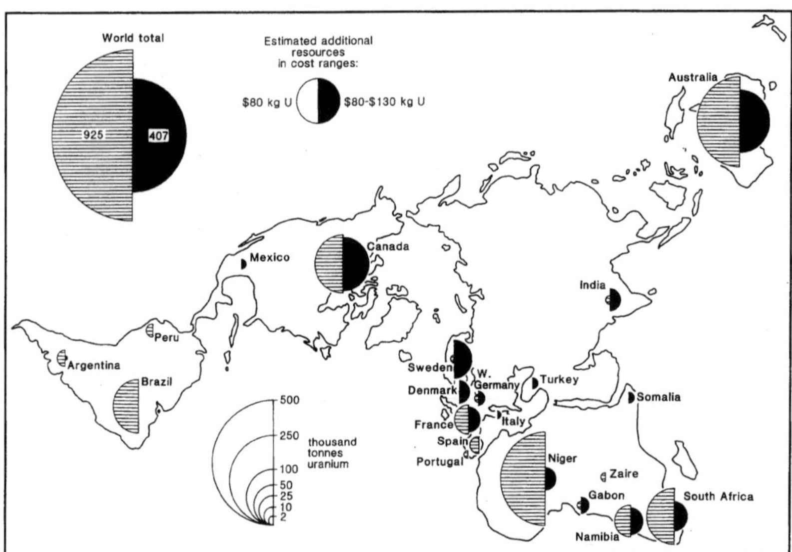

Figure 7.2 Distribution of reasonably assured and estimated additional uranium resources in the world outside Communist areas, 1985. EAR2 resources are not shown on these maps and this depresses US resources by around 500,000 tonnes U in the lower cost category
Source: OECD/NEA/IAEA 1986: 22, 24, Tables 1, 2

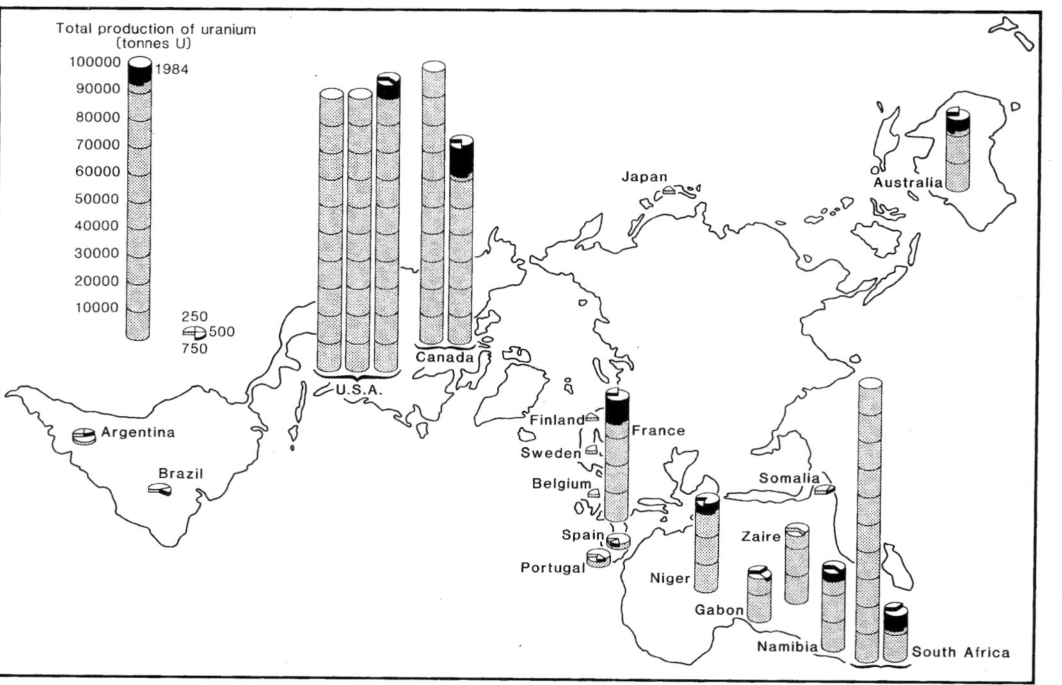

Figure 7.3 Uranium production in 1984 and cumulative production to the end of 1984. The diagram puts the past contribution made by the USA into perspective
Source: OECD/NEA/IAEA 1986: 24, 46

information for the USSR, Eastern Europe or China. Estimates of uranium resources in the centrally planned economies made for the 11th World Energy Conference 1980 were of 500,000 tonnes U in the low cost RAR category and 1.5 million tonnes U in the total low cost EAR, mostly in the USSR, the GDR and Czechoslovakia. China may have 800,000 tonnes U.

Given these facts, it is clear that Australia (22 per cent), the USA (17 per cent), South Africa (15 per cent) and Canada (9 per cent)account for two-thirds of the West's known RARs. Niger's resources in the cheapest category exceed Canada's, as do those of Brazil, and Namibia also has major resources. Useful but smaller RARs occur in France, India, Sweden, Spain, Denmark, and Algeria. In production terms, outside the 'big four' and Namibia, only France and Niger match Namibia's annual output of uranium, and among the other WOCA producers only Gabon approaches an annual output of 1,000 tonnes U. It is also clear that production has not always matched resources; Australia, for example, with the world's largest concentration of resources, has been responsible for a much smaller proportion of cumulative world production (see Table 7.1 and Figure 7.3). France is Western Europe's most important producer (Bessines, Ecarpière and Lodève), but 80 per cent of Western Europe's known uranium reserves occur in a layer of bituminous shale occupying a 500 km^2 triangle between Skövde, Tidaholm and Falköping in the Västergötland district of Sweden.

Australia

Uranium mineralization was first discovered in Australia at Carcoar in New South Wales in 1894, in South Australia at Radium Hill in 1904 and at Mount Painter in 1910. However, Australia's first uranium mine, the Mary Kathleen mine in Queensland, was not opened until October 1958 (it was closed in 1982). Since 1970 very large deposits have been discovered, increasing resources in the lowest cost range to 463,000 tonnes, and there is always a strong probability that this figure will increase. The recent Olympic Dam find at Roxby Downs in South Australia, a copper–gold–uranium–silver association, is thought to harbour twice as much uranium as any other known find in the world. There are eight other significant discoveries so far unmined and in 1988, despite the plentiful reserves, only one mine, Ranger in the Northern Territory, was in production. The Narbarlek ore was all extracted in 1979 ready for subsequent milling (Figure 7.4).

Australia has not developed its uranium resources in a systematic way, and does not have a reliable supplier image except, perhaps, in the FRG and Japan who have bought much of Australia's output since 1977. Periodic slumps in uranium prices have not helped, but the main reason is that uranium mining has become a highly politicized issue. A large proportion of the Northern Territory deposits are under traditional Aboriginal tribal

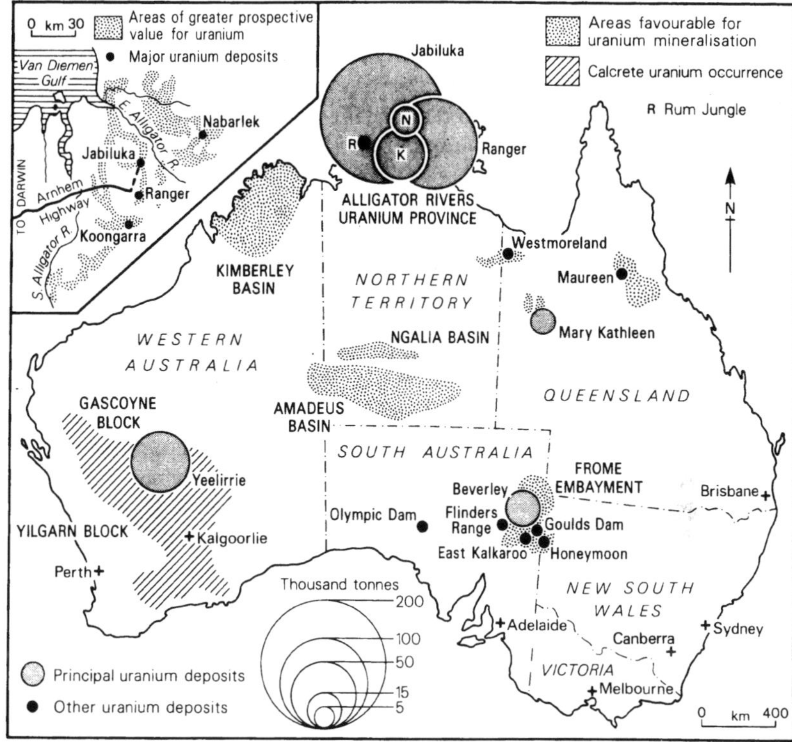

Figure 7.4 Uranium resources in Australia. Mary Kathleen is a pyrometasomatic deposit, the Westmoreland deposits are pitchblende in sheared sediments and the Rum Jungle–Alligator resources are in metamorphosed Lower Proterozoic sediments. In arid areas attention is being given to the calcrete deposits which fill dry river channels and where uranium is localized near the water-table under the present cycle of erosion, e.g. Kalgoorlie–Wiluna. The Olympic Dam resources amount to around 1.2 million tonnes
Source: Mounfield 1981

grounds, and the difficulty of reconciling traditional beliefs and way of life with modern mining activity has caused the Aboriginals to become politically active in attempts to protect their fragile environment (Fox 1976–7; Mounfield 1981). They have received strong support from the Australian Council of Trade Unions (ACTU). Moreover, the Australian Labour Party (ALP) is split on the issue of uranium mining. The party leadership has been pro-uranium and has declared for development of the Olympic Dam deposits, but formal ALP policy resists uranium mining and would consider export applications only when the uranium is 'mined incidentally to other minerals'. This has been taken to mean an exemption for Olympic Dam, even though its ore deposits might be uneconomic to mine if it were not for their uranium content.

USA

The 1901 edition of *Webster's Dictionary* defined uranium as 'a worthless metal not found in the United States'. However, from the late 1940s intensive prospecting and exploration occurred and by 1960 most of the presently known outcropping uranium deposits had been revealed. The Colorado Plateau, the central Wyoming Basin and the Gulf Coastal Plain of Texas constitute the most important uranium provinces, accounting for about 82 per cent of the RARs in the US$80/kg U category. Since 1983, the resource categories have undergone changes mainly because higher costs of mining and milling have moved some of the resources to higher cost categories (see Table 7.1). Young organic sediment uranium deposits provide potential for significant additions to the resource base.

The peak production years in the USA were from 1960 to 1962, when two dozen mills were in operation treating 7.25 million tonnes of ore to produce 13,000 tonnes U. There was a considerable reduction from 1962 to 1968 in the number and production rates of the active uranium mills. By 1957 it was apparent that very large ore reserves had been developed and in 1958 the USAEC withdrew its offer to purchase uranium from any ore reserves developed thereafter. This led to the shutdown of mills after contracts expired and to 'stretch-out contracts' that were extended through

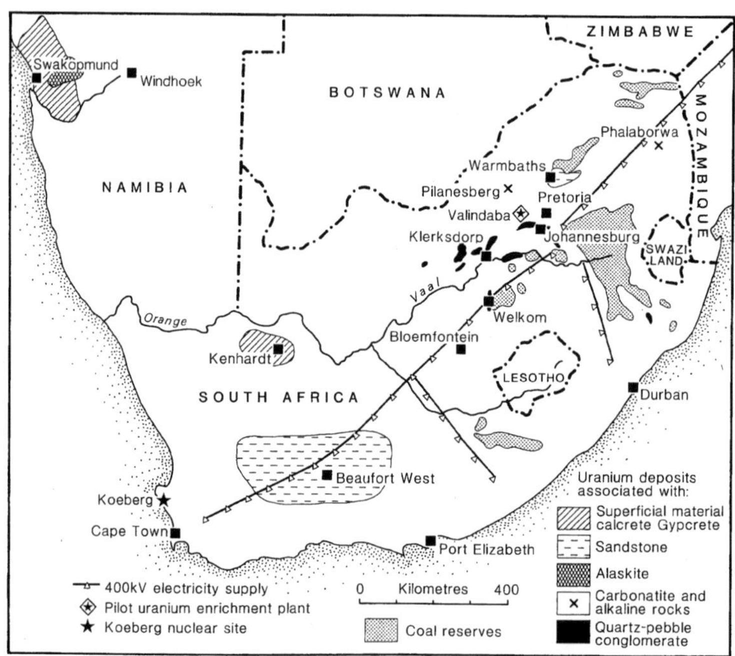

Figure 7.5 Main areas of uranium mineralization in South Africa and Namibia

1970. Thus production declined substantially in the early 1960s, but recovered in the 1970s to a record production in 1980 of nearly 17,000 tonnes U, with twenty-two conventional mills, eleven solution mining and milling facilities, eight plants to recover uranium from phosphoric acid and two plants for leaching uranium from copper oxide liquors. When the industry realized that no more nuclear power plants were being ordered and the market was again deflating, production again drastically declined to only 5,722 tonnes in 1984. By the beginning of 1985 only eight conventional mills were operating and fourteen had ceased production. Many of the unconventional uranium recovery plants were also inactive. Uranium mine and mill projects were abandoned and oil companies lost interest in large-scale commitment to uranium. In the late 1980s a large stockpile of uranium was being worked through by the toll enrichment operations of the three federal government enrichment plants.

Southern Africa

Southern Africa's reasonably assured resources are mostly in the Witwatersrand Basin and Namibia (Figure 7.5).

The South African ore grade is commonly low, averaging only 0.025 to 0.03 per cent, but the producers have certain distinct advantages. In effect, the uranium is produced as a by-product of gold mining, since most of the famous South African gold mines also produce uranium. As auriferous ores contain uranium initial steps are taken by normal processing methods to remove the gold, and thereafter the residual slimes are delivered to plants for the extraction of uranium. Thus only these treatment costs need be charged against uranium receipts. South Africa now has its own enrichment plant and is widely assumed to have nuclear weapons capability also.

The Rössing mine in Namibia was opened in 1976 solely to provide uranium, mainly to the UK. It is a very large mine indeed, with a capacity of up to 4,250 tonnes U per annum. Namibia is a former League of Nations mandated territory illegally administered by South Africa from 1919 to 1988, and the Rössing mine has been a subject of controversy since it was first opened. Had it not been for the Rössing uranium, a money spinner of consequence for Rio Tinto Zinc, it is possible that the UK might have done more to persuade South Africa to end its former illegal occupation. The International Court of Justice declared trade with Namibia contrary to international law in 1971 because of South Africa's occupation.

Canada

The uranium industry in Canada began in 1933 with the exploitation for radium of pitchblende deposits found in 1930 at Port Radium in the Northwest territories. The bulk of known resources occur in quartz–pebble conglomerates in the Elliott Lake–Blind River district of Ontario, where the mineral was discovered at several localities between 1953 and 1955 (Brown 1967). Nearly half of Canada's uranium output currently comes from three production centres in this area. Other uranium resources occur as vein-type deposits in the Beaverlodge areas of Saskatchewan, British Columbia, the Northwest Territories and Newfoundland.

At the industry's peak in 1959, twenty-three mines and nineteen treatment plants were operating in five producing districts to provide 12,200 tonnes U. By 1966 there were only four mines left, producing a total of less than 3,000 tonnes U. The market upturn of the 1970s enabled five primary producers to produce 11,170 tonnes U in 1984, of which 9,693 tonnes, valued at Can $916 million, was shipped abroad.

Canada has adopted a nuclear power programme which uses a technology based on natural uranium fuel and is conscious of the military uses of enriched uranium. Although uranium resources, like other minerals, are owned by the provinces, the Canadian Atomic Energy Control (AEC) Act gives the federal government jurisdiction over Canada's uranium industry and the export of uranium is closely controlled. In 1977, for example, the government imposed a freeze on uranium exports to the EC countries and Japan, consequent upon what is considered to be its failure to obtain what it considered to be satisfactory safeguard agreements. A crucial element was whether customers should be free to use spent fuel for reprocessing without government approval, an issue on which agreement was not reached until January 1978. Nevertheless, Canada's uranium export policy has been developed with a view to maintaining the country's role as a reliable uranium supplier to its industrial customers. Canadian producers such as Rio Algom Ltd are free to negotiate their own contracts within the context of federal government export policy.

THE URANIUM MARKET: THE DEMAND SIDE

There has never been a generally recognized international market price for uranium; there is a great deal of elasticity in uranium prices, and the keyword in the uranium market in terms of both production and prices has been volatility. Post-war production rose rapidly to a peak in 1959, when military demand was at its highest, and then fell equally rapidly in the 1960s, turning into a slower upward trend after 1967 (Figure 7.6). The industry's troubles between 1960 and 1967 stemmed from the effects of overproduction and unfulfilled expectations about nuclear power. At the end of 1957 and in the early months of 1958 it was estimated that the nuclear power programmes of the UK and the Euratom countries (then France, the FRG, Italy, Belgium, the Netherlands and Luxembourg) would need up to 27,000 tonnes of uranium oxide by 1967 to provide fuel for 6,000 MWe of installed generating capacity in the UK and 15,000 MWe in the Euratom countries. It was assumed that the USA would use at least 30,000 tonnes or civil and military purposes and that nuclear power

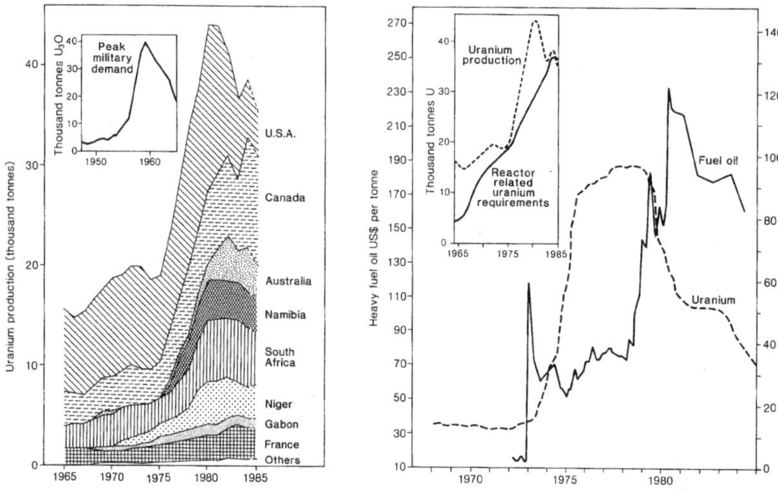

Figure 7.6 Uranium production and prices. The left-hand graph shows production trends to 1985 and is adapted in part from OECD/NEA/IAEA (1986). The right-hand graph compares trends in the price of uranium and fuel oil from the 1970s to 1985

stations in other countries might require a further 3,000 tonnes. In sum this amounted to a demand well in excess of the anticipated total output for 1959. By the end of 1958, however, serious doubts appeared about these projections, and by 1960 the misgivings were fully confirmed by drastic pruning of the UK and Euratom nuclear power programmes to 2,000 MWe and 5,000 MWe respectively, by technical advances in the use of uranium fuel, which meant that considerably less was needed to produce a given amount of power, and by the decision of the USA to ban imports of uranium and to restrict its demands to limited contracts with domestic suppliers. This cyclical 'stop–go' trend in the uranium market was to become typical as OPEC oil price rises aroused interest in nuclear power one year and the general depression in the world economy or a notable reactor accident decreased it the next.

Trends in output from 1968 to 1973 (see Figure 7.6) were dominated by antipathy between Canada and the USA over the handling of the uranium market. The Canadian Government held the view that the problems of its uranium industry, reflected in oversupply and low prices, were compounded by the way in which the US Government had closed its domestic market to foreign-produced uranium and simultaneously had moved uranium from the US stockpile into the international market. Moreover, the USA imposed conditions on foreign users of US enrichment plants and US companies were competing aggressively for sales outside their protected domestic market. By 1971 prices had been driven down to half those being paid at peak production and in 1972, on the initiative of the Canadian Government, a group of major uranium mining organizations, excluding those of the USA, held a series of meetings aimed at raising the world price of nuclear fuel. Australian, Canadian, French, South African and UK companies were represented at discussions held with the stated objective of 'achieving an orderly regulation of the uranium market'. The Canadian Government directed its Atomic Energy Control Board to reject any export order for uranium at prices below those called for by the 'producers' club' and these minimum price directives were maintained until early 1975.

It is difficult to judge the effects of this marketing arrangement among the leading non-American producers. Pricing clauses in contracts are complex and contract prices are often kept secret. The spot price of uranium oxide rose from less than US$11/kg in 1971 to US$14/kg in 1973 and to US$28/kg in 1974. The average contract price for 1980 deliveries increased from US$24/kg in 1973 to US$43/kg in 1974. It is certain, however, that the massive OPEC oil price increase of 1973 had a much more substantial effect on uranium prices than the activities of the cartel; from 1973 to 1975 uranium prices went up as fast as those of oil (see Figure 7.6). The 1974 price for deliveries scheduled for 1980 rose to $78/kg; the 1976 price was also at this level, and in 1976 the Pennsylvania Power and Light Company set a new record by entering into a contract with United Nuclear for delivery of oxide in 1980 at US$123/kg. This price was exceptional, however. In mid-1977 there was a general feeling that prices had been inflated by overreaction on the part of producers to their previous loss-making situation and that a fair price for delivery in the early 1980s would be about US$66/kg compared with the US$88–90/kg then being quoted. In the event, as Figure 7.4 shows, prices declined rapidly after 1978 and did not follow the second OPEC major rise in oil prices in 1978–9. 1979 was the year of the Three Mile Island nuclear power station accident in the USA. Uranium oxide prices averaged US$53/kg in 1983, US$37/kg in 1984, and were below US$33/kg by mid-1985. Throughout 1987 the spot price was stable at US$44/kg and hovered around US$40–45/kg during 1988–9.

The volatility of uranium prices has little effect on the cost of producing nuclear power. In a PWR, uranium fuel is around 4 per cent of operating costs while the capital costs are being amortized, and the enrichment of uranium is another 6 per cent. A thousand megawatts of PWR electricity working at full load requires around 350 tonnes of uranium a year; 1,000 megawatts of base load electricity from a coal-fired plant may require 2.5 million tonnes of coal which means that transport is a major part of the costs. The increase in uranium prices by a factor of eight in the 1970s had little effect on the cost of nuclear electricity, while the quadrupling of oil prices transformed the economics of power and transport. More important has been the effect of depressed prices on exploration and investment in new mines. It takes around eight years from the discovery of a uranium deposit to commercial operation, and twelve to fifteen years from the start of exploration in a new area to production. Uranium exploration in 1988 was lower than it had been for a decade, and most of the exploratory drilling was concentrated on evaluation of the extent of existing deposits or previously discovered occurrences.

As prices fell, so did uranium production, with a slight and short-lived resurgence in 1984 as a result of increased output from Australia's Ranger operation and the start of production at Canada's giant Key Lake project, the world's largest production centre (4,600 tonnes U per annum). The 1985 output of 37,000 tonnes U may rise to around 55,000 tonnes in 1995, but production capability is 15–20 per cent in excess of demand. Indeed, since at least 1970 the uranium industry outside the Communist Bloc has consistently produced at levels exceeding reactor requirements, with the result that stockpiles equivalent to at least three years of reactor requirements have accumulated. It is likely to be 1995 or thereabouts before more uranium production capacity may be needed.

The possible impact of fast reactors on uranium fuel

The production of plutonium is invariably part of the nuclear fuel cycle, and any degree of commitment to the nuclear fuel cycle involves the production of plutonium. Thus the question arises as to what to do with the plutonium. A once-through fuel cycle means that it is left in the fuel element and housed in interim storage facilities, including cooling ponds at the power stations, pending the building of long-term repositories. A once-through cycle using slightly enriched fuel provides a fuel utilization of only 0.3 per cent or thereabouts, and so potentially useful fuels, uranium as well as plutonium, are treated as waste products. With reprocessing, the uranium that is separated out can be recycled to produce new fuel and the plutonium can be stored. However, plutonium has some value as a fuel in ordinary LWRs; indeed, as soon as it is produced in a thermal reactor it immediately begins to incinerate itself. Thus a third possible fuel cycle involves the recycling of both uranium and plutonium through thermal reactors, and the added plutonium can save up to 15 per cent of the uranium fuel. Reprocessing offers the chance of reducing the uranium fuel requirement by about 23 per cent compared with the once-through fuel cycle, thermal recycling with uranium and plutonium increases this figure to 39 per cent, but 'the most natural step with plutonium is to use it in a fast reactor' (Marshall 1979). An established fast reactor cycle would enable dramatic savings in the consumption of uranium; a 60–70 per cent fuel utilization rate would effectively increase nuclear resources by a factor of about 200, and because the fast reactor produces plutonium, the ratio might be even better than that.

Figure 7.7 shows an attempt by OECD/NEA/IAEA to demonstrate

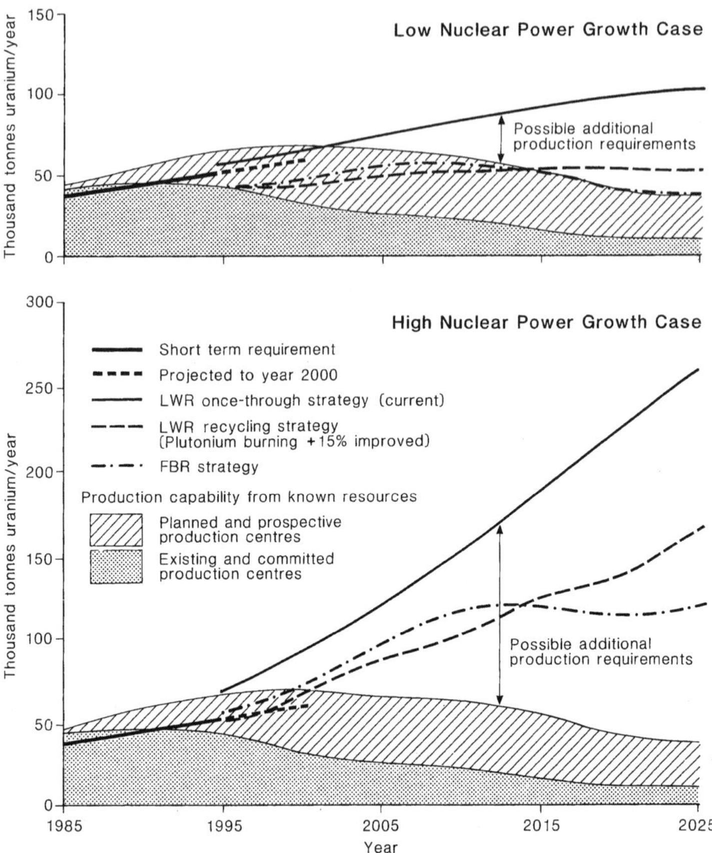

Figure 7.7 Long-term projections of the effect of alternative growth rates in nuclear power on uranium production capabilities in the world outside the USSR and China
Source: Redrawn from OECD/NEA/IAEA 1986: 21–2

the possible significance for uranium fuel demands of a successful widespread fast breeder strategy. It must be emphasized, however, that there are considerable uncertainties involved in producing this kind of extrapolation. In the first place, fast reactors are some way from full commercial development. In late 1988 there were indications that the UK would delay or cancel large parts of its fast reactor programme and the world's first commercial-scale fast breeder at Creys Malville had been out of action for 2 years. Second, the diagram postulates a complete turn-around in the historic uranium demand–supply situation, with possible shortages of uranium appearing from about the year 2000 onwards. There are other uncertainties also, including nuclear power station ordering and completion rates in the post-Chernobyl period. If uranium does become scarce and development problems with FBRs are solved, they could form the basis for viable nuclear power programmes, but this point would come at different times in different countries.

MINING AND MILLING PROCESSES

Uranium mining methods include open-pit as well as virtually all conventional underground techniques. Milling is essentially a leaching precipitation operation and the product is an oxide or salt concentrate containing 60–85 per cent U_3O_8 (uranium oxide) by weight. The basic steps are ore preparation, leaching and product recovery. The ore is first crushed and ground in wet rod-and-ball mills to produce a water slurry. Uranium is leached out of the slurry with acid or alkali depending on its nature, and the uranium solution is separated from the slurry to leave sandy waste residue (tailings). Uranium is extracted from the liquor by an ion exchange technique or by solvents. It is precipitated and collected on a rotary vacuum filter, dried in a kiln and calcined in a furnace to form the ore concentrate which because of its bright yellow colour is known as 'yellowcake'. These processes are summarized in the flowsheet in Figure 7.8, but flows at particular mills may deviate somewhat from this generalized sequence according to the technology employed. The objective of all mills, however, is to produce the uranium concentrates which provide the feedstock for the subsequent processes that lead to the production of high purity uranium for fuel elements, and which are economically transported over long distances because of their high uranium content. Table 7.2 provides a summary view of the world's uranium ore processing facilities, excluding the CMEA countries.

After being refined near the mines and mills the product is shipped to a conversion plant. Here, to reach the high level of purity required for nuclear fuel, the ore concentrate is dissolved in 99.95 per cent pure nitric acid.The uranyl nitrate in turn is converted to volatile uranium hexafluoride or 'hex' which is used in the enrichment process.

If enrichment is not required, for example for heavy water reactor fuel, uranium dioxide is produced from the uranyl nitrate and transported directly to a fuel element manufacturing plant. At these plants fuel elements are produced from both non-enriched and enriched uranium, as required by the customer.

Health issues in mining and milling

The earliest commercially exploited source of uranium, the Joachimsthal pitchblende, comprised a small part of the metalliferous resources of the Erzebirge (Ore Mountains) of Central Europe, extending from Joachimsthal to Schneeberg in Germany. For centuries underground workers in the

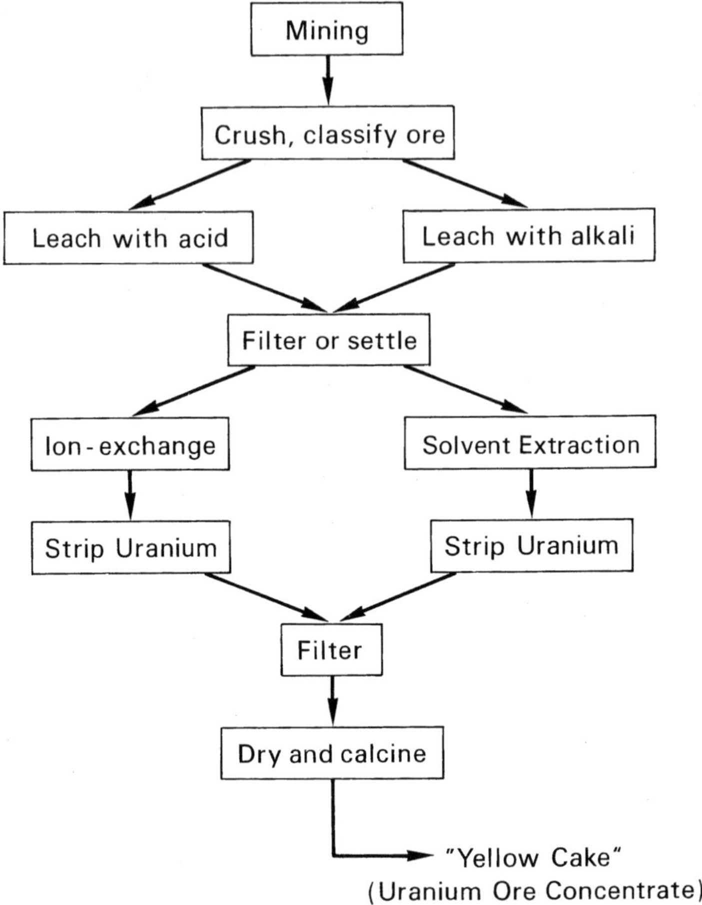

Figure 7.8 Simplified flow diagram of the first stages of uranium ore processing from mining to the production of U_3O_8. The result of the 'milling' process shown by this diagram is a dry powder containing 70–90 per cent by weight of uranium oxides with a chemical formula equivalent to U_3O_8. When this is in the form of ammonium diuranate, it is referred to as 'yellow cake'. This forms the raw material for the fabrication of nuclear fuel
Source: Hill 1974: 6

lead and silver mines of this region suffered from a lung disease, usually fatal and locally called *Bergkranheit* or mountain sickness, which was not correctly identified as lung cancer until 1879.

Wherever uranium exists in rocks and ores the noble gas radon appears in the natural background radiation environment, and for most practical purposes uranium-laden ore can be regarded as a source of this gas, which has a strong tendency to diffuse out into the air. When the air is breathed

Table 7.2 Uranium ore processing capacities by country and status 1987

Country	*Operable (tonnes U/ year)*	*Under construction (tonnes U/ year)*	*Planned (tonnes U/ year)*	*Shut down or on stand by (tonnes U/ year)*	*Construction deferred (tonnes U/ year)*
Argentina	240	–	500	115	–
Australia	4,500	2,500	–	700	–
Belgium	50	–	–	–	–
Brazil	420	1,000	–	–	–
Canada	15,060	–	4,600	–	–
China	1,100	–	–	–	–
France	4,000	–	–	–	–
FRG	–	–	–	125	–
Gabon	1,500	–	–	–	–
Greece	150	–	–	–	–
India	200	–	–	–	–
Japan	–	–	–	50	–
Mexico	–	–	–	–	400
Morocco	470	370	185	–	–
Namibia	4,000	–	–	–	–
Niger	4,600	–	–	–	5,000
Pakistan	30	–	–	–	–
Portugal	170	–	200	–	–
South Africa	8,300	–	–	2,050	–
Spain	215	–	615	–	–
USA	9,020	–	–	18,710	–
Yugoslavia	120	–	–	–	–
Total	54,145	3,870	6,100	21,750	5,400

Source: Nuclear Engineering International, *World Nuclear Industry Handbook, 1988*: 19 and *1989*: 17

in, the inert radon gas itself is rapidly exhaled but the decay products, or radon daughters, have important radiobiological consequences because two of them at least are energetic alpha particle emitters which tend to lodge in the lung mucosa and induce cancer.

The problem received publicity and attention in the 1960s when it was found to be responsible for a high incidence of lung cancer among US uranium miners, and during the 1970s frequent reports appeared of similar situations in other mines elsewhere in the world (Archer and Wagoner 1973; Mays 1973; Budnitz 1974; Cohen 1982a). Described at one stage as a problem of epidemic proportions, it was brought under control by persistent efforts to improve mine ventilation and by legislation on health standards for mine workers.

Health hazards in this part of the nuclear fuel cycle are not limited to the mines. In the past, waste material from mining and milling was simply dumped on the surface at the mine and mill sites. Gradually, consciousness of the environmental impact of uranium mill tailings grew to the point where the heaps of waste accumulated by four decades of ore processing became widely recognized as an important public health issue (Cohen 1982b). The Nuclear Energy Agency of the OECD regards the management of the mining and milling wastes as sufficiently significant to have established a co-ordinating group from eleven countries to apply radiological protection principles to their management (OECD/NEA 1984d). It is the residual uranium content (tails assay) which determines the long-term level of radioactivity of the tailings, and so mill operators are being urged to remove as much as possible of the uranium values. It has been predicted, however, that about 500 GWe of nuclear generating capacity will be in service world-wide soon after the year 2000 and that the total uranium needed by then would be about 2 million tonnes. If this uranium were all produced from 0.1 per cent uranium ore, some 2,000 million tonnes of mill tailings would be left behind for disposal. The accumulation rate in OECD countries in the mid-1980s is 25–30 million tonnes per annum, and the total accumulated by 1984 was 300 million tonnes. To this must be added unknown quantities stemming from mining and milling in the USSR, Eastern Europe and China.

The hazards associated with mine waste and mill tailings are threefold. First, there is atmospheric dispersal of radon-222 and its daughters, which is short lived but capable of irradiating lungs through alpha emitters. Second, there is the possibility of water dispersal of soluble tailings material by leaching. Third, wind action can disperse waste dust. Under good management wastes are no longer being dumped but are being disposed of in engineered valley dams, cuts and specially excavated pits and by back-filling to give better containment. In 1983 the average tailings stabilization cost in the southwestern USA was US$1.75 per cubic yard but varied between 90 cents per cubic yard for recontouring to a high of US$26.40 for the provision of rock cover (De Vergie 1983).

The consequences of residual dumps from previous mining and milling operations are having to be dealt with also. Strip mining at Rum Jungle in Australia has ruined thousands of hectares of countryside through effluent runoff. In the USA leaching from dumps adjacent to the Colorado River and its tributaries, used for irrigating food crops over one-twelfth of the USA, has had to be controlled. In 1978 there were twenty-two sites in eight western states which still had tailings dumps. Salt Lake City, Durango, Maybell and Rifle have all had large dumps nearby offering a potential danger to public health. In Grand Junction, Colorado, over 200,000 tonnes of waste were used in laying foundations for the city's streets, drives, houses, offices and swimming pools built during the 1950s and 1960s. The material was cheap and plentiful, and not until 1966 did the state government prohibit its use for construction purposes. In 1972 a programme was begun to move the tailings in Grand Junction and to dig out the material from the 'hot spots' in the town's 740 contaminated buildings, but it proved difficult to hire contractors to do the work and even by 1985 decontamination was not complete. A large programme of remedial action at tailings sites began only in 1983 as a result of federal legislation. In the Uranium Mill Tailings Radiation Control Act 1978, the US Congress instructed the Environmental Protection Agency to identify the location and hazards of uranium mine wastes associated with 63 surface mines and 256 underground mines in the seven western states.

It must be emphasized, however, that the health hazards in mining and waste disposal affect *particular localities* in the southwestern USA and not the seven uranium-rich states as a whole. J.H. Fremlin has referred to an extensive study of cancer rates in the USA made by the Argonne Laboratory in an attempt to find the effects of variations in background radiation on cancer rates throughout the USA. In the investigation the age-corrected cancer rate and the mean natural background radiation of each state was recorded, with the results shown in Figure 7.9. Professor Fremlin has argued from the evidence of the diagram that there is a strong implication that the seven western mountain states owe their lower cancer rates to the presence of other kinds of carcinogen in the more densely industrialized states in the east. 'If all cancer rates in the USA could be reduced to the level of the high-natural-background mountain states, 24,000 deaths a year would be avoided' (Fremlin 1986).

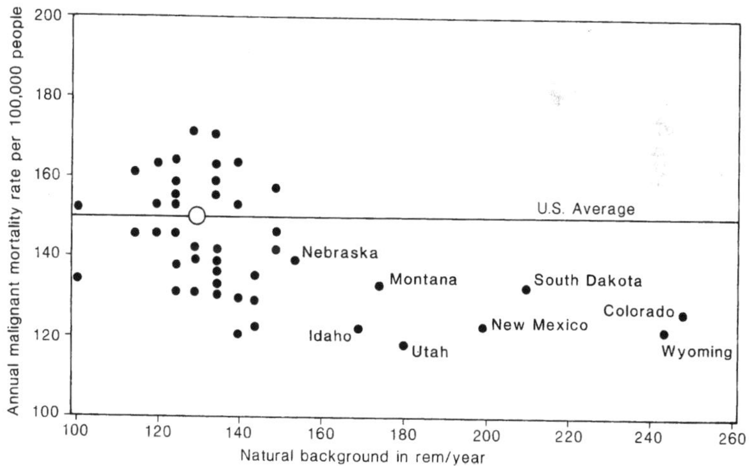

Figure 7.9 Malignant mortality rates for the US white population from 1950 by state and natural background. Each dot on this graph represents a US state; those with high levels of natural background radiation and uranium resources are individually named and clearly are not those with the highest malignant mortality rates

Source: Fremlin 1986, based on a survey by the Argonne Laboratory

URANIUM ENRICHMENT

Different kinds of fuel elements for power reactors are available, but nuclear power stations using enriched uranium fuel dominate the world's nuclear generating capacity and the bulk of uranium fuel is in the form of enriched uranium dioxide. Different reactor designs call for uranium of varying degrees of isotope enrichment, ranging from natural uranium containing 0.7 per cent U-235 (CANDU reactors) to enriched uranium containing in the neighbourhood of 90 per cent U-238 (plutonium product). Thus, uranium enrichment has a key role in the nuclear fuel cycle, but is one of the most difficult of nuclear power technologies. It accounts for 40 per cent of the fuel cost (see Introduction, Figure 0.2) but enrichment is not a tangible product; it is a physical change wrought in uranium by increasing the proportion that is fissile.

The object of enrichment

Natural uranium contains three isotopes, namely U-235, U-234 and U-238, occurring as 99.27 per cent U-238. 0.72 per cent U-235 and 0.0006 per cent U-234. However, since fission comes about from the bombardment of U-235 by neutrons, and not by the bombardment of U-238, for a chain reaction to occur it is essential that, on average, at least one of the two or three neutrons emitted at fission will encounter a U-235 atom and produce another fission and release more neutrons. Thus, uranium which is enriched by increasing the proportion of U-235 at the expense of U-238 is a more fissile and more useful material for fuel elements (see Chapter 1).

For enrichment to take place, separation of the isotopes has to be achieved, but U-235 and U-238 have the same chemical properties and thus cannot readily be separated by chemical processes; almost all the processes used depend upon the small difference in their mass (Table 7.3). Four different types of technology are available: gaseous diffusion, which has been the process used for most uranium enrichment to date; the gas centrifuge process, which is well proved technically and is now a real competitor for gaseous diffusion; two versions of an aerodynamic jet nozzle process, one developed in the FRG and the other in South Africa; and a commercially untried process which is based on the preferential excitement by a laser of one isotope compared with the other. In all these processes it is necessary first to convert the uranium oxide into uranium hexafluoride (UF_6) which is the simplest and most suitable gaseous compound of uranium.

Gaseous diffusion

Gaseous diffusion depends upon the fact that, since the average energies of the light and heavy isotopes in gaseous form are the same, the light component

Table 7.3 Uranium enrichment capacity operable or under construction in 1988

Country	*Facility location/ name*	*Owner/ operator*	*Process*	*Capacity (thousand SWU/year)*	*Status*	*Year of first operation*
Argentina	Pilcaniyeu	CNEA	GD	20	OP	1987
				100	UC	
Brazil	Resende	Nuclebras	JN	10	OP	1987
China	Lanchow	–	GD	80	OP	1980
France	Tricastin	Eurodif	GD	10,800	OP	1982
	Pierrelatte	Comurhex	GD	600	OP	
FRG	Gronau	Urenco	Cen	400	OP	1988
	Karlsruhe	Steag	JN	50	OP	1986
Japan	Hyuga	Asahi Chemical Industry	Chemex	2	OP	1986
	Ningo-Toge	PNC	Cen	75	OP	1981
	Rokkasho-Mura	JNFS	Cen	150	UC	1991
Netherlands	Almelo	Urenco	Cen	1,120	OP	1979
Pakistan	Kahuta	–	Cen	5	OP	1984
South Africa	Valindaba	UCOR	JN	300	OP	1982
UK	Capenhurst	Urenco/ Centec	Cen	770	OP	1979
USA	Oak Ridge	Exxon	GD			1945
	Paducah	DoE	GD	9,430	OP	1954
	Portsmouth	DoE	GD			1956
USSR	Siberia	State Committee	GD	10,000	OP	1982

Source: Nuclear Engineering International, *World Nuclear Industry Handbook 1988* and *1989*

Notes: (a) Abbreviations: Cen, centrifuge; Chemex, chemical exchange; GD, gaseous diffusion; JN, jet nozzle; Op, operational; UC, under construction.

(b) Over the past decade there has been retrenchment in uranium enrichment. Plant has been shut down (including 7,700 thousand SWU/year in the USA and 650 thousand SWU/year in the UK in 1988–9) or put onto standby, and construction plans have been cancelled or deferred.

travels faster. This fact is utilized to achieve separation by pumping UF_6 gas through a porous barrier. Since the lighter component moves faster, if the holes in the membrane are the correct size, the U-235 molecules will move through more readily than the U-238 molecules. However, the degree of enrichment is very slight, being proportional to the square root of the ratio of the molecular weights of the two molecules, and after the gas has passed through the membrane it is only about 1.0049 times richer in U-235 than it was on the high pressure side. Thus, to achieve useful enrichment levels (typically 1.5–3.5 per cent for reactor use) it is necessary to repeat the process many times in a cascade of separation stages. The lighter fraction from one stage is fed to the next stage above in

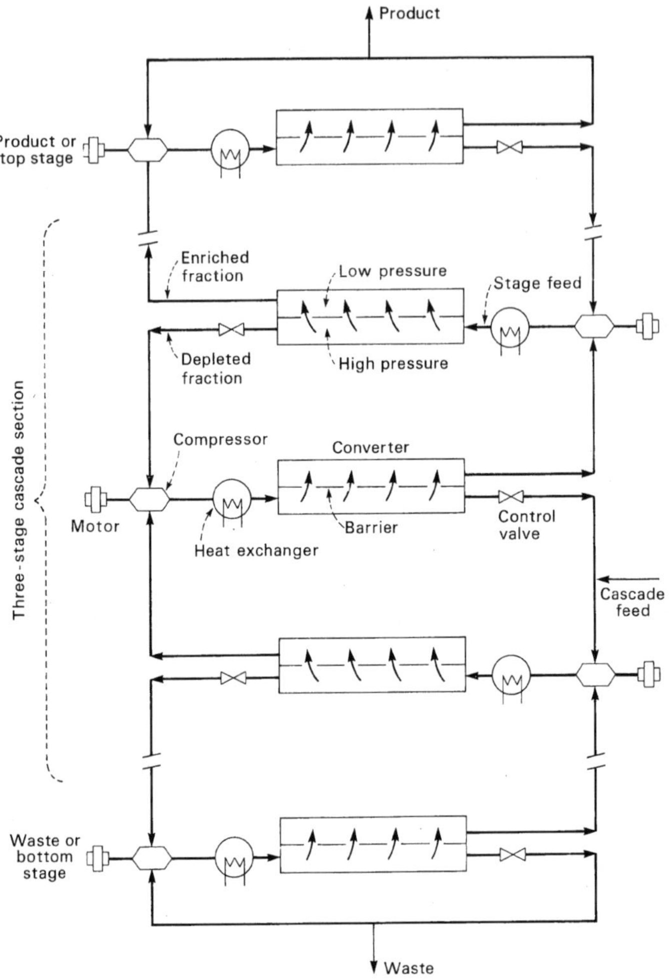

Figure 7.10 A cascade of gaseous diffusion stages
Source: After Greenwood *et al.* 1976: 43

the cascade, whilst the heavier fraction is passed to the stage below (Figures 7.10 and 7.11(a)). In addition to a product enriched in U-235, the plant also produces waste or 'tails' depleted in U-235 which is stored pending the time when it may be used to fuel fast reactors.

The degree of enrichment and the tails assay required determine the number of times that the operation is performed. For example, if a plant is being operated to produce uranium enriched to 3.0 per cent U-235, the level required in LWRs, at a tails assay of 0.20 per cent U-235, about 1,300 stages are required. The same tails assay with a weapons-grade

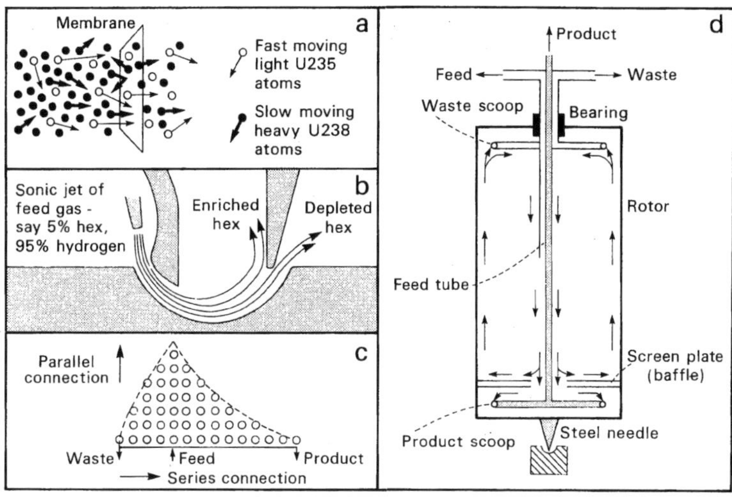

Figure 7.11 Enrichment by gaseous diffusion and by centrifuge. (a) Enrichment by gaseous diffusion. Desirable U-235 atoms are marginally lighter than unwanted U-238. Light atoms of gas traverse a membrane faster than the heavier atoms and thus can be separated from them. The process needs to be repeated through many membranes, and a small enrichment plant uses as much power as a large town. (b) Enrichment by stationary-walled centrifuge. (c) Shape of a centrifuge cascade. (d) A counter-current centrifuge

enrichment of 90.0 per cent requires 3,912 stages. Thus the capacity of gaseous diffusion plants is measured by the amount of separative work they can perform, expressed in separative work units (SWU) per annum, measured either in kilograms or tonnes. For example, the production of 300 tonnes of uranium per annum at an enrichment of 2.0 per cent U-235 and using 0.25 per cent waste concentration (tails assay) would require 570 tonnes per annum of separative work and 1,150 tonnes per annum of natural uranium as feed. However, a plant of this separative capacity could equally well carry out other separations requiring the same total amount of separative work; for example, still keeping a waste concentration of 0.25 per cent, the plant could produce either 150 tonnes per annum at 3.0 per cent U-235, or 550 tonnes per annum at 1.5 per cent U-235. If the waste concentration were raised to 0.35 per cent, the product rate at 2.0 per cent would increase to 375 tonnes per annum.

Until the early 1980s the USA dominated the world's gaseous diffusion enrichment capacity with three enormous plants at Oak Ridge, Tennessee, Paducah, Kentucky, and Portsmouth, Ohio. These were built between 1945 and 1956 but were substantially improved and uprated in 1981. They represent a massive capital investment and have been immensely significant in the world's enriched uranium supply system. Until the early 1980s enrichment plants elsewhere in the world could be counted very easily on the fingers of one hand. The UK had a diffusion plant at Capenhurst, opened in 1953 to produce weapons-grade material and enlarged in 1966 to accommodate the civil nuclear power programme. Prior to its closure in 1982 it was capable of 400–600 SWU/year, France had separative capacity at Pierrelatte and the USSR had gaseous diffusion capacity in Siberia. China may have had an early plant in Lanchow. But virtually all productive capacity for uranium isotope separation for civilian applications was in the USA where the three plants had a maximum output capacity of 17 million SWU(kg)/year. The location and size of enrichment plants in 1988 is shown in Table 7.3. In 1988 the US factories accounted for only 39 per cent of the non-Communist world market for enrichment operations; Eurodif and Urenco took 54 per cent.

The overheads of gaseous diffusion plants are so high that it is not economically attractive to run them much below their full capacity. The operating costs are dominated by the cost of the electricity used in the process; for example, the three US gaseous diffusion plants required around 7,400 MWe of power at or near full output.

Stationary-walled centrifuge or jet nozzle process

The idea of the jet nozzle process is to blow a gas mixture of UF_6 and hydrogen (or helium) through a nozzle directed against a concavely curved wall. Centrifugal forces leave the heavier U-238 isotope nearer the wall than the lighter U-235 isotope, so that a 'skimmer' at the end of the curve separates the gas into enriched and depleted streams (Figure 7.11(b)). The process was devised by Professor E.W. Becker in West Germany and has been taken up in modified form by the Uranium Enrichment Corporation of South Africa (UCOR).

The stationary-walled centrifuge is an energy-hungry device, needing even more energy per unit of output than the gaseous diffusion process, but UCOR uses coal-generated electricity which is cheaper in South Africa than in Europe. A pilot plant at Valindaba has been followed by a commercial enrichment plant at the same site, with a capacity of 300 SWU (tonnes)/year, primarily with the object of supplying South Africa's first commercial nuclear power station, a 2,000 MWe plant at Koeberg near Cape Town. In essence, in the South African context, enrichment constitutes a convenient way of storing electricity on the Rand for shipment to the coast.

Gas centrifuges

The concept of separating isotopes by means of gravitational or centrifugal fields underlies the theory of the counter-current centrifuge (Figure 7.11(d)). The gas centrifuge is a cylinder containing UF_6 gas and rotating very rapidly. The gas at the periphery, enriched with the heavy isotope, is swept to one end of the machine whilst that in the centre, near the axis, is swept to the other end. This isotopic concentration difference along the length of the machine enables the enriched and depleted streams to be taken off at opposite ends. The separative work output of a cylindrical centrifuge depends on the acceleration at the periphery of the centrifuge (which depends on the radius and the rotation rate) and on the length of the centrifuge, increasing with both. Separation factors are much larger than in the case of diffusion plants and this means that enrichment can be accomplished in far fewer stages. Partly for this reason, the energy required per unit of separative work is only one-tenth of that needed with the diffusion method. The centrifuges have to be operated in a cascade that must be assembled from identical machines with many in parallel at the feed point and fewer in each successive step towards the product and tails ends (Figure 7.11(c)). Even a small commercial plant needs several thousand centrifuges and the commercial viability of the process is strongly influenced by their failure rate.

Gas centrifuge development has been successfully pursued in the USA, Japan and particularly in Western Europe, but less successfully in Pakistan. The West European effort has been led by the UK, the FRG and the Netherlands, and, whilst each country initially worked on its own to further the technology, an agreement was signed between the 'troika' at Almelo in March 1970 to work together on further development. The agreement brought into existence two companies – The Prime Contractor and The Enrichment Organization. The Prime Contractor, sited in the FRG, is concerned with the development, design and manufacture of centrifuges. The other company, now called the Uranium Enrichment Company (URENCO) is based in Marlow, Buckinghamshire, in the UK, and exists to purchase plants from The Prime Contractor, operate them through subsidiaries and market the enriched uranium. The shareholders of URENCO are British Nuclear Fuels Ltd (BNFL), Ultra-Centrifuge Nederland N.V. (UCN) and Uran-Isotopentrennungs-GmbH (URANIT). Although the market for enriched uranium has not matched the forecasts of the early 1970s, in 1987 URENCO held orders worth £2,500 million with the home markets and with the USA, Sweden, Switzerland and Brazil.

Laser enrichment

Laser enrichment is the most recent of the four technologies and is still in the experimental–developmental stage.

Isotopes can absorb light but at slightly different energy levels according to their class. Thus the laser enrichment process depends on the selective excitation of either U-235 atoms or $^{235}UF_6$ molecules in gaseous form by a sharply tuned laser. The gas is then irradiated with another light source, which need not be so sharply tuned. This causes ionization of the excited atoms or molecules and the other class of isotope that had not absorbed the light at that particular level is left behind. The process may achieve nearly complete separation in a single stage, and the energy requirements are likely to be two or three orders of magnitude lower than for gaseous diffusion. Moreover, laser isotope separation technology would have a major impact on world uranium requirements because optimal U-235 enrichment tails could then shift to much lower values, perhaps at around 0.1 per cent, which would substantially reduce natural uranium needs to produce enriched uranium. Most of the experimental work to date has been conducted by US companies such as Exxon Nuclear, Avco and Jersey Nuclear a subsidiary of Standard Oil; France and Japan are also studying the technology.

Supply and demand for enrichment

The uranium enrichment industry has been badly hit by the slow growth of nuclear power and there seems to be no likelihood of a shortage of enrichment capacity for several decades. The US Department of Energy has the advantage of plants which worked off most of their capital cost long ago and are being required neither to make a profit nor pay tax. They have also proved themselves as reliable suppliers to US utilities. In 1985, however, the USA cancelled plans to build a gas centrifuge plant at Portsmouth, Ohio, and gaseous diffusion plants are finding it difficult to compete internationally. They are energy hungry, although URENCO claims that its latest production machines consume only a twentieth of the electricity used by the diffusion process, and they are allowed to enrich only a client's uranium. France, on the other hand, offers enriched uranium for sale. France is the leading participant in EURODIF, centred on the massive gas diffusion plant at Tricastin in which Italy, Belgium and Spain are also partners. France uses EURODIF and agreements with the USSR to underwrite guarantees of supplies of enriched fuel for French-type PWRs built under contract outside France. Tricastin was built in anticipation of a scale increase in civil nuclear power which has not materialized, and because in the early 1970s the Americans began to raise prices and set stiff contractual conditions which drove European customers to seek other means of supply. However, since 1979 Tricastin has had to work at well below its full capacity for long periods.

US dominance in the enriched uranium market until the late 1970s was felt world-wide on occasion, especially when its near-monopolistic position in enrichment was used as a political lever. This was true particularly in Western Europe. Virtually every new nuclear power station commissioned in Western Europe during the 1960s and 1970s committed the continent to increased foreign fuel contracts and increased dependence on the USA for enrichment. In the mid-1970s many Western electrical utilities were turned away from the door by the USAEC, whose enrichment queue they were in, because of uncertainties about supplies for home power stations. Alternative sources were sought including the USSR, and the need for increased enrichment capacity in Western Europe was recognized. Under Euratom rules, the EC's supply agency vets purchases of enriched uranium by member states. The fact that in Europe two commercial enrichment organizations, URENCO and EURODIF, have emerged, represents past uncertainty within the EC regarding the best technology to develop, but the fact that each involves several countries is because the financial, technological and production infrastructure investments involved are so heavy and the market uncertainties so real that no one country has been willing to develop an independent enrichment system. The enrichment industry has also been complicated by the vigorous entry of the USSR into world and West European markets through its state company Technsnabexport. Virtually every West European country with nuclear power plants has entered into contracts for uranium enrichment with the USSR at one time or another.

FUEL ELEMENT MANUFACTURE

The milling, refining, conversion and enrichment processes constitute a sequence of feed materials operations providing the raw material for nuclear fuel production at fuel element manufacturing plants such as Richland, Washington State in the USA, Malvesi, near Narbonne, in southern France, Pierrelatte in the Rhone valley, Hanau and Lingen in the FRG, Springfields, near Preston, in the UK and others. Refining and conversion capacity is heavily concentrated in Canada, France, the USA, the USSR and the UK. The value and importance of the conversion and fuel-fabricating facilities should not be underestimated; for example the fuel produced by the UK Springfields factory alone in a full working year is the equivalent in energy terms of about 5 million tonnes of oil or 8–10 million tonnes of coal.

Enriched uranium hexafluoride is chemically converted to pure uranium dioxide powder which is then pressed into pellets and fired in a kiln to produce a dense ceramic fuel capable of withstanding high temperatures and retaining gaseous waste products. The fuel pellets are packed together and then sealed in tubes of a zirconium alloy which resists corrosion by water. These loaded tubes, called fuel pins, are put together in a lattice of fixed geometry termed a fuel assembly. A similar procedure is adopted for non-enriched fuel for CANDU reactors and for the fuel for advanced gas cooled reactors, except that, in the latter case, stainless steel is used in place of the zirconium alloy to contain the fuel because it has a better resistance to the corrosive action of the carbon dioxide reactor coolant.

SUMMARY

Nuclear power reactors are only part of the world's nuclear power industry; they rest upon an international infrastructure of uranium mines and mills, enrichment plants and fuel manufacturing factories which together constitute the front-end of the nuclear fuel cycle. Even so, the fuel supplies for nuclear power stations are only part of the total cost of nuclear power, and so the impact of volatility in the uranium market has been felt mainly among uranium producers rather than among electrical utilities.

The uranium market is highly politicized, partly, perhaps, because the material is linked almost exclusively to nuclear power and has its future determined so much by one market. This symbiotic relationship has been sensitized by the fact that cancellations of nuclear reactors since 1980 have exceeded the number of new orders placed, and the price of uranium has fallen sharply partly because of this and partly because of the arrival of new large low cost production centres such as Ranger and Key Lake. A further problem is that supplies are often tied into firm contracts for the enrichment processes needed to make the uranium usable in reactors, and these contracts run ahead of utilities' needs, creating additional stockpiles of prepared material. Some US power utilities have been selling material from their own stock, undercutting producers, and sometimes producers have bought from the utilities to sell to other users. To the impartial observer, there does not seem to be much sign in all this of an orderly uranium market.

A depressed uranium market in the 1990s could create problems for the future if it means that new ore bodies are not sought out and new mines not brought into production. Over the short and medium term it is not the absolute size of uranium reserves that matters on the supply side as much as the production capability of the suppliers, who will not commit themselves to investment in expensive new capacity with the long lead times involved without good assurances on uranium demand. New methods of mining have been attempted in order to reduce costs, but, faced with low prices, market uncertainties caused by delays and cancellations in reactor construction and upward pressures on production costs, many uranium mining and milling operations have been shut down since 1980. Accompanying these trends has been a shift in the geography of ore extraction and oxide production away from the USA, whose output fell from 10,300 tonnes U in 1982 to 5,700 tonnes U in 1984, towards Canada and Australia and, to meet the expanded French nuclear power programme, Niger, Gabon and France itself.

Uncertainties in the uranium market are compounded by ambivalent attitudes to uranium mining in Australia, the leading uranium-rich WOCA country. Uranium is the country's tenth largest export earner, a long way behind coal, meat and wheat, but ahead of copper, zinc and dairy products. The Australian government operates a very tough set of conditions on exports, but its vacillations in policy towards uranium mining have prevented the country from earning the image abroad of a reliable supplier. Australia, however, may still be able to take the lead in establishing proper and enforceable international safeguards for the mining and use of uranium, with the ultimate sanction of withdrawal of supplies as a lever to use against those who err. In June 1983 France was told by the Australian government that deliveries of Australian uranium were suspended until it stopped nuclear weapons tests in the South Pacific Ocean.

The issue of nuclear proliferation and safeguards, coupled with policies of national energy autonomy and export policies in general, can profoundly influence the availability of uranium to customers outside the country of origin. In determining export policy Canada takes account of the domestic nuclear power programme (a domestic reserve requirement has to be satisfied before any exports are permitted), pays considerable attention to non-proliferation and safeguards issues and gives weight to economic considerations (there is a requirement that the uranium should be exported in as processed a form as possible, and another that prices upon delivery must provide for protection of the producers' investment). Similarly, enrichment facilities have served a political purpose from time to time.

In terms of environmental and public health impact, the major problems at the front-end of the nuclear fuel cycle have been associated with the health of miners and the treatment of ore waste and mill tailings. Gradually, better mine ventilation has made the mines safer, and improved procedures are being implemented for waste management. Work still remains to be done, however, to remedy the impact of past over-casual mining and milling operations and to control present and future operations effectively.

NOTES

1. Different terms are used in reports to measure uranium resources. In the past it has been conventional to express ore and mill product assays in terms of uranium oxide (U_3O_8) content, even though the uranium may be present in different chemical forms. One kilogram of U_3O_8 contains around 0.85 kg of uranium, and confusion can arise if an assessment of resources based on uranium itself is compared with another based on oxide or concentrate. Nowadays the common practice is to use the uranium (U) figure.

FURTHER READING

Bowie, S.H.U. (1979) 'Theoretical and practical aspects of uranium geology,' *Philosophical Transactions of the Royal Society, Series A* 291.

OECD/NEA/IAEA (Organization for Economic Co-operation and Development/Nuclear Energy Agency/International Atomic Energy Agency) (1988) *Uranium Resources, Production and Demand*, Paris: OECD.

Moss, N. (1981) *The Politics of Uranium*, London: Andre Deutsch.

8 The nature of the hazards: radiation and its biological effects

Like earthquakes, floods, tornadoes and volcanic eruptions, radioactivity has always been present in the human environment. Human life, therefore, has never been free from exposure to ionizing radiation, both from naturally radioactive elements present in the earth's crust ever since its formation (more than sixty radionuclides are known to occur naturally in the earth's environment) and from bombardment of the earth's gaseous envelope by cosmic rays originating outside the atmosphere. Such natural or *background radiation* provides the largest source of radiation for human populations (Table 8.1).

Establishing the level of risk associated with nuclear power stations and related phenomena such as mill tailings, fuel reprocessing plants and stores

Table 8.1 Contributions to average radiation exposure (in millisieverts)

UK		*Source*	*USA*
1988	*1984*		*1987*
1.2	0.7	Radon from earth materials (98 per cent from indoor exposure)[a]	2.0[b]
0.35	0.40	Terrestrial gamma rays[a]	0.28
0.30	0.37	Ingested natural radionuclides[a]	0.39
0.30	0.25	Medical exposures	0.53
0.25	0.30	Cosmic rays[a]	0.27
0.10	0.10	Thoron from earth materials[a]	–
0.01	0.10	Fallout (including Chernobyl)[a]	0.0006
0.01	0.01	Miscellaneous sources	0.05–0.13
0.005	0.009	Occupational exposure	0.009
0.001	0.002	Radioactive effluent discharges	0.0005
Total			Total
2.5	2.2		3.6

Source: Clarke and Southwood 1989
Notes: [a] Natural sources.
[b] Includes thoron.

for radioactive waste depends upon assessment of the extent to which any *additional* amounts of radiation produced by the nuclear fuel cycle may constitute a health hazard in the normal operation of particular parts of the cycle and under accident conditions. This task is complicated by disagreement in the scientific community regarding the significance for human health of low levels of radiation. It is accepted that a dose of 20 Sv to the human body over a short time will cause certain death, a dose above 5 Sv will cause gastro-intestinal radiation sickness and damage to the lens of the eye, a dose above 4 Sv will carry a 50 per cent risk of death, a dose of about 2 Sv will cause skin burns and bone marrow radiation sickness and a dose over 0.1 Sv will cause temporary blood count depression, temporary sterility and mental and growth retardation in young children. But what about doses below 0.1 Sv? What is a dose? What is a Sievert? What is a 'short' time?

It is necessary to answer these questions briefly by defining terms used by radiobiologists and physicists to measure and describe radiation and the damage it may cause.

RADIATION TERMS

Different types of unit are used to measure radiation and its effects: some types of unit measure the quantity of radioactivity emitted by a substance and others the biological effect produced on exposure to radiation, and the units have changed with the introduction of Système International (SI) units. The *becquerel* (Bq) is the SI unit for measuring the quantity of radiation in terms of the number of radioactive disintegrations taking place in a material and is defined simply as the radioactivity produced by the disintegration of one atom per second in any amount of a radioactive substance. The unit is named after the French physicist Henri Becquerel who discovered the phenomenon of radioactivity in 1896. The previous unit used was the *curie* (Ci), named after the nineteenth century French physicist Marie Curie. The curie is defined as the radioactivity corresponding to any mass of a radioactive material disintegrating at a rate of 3.7×10^{10} per second (the rate for 1 g of radium-226). For example, the radioactivity of a substance would be 10 Ci if measurements indicated that its atoms were disintegrating at a rate of 37×10^{10} disintegrations per second, a figure ten times greater than that of 1 g of radium.

Becquerels and curies are used to measure the amount of radioactivity from a given mass of material. The same mass of two materials which have different activities will emit different amounts of radiation. Conversely, a given amount of radioactivity, for example 1 millicurie (1 mCi), is produced by 1 million tonnes of seawater or by a piece of iodine-131 the size of a pinhead. These units do not indicate the type of radiation emitted by the decaying atoms or the energy of the radiation. Ionizing radiation is produced when one or more of the protons, neutrons and electrons which make up the atoms of all elements are released from an unstable mother element or *radionuclide.* The principal emissions are alpha (α) particles (the nuclei), beta (β) particles (high energy electrons) and gamma (γ) rays (high energy electromagnetic waves).[1]

Ionizing radiation has the capacity to be a cancer-inducing agent by damaging the chromosomes of tissue cells. Whether or not damage is caused, and the extent of the damage, depends upon the intensity of the radiation, the type of radiation and the time over which it is received. The intensity of the radiation received is proportional to the activity of the source and the reciprocal of the square of the distance of the receiver from the source. Thus direct contact with a source of radioactivity, either via the skin or ingested, is likely to be particularly harmful.

Before the introduction of SI units the amount of ionization produced in tissues by the absorption of radiation energy, the *radiation dose,* was measured in *rads* (*r*adiation *a*bsorbed *d*ose), where 1 rad equals 0.01 joules of energy absorbed per kilogram of tissue, and this unit is still widely used. The official SI unit for this purpose, however, is the *gray* (Gy) and 1 Gy = 100 rad (i.e. 1 Gy equals 1 joule of energy absorbed per kilogram of tissue). The unit is named after L.H. Gray who was a pioneer in radiobiology. Absorbed dose is, in effect, the amount of energy dissipated, usually in the form of heat, when radiation is absorbed in a mass of material.

It is generally assumed that a whole-body radiation dose of 4 Gy (400 rads) will result in a 50 per cent chance of death within a few days. Doses below 1 Gy may cause radiation burns, sickness and hair loss but are generally not lethal.

Biological damage depends also on the nature or type of ionizing radiation; for example, alpha particles can be eighty times more harmful than gamma rays or beta particles. The gray and the rad are satisfactory units to use when studying the effects of radiation on living tissue provided that their use is confined to X-radiation, gamma radiation and beta radiation. Under such circumstances the biological damage caused by these types of radiation is very closely related to the energy which they deposit in living tissue. However, such a relationship does not hold for the much heavier neutrons, protons, alpha particles and fission fragments, and it is necessary to introduce a correction factor when studying the biological effects of these types of radiation. The correction factor used is known as the *quality factor,* or the Q factor, and has the following values for the type of radiation indicated:

$Q = 1$ for electrons (beta particles), X-rays and gamma rays;
$Q = 10$ for protons and fast neutrons;
$Q = 20$ for alpha particles and fission fragments.

When the Q factor is incorporated into the rad or becquerel as a multiplying factor, the absorbed dose is said to have been converted into a biological dose equivalent; before the use of SI units the converted unit was known as the *rem* (*r*oentgen *e*quivalent *m*an). Thus 1 rem = 1 rad × Q.

At relatively low and intermediate dose levels the rem is a very valuable unit in radiology because it indicates the biological effectiveness of all types of nuclear radiation, including delayed effects such as cancer and leukaemia. At high dose levels, however, the application of the Q factor has been demonstrated to be inappropriate and the rad or gray is used, together with a specification of the type of radiation.

The sievert (Sv), named after the Swedish scientist Rolph Sievert, is the SI equivalent and replacement for the rem as the 'dose equivalent unit' and 1 Sv = 100 rem.

The sievert, rem, rad and gray are related in the following way:

$$\begin{aligned} 1 \text{ Sv} &= 100 \text{ rem} \\ &= 100 \text{ rad} \times Q \\ &= 1 \text{ Gy} \times Q \end{aligned}$$

The radiological significance of different levels of radiation are indicated in Tables 8.2 and 8.3.

CONCEPTS OF DOSE

In any given period of time an individual may be subject to radiation from several sources (see Table 8.1). The *committed effective dose equivalent* is a composite figure in which all the doses from different types of radiation to different body organs are combined together according to their capacity to

Table 8.2 Harmful effects of ionizing radiation on particular parts of the human body

Aspect	*Syndrome*		
	Central nervous	*GI tract*	*Haemopoietic*
Critical organ	Brain	Small intestine	Bone marrow
Syndrome threshold (Gy)	20	3	1
Latent period	15–20 min	3–5 days	2–3 weeks
Death threshold (Gy)	50	10	3
Time of death	2 days	3–14 days	3–8 weeks
Symptoms	Tremors, convulsions, ataxia	Anorexia, vomiting, diarrhoea, electrolyte loss, dehydration	Fever, dyspnea, leukopenia, thrombopenia
Radiopathology	Oedema (CNS), encephalitis	Depletion of the intestinal epithelium, infection, neutropenia	Bone marrow, atrophy, pancytopenia infection, haemorrhage, anaemia

Source: Windeyer 1974: 37

Table 8.3 Effects of exposure of one million people of all ages to 1 mSv each of whole-body radiation (assuming simple proportionality between dose and frequency)

Organ exposed	*No. of genetic effects*		
Gonads	*6 (all generations)*		
	No. of cancers		
	Fatal	*Curable*	
Bone marrow	2	0.1	(leukaemia)
Lung	2	0.1	
Breast	5	3	(in women; none in men)
Thyroid	0.5	10	
Bone	0.5	0.2	
Other organs	5	2	
Total			
Genetic effects	6 (av. value; fewer for women than for men)		
Fatal cancers	15 in women,	10 in men	
Curable cancers	15 in women,	12 in men	

Source: Pochin 1983

cause fatal cancers. This quantity is often simply called the *dose.* The doses to different organs are weighted in such a way that the effective dose equivalent would give rise to the same risk of inducing fatal cancer or serious genetic defects in the first two generations as a whole-body dose equivalent resulting from uniform external irradiation which is numerically equal to the effective dose equivalent. For example, an effective dose equivalent of 5 mSv would give rise to the same risk as a dose equivalent of 5 mSv to the whole body from uniform external irradiation.

The *collective dose* or *effective collective dose equivalent* is a way of measuring or estimating the irradiation of a population group. It is the sum of the products of the individual dose equivalents and the number of individuals in each exposed group in a population. The *collective dose commitment* is the dose commitment multiplied by the number of individuals in the specified population. It is commonly expressed in units of man sieverts (man Sv) or man rems.

The *dose equivalent commitment* is the infinite time integral of the dose equivalent rate per capita for a specified population for any particular decision, practice or operation. The exposed population is not necessarily constant in number.

Figures published by the International Commission on Radiological Protection (ICRP) indicate that the average risk of developing fatal cancers from exposure to radiation is about 1 in 100 per sievert per year for all members of the population. This represents an average figure which accounts for the increased sensitivity of certain members of the population to specific types of cancer, and the greater resilience of many others. These ICRP figures have been adopted by national agencies to specify limits to doses from exposure to radiation for workers handling radioactive materials and for members of the public under normal conditions. In the UK, for example, the current limits which apply are 50 mSv per year for employees in the nuclear industry (occupational dose limit) and 5 mSv per year for members of the public. These dose limits represent risks of 1 in 2,000 and 1 in 100,000 of developing fatal cancers, but at the time of writing the radiation exposure rates are being reviewed to take account of new evidence from the monitoring of survivors of the Nagasaki and Hiroshima atomic bomb attacks during the Second World War. Recent reports suggest that the risks from low level exposure may be about three times higher than previously thought (personal communication, E.A. Davis, 1990).

SPATIAL VARIATIONS IN BACKGROUND RADIATION

Present dose levels to humans from naturally occurring radiation generally lie between 1 and 3 mSv/year but, as a result of differences in geology, altitude and latitude, the variation from place to place is quite wide. The contribution from *cosmic radiation* at 50 °N at a height of 5,000 feet (1,524 m) is 60 per cent greater than at sea level. The variation with latitude is less. At sea level the cosmic ray intensity at the poles is 12 per cent greater than at the equator. There is a somewhat larger latitude effect at higher altitudes, but even at 10,000 feet (3,048 m) it is only about 50 per cent greater at the poles than at the equator.

Radon from earth materials and beta and gamma radiation from radionuclides in rocks and soils constitute the major part of *terrestrial radiation.* The major contributors are the potassium (K-40), uranium (U-238) and thorium (Th-232) decay series in the approximate ratio of 2:1:2. Terrestrial gamma radiation exposure is strongly influenced by the nature of the geology and terrain in particular localities. Over large fresh-water lakes, for example, there is very little, but high values are observed over acid igneous rocks such as granite. Radiation from terrestrial gamma sources is also influenced by climatic conditions; probably the most important effect is shielding by snow cover and by moisture in the soil after heavy rains. Published data on the beta contribution to the terrestrial dose vary at around 0.03–0.05 Gy/year. The beta contribution to a genetic dose is less than this because critical body organs are shielded by body tissue.

Table 8.1 shows the average level of background radiation in the UK to be 2.5 mSv/year (Clarke and Southwood 1989). The mean exposure value in the USA is 3.6 mSv/year, but with a variation of 0.8 mSv/year between low dose level localities in the coastal states from Texas to New Jersey and higher level localities in the Colorado Front Range. The Kerala coast of India, where monazite is a common geological material, contains some

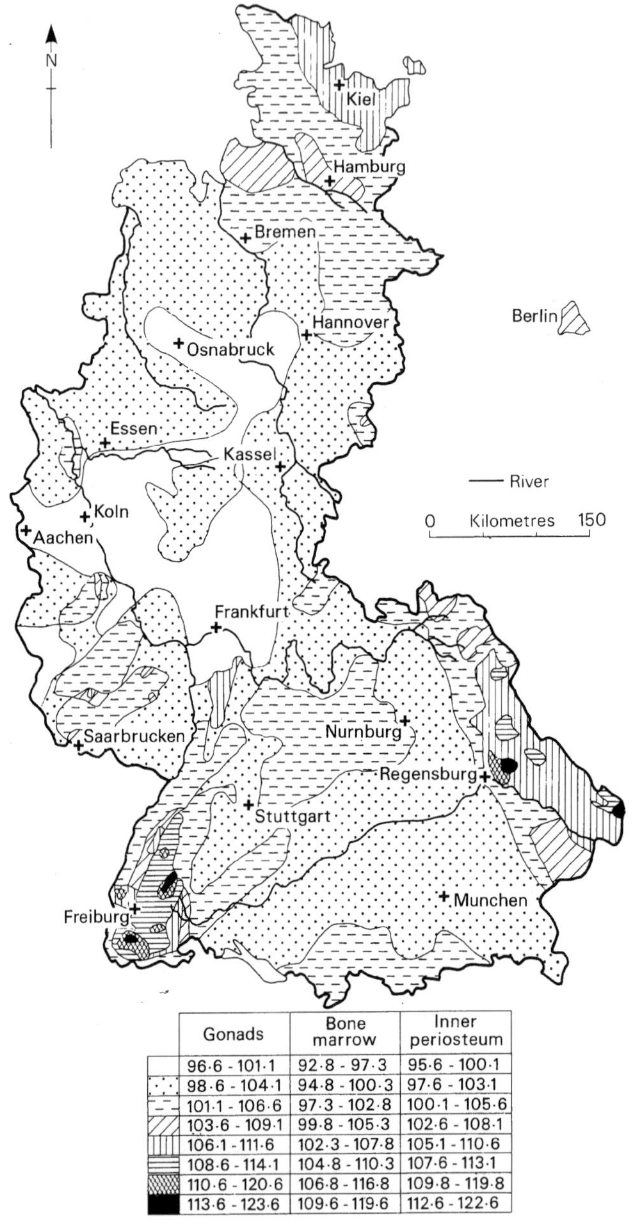

	Gonads	Bone marrow	Inner periosteum
	96·6 - 101·1	92·8 - 97·3	95·6 - 100·1
	98·6 - 104·1	94·8 - 100·3	97·6 - 103·1
	101·1 - 106·6	97·3 - 102·8	100·1 - 105·6
	103·6 - 109·1	99·8 - 105·3	102·6 - 108·1
	106·1 - 111·6	102·3 - 107·8	105·1 - 110·6
	108·6 - 114·1	104·8 - 110·3	107·6 - 113·1
	110·6 - 120·6	106·8 - 116·8	109·8 - 119·8
	113·6 - 123·6	109·6 - 119·6	112·6 - 122·6

Figure 8.1 Average natural radiation exposure (mrem/year) in the FRG
Source: Von Bonka 1974: 149
Note: The millirem units were used in the original map and are therefore retained here. The reader wishing to transfer to SI units is reminded that 100 mrem = 1 mSv

population groups subject to doses of over 12.0 mSv/year. Apart from a possible excess of Downs Syndrome, health studies carried out by 1981 had shown no adverse medical effects at Kerala but the lack of medical records for the 70,000 inhabitants has made research difficult (Rose 1982). The mean level of natural exposure in the FRG has been calculated to be about 1.04 m/Sv but, as Figure 8.1 shows, some parts of Baden-Württemberg, to the north of Basel and Waldschut and to the west of Wolfach, are subject to external radiation exposure levels of over 2.0 mSv/year (Von Bonka 1974)

The data used for the construction of Figure 8.1 include a calculation for radiation arising from house building materials, for such materials can introduce another element of place-to-place variation in exposure levels. The level of radiation experienced by a Tokyo family living in a wooden house would be half that experienced by a family living in a brick house in London, and an apartment in a concrete Swedish high rise block may offer seven times the level of the Tokyo home. In the UK in 1986 the National Radiological Protection Board (NRPB) conducted house surveys in Devon and Cornwall, where there are many houses built from and on granite. The surveys revealed that the radon doses received in over 20,000 houses, given normal occupation through the year, could exceed 400 Bq per cubic metre of air. The NRPB recommended that this should be an action level, i.e. that occupants should be warned to take some remedial action, and then in 1990 suggested that an even more stringent action level of 200 Bq should be adopted. Appropriate action includes improved sealing to the ground floor and pumps installed close to the ground to remove radon.

Individuals are also subject to radiation from medical X-rays (average 0.53 mSv/year in the USA) and from air travel (perhaps 0.01 mSv/year on average in the USA).

Clearly, therefore, people throughout the world are subject to varying levels of radiation from a variety of sources. The questions remain whether this has any public health significance and whether the additional increments of radiation arising from nuclear power have any significance for the health of individuals and populations. The fact that the ICRP recommends an action level for natural or background radiation at 25 mSv/year, and that this is exactly 25 times greater than the allowable lifetime public dose from artificial sources, implies affirmative answers, but the issues involved are quite complex. To measure the significance of this figure, it is helpful to keep in mind the limit for radiation workers in the UK of 50 mSv/year and that 5,000 mSv in a short period of time would prove fatal within a few days or weeks.

HARMFUL EFFECTS OF IONIZING RADIATION

It is recognized that there can be damaging effects on an individual who has been exposed to radiation and that this class of damage, called the *somatic effects*, may become evident either almost instantaneously with the irradiation, after several weeks or, in the case of cumulative small doses, after several years. The effects may be local, if only a small part of the body is exposed, or general, if the whole body or a large part of it is exposed. They include local effects on tissue, functional disorder, leukaemia and other forms of cancer (see Tables 8.2 and 8.3).

The mechanism through which somatic effects are produced is not fully known but radiation must pass through the nucleus of a cell to cause cell death and cancer. The blood-forming tissues are particularly sensitive, but different substances may become localized in different parts of the body. Radium, for example, is a bone-seeker, whereas iodine is concentrated in the thyroid gland. There is extensive medical evidence that embryos in the female womb and children are more susceptible to radiation injuries than adults, and that even low doses may induce both developmental disorders and malignant changes in embryos. It was for this reason that one of the early emergency measures taken in 1979 at Three Mile Island was the evacuation of pre-school-age children and pregnant women from within a 5 mile (8 km) radius of the plant.

Sooner or later somatic effects are seen in the exposed individual. The hereditary or *genetic effects* of ionizing radiation are those which, while not being expressed in the irradiated individual, are transmitted to his or her progeny through alterations in chromosomes within the genetic material. The ovaries and testes are particularly susceptible to direct radiation exposure. Changes in the structure and number of chromosomes, and mutations of genes, can be induced by exposure to ionizing radiation and it is generally accepted that such changes can reduce fertility, increase the likelihood of stillbirths and induce congenital malformations. Doses as low as 0.01–0.1 Gy are thought to be capable of inducing minor genetic changes such as eye-colour anomalies. It is estimated that in the UK there is a natural incidence of about 4 per cent of all births that are affected by hereditary disorders and it is thought possible that some of these abnormalities are due to the effects of background radiation (Windeyer 1974).

Because they cause hereditary damage, genetic effects have the ability to make themselves felt through whole populations, and significant exposure rates are habitually averaged over whole populations. The investigation and evaluation of genetic effects in humans has many inherent difficulties and much of the available information comes from observations on mice or from experience with large doses delivered over a short period of time, as from the effects of the atomic bomb explosions in Japan or the treatment of patients by radiotherapy. There is much less experience of the same dose spread out over a longer period and the cumulative effects of multiple small doses spread over a period of years. Careful international monitoring of the populations affected by the Chernobyl accident in 1986 could ultimately help to fill this gap in scientific knowledge (Konstantinov and González 1989).

Some of the harmful biological effects of exposure to ionizing radiation, such as skin cancer, cataracts and impairments of the function of heavily irradiated organs, became known within a short time of Röntgen's discovery of X-rays in 1895. Information about genetic effects became available from 1927, and the effects of intake by inhalation and ingestion through food-chains have been studied since the mid-1950s. As a result, data are available on the fractions of different radionuclides discharged into air, water, soil or rock which may be expected to return, after different periods of time, to sources of human food or, by atmospheric transmission or resuspension, to human intake by inhalation or, less commonly, to human exposure by external irradiation following deposition on the ground. These data make it possible to make a judgement of the risks to human health that may result from introducing into the human environment a man-made source of radiation, such as a nuclear waste depository, fuel reprocessing plant or nuclear power station, so that a quantitative assessment can be made of the degree of hazard or risk that the facility may represent (González and Webb 1988; Probert and Tarrant 1989).

Sir Edward Pochin has used a simple and effective diagram to indicate the steps involved in the process of assessment (Pochin 1985):

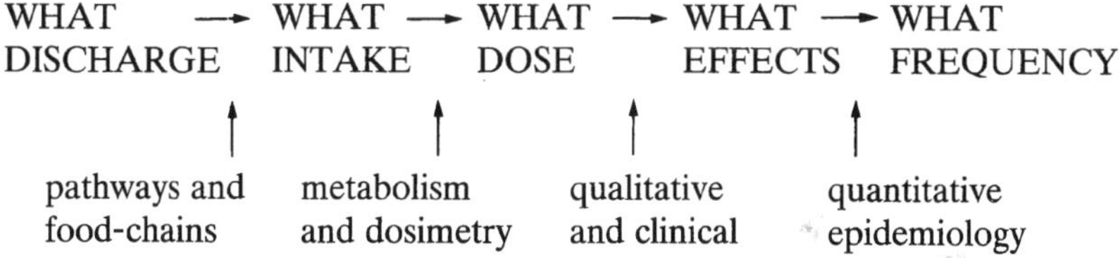

The decision-making process involves answering the following questions.

(a) What amounts of different radioactive materials (what activities of various radionuclides) are released to the environment either by direct discharge or by ultimate escape from containment?
(b) What fractions of these discharged activities may return to man and, in particular, may be taken into the body.
(c) What amounts of radiation – what doses – may they then deliver to body tissues?
(d) What kinds of effect may such doses cause?
(e) What is the frequency with which harmful effects may be induced by these doses?

Answers to these questions for different kinds of man-made radiation sources provide the basis for standards used in the design, construction and operation of nuclear facilities, and for estimating the radiological risk associated with particular practices, e.g. in the management of radioactive waste. It has not proved possible, however, to answer all the questions in a straightforward and unequivocal way. There are areas where the process of translating answers to the questions into standards has been contentious because the answers to the questions are not clear and the extent of risk therefore remains problematical. Nowhere is this clearer than in the question of assessing the significance for human health of low levels of dosage.

LINEARITY, NON-LINEARITY AND THE *DE MINIMIS* CONCEPT

During the early days of the nuclear power industry it was often argued that exposure of individual human beings or human populations to ionizing radiation at levels near those of the natural background, especially if they were within the variation of the natural background from place to place, represented a trivial or non-existent risk compared with other hazards of life, and that limits could be established below which additional radioactivity could be introduced to the environment with little or no adverse effect on individuals or populations. In 1958, for example, an expert panel of the UK's Radioactive Substances Advisory Committee reported that:

> It is not in the nature of radioactivity that the economic value of the useful materials involved can be realized without some release of ionizing radiations to the general environment. But this release can be controlled *within safe limits.* (Emphasis added)

This argument assumed, however, that the effect of the natural background itself was trivial and negligible.

The United Nations Scientific Committee on the Effects of Atomic Radiation (UNSCEAR) was formed in 1955 as a reflection of growing unease over the possible consequences of the exposure of mankind to ionizing radiation, a concern intensified in 1954 by radiation injury to Japanese fishermen aboard the fishing vessel *Fukuryu Maru*, by the exposure of British personnel in Australia in weapons tests just after the Second World War and, in subsequent years, by increasing levels of world-wide fallout from nuclear weapons tests. The committee is required to collate and assemble radiological information furnished by governments, members of the United Nations or specialized agencies relating to observed levels of ionizing radiation and reports on scientific observations and experiments relevant to man and his environment. A warning issued by the Committee in 1962 stated:

> It is clearly established that exposure to radiation even in doses substantially lower than those producing acute effects may occasionally give rise to a wide variety of harmful effects ... which in some cases may not be easily distinguishable from naturally occurring conditions or identifiable as due to radiation. Because of the available evidence that genetic

> damage occurs at the lowest levels as yet experimentally tested, it is prudent to assume that some genetic damage may follow any dose of radiation, however small.

Thus, *the cautious assumption is that damage from radiation has no threshold, and that each additional increase of dose may bring an increased risk.* This linear relationship between dose and effect is shown diagrammatically in Figure 8.2. The relationship has been challenged by those who maintain that there is a threshold of dose below which no lasting effects will occur. However, the weight of scientific opinion is that the correct relationship between risk and dose is likely to be linear–quadratic for gamma rays and linear for alpha particles and neutrons, but there is an articulate minority that argues for a square root relationship giving proportionately higher risks at low doses (Roberts 1987).

If the dose–response relationship for radiation were very non-linear, small doses would have a much smaller effect, per unit dose, than large doses. There would then be some justification for downgrading the significance of small doses – of assuming a threshold – when there was little risk of a combined effect from many small doses. However, the interest is in the increments of dose and risk from nuclear installations and waste which are added to existing levels arising from natural background radiation. Also, the *working hypothesis* in radiation protection is that of linearity, i.e. proportionality between dose and response, and discussion then concentrates upon the level of risk that is acceptable as entire populations and individuals become exposed to *additional* radiation sources arising from the growth of nuclear power. Thus, any dose, no matter how small, is assumed to have associated with it some degree of risk, and any decision to disregard some portion of a dose carries with it the implication of ignoring some extra health detriment. The situation is clarified, perhaps, if one kind of relationship is represented numerically, as in Table 8.3.

The cancers induced by radiation do not differ in clinical behaviour or appearance under the microscope from cancers of the same types which occur 'naturally' in different body organs. Therefore the number of cancers induced by radiation must be determined by comparing the numbers of cancers developing in a population who have been exposed to a known dose of radiation, such as the victims of Chernobyl, with the observed or expected numbers in a population not so exposed but similar in all other respects.

It remains common practice to use the level of background radiation as a basis of reference with which additional increments of man-made radiation may be compared. This may seem odd in view of the wide variations in background radiation levels from place to place, but it is argued that if the man-made dose levels are kept well below the natural background level any genetically significant effects will neither differ in kind from those experienced throughout human history nor exceed them in quantity.

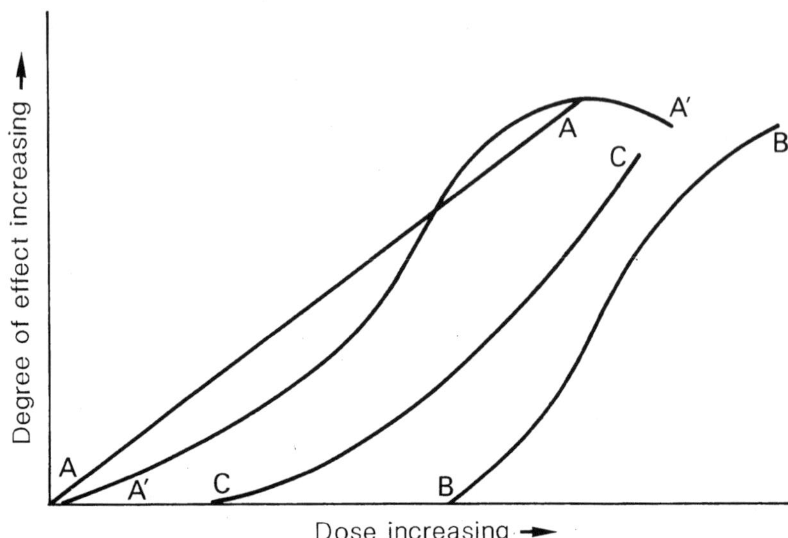

Figure 8.2 Alternative postulated dose–effect relationships at low levels of radiation dose. After a whole-body dose of 100 mSv to an adult human being there is about one chance in a thousand that a cancer has been induced which will eventually kill unless the individual dies of something else first. This proportionality is well established for doses from some tens of millisieverts upwards; it is virtually impossible to establish the relationship between lower doses and ill effects (because of the lack of evidence this area has been called 'trans-scientific').

The curves shown here attempt to extrapolate the response to low levels of radiation as presented in ICRP publications. The linear or curvilinear (quadratic) relationships (A and A′) might be expected on physical or molecular grounds. Curve B with a well-defined threshold might be expected on biological grounds. Curve C with a long 'tail' might be expected in populations such as man with a large range of sensitivities. Although there is general acceptance of the cautious stance of using an assumption of linearity for setting safety standards, there has been considerable discussion of the relative merits of linear and quadratic relations (Hubert 1983). Adopting a purely quadratic relationship would lead to much lower risk estimates but the debate loses much of its sting in the context of quantification of the effects of radiation.

It is important to recognize that, under the assumption of a linear no-threshold dose–effect relationship, the total number of cancers induced by a given *population dose equivalent* is independent of the detailed distributions of age, sex, dose and radiosensitivity of individuals. Thus, for example, if one individual in a hundred were unfortunate enough to be fifty times more radiosensitive than the 'average', then ninety-nine people would have to be, on average, correspondingly less sensitive by a factor of two

Source: Adapted from Dunster 1985

POSSIBLE PATHWAYS FROM A RADIATION SOURCE TO MAN

Radionuclides in the earth's crust reach the earth's surface in springs and other ground-water discharges and are then redistributed by surface water movements. They may finally reach human beings either directly, through drinking water, or indirectly, via food-chains. Man-made sources of radiation may emit radiation through discharge gases sent out through chimney stacks, through liquid discharges, through ground-water attacks on radwaste in trenches or temporary repositories, through controlled or accidental discharges at nuclear power stations and fuel reprocessing plants or through uncovered uranium mill tailings.

The principal pathways by which radioactive materials present in the environment may reach human beings are shown in Figure 8.3, and it is clear that potentially harmful exposure to radioactive hazards may be achieved not only by *direct radiation* but also through the *ingestion* and *inhalation* of radioactive materials. In the mining and milling of uranium ore, for example, dust particles small enough to be taken into the lungs constitute a potential health hazard not only because many of the minerals in the ore contain silica, capable of causing silicosis of the lung, but also because the hazard is aggravated by the radioactive content of the ore (see Chapter 7). With growing understanding of the nature and functioning of ecosystems, less obvious pathways have received scientific recognition and with this realization *critical groups* within human populations have been identified. Such groups of individuals, because of their location, occupation or diet, can be assumed to be exposed to a greater level of risk than the population as a whole. Table 8.4 identifies critical groups in the UK. With the

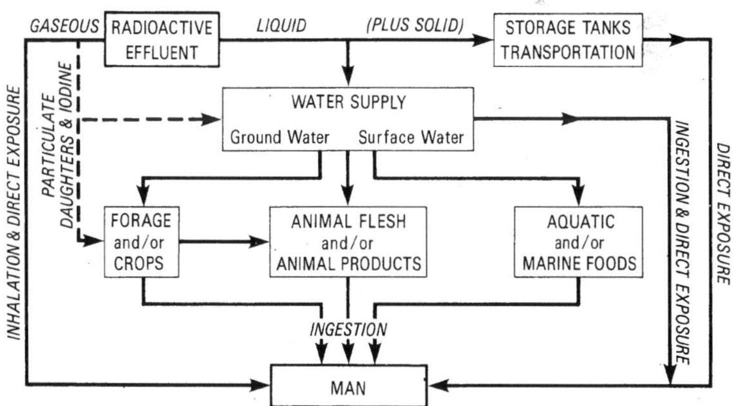

Figure 8.3 Generalized pathways of exposure to man from the operation of a light water reactor

Source: US Nuclear Regulatory Commission, Final Generic Environmental Statement on the Use of Recycled Plutonium in Mixed-Oxide Fuel in Light-Water Cooled Reactors, Washington, DC

Table 8.4 Summarized estimates of public radiation exposure from selected nuclear plant discharging liquid radioactive waste in the UK, 1983 and 1984

	Key radiation exposure pathway	*Estimated exposure of critical group*[a] *(mSv)*	
Establishment	*(critical group)*	*1983*	*1984*
BNFL plant Sellafield	Fish and shellfish (local fishing community)	2.25[b] (45%)	0.84[b,e] (16.8%)
UKAEA plant Dounreay	Handling of fishing gear (local fishermen)	0.2[c] (4%)	0.3[c] (6%)
CEGB and SSEB plants			
Berkeley & Oldbury	Fish and shellfish (local fishing community)	0.005[c] (0.1%)	0.001[c] (0.02%)
Bradwell	Fish and shellfish (local fishing community)	0.02[c] (0.3%)	0.01[c] (0.2%)
Dungeness	Fishing consumption (local fishing community)	0.005 (0.1%)	0.02 (0.04%)
Hartlepool	Fish and shellfish (local fishing community)	0.02[d] (0.4%)	0.02[d] (0.4%)
Heysham	Fish and shellfish (local fishing community)	0.55[d] (11%)	0.45[d] (8.6%)
Hinkley Point	Fish and shellfish (local fishing community)	0.005 (0.1%)	0.005 (0.1%)
Hunterston	Fish and shellfish (local fishing community)	0.1[d] (2%)	0.05[d] (1%)
Sizewell	Fish and shellfish (local fishing community)	0.01[c] (2%)	0.01[c] (2%)
Trawsfynydd	Fish consumption (local fishing community)	0.25 (5%)	0.32 (6.4%)
Wylfa	Fish and shellfish (local community)	0.15[d] (3%)	0.2[d] (4%)

Source: House of Commons 1986: lxviii

Notes: [a] Percentage of ICRP 5 mSv dose limit for members of the public in parentheses.

[b] Calculation based on NRPB advice using enhanced gut uptake factor for plutonium.

[c] Partly due to discharges from Sellafield.

[d] Mainly due to discharges from Sellafield.

[e] 0.84 mSv estimate of committed effective dose equivalent based on reduced consumption of molluscs owing to closure and avoidance of beaches following November 1983 incident. Estimate for 1984 would be 2.0 mSv (i.e. 40 per cent of ICRP 5 mSv limit) if mollusc consumption in 1984 had resumed 1983 levels.

exception of the BNFL plant at Sellafield, estimated doses received by the 'critical groups' are well within ICRP–NRPB radiological protection objectives of 5 mSv/year and in 1984 were below 1 mSv/year. This is as it should be. Uncertainties about the precise behaviour of radionuclides in the environment and about their effect on health mean that a cautious approach is necessary.

Models of the processes by which radionuclides progress through pathways to man attempt to estimate radiation doses by evaluating dilution in surface and ground-water bodies and possible radionuclide concentration in certain biosphere 'compartments' (e.g. the capacity of oysters to reconcentrate zinc-65). Such a system of compartments is shown in Figure 8.4.

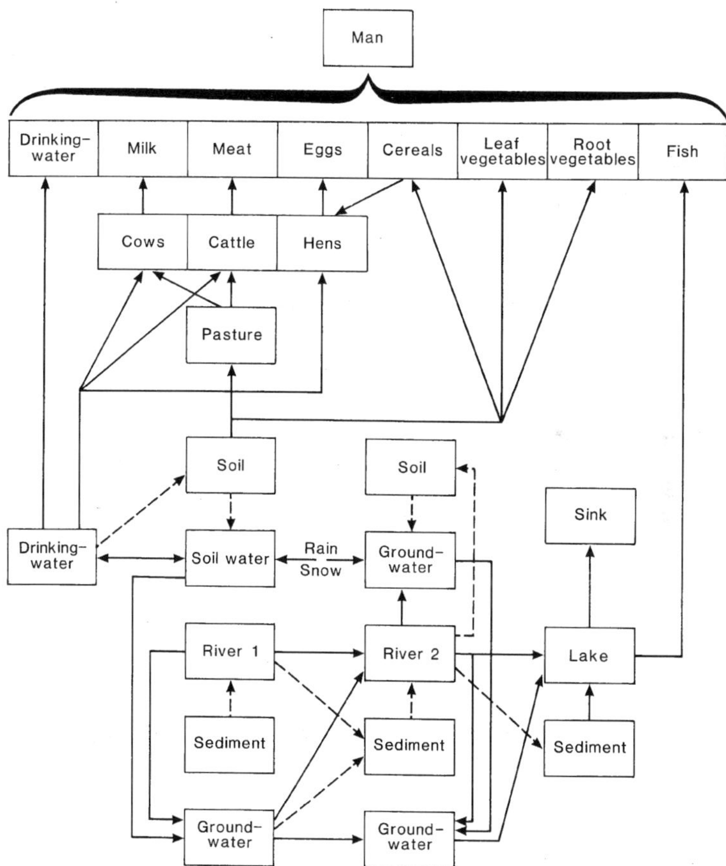

Figure 8.4 Typical compartments used in studies of biosphere movements of radionuclides
Source: Nagra 1985, reproduced in Hill 1987

A food-chain is constructed for each compartment, and when a full set of such chains has been constructed the change in radionuclide concentration is calculated along them by using concentration ratios between various links (e.g. water to grass, grass to cow, cow to milk or beef, milk or beef to man). Finally, doses are calculated from the activity concentrations in water and in the individual foodstuffs, and the quantities of food consumed, using dose conversion factors based on human metabolism (Hill 1987). These factors take into account total body mass and the mass of each organ, radionuclide partitioning between different body organs, radionuclide retention in the body (biological half-life) and differing radiosensitivity of specific body organs.

There are variations in the potential take-up according to soil type and food sources. In localities where the soil is podzolic and in areas of peat bog relatively high transfer of caesium-137 (Cs-137) into man is possible. In such a situation, and with a rural economy using milk and vegetables grown locally, the critical product for the local population is milk, and children become the critical population group (Marei 1972; Pearce 1988).

PLUTONIUM PROBLEMS

With increasing use of plutonium (Pu) in the fuel cycle, the environmental distribution of plutonium and the possibility of its entrance into man become significant. Some authorities have forecast a fourfold increase in the annual production rate of Pu-239 in the USA alone, from 20,000 kg in the period 1970–80 to 80,000 kg in the period 1990–2000. There is a potential threat from plutonium under both normal operating and accident conditions in the handling of effluent discharges at nuclear power stations, in the fabrication and transport of fuel elements, and at fuel processing plants. Plutonium, like radium, concentrates in the bone and produces similar types of bone lesions, but it is much more long lived and toxic.

W.H. Langham's diagrams show schematically the conceivable plutonium transmission routes to man (Figure 8.5). Langham (1972) argues that because of the element's sensitivity to detection in air, water and other parts of the environment, and because of the magnitude of the discrimination factors along the food-chain (minimum discrimination of the order of 10^8), it is 'almost inconceivable that environmental contamination would be allowed to approach harmful levels from ingestion'. He also adds, however, that the discrimination factors (labelled Df in the diagram) are order-of-magnitude estimates only and that a better understanding of the transport of plutonium along aquatic chains is desirable.

Two modes of direct exposure of man by inhalation of atmospherically suspended plutonium are envisaged (Figure 8.5): the first is inhalation of particles from the primary contaminating source prior to surface deposition, and the second is inhalation of particles resuspended in the atmosphere from the contaminated surface subsequent to deposition. In this

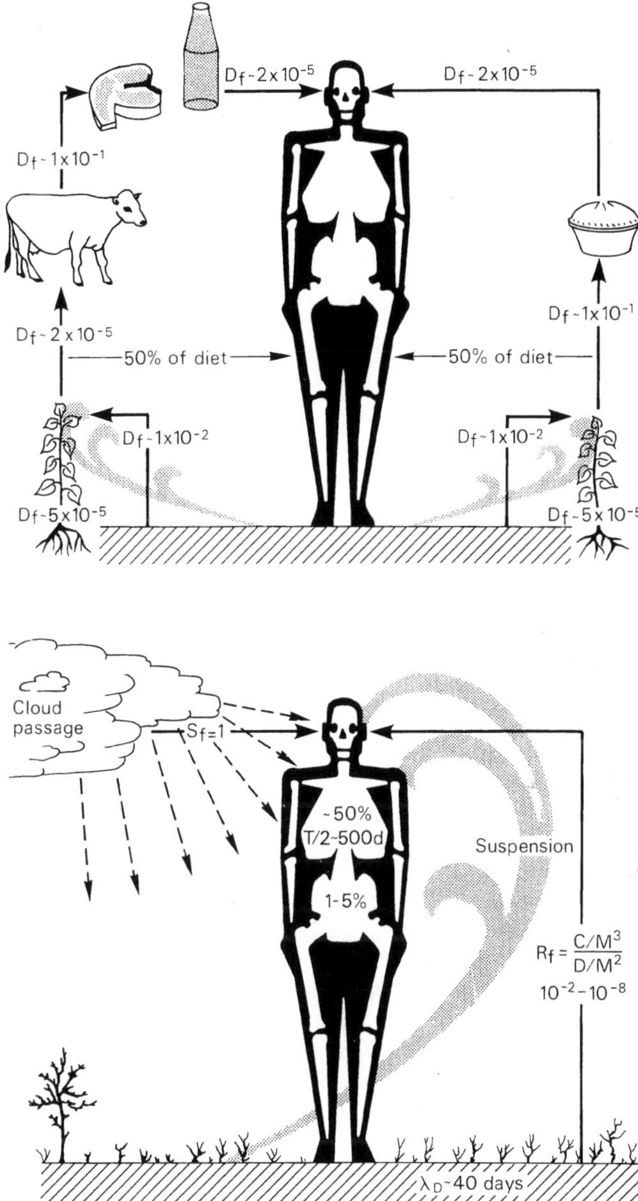

Figure 8.5 Transfer of transuranium elements to man: (a) the steps by which the transuranium elements might be transferred from soils to man; (b) direct exposure of man via inhalation of atmospherically suspended plutonium
Source: Langham 1972

diagram Sf is the suspension factor and Rf is the resuspension factor which, writes Langham (1972), is 'dependent on a staggering number of inter-related variables involving ill-defined phenomena that within themselves vary from place to place and with time'. Such variables include the nature of the contaminated surface (soil type, vegetation cover) and local micro-meteorology (turbulence, wind velocity, rainfall). The estimation of exposure from the primary contaminating source is easier, but even this includes variables such as particle size, inhalation length, length of time of exposure and chemical form of the plutonium (Langham 1972).

THE HOT-PARTICLE PROBLEM

A problem which is particularly relevant to breeder reactors, which in future power programmes would be fuelled with plutonium, to the use of plutonium to enrich the fuel cycle of some of today's LWRs and to the reprocessing of irradiated fuel elements is the so-called 'hot-particle problem'. Hot particles are submicron-sized particles of radioactive substances which, because they are insoluble and emit alpha particles, could become trapped in tissue and deliver an extremely high radiation dose over a very small volume. Typically, such particles might become lodged in a small cut, in a hand, for example, when moving contaminated containers, or deep inside the lung. The crucial question is whether individually such particles pose a hazard which is much greater than allowed for in the existing radiation protection standards which average out the dose over whole organs or the whole body. The tissue surrounding a trapped particle accumulates a dose many thousands of times that of the average 'whole lung' dose from the same particle and it is important to know whether this matters more in terms of tumour incidence than would seem to be the case if the radiation dose were 'averaged out' over the whole lung. Intensive experimental investigations with laboratory animals have demonstrated that minute quantities of plutonium can cause malignant growth in the tissues, and the maximum permissible skeleton plutonium burden is set at less than a billionth of a gram. Such an apparently stringent standard underlines the fact that plutonium is perhaps the most hazardous material known to man, but since 1942, through nuclear power and weapons programmes, plutonium isotopes have been produced in quantity. The fuel for fast breeder reactors is largely plutonium, and it is argued in some quarters that the increased quantities of fuels being reprocessed for the nuclear industry must lead to an increased likelihood of the accidental release of plutonium to the environment. The question is whether the risk involved can be assessed, weighed and reasonably balanced against the benefits of electricity from fast breeders. Because of the potential dangers associated with plutonium the Flowers Commission concluded in 1976 that the UK should not rely for electricity supply on a technology involving plutonium unless there is no reasonable alternative (Flowers Commission 1977).

RADIOBIOLOGICAL PROTECTION STANDARDS

The evidence of damaging biological effects from exposure to radiation has resulted in the setting up of national and international protection organizations. The first of these was an unofficial British X-ray and Radium Protection Committee, formed in 1921 to formulate standards for the protection of those occupied in the medical application of X-rays and radium. Other national organizations followed, such as the American Roentgen Ray Society, and in 1928 the International Congress of Radiology formed the International Committee on X-ray and Radium Protection which, in 1950, evolved into the International Commission on Radiological Protection. This is, perhaps, an archetypal quasi-official non-governmental organization, but it has gained world-wide status as the recognized authoritative body for recommending standards of radiological protection, and most national bodies having responsibility for radiological protection fix their own standards by reference to those recommended since 1951 by the ICRP (González 1983; H. Smith 1988). Since 1955 the ICRP has produced recommendations regarding maximum permissible doses for exposures of radiation workers to controllable sources, for individual members of the public and for whole populations, but it has set its standards on the assumption that there is no valid proof of the existence of thresholds. This position is made quite clear in paragraph 29 of ICRP Publication 9 which states:

> A basis of the Commission's recommendations is the cautious assumption that any exposure to radiation may carry some risk for the development of somatic effects, including leukaemia and other malignancies, and of hereditary effects. The assumption is made that, down to the lowest levels of dose, the risk of increasing disease or disability increases with the dose accumulated by the individual. This assumption implies that there is no wholly 'safe' dose of radiation. The Commission recognizes that this is a conservative assumption, and that some effects may require a minimum or threshold dose. However, in the absence of positive knowledge, the Commission believes that the policy of assuming a risk of injury at low doses is the most reasonable basis for radiation protection.

Despite this rightly cautious stance, the ICRP has grasped the nettle of setting standards, and of changing them from time to time in the light of new thinking (Bonnell 1983).

Radiobiologists generally distinguish between those harmful radiation effects in which the severity of the effect varies with the dose, as for skin damage, and for which there may therefore be a dose threshold (non-stochastic effects) and those for which the probability of the effect – the chance of suffering an effect – but not the severity, depends on the dose and increases with increasing dose (stochastic effects).

The objective of radiation protection is to prevent detrimental non-stochastic effects and to limit the probability of stochastic effects *to levels deemed to be acceptable to society.* For this purpose the ICRP since 1977 has suggested three principles to help define acceptability (ICRP Publication 10). These are as follows:

(a) no practice shall be adopted unless its introduction produces a positive net benefit ('justification of practice');
(b) all exposures should be kept as low as reasonably achievable, economic and social factors being taken into account (optimization of protection, often referred to in technical literature by the acronym ALARA);
(c) the radiation doses to individuals shall not exceed the dose equivalent limits recommended for the appropriate circumstances by the ICRP.

In setting standards, and in defining acceptability, the notion of a *de minimis* dose has come under discussion. This is the level of additional radiation dose arising for an individual exposed to an artificial man-made source below which the resulting increment in radiation risk is entirely negligible to the exposed individual. This may seem to be the notion of a threshold in a new guise, but it is an attempt by the nuclear industry to persuade regulatory authorities that in their decisions regarding risk and safety there is something that is beneath concern, even though there is an acknowledged detriment. The argument used for a cut-off is a legal or quasi-legal one, that '*de minimis non curat lex*' or 'the law does not concern itself with trivialities'. However, this may be a corruption of the more accurate term '*de minimis non curat practor*' or 'the prosecutor is not concerned with triviality' (Clarke 1985; Lindell 1985). Acceptance of such a principle would mean that in designing and building facilities the industry would not need to take safety measures against things beneath concern. The notion does not challenge linearity, merely the significance of linearity at very low and infrequent dose levels.

DOSE LIMITS

The ICRP has been firm in its caution when issuing guidance on dose limits for occupational and public exposure to man-made radiation sources other than medical exposures. These recommendations form the basis of a European Directive on Radiation Protection. The dose limit relates to the sum of the annual 'effective dose equivalent' from external irradiation and the 'committed effective dose equivalent' from intake of radioactive materials in the same year.[2] It is applied to the average member of a critical group.

Exposure limits for radiation workers in the vicinity of controllable radiation sources are set at a level where there is 'a low probability of radiation injury without undue restriction of the uses and benefits of ionizing radiations'. The current exposure limit for workers (occupational exposure) is set at 50 mSv/year (or 5 rem/year). This is based on an ICRP estimate of the annual risk of death of 10^{-2} per sievert, which produces a radiation risk comparable with the fatal accident risk in an 'averagely safe' industry. The gonads and red bone marrow are considered critical organs when the whole body is exposed uniformly.

The dose limits considered appropriate for members of the public are one-tenth of those for workers. This takes account of the fact that radiation workers are adults, working under medical supervision, wearing personal monitors or film badges and whose received doses are recorded in personal files, while members of the public include all age groups and are not monitored. The dose limit is intended to provide standards for the design and operation of radiation sources such that individuals in a critical group of the public are unlikely to receive more than a specific dose under given conditions. The currently recommended dose limit is 5 mSv/year (500 mrem/year), but actual dose equivalents to members of critical groups are commonly much lower than this.

The additional risk of death for a group of men and women uniformly exposed over a lifetime at an average level of 1 mSv/year is estimated to rise from zero in the first few years to about 1 in 100,000 after several decades. The average additional risk of death to the whole population is believed to be less than 1 in 1 million (ICRP Publication 26, 1977).

Because it is not possible to measure a committed dose equivalent directly, ICRP provide secondary limits expressed as annual limits of intake in becquerels per year. These can be further developed to give derived limits, e.g. for the concentration of radioactive materials in foodstuffs such as meat, fish or milk. These derived limits are then specific to a critical group of known habits and food consumption in a known environment, although generalized derived limits covering any reasonable behaviour patterns and environmental conditions can also be developed.

AN EXAMPLE OF THE APPLICATION OF DOSE LIMITS

These secondary and derived limits are appropriate as a basis for control procedures, for monitoring the environment around nuclear installations under normal operation and after an uncontrolled accidental release of radionuclides at an installation. In the UK, for example, the Ministry of Agriculture, Fisheries and Food, in conjunction with the Welsh Office, has undertaken routine monitoring of radioactivity in food and agricultural materials around each of the nuclear sites in England and Wales since 1 January 1986. When news came through of the Chernobyl accident in the USSR on 28 April 1986 the Ministry of Agriculture began the daily testing of milk from southern England. The sampling effort was initially devoted to milk because of its sensitivity to atmospherically deposited radionuclides. This was extended to a countrywide programme of monitoring food and agricultural produce in England and Wales once radioactivity had been detected. Initial efforts were directed at identifying the parts of the country

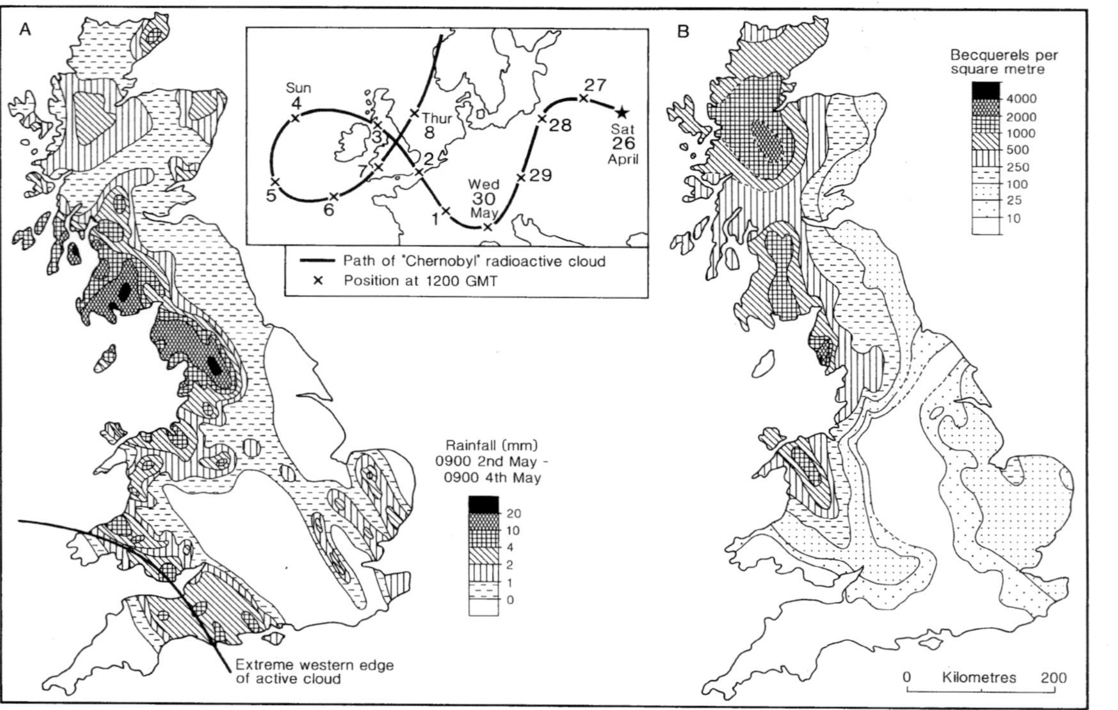

Figure 8.6 (a) The Chernobyl cloud and rainfall in the UK (09.00 h 2 May to 09.00 h 4 May 1986); (b) ground contamination by Cs-137 in the UK following the Chernobyl accident

Sources: (a) Smith 1986; Smith and Clark 1986. (b) Institute of Terrestrial Ecology (National Environment Research Council)

Note: The computer-generated map (b) is based on samples at 500 locations taken 10 days after the peak deposition on 3 May 1986. It was first published in the *Guardian* newspaper on 25 July 1986. The pre-Chernobyl background levels of Cs-137 on vegetation ranged from below 5 Bq/m^2 to around 20 Bq/m^2.

Table 8.5 Examples of sheep meat readings (Cs-137 plus Cs-134) taken in 1986 in selected regions of the UK (maximum readings within daily samples)

Locality	*Date*	*Reading (Bq/kg)*	*Date*	*Reading (Bq/kg)*	*Date*	*Reading (Bq/kg)*
Allerdale	6 Jul	1,024	14 Jul	2,238	26 Aug	1,432
Copeland	29 Jun	1,822	12 Jul	3,691	16 Oct	2,745
Eden (S.Lakeland)	1 Jul	1,268	22 Aug	2,553	15 Sep	1,165
Clwyd	1 Jul	1,631	22 Jul	3,246	24 Jul	413
Gwynedd	19 Jun	1,630	21 Aug	4,950	13 Aug	2,422

Source: MAFF/Welsh Office 1987

and types of produce most at risk (MAFF/Welsh Office 1987). Subsequently, when the pattern of deposition had become evident (Figure 8.6), efforts were concentrated in the regions where the highest levels of deposition had been found and on those feedstuffs most affected. Results were expressed in becquerels per kilogram fresh weight for meat or in becquerels per litre for milk.

The monitoring of sheep meat samples was late. It did not begin until mid-May, and by mid-June it became clear that in some parts of the country the caesium levels in sheep intended for market would exceed the action level of 1,000 Bq/kg by the time animals were due for slaughter (Table 8.5). The Government therefore introduced a ban on the movement and slaughter of sheep in extensive parts of Cumbria and North Wales. Extensive monitoring of sheep within these areas followed, and farms were released from restrictions only when total caesium activity in sheep fell below 1,000 Bq/kg. By June 1988 more than 100,000 sheep from over forty farms were still subject to government restrictions, together with 300,000 more in Cumbria, Scotland and Northern Ireland. Moreover, independent research by Dr Kenton Morgan, based on fieldwork in 1986–7, identified high contamination levels in sheep from around Cullompton, Devon, and Creech St Michael and Taunton in Somerset. The implication of his findings is that the monitoring by the Ministry of Agriculture may not have thoroughly identified all the contaminated areas (*Guardian*, 5 June 1988). The UK has now set up the Rimnet System to monitor any future incidents (Figure 8.7)

The Chernobyl explosion injected 1.85×10^{18} Bq of radiation into the atmosphere over an 11 day period from 26 April. Most of this was emitted early on, and most stayed in the lower atmosphere. The radiation that affected the UK came from a 'finger' of the initial cloud that first moved south to Italy and then back to the northwest over France to arrive in Kent on 2 May. Little or no rain had affected the cloud in the first seven days. On 2 and 3 May a depression ran across Britain from the southwest, giving thundery outbursts of rain over most of the country. Ahead of the cold

Figure 8.7 The UK Rimnet system. Following Chernobyl, the radioactive incident monitoring network (Rimnet) in the UK was announced by the Department of the Environment in 1988. Its purpose is to detect overseas incidents as a secondary system to that provided by an international system of data exchange also set up after Chernobyl. The Rimnet sites are mainly ones that were already in existence for other official or quasi-official purposes, e.g. meteorological stations at RAF bases and UKAEA establishments, and were capable of being provided with the equipment needed to measure dose rates. Some will also measure rain. Rimnet information on an incident will be fed into a central data base for analysis

front was a low-level pre-frontal jet stream that rushed the polluted air up eastern England, across the Pennines and so to Cumbria where heavy rainfall, orographically enhanced, washed out the radiation (Smith 1986, 1989; Smith and Clark 1986). This was the first such wash-out to occur. It is the coincidence of the trajectory crossing Cumbria *and* the precipitation that accounts for the peak of Cs-137 deposition in Cumbria shown in Figure 8.6(b) and that peak has nothing at all to do with the presence nearby of Sellafield. It could just as easily have happened over Italy, earlier in the week.

It is clear from Figure 8.6(b) that the highest deposition of Cs-137 occurred in the upland areas of the north and west. There is a sharp gradient in England and Wales between the low deposition in the Midlands and south of England and that in the north and west, marking the boundary between those parts of the country subject to heavy rainfall while the cloud of radioactivity was passing over.

THE ACCEPTABILITY OF RISKS

In suggesting dose limits for radiation workers and the public alike, whilst simultaneously accepting the possibility that dose threshold limits for stochastic effects of radiation may not exist, the ICRP accepted a need to balance costs and benefits. If any exposure to radiation has to be considered potentially detrimental, it follows that the possible risks must be outweighed by the benefits to society.[3] It has been argued that the balancing of gain and hazard is inherent in the whole philosophy of radiological protection, in diagnostic medical exposures and in therapeutic radiology. However, there is a significant difference between radiobiological exposures to individuals in medical science and exposure to ionizing radiation as a result of the nuclear fuel cycle. In the former case choice is voluntary – individuals can make a choice for themselves whether or not to risk the dose – whilst in the latter instance the choice is involuntary and made for them, and for subsequent generations, by individuals and agencies who may be accountable but who are largely outside their control. The ICRP is well aware of this, and recommends that the whole population dose should be kept to the minimum amount consistent with necessity. It has suggested dose limits at levels where it is considered that the risk would be acceptable to the individual and to society, but since 1973 the ICRP has strongly emphasized the importance of the ALARA principle.

The reduction of doses from man-made radiation sources to a level as low as reasonably available can be carried out by using differential cost – benefit analysis. Figure 8.8 shows how this works. The cost of the radiation health detriment saved by reducing the doses to individuals is compared with the cost of increasing the level of protection, e.g. by some form of engineered safeguard. The optimum level of production is achieved when the next level spent on protection exceeds the value of health detriment

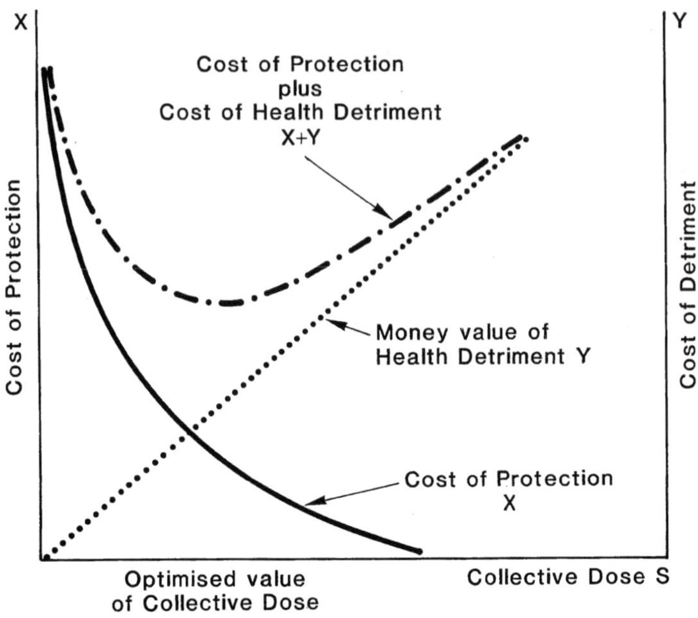

Figure 8.8 The optimization of radiation protection
Source: OECD/NEA 1984: 8

thereby averted (Clarke 1985). Some authors have suggested that to extend the ALARA principle into quantitative ideas of optimization of protection requires a more flexible and comprehensive mechanism than cost–benefit analysis. The argument is that such an aid to decision-making must be capable of handling trade-offs between items such as occupational exposures, probability and consequence of accidents, routine public exposure and cost (Webb and Dunster 1985).

The question of the level of risk that is acceptable or not acceptable is central to the issue of public attitudes to nuclear power. This is a subject which has generated a large literature, and only a brief summary can be attempted here.

At the extremes, two polar positions can be identified. Radical opponents of nuclear power have taken the ICRP assumption of risk of injury at low doses to justify the stance that, if there is no safe level of radiation, there can be no permissible additional levels of man-made radiation. This viewpoint has been expressed by some of those vehemently opposed to nuclear power, such as J.W. Gofman, A.R. Tamplin and E.J. Sternglass. A fairly typical example is contained in a piece written by W.C. Hueper in the November 1971 issue of the journal *Health Physics*:

> Neither the epidemiological data relating to environmental cancer risks involving man nor available data on experimental animals exposed to

environmental carcinogens[4] under any conditions of exposure, permits valid calculations, estimates, or educated guesses at a 'safe' dose of proven harmlessness.... The general population consists of individuals of diverse genetic backgrounds ... the adoption of a safe dose represents an irresponsible gamble with the health and life of the people left without vote in such fatuous decisions.

(Hueper 1971)

The alternative view is that no human activity can be regarded as absolutely safe and that society can only reasonably ask a technology such as nuclear power to be safe *relative to other things which are regarded as satisfactory, and relative to the benefits which may accrue from accepting the risk.* Such a view is exemplified by C. Starr and M.A. Greenfield who, in the July–August 1973 number of *Nuclear Safety,* wrote:

A study of the public acceptance of mortality risk arising from involuntary exposure to sociotechnical systems such as motor-vehicle transportation indicates that our society has accepted a range of risk exposures as a normal aspect of our life. Figure [8.9] shows the relation between the per capita benefits of a system and the acceptable risk as expressed in deaths per exposure year (i.e. time of exposure in units of a year). The highest level of acceptable risks which may be regarded as a reference level is determined by the normal USA death rate from disease (about 1 death per year per 100 people). The lowest level for reference is set by the risk of death from natural events – lightning, flood, earthquakes, insect and snake bites, etc. (1 death per year per million people). In between these bounds, the public is apparently willing to accept involuntary exposure, i.e. risks imposed by societal systems and not easily modified by the individual, in relation to the benefit derived from the operation of such systems. The position of nuclear power plants is well within the acceptable limits range.

(Starr and Greenfield 1973)

Starr and Greenfield published this opinion thirteen years before the Chernobyl accident. In an analysis of some of the consequences of that accident Dr James Corbett of the UK's CEGB Berkeley Nuclear Laboratories has put the cost–benefit argument as follows:

We now believe that between 10,000 and 40,000 people in the USSR will die of cancers initiated by the Chernobyl release. Large as these numbers are, they are taken randomly from a population of 250 million who share the benefit of the Soviet nuclear power industry. The average radiation dose on which the figures are based is between 5 and 20 mSv, equivalent to a few years' natural background. Since such doses are commonly accepted as a certainty in return for better living conditions, the Soviet people as a whole do not seem to have paid a high price, in terms of risk, for their nuclear power.

(Corbett 1987)

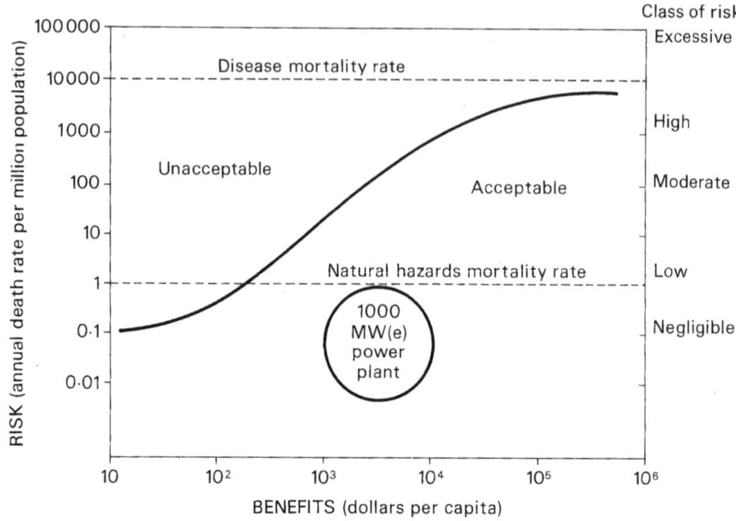

Figure 8.9 Risk–benefit pattern for involuntary exposure
Source: Starr and Greenfield (1973) 'Public health risks of thermal power plants', *Nuclear Safety* 14 (4): 273

A refinement to the estimation of risk is that of risk accountancy. Inhaber (1978) has argued that if numerical estimates of risk are to be attached to particular energy machines, such as power stations, all the risks associated with the production of a particular amount of energy should be added together. Under such a system the analysis starts with the risks involved in obtaining and processing the materials out of which the power station is made, and includes all the subsequent stages in their use up to the point where the energy is produced (Table 8.6).

In the UK, Lord Rothschild (Royal Society 1981) has suggested that the risk of an atomic explosion at a nuclear power station is zero, that the risk of death from the escape of radioactive substances within a 25 mile (40 km) radius of such a power station is less than 1 in 1 million per year, that the mortality risk from one-man sabotage at a nuclear power station ranges from 1 in 1,000 to 1 in 1 million per year, that the risk from a

Table 8.6 Estimated deaths for a specified energy output (10 GW years)

Coal	50–1,600
Oil	20–1,400
Wind	230–700
Solar (space heating)	90–100
Uranium	2.5–15
Natural gas	1–4

Source: Inhaber 1978

Bader–Meinhof type of suicide attack on a nuclear establishment is 1 in 100 per year and that the risk of death from intentional attack from the air is between 1 in 100,000 and 1 in 1 million per year.

Calculations of the type made by Inhaber, Corbett, Starr and Lord Rothschild are based on the principle of averaging individual risks over a large population, and on the equally well-established practice of measuring public risk in terms of fatality or mortality rates. Such rules of thumb simplify the calculations, but the release of radioactivity from an accident such as core meltdown in a PWR would affect many thousands of individuals in particular areas, including individuals not born at the time of the accident. If, for example, a core meltdown had occurred at the Three Mile Island nuclear power station in March–April 1979, the actual somatic and genetic damage would not have been averaged and equally shared throughout the population of the USA. It would have been geographically concentrated, first on the 600,000 people living within an approximate 20 mile (32 km) radius of the plant and second on that proportion of the 600,000 living downwind. Moreover, the effects would have been felt most by pregnant women and young children. In the Chernobyl incident, 135,000 people evacuated from the worst-affected zone had already received on average an estimated dose of 120 mSv from external irradiation, which is likely to have induced 160 potentially fatal cancers within this group. Doses to the thyroid may cause 3,000 non-fatal cancers of the thyroid among the evacuees, and Soviet documentation does not exclude the likelihood of hyperthyroidism in young children (Corbett 1987).

Risk calculations are also made on a 'risk per annum' basis, but the life of a nuclear power station may be twenty five years or more and there can be no realistic assumption that any power-producing machine will be in exactly the same state from one year to the next. Thus averaging out the risk per year may be as problematic as averaging it out through whole populations especially if, as chaos theory suggests, the development of an emergency at a reactor may be very sensitive to the condition of the reactor at the initiation of a sequence of events.

Calculations based on mortality rates often do not take account of the disabilities involved in the less publicized morbidity rates that may be incurred by both somatic injury and genetic damage transferable from one generation to the next. In terms of human, social and economic values, the importance to public welfare of an increased morbidity rate from nuclear power, if such were proved to occur, might be greater than real or calculated mortality rates. For example, the annual number of road accidents in Western countries is often quoted with alarm, but the disabling injuries, hundreds of times as many, which may have an equal or greater social importance are less widely publicized. Moreover, the calculations of risk depend on an implicit assumption that society in the future will always have the technical knowledge and economic resources to maintain the safety precautions currently regarded as desirable.

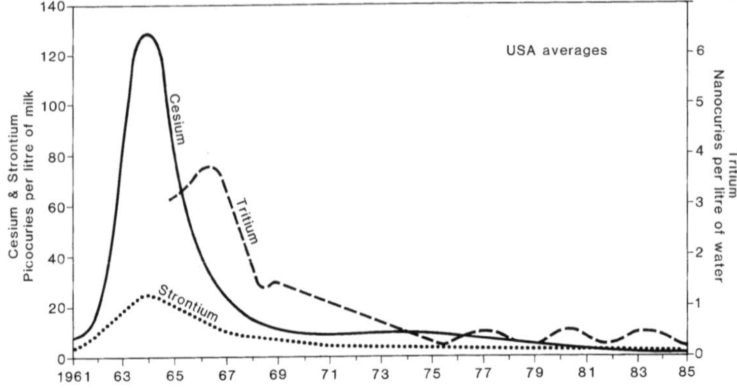

Figure 8.10 Radioactive fallout from weapons tests (US averages) 1961–85
Source: *The Economist*, 29 March 1986

ENVIRONMENTAL MONITORING AROUND NUCLEAR PLANTS

A few years ago it was suggested that monitoring of the environment around nuclear power stations in the UK had degenerated into a ritual performed according to ancient traditions, using inaccurate detectors and oversimplified uncritical interpretation of the results. However, knowledge of significant pathways from a radioactivity source to man, and the instruments and methods used for monitoring the surrounding environment of nuclear plants and for analysing the results, have all improved markedly since the early days of the industry (Clarke and Thompson 1978).

In the immediate vicinity of a nuclear power station it is normally difficult to detect any radiation attributable to the station itself, even with modern instruments. The natural background rate fluctuates for many reasons. Changes in barometric pressure affect the amount of the radioactive gas radon emerging from below ground, cosmic radiation levels vary unpredictably, heavy rainfall may wash down natural radioactive decay products from the atmosphere and clouds of radioactive material are still passing round the northern hemisphere as the result of weapons trials conducted in the atmosphere by some nations (Figure 8.10).

Early instruments were simply inadequate for the task of distinguishing any additional radiation due to nuclear installations from this background 'noise', and this soon became apparent as regulatory agencies imposed increasingly stringent monitoring standards. In the USA, for example, until 1971 monitoring programmes at nuclear power stations were concerned primarily with ensuring that integrated exposure levels outside the plant did not exceed the then required Federal Standard of 500 mrad/year. For that level of monitoring, thermoluminescent or film dosimeters were usually used. However, in 1971 the 'lowest practicable' levels of external gamma radiation from nuclear effluents at the boundary of LWR sites were made more demanding, at 10 mrad/year. This increment of radioactivity imposed on a variable natural background of the order of 100 mrads proved almost impossible to measure with passive devices, and so high pressure ionization chambers were introduced. These provided a means of analysing large numbers of samples for iodine-131 (I-131) and other gamma-emitting radionuclides.

Approximately 30 per cent of the iodine taken into the body is taken up by the thyroid gland, and the physical characteristics of I-131 make it the most important single iodine isotope from the viewpoint of general population exposure from the ingestion of foodstuffs. In the early days of the nuclear power industry there was only a limited capability to process large numbers of environmental samples. This problem was solved by the development of gamma ray spectroscopy using multichannel analysers. Combined with sensitive digital dosimeters, recording on magnetic tape left at permanent monitoring sites located around a nuclear power station site, these have made sensitive and accurate environmental monitoring possible.

The public health significance of radio-iodine and of the route to man through the forage–cows' milk food-chain were not recognized during the early days of nuclear weapons testing, but had been identified in time to be used for counter-measures in the first major nuclear reactor accident to threaten public health (Nair *et al.* 1986). The first use of protective action for I-131 in milk occurred in October 1957 after the fire in the Windscale (Sellafield) reactor in the UK (see Chapter 10). Scientists knew that radio-iodine settling on grass and rapidly entering cows' milk would be a key route by which radioactivity from the fire could reach local residents. However, they had not set a limit for the level of radioactivity that should be allowed before the milk must be thrown away. The limit decided on, arbitrarily and after brief discussion but on the best judgement to hand, was 0.1 μg of I-131 per litre of milk (Pearce 1988). The milk collected from farms up to several miles from the plant was destroyed to prevent high thyroid doses. Such protective control measures can reduce the exposure to radiation that would occur from the future ingestion of foods contaminated with radioactive materials, but to ensure timely action it is necessary to have an effective environmental monitoring system, an ability to use the data it provides, procedures for prompt action to be taken if required and a relatively low surrounding population density to facilitate evacuation measures.

LEUKAEMIA 'HOT SPOTS' AROUND NUCLEAR INSTALLATIONS

Efforts to demonstrate unhealthy radiation effects from nuclear power installations began in earnest with highly publicized articles and books written by well-qualified US scientists in the late 1960s (Tamplin 1969; Gofman and Tamplin 1970; Sternglass 1973). The statistical basis of some of this work proved sufficiently uncertain for it to be discounted in conventional scientific circles (Yulish 1973; Gerende *et al.* 1974). More recently, however, the nuclear power industry has come under sustained attack in the UK in relation to alleged leukaemia 'hot spots' or clusters in the vicinity of nuclear installations. These have been identified around Sellafield (Cumbria), Hinkley Point (Somerset), Dounreay (Thurso) and near Aldermaston and the Burghfield weapons factory in Berkshire. A particularly protracted dispute has developed over the public health risks that may be associated with spent nuclear fuel reprocessing facilities at Sellafield.

The Sellafield site employs around 6,000 people and is an important part of UK nuclear industry. It was acquired in 1947 by the UK Atomic Energy Authority to produce weapons-grade plutonium. Civilian nuclear power functions were added later and are now run by BNFL. The site, on the Cumbrian coast, is divided into the Windscale Works and the Calder Works, by which names it was more commonly known prior to 1981. The principal activities are the reprocessing of irradiated fuel and the conditioning and storage of radioactive wastes. Low level liquid wastes from these operations are discharged under authorization into the eastern basin of the Irish Sea by pipeline (see Chapter 12, Figure 12.9).

New production facilities are enabling better control and higher standards, but for several years the discharges of alpha and beta emitters in liquid discharges from the BNFL site were significantly higher than those from similar plants in other countries (Black Report 1984). The plant has also been bedevilled by a history of accidental radwaste spillages and emissions and has become the focus of attention by anti-nuclear pressure groups such as Greenpeace and Friends of the Earth. The issue came to a head when a television programme produced by Yorkshire TV and broadcast nationally on 1 November 1983 alleged that there had been an excess of young people with leukaemia and other cancers in the Sellafield locality. The Black Committee, set up by the Government to investigate the statements made in the programme, concluded that the suggestion of a causal relationship between an increased level of radioactivity and an above-average experience of leukaemia in the vicinity of Sellafield, while possible, was not proven (Black Report 1984). Other workers have suggested that the relevant scientific disciplines have been unable conclusively to confirm or reject suspected causal links between observed cases of leukaemia and past discharges from Sellafield. They cannot confidently identify or discount other possible causes (MacGill 1987). The verdict therefore remains uncomfortably open. In June 1988 the Committee on Medical Aspects of Radiation in the Environment (COMARE), set up by the UK Government in 1985 as a result of a recommendation by Sir Douglas Black's Sellafield inquiry, reported on cases of childhood leukaemia round Dounreay registered between 1979 and 1984. Six cases were reported, where statistically one would have been expected, and the Committee indicated that the likelihood of that cluster happening by chance was very low. The Committee Chairman, interviewed for television on 8 June 1988, said '... there's something happening there that we don't understand and we need to get to the bottom of it'.

Research directed at 'getting to the bottom of it' has continued and on 17 February 1990 the *British Medical Journal* contained the results of work by six scientists revealing a marked correlation between high exposure to radiation of men *working in the Sellafield plant* and leukaemia or non-Hodgkin lymphoma in their offspring (Gardner *et al.* 1990a,b). This was the first discovery of an association between fathers' exposure to ionizing radiation and cancer in their children. Previous studies had shown that a single dose of diagnostic X-rays of between 5 and 50 mSv administered to mothers expecting twins raised the risk of leukaemia among those twins by 15–20 per cent compared with non-irradiated twins (Beir 1990). Among survivors of Hiroshima and Nagasaki the rate of mortality from leukaemia was raised only at high doses of 400 mSv and above and an increased incidence of leukaemia among the survivors' children conceived after the explosion of the atomic bombs has not been discovered.

The UK's Medical Research Council (MRC) Environmental Unit collected the hospital records, family histories and habits of fifty-two cases of leukaemia, twenty-two cases of non-Hodgkin lymphoma and twenty-three cases of Hodgkin's lymphoma of patients under 25 born in West Cumbria and diagnosed there between 1950 and 1958, together with similar data for 1,001 healthy control cases matched for sex and date of birth and taken from the same birth registers as the patients. The team, led by M.J. Gardner of the MRC's Environmental Epidemiology Unit at Southampton University, assessed the risks to children born within a radius of 5 km from Sellafield and at increasing distances from the plant. Five cases of leukaemia and two of non-Hodgkin's lymphoma occurred in the inner circle. The fathers of all five cases of leukaemia and one of the cases of lymphoma were employed at the Sellafield plant. There was a fall-off in the incidence of these cancers to about one-third or less on moving further away from Sellafield.

Table 8.7 shows that the fathers' exposure amounted to between 2.5 and 15 times natural background radiation (2.5 mSv/year). Exposure was spread over many years but statistical analysis showed the strongest correlation with fathers' exposure to more than 10 mSv during the six months before conception of the children who contracted leukaemia. This coincides with the period of two to three months that it takes human sperm to grow and mature.

Other work has shown no excess deaths from leukaemia among the Sellafield workers themselves (Smith and Douglas 1986). There remains a possibility that the cause of the children's leukaemia at Sellafield lies less with their father's irradiation from external sources at the nuclear plant

Table 8.7 Exposure to ionizing radiation of workers in the Sellafield plant (UK) whose children died of leukaemia or non-Hodgkin lymphoma

Case	*Total dose (mSv)*	*Total dose compared with natural background*	*Cause of death*
1	102 in 7 years	6×	Leukaemia
2	162 in 6 years	11×	Leukaemia
3	188 in 7 years	11×	Leukaemia
4	97 in 13 years	5×	Leukaemia
5	370 in 10 years	15×	Leukaemia
6	97 in 15 years	2.6×	Non-Hodgkin lymphoma

Note: Leukaemia is a malignant multiplication in white blood cells. Lymphoma is the name for various cancers primarily affecting lymph nodes. Hodgkin's lymphoma or Hodgkin's disease is characterized by enlargement of the lymph glands and of the lymphoid tissue of the liver and other organs, by anaemia and by the appearance of giant white blood cells. Non-Hodgkin lymphoma is a collective term for all other lymphomas.

than from internal contamination with ingested or inhaled radionuclides, including plutonium and tritium. The Gardner report states that 'data on internally incorporated radionuclides will be analysed when these become available', but its findings do support the hypothesis that exposure of fathers to ionizing radiation before conception is related to the development of leukaemia in their offspring.

After the publication of the Gardner report the four labour unions representing manual workers at Sellafield asked that exposure in any six month period be cut to half the level suggested in the report as likely to lead to genetic changes in workers' sperm. Early in 1990 a similar medical study was under way at Aldermaston, where the main problem may be the ingestion of radioactive dust rather than external radiation.

SUMMARY

Decision-makers in the nuclear power industry have accepted the need to form an estimate of risks to human health that may result from the activities of their industry. Numerous attempts have been made to express these estimates in a quantitative way so that the significance of the hazard can be seen and reviewed in terms of both its nature and its size. It may well be that modern science knows all that needs to be known for adequate protection against ionizing radiation (Taylor 1980). The safety record of the industry is good compared with industrial operations of comparable size, and operational risks of most parts of the nuclear fuel cycle are tolerably small. But the risk as publicly perceived is not necessarily a proper reflection of the actual risk. The result is puzzlement on one side, when rigorous standards seem insufficient for public acceptance, and apprehension on the other, when risk assessments are either distrusted or only partially understood.

Radiation hazards are not easy to understand and radiation can be detected and measured only with special equipment. Without such equipment it cannot be seen, heard or felt and is therefore outside normal human sensory experience. The effects of radiobiological damage can be chronic and delayed as well as immediate. It is not surprising, therefore, that there has been public concern about what is perceived as a subtle and insidious environmental hazard. Figure 8.11 is a diagram well known to geographers involved in hazards research. In the case of nuclear power a degree of biological repair to radiation damage sometimes does exist. Cultural absorption of the technology has been mixed and uncertain. The public intolerance threshold is quite low but varies from country to country, and the process of purposeful adjustment has likewise been patchy.

It is not possible, with any technology, to eliminate all risk of harm, but people should not be exposed to hazards without their knowledge or consent. Few people in the UK would support a move to ban motor vehicles, yet 6,000 people are killed in road accidents each year. Where benefits are substantial and widely shared through society, and the risk to any individual is comparatively small, it is generally accepted in practice. It is therefore necessary for opponents of nuclear power not simply to assert the riskiness of nuclear power but to demonstrate that the level of risk associated with it is unacceptable to an informed public opinion. The Black Report in 1984 indicated that fatal accidents averaged over broad areas of industry traditionally considered dangerous in the UK (quarries, mines, railways and the construction industry) lie between 1 and 3 in 10,000 annually, while in manufacturing industry as a whole the level approaches 3 in 100,000. The authors of the Report pointed out that in 1983 a working party of the Royal Society estimated that most people are willing to accept a risk of one chance in a million of dying in any one year from an environmental hazard, and would be reluctant to spend money or time on reducing such a risk further. Neither the nuclear fuel cycle as a whole nor any of its constituent parts have been demonstrated to exceed this level of average risk to populations. In the UK on the basis of the national incidence rate, nine leukaemias would be expected over ten years in a typical town of 10,000 people, and seven of these would lead to deaths. The number expected from average background radiation using ICRP risk factors would be 0.20, and the number expected from a nuclear power station giving an increment over the background level of 0.04 mSv/year to the local community would be 0.008 (Taylor 1987). The Sellafield case, however, underlines the need for continued vigilance in the exposure of radiation workers.

It remains extremely difficult for members of the public to balance the benefits of nuclear power against its risks and hence to determine through due political processes the acceptable level of the spread and development of the technology. Moreover, the public is not well informed: many people

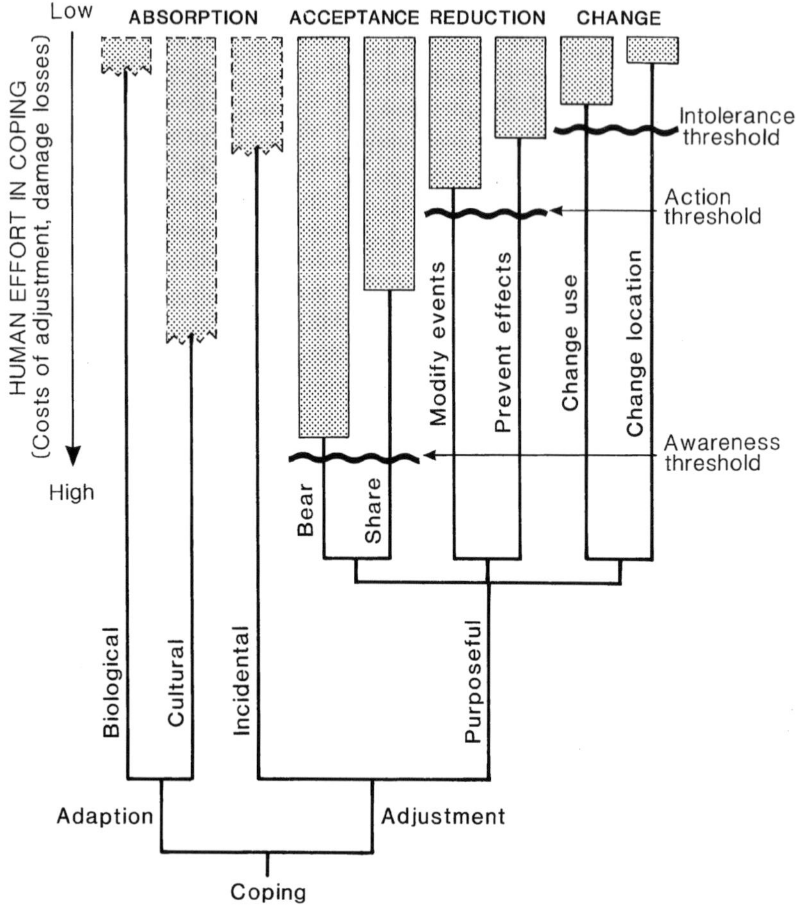

Figure 8.11 Modes of coping with hazard
Source: Burton *et al.* 1978: 205

quite erroneously think that accidents cause more deaths than disease (Slovic *et al.* 1980). For most people the genetic hazards of radiation damage constitute a fearful and emotionally charged issue, and thus it becomes relatively easy for extreme anti-nuclear lobbyists to portray nuclear power as a genetic mortgage likely to bankrupt future generations. Public unease is hardly assuaged by statistical exercises designed to balance the risk of nuclear power against other hazards, or by attempts to justify nuclear power quantitatively against energy policies involving other fuels. The techniques of risk assessment applied to nuclear installations have become increasingly sophisticated and expensive. Criteria have been published to guide acceptance by regulatory authorities. Risk estimates can be tested for consistency and with various arrangements of the basic model. However, no simple measure of risk is applicable in all situations; a hazard with serious consequences is only acceptable if the risk is very low indeed, and public perceptions are concentrated on probable consequences rather than calculations of the frequency of occurrence of an accidental release of radioactivity. People are less concerned about seemingly abstract notions of acceptable risk and more about the quality of the management responsible for their safety and about the processes by which safe conditions are agreed and then attained; it is a matter of trust (Roberts 1987). The public may not be wrong in this. The Three Mile Island and Chernobyl accidents both showed that regulations and good equipment cannot guarantee safety. Yet it is paradoxical that ionizing radiation, continuously part of the environment since terrestrial life began, 'is far more feared than the whole class of chemical pollutants to which humanity is exposed for the first time' (Fremlin and Wilson 1982).

NOTES

1. Beta rays (or particles) are high energy electrons and are therefore negatively charged. If they are stopped suddenly their kinetic energy is converted to heat and electromagnetic radiation. Gamma rays are a form of electromagnetic radiation; they can be emitted by radionuclides in association with the release of particulate radiation. Alpha particles (or rays) are helium atoms without their two outer electrons. They therefore carry a net positive charge.
2. In determining an estimate of the stochastic risks, derived from reviews by the UNSCEAR and the Biological Effects of Ionizing Radiations Committee (BEIR) of the US National Academy of Sciences, it was assumed that each organ or tissue contributed a certain fraction of the total risk (estimated to be 1.65×10^{-4} per rem, or 1.65×10^{-2} per sievert in the SI system of units) following irradiation of the whole body. The effective dose equivalent weighted for susceptibility to the harming of different tissues was used to estimate the contribution of the organ or tissue dose to the whole body equivalent dose. In November 1990 the ICRP reduced the average maximum exposure for nuclear power workers (occupational exposure) to 20mSv/year.
3. The word 'detriment' is used in the ICRP system as a measure of the harmful effects that could occur as a result of exposure to radiation. These are defined as health effects to the individual or descendants of the individual or as other effects, such as controls on contaminated property. The effects on health are called 'the detriment to health' and detriment is defined as the mathematical expectation of harm from a radiation source. In evaluating the health detriment both the probability and severity of harmful effects are taken into account.
4. Carcinogens are chemical, physical and parasitic agents of natural and man-made or synthetic origin which are capable of producing cancers in human and animal organs and tissues.

FURTHER READING

Conrad, J. (ed.) (1980) *Society, Technology and Risk Assessment*, London: Academic Press.

Horrill, A.D. and Lindley, D.K. (1990), 'Monitoring method based on land classification for assessing the distribution of environmental contamination', in *Environmental Contamination following a Major Nuclear Accident*, Vienna: IAEA.

Pentreath, R.J. (1980) *Nuclear Power, Man and the Environment*, London: Taylor and Francis.

Pochin, Sir Edward (1983) *Nuclear Radiation: Risks and Benefits*, Oxford: Clarendon Press, Monographs on Science, Technology and Society No. 2.

Roberts, L.E.J. (1987) 'The risk factor: acceptance and acceptability'. *Nuclear Energy* 26 (6): 349–59.

Smith, F.B. and Clark, M.J. (1989), *The Transport and Deposition of Airborne Debris from the Chernobyl Nuclear Power Plant Accident with Special Emphasis on the Consequences to the United Kingdom*, Meteorological Office Scientific Paper No. 42, London: HMSO.

9 Money costs of nuclear power and their locational effects

Given that the arguments about hazards and risks of the sort outlined in Chapter 8 are difficult to resolve in public policy, because they are so emotively charged, it becomes tempting to suggest that the money costs of nuclear power may provide a better theatre for rational argument. They could, perhaps, provide quantifiable evidence for or against building more nuclear power stations. Yet costs and safety are connected. Public concern over the safety of nuclear power plants in many countries has resulted in higher levels of expenditure on engineered safeguards than might otherwise have been incurred.

By the mid-1980s, in a number of the world's energy markets, nuclear generation costs were lower in real resource terms than for fossil fuels, a situation helped by the escalation in the real cost of fossil fuels up to the mid-1980s when prices fell again and reduced nuclear power's advantage. The USA and the UK, through difficulties in completing projects to time and cost, were the main countries where cost benefits for nuclear power were not obvious. Historic costs, therefore, are not an adequate guide for future investment, and the decision whether to build any kind of power station on a utility's system should depend upon estimates and predictions of *future* circumstances. The degree of accuracy that may be expected from such forecasting depends upon the skill and luck of the forecaster and upon how far into the future predictions have to be made. Power stations are expected to have long working lives; usually they are designed for an operational life of thirty-five years but may be used in different ways over this period. Amortization periods vary internationally between thirteen and thirty-five years and discount rates range from 4 to 9 per cent (Table 9.1).

The construction period for a large station may be six to ten years, and the forecasting period may add up to half a century. By 1982 the overall lead time required to design, license and build a nuclear power station in the USA had stretched to such a point that it was longer than the time period within which a utility was able to plan reliably (Huggett 1982).

Projections into the future involve assumptions about construction costs, inflation and interest rates, plant availability, load factor, operating life and

Table 9.1 Economic parameters used in national calculations of costs

Country	*Construction time (years)*		*Plant lifetime (years) (Amortization/Technical)*		*Discount rate (%)*
	Nuclear	*Coal*	*Nuclear*	*Coal*	
Belgium	8	5	20/35	20/35	8.6
Canada					
Central	6	7	40/40	25/35	4–5
East	7	3	30/40	30/40	4.76
West	–	6	–	30/35	6
Finland	6	4	25/30–50	25/30–50	5
France	6	4	25/30–40	25/30	8
FRG	6	4	20/–	20/–	4.5
Italy	8	6	13/25–30	13/25–30	5
Japan	8	6	16/–	15/–	5
Netherlands	7	6	20/30	15/25	4
Spain	9	5	25/30	25/30	5
Sweden	–	5	–	25/25	5
UK	6.5	5	40/40	45/45	8 (1989)
USA	7	5	30/40	30/40	5

Source: OECD/NEA/IEA 1989: 66

operating requirements of the system, fuel cost escalation, changes in safety legislation and insurance arrangements, income from plutonium sales and plant decommissioning and clean-up costs. Additional considerations include taxes on fuel oil, inter-industry agreements regarding the level of coal-burn in power stations and interest and exchange rates.

The history of nuclear power development is punctuated with cost comparisons, many of which have been rendered entirely redundant by unexpected increases in the costs of labour, materials and capital, quite apart from unforeseen delays in construction. Thus, while it cannot be denied that economic assessments provide a useful guideline in the decision-making process, experience indicates that such assessments cannot be sufficiently accurate to justify their adoption as the sole criterion of choice.

FACTORS INVOLVED IN ESTIMATING ELECTRICITY GENERATING COSTS

The money costs incurred in producing electricity are normally considered up to the point at which the electricity enters the transmission system. In the USA this point is called the power station *busbar*. It is the point beyond the high voltage terminal of the generator in the power station complex but prior to the switchyard and distribution facilities into the transmission grid.

There are three main ways of calculating and presenting generating costs:

(a) as an annual cost per kilowatt hour, relating to a share of capital with operating and fuel costs appropriate to a particular year spread over the units sent out in that year;
(b) as a levelized cost per kilowatt hour, which allocates lifetime production costs over lifetime output of electricity;
(c) as system costs which examine the costs of operating a whole electricity network against alternative assumptions involving only partial costs, e.g. generating costs only.

Each method yields different numerical answers, but it is necessary first to consider the cost factors which appear in all three methods (Figure 9.1).

Electricity generating costs are normally expressed in money terms per unit of net output, e.g. in mills (tenths of a cent) per kilowatt hour in the USA and in pence per kilowatt hour in the UK. Certain major items determine this single parameter.

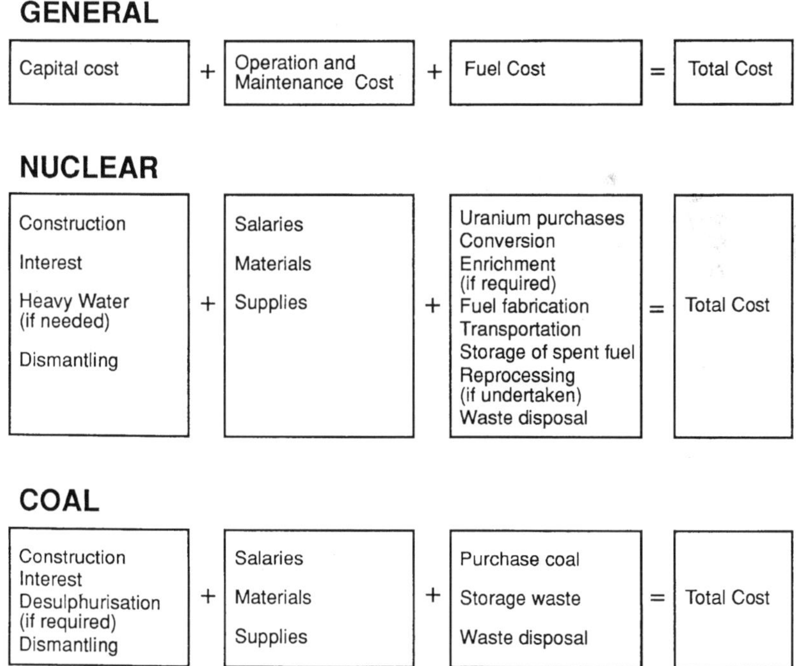

Figure 9.1 Basic components of generation costs
Source: OECD/NEA 1983a: 12

(i) The first major item is the *capital investment cost* of the power station, normally expressed as a unit cost per kilowatt of installed design capacity. The conversion of a capital cost per kilowatt figure into the fixed-cost component of generating cost per kilowatt hour requires explicit introduction of two additional elements.

(a) The *load factor* (UK) or *plant factor* (USA) can be defined as the ratio between the energy that a power station has produced during a period of operation and the energy that it could have produced at maximum capacity under continuous operation during the whole of the period. The load factor represents the degree of utilization of an electric power plant and, because it is desirable to spread fixed costs over the largest possible output, the higher the load factor, the lower becomes the total generating cost. The average annual and lifetime load factors for countries with four or more operating reactors of 150 MWe gross or over in 1987 are shown in Table 9.2. The term load factor or plant factor should not be confused with *operating factor*, a term used in the USA to indicate the ratio between the number of hours a power plant is in use (on-line) and the total number of hours occurring in a reference period. The *unavailability factor* is another US term used to indicate the amount by which the available capacity is lower than the maximum capacity. The two reasons for plant unavailability are (i) breakdowns (unplanned outages) and (ii) maintenance and refuelling, for some plants which cannot be refuelled on load (planned outages).

Table 9.2 Nuclear power station load factors during 1987

Country	*Annual average load factor (% for year to end June 1987)*	*Average cumulative load factor (% for year to end June 1987)*	*No. of reactors*	*Total nuclear capacity (MWe gross)*
Finland	87.1	79.3	4	2,400
Switzerland	82.1	79.7	5	3,079
Sweden	79.1	66.9	12	10,115
South Korea	76.5	69.2	6	4,819
Belgium	75.4	78.0	7	5,718
Japan	75.1	65.4	34	25,846
Spain	74.0	64.8	7	5,655
Taiwan	72.8	61.9	6	5,144
Canada	71.4	78.7	17	11,896
FRG	71.1	66.3	18	18,475
France	66.4	62.8	46	43,663
USA	58.8	56.6	96	89,259
UK	54.3	48.3	29	12,139
India	45.5	44.3	6	1,330

Source: Nuclear Engineering International, *World Nuclear Industry Handbook* 1988:19

A concept similar to the load factor is the *system load factor*, defined as the ratio of average load to peak load. This is determined by the demand characteristics of a utility's customers. Generally speaking, the greater the proportion of industrial customers, the higher is the system load factor, because such users tend to have a more constant round-the-clock demand than residential premises, shops and offices. Thus the base load is increased by a high proportion of industrial users. This fact makes the use of population size as an estimate of load in particular markets very much an approximation. It must also be recognized that there is a significant difference between the costs of generating the electricity and the price of electricity to the consumer. The latter includes charges for transmission and distribution, as well as charges to cover overheads and profit margins. Prices to different categories of customer can be set to take account of their pattern of electricity use, and prices to a specific consumer can be set to encourage or discourage use at particular times of the day or year.

(b) The rate of fixed charges on capital cost is the second major element that must be introduced into cost calculations in order to translate capital costs into a component of generating costs. This also is a compound item, including provision for repayment of the capital required for the plant over its lifetime, interest payments, plant insurance and the like. If the utility is owned by shareholders, a common situation in the USA, the figure has to include reasonable earnings (profits) on the equity capital. If the power station is part of a nationalized electricity production and distribution system, it is possible that central government may regard the rate at which future gains from an investment are discounted to be unduly high from the standpoint of the community as a whole. Government may then apply a test rate of discount such as that introduced in the UK in 1967 for public sector investments. The rate was, and is, defined in real terms, i.e. without regard to prospective inflation, and therefore it differs from the actual market rates at which the organization concerned can borrow money. When it was first introduced the test rate was fixed at 8 per cent per annum, but in 1969 it was raised to 10 per cent, where it remained until early 1978. In a White Paper on the nationalized industries in the UK which was published then, a uniform rate for the public sector was no longer prescribed, but the minimum rate to be used became 5 per cent per annum. The rate rose to 8 per cent per annum in April 1989.

(ii) Fuel and other operating costs such as fuel transport and storage are the complement to fixed costs in calculating total generating costs. Many nuclear industry cost assessments have argued that nuclear fuel costs have remained nearly constant in real terms over considerable periods compared with rising costs of fossil fuels. In the UK, Magnox fuels with 5,000 nwd/tonne burnup are more expensive than AGR (24,000 mwd/tonne) so that fuel costs per kilowatt hour have dropped. Uranium prices and enrichment costs have fallen considerably during the 1980s. There have been some big rises in back-end costs but these are a minor component of PWR fuel costs.

CAPITAL COSTS

The generating cost per kilowatt hour of base load coal-fired or oil-fired stations is shared between capital and running costs in the ratio of 1:2, whereas the cost of a nuclear power station is broken down in an almost exactly inverse ratio at 2:1; for hydroelectric stations the ratio may be as high as 4:1 or 5:1. Hence coal-fired electricity generating costs are very sensitive to delivered coal prices and transport costs, and the capital-intensive nature of nuclear power stations makes them particularly sensitive to inflation during construction. Consequently the increasingly long lead times needed for nuclear power plant construction shown in Table 9.3 assume particular significance, especially as they have far exceeded increases shown by other kinds of electrical power plants.

For very similar reasons nuclear power station costs are very sensitive to the load factor at which plants operate, as Figure 9.2 shows.

Table 9.3 Trends in average power plant lead times in International Energy Agency (IEA) areas: years taken from ordering to commissioning

	1965–9		*1970–3*		*1974–9*		*1980–4*	
IEA North America								
Nuclear	5.0	(7)	5.6	(27)	8.7	(41)	12.5	(29)
Hydroelectric	3.4	(13)	5.0	(20)	5.5	(38)	6.0	(12)
Solid fuels	4.0	(106)	4.7	(99)	5.6	(129)	6.8	(104)
Oil/gas	3.4	(69)	4.2	(55)	5.5	(68)	8.4	(5)
IEA Europe								
Nuclear	6.0	(27)	6.4	(15)	6.5	(19)	11.1	(27)
Hydroelectric	3.8	(85)	4.0	(70)	5.0	(69)	6.1	(30)
Solid fuels	4.2	(55)	4.6	(53)	5.8	(39)	5.4	(34)
Oil/gas	3.3	(55)	4.1	(71)	4.6	(65)	8.0	(32)
IEA Pacific								
Nuclear	–	–	5.0	(4)	6.3	(17)	10.2	(6)
Hydroelectric	4.2	(43)	4.3	(20)	5.0	(14)	5.8	(23)
Solid fuels	4.0	(27)	5.6	(14)	6.3	(11)	6.2	(19)
Oil/gas	3.7	(40)	3.4	(67)	5.3	(43)	5.6	(14)

Source: Power Plant Register, Science Policy Research Unit, University of Sussex, UK, quoted in OECD/NEA 1985:77

Notes: (a) Numbers of generating units completed in parentheses.
(b) Units below 30 MWe excluded.

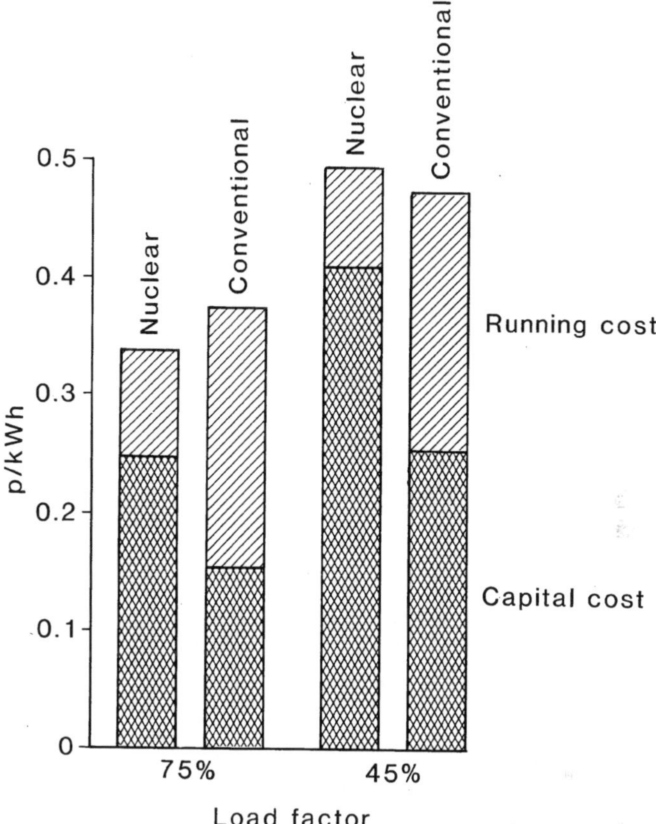

Figure 9.2 Breakdown of nuclear and conventional generating costs at different load factors
Source: Searby 1971

The status of a new power station in a utility's generating system and the way in which it affects existing stations is relevant to economic projections. Utilities possess plants of varying size and economic efficiency and marginal running costs are relevant to their merit order. The newest, largest and lowest fuel cost stations will be used to supply base load, the next most economic plants will be used to satisfy intermediate load, and the older stations and pumped storage hydroelectric plants will be used to meet peak loads and may often be producing electricity that is sold at a loss (Figure 9.2). Thus each power station has a merit position in relation to other power stations in a system, and the economic concern when a system builds a new power station is its position in the merit order and the position of stations that it displaces. As more nuclear stations are added to a system, existing coal- or oil-fired plants tend to be pushed into marginal roles to satisfy intermediate or peak loads and thus burn less fuel. In order to make an economic assessment on a common basis between two stations with different operating regimes, allowance has to be made for the fact that the deficiency of electricity generated from a low meit order station will have to be made up from stations elsewhere in the system.

Although nuclear power plants in general have lower running costs than coal-fired stations, they also have a lower load factor. This is partly because the shutdown periods of nuclear power plants may be longer and more frequent, not all are designed for on-load refuelling, partly because complexities in the design of their operating systems may result in not infrequent breakdowns. Whereas in the past system planners assumed that a nuclear power station might generate power for 80 per cent of the time, it is now thought prudent to assume only 65 per cent, compared with 70–75 per cent for conventional coal-fired plants (Fishlock 1982). A reduced load factor has a much greater effect on the generating costs of a capital-intensive nuclear station than on those of a conventional plant. In France, where load following is extensively used to give the nuclear units comparable flexibility with that of coal- and oil-fired plants, the 900 MWe units had an average availability of 78.6 per cent in 1987 and 83 per cent in 1986 compared with a design figure of 71 per cent (*Nuclear Engineering International*, 'Datafile France' 1988). In the UK, the availability of the AGRs in the late 1980s was a lamentable 36 per cent.

Given the nature of the variables involved and the complexity of reconciling them into practical forecasting models, it is not surprising that the record of estimating and predicting costs for nuclear power plants has been patchy, with the estimated and actual costs often differing by a factor of two or more. It is, perhaps, a little more surprising that calculations for the same nuclear power station can also differ widely. For example, the USAEC estimated the capital cost of Peach Bottom 2 at US$525/kWe whereas Commonwealth Edison came to a figure of US$331/kWe (King and Yang 1981). Such differences stem from the use of different assumptions, criteria, viewpoints and costing procedures. Figure 9.3 shows that in the USA the capital cost for nuclear plants from the mid-1960s to the late 1970s was about three times greater than the predicted cost by the seventh year interval. It would therefore seem wise not to attach too much weight to prognostications that cover more than a decade from the forecasting date.

RUNNING COSTS

For fossil-fuelled plants fuel supplies constitute the largest single item under the heading of works costs or running costs. The proportion of fuel costs to generating costs in most utilities is generally two-thirds at a conventional power station and one-fifth at a nuclear plant (Rawstron

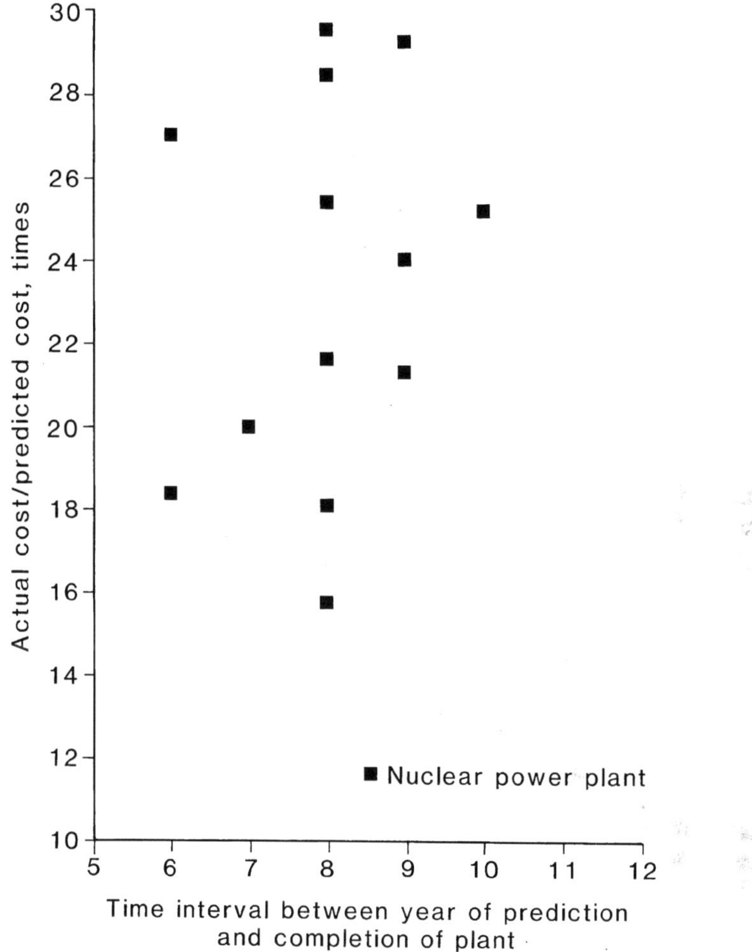

Figure 9.3 Deviation of cost predictions for nuclear power plants in the USA
Source: King and Yang 1981

1951 and 1955). A 1,000 MWe coal-fired station operating at 75 per cent load factor to provide base load current could consume up to 5 million tonnes of coal a year, depending on the fuel's calorific value. The refuelling of a reactor at a nuclear plant of comparable size involves no more than one or two lorry or railway wagon loads of fuel elements each year. Thus, in relation to fuel supplies, a nuclear power station is potentially much more locationally mobile than its coal-fired equivalent with its demanding requirements for coal transport. The competitive position of the coal-fired station can be significantly influenced by the arrangements that are made for transporting coal. In the UK, for example, 75 per cent of the coal reaching power stations is normally carried by rail, with road transport taking 10–12 per cent, coastal vessels 10 per cent and conveyor belts and canal barges each carrying around 2 per cent. Under a fifteen year agreement that began in 1976, the CEGB undertook to use the railways to carry coal from all its pits with rail connections. Before the year-long National Union of Mineworkers strike in 1984–5. British Rail moved 1.3 million tonnes of coal a week to power stations. However, the strike and sympathetic industrial action by rail-workers reduced this to 175,000 tonnes a week, carried in an average of thirty-five trains a day. During the strike coal proved to be 20–25 per cent cheaper to move by road in situations where the road links between the pits and the power stations are markedly shorter than the rail connections. But in the UK the average rail journey for fuel is much shorter than in the USA, the USSR or China.

As a general rule, fuel prices are quoted to include the costs of delivery to power stations. In other words, they include the cost of transport by rail, water or road, but not the cost of unloading, stocking or handling at the power station itself. Even when the prices of coal or lignite are lower than those of oil or nuclear fuel, their use requires more equipment for handling the fuel and the ash they produce. Indeed, the use of fuel with a high ash content raises the additional problem of ash disposal, and the solution of this problem may give rise to expenditure offsetting the advantages of low coal prices.

Although nuclear fuel is a much smaller proportion of total generating costs than the fuel for any fossil-fuelled station, large investments have to be made by such capital-intensive services as uranium enrichment, reprocessing of spent fuel and disposal of radioactive wastes. There are two broad alternatives available to electrical utilities for financing nuclear fuel supplies: ownership or some form of leasing. A major advantage of ownership is that it allows flexibility in each step of the nuclear fuel cycle and enables the utility to seek the best buy for each step, as well as allowing it to follow the most advantageous method of disposing of residual values in spent fuel. Leasing fuel reduces the immediate requirement for working capital, but the lease arrangements available to utilities for financing nuclear fuel vary widely as to the services provided and terms of payment, especially in the USA. One example may serve as an illustration. In 1973 the Commonwealth Edison Company entered into a sale and lease-back arrangement under which the initial fuel load at its Dresden No. 3 reactor was sold for its tax basis and immediately leased back. The lease payments were made on the basis of the rate of burnup of the fuel in the reactor. The spent fuel assemblies were purchased by the company for their salvage value (Bachert 1973).

The question of the value of spent fuel is not easily answered. A utility may dispose of the plutonium in spent fuel in three ways. It may be used to help fuel existing power stations, it may be stored to use more efficiently in future plants, or it may be sold. For a time a ban was imposed on commercial reprocessing of fuel in the USA but was lifted in 1981. However, the US still has no civil plants for spent fuel reprocessing and no utilities are contemplating reprocessing at present.

Insurance costs for nuclear reactors are often held down artificially by a legislated ceiling on liability. In the USA this ceiling is about 4 per cent of the government's most recent damage estimates of the worst accident scenario, and critics say that the government's damage estimates are unreasonably low and that if nuclear power, like other technologies, were forced to pay insurance premiums in line with its risks, the costs per kilowatt hour sent out would be substantially higher (Green 1974).

In the early days of the nuclear power industry there was considerable reluctance on the part of insurance firms to expose their funds to the very heavy losses that could possibly arise from the operation of a nuclear reactor and from the products of nuclear fission. The concept of pooling was adopted at a very early stage. Under this collective arrangement, instead of insurers dealing with each risk as it comes along, the amount that they are prepared to commit annually for any one nuclear installation is determined by them in advance. This is known as their retention or 'line' and it is fixed on a strictly net basis. Thus they undertake not to obtain any outward insurance with respect to their share but, equally, they are protected against having to add to their commitment by receiving inward reinsurance relating to the same business. The total of the contributions made by the participating insurers at the beginning of each year constitutes, in aggregate, the capacity of the national pool concerned. Such pools exist in the USA, Sweden, the UK, Belgium, Denmark, Finland, France, the FRG, Italy, Norway, Switzerland, Netherlands, Austria and Japan.

In the USA the 1954 Atomic Energy Act allowed for private ownership and operation of nuclear reactors but at that time the competitive position of nuclear power had not been established and utilities did not know when it would become profitable. The reactor suppliers and the operators were unwilling to take on the additional risk of a catastrophic accident which conceivably could bankrupt the companies concerned. To meet this need, the Price–Anderson Act was passed in 1957. It was designed to protect the public and USAEC licensees and contractors against risks associated with the use of nuclear power by limiting the amount of liability protection to US$560 millions. The Price–Anderson Extension Act, amended in 1975, began the phasing out of the federal government's indemnification of commercial reactors, although non-profit-making research and development reactors remain covered to a US$560 million liability limit.

There has been much discussion about whether and to what extent Price–Anderson indemnification has been a subsidy for nuclear power in the USA. It removed a stumbling block to the development of the nuclear power industry but it seems likely that without Price–Anderson the utilities would have had to purchase extra liability insurance and the price of nuclear power would have been higher as a result.

ALTERNATIVE COSTING PROCEDURES

The use of different accounting procedures to make economic comparisons between power stations, existing or proposed, raises important problems in economic–geographical analysis, just as it does for power system planners themselves. Calculations on historic cost (HC), discounted cash flow (DCF) and revenue requirements (RR) bases are used in the industry, but not infrequently the *results* of one set of calculations may be quoted and compared with a result for another project to which quite different accountancy procedures have been applied, without the methods used being specified at all clearly. In short, it is always possible by using widely different costing procedures to come to widely different results in estimating the cost of electricity to be produced by any power station, be it nuclear, coal- or oil-fired or hydraulic. As long ago as 1963 the United Nations Organization warned:

> No valid conclusions can be drawn from a superficial comparison of the cost of electric power produced by conventional thermal power stations. For a proper study it is essential to compare the various cost components and to make sure that they are homogeneous, i.e. that they have been worked out according to the same rules and for the same utilization factor of the installed capacity.
>
> (UNO 1963)

Yet, nearly a quarter of a century later, P.M.S. Jones was moved to complain that:

> It is ... pointless to try to take cost data from individual countries' experience and simplistically to transfer it to others using nominal exchange rates, or to attempt to produce generic or typical costs ... many comparisons or assessments appearing in the literature, some from highly reputable agencies, do use converted data without recognising the potential magnitude of the errors they could incur.
>
> (P.M.S. Jones 1987)

By not ensuring strict comparability, pro-nuclear and anti-nuclear camps are both liable to quote with complete assurance apparently quite definitive but widely differing figures and make authoritative forecasts on generating cost trends, claiming accuracy to one-hundredth of a cent or penny per kilowatt hour to be sent out many years hence, apparently undeterred by the level of uncertainty that enters into the calculations. It is perfectly possible, moreover, for an ideological or career commitment to a particular power generating form to be held by individuals and groups who work for it, have invested in it or have fallen in love with it. It is not impossible for this to be translated into a preference for calculations and forecasts favouring that way of producing electricity, especially if there has been previous heavy political and technological commitment by the power utility as employer, or government as stage-manager, to that form. It is, perhaps, less easy to accept that a widespread conspiracy exists amongst electrical generating utilities widely scattered throughout the world consistently to manipulate the results of generating cost comparisons between fuels in favour of nuclear power. But a trend of regression towards a mean – a commonly held view created by shared exogenous pressures – may be more credible. Many states in Western Europe, for example, turned to nuclear power after the OPEC oil price increases of the mid-1970s, but those who went along the nuclear path, such as France, and who were able to claim cost advantages in favour of nuclear power found that the collapse of oil prices in 1986 markedly altered the arithmetic. Nevertheless, the nature of the arithmetic – the way in which the calculations are made – has to be the starting point.

COMPARISON OF COSTING PROCEDURES

The DCF and RR methods are based on two fundamental principles: (i) money has a time value, i.e. under most circumstances a sum of money in the hand is worth more to the holder than the same sum in real terms in the future because the holder can use the money in the intervening period, and (ii) a venture is solvent when receipts and disbursements balance out (the balancing process is defined as the equality of the present worth of net cash inflows and net cash outflows). The difference between the methods is in the manner that they are used or in the quantity that they are supposed to compute.

Use of the DCF method begins with all known cash flows, such as in a completed project, and then involves searching for the discount rate that allows the inflow and outflow streams to be equivalent. Such a resultant rate is also called the Internal Rate of Return (IRR). A direct comparison of such a rate with the owner's experience will help determine whether the project is financially desirable. In general, the alternative that yields a higher discount rate is the more attractive of two options, other things being equal.

The RR method starts out at the opposite end. It assumes the owner's exact expectation of return on his capital and then proceeds to calculate the minimum he must obtain by selling the products. This figure must be sufficient to cover all the owner's operating costs, taxes, required return and the recovery of his capital. The resulting revenue can then be converted to the minimum, or 'bare bones', sale price of the product. A judgement can next be made to see whether the minimum price can survive the market-place. When two options are available, the alternative that leads to a lower 'bare bones' price of the product is the financially more attractive alternative, assuming that all other factors such as quality, time constraints and quantity of product are similar.

In recent years, the economic analysis of energy technologies has assumed an important role in both government and power utility circles.

Phung (1980) has pointed out that in the USA both the DCF and RR methods are frequently used, not only in their traditional roles as means to compare *similar* and/or compatible alternatives, but also in an *absolute* manner. Phung comments: 'In this new role, the methods would not be suitable unless the meaning and limitations of each sub-element of the methods is well understood and qualified. The lack of such understanding may lead to doubtful, sometimes even deceitful results.' Phung provides two tables to illustrate the problem. Table 9.4 shows the estimated investment, revenue and cost streams of two projects: project A lasts five years, while project B lasts six years. Which project is economically more attractive? Table 9.5 shows the possibilities for projects C and D. These projects differ in almost all respects: tax rates on profit, *ad valorem* charges, capitalization, useful life-cycles and operating costs. Which project is economically more attractive? How may projects A and B be compared with projects C and D? Phung points out that cost comparisons of this type are met daily by planners and that in the area of energy production, where there are many technologies, their inter-comparison can be 'very challenging'. He argues that the practising cost engineer will identify immediately that a good way to compare projects A and B is to calculate their respective internal rates of return. The project that has the higher rate (project B) would be more attractive. The DCF method has been used because all relevant cash flow streams are given and there is no pattern in each stream. As far as projects C and D are concerned. Phung argues that a good way to make a comparison is to calculate the minimum unit cost of

Table 9.4 An example of applying the discounted cash flow method in alternative project comparison

Year i	*Investment* I_i *(US\$)*	*Operating costs* O_i *(US\$)*	*Income tax* T_i *(US\$)*	*Ad valorem charges* II_i *(US\$)*	*Gross revenue* R_i *(US\$)*
Project A					
0	1,000,000	0	0	0	0
1	0	350,000	40,000	15,000	600,000
2	0	350,000	50,000	20,000	700,000
3	0	400,000	60,000	20,000	800,000
4	0	450,000	60,000	25,000	850,000
5	0	500,000	60,000	25,000	850,000
Project B					
0	1,500,000	0	0	0	0
1	0	500,000	60,000	40,000	900,000
2	0	550,000	80,000	40,000	1,000,000
3	0	550,000	80,000	45,000	1,110,000
4	0	600,000	90,000	45,000	1,200,000
5	0	650,000	90,000	50,000	1,300,000
6	0	750,000	80,000	50,000	1,300,000

Table 9.5 An example of applying the revenue requirement method in alternative project comparison

	Project C	*Project D*
Service life (years)	5	6
Annual production (MMBtu)	300,000	500,000
Beginning of life investment (US$)	1,000,000	1,500,000
Salvaged value	None	None
Depreciation[a]	SYD over 5 years	SL over 6 years
Capitalization		
Bond fraction	0.5	0.6
Bond rate (%)	10	10
Stock fraction	0.5	0.4
Stock rate (%)	18	20
Effective income tax rate	0.50	0.48
Ad valorem rate	0.02	0.03
Operating cost (US$/year) estimated at $i = 0$	300,000	500,000
Operating cost escalation rate (%/year)	12	10

Source: Phung 1980: 1054
Note: [a] SYD, sum-of-the-years digits schedule; SL, straight-line schedule

production. The project that leads to a lower cost of each unit product (project D) is more attractive. The RR method has been used in this case.

The discussion of costing procedures to this point makes it clear that there are real problems involved in comparing the competitive economic position of any particular means of electricity generation when the methods of economic evaluation vary internationally, through time and from utility to utility. The way is left wide open for 'creative accounting'. Summary comparisons between projects are frequently made, with average electricity generating costs being calculated by adding the annual expenditure on operations and fuelling to that in a sinking fund for the amortization of capital and then dividing the result by the annual amount of electricity generated. Economic comparisons made in such a simple and straightforward way may prove no less likely to be correct than more complex analysis, but it would seem an improvement to operate on a 'present worth' basis, under which the utility assumes that the worth today of a sum of money received in several years' time at a particular discount rate is the amount that it would have to find now, and invest at the same rate of interest in order to provide it with that original sum of money at a future chosen date, generally the date when the power station is ready to take its first commercial load. This has been the way of making detailed calculations of generating costs in the UK since the 1960s (Berry 1966). Regardless of the means of calculation, however, it must be remembered that forecasts are no more than approximate indications of what *might* happen in the future, whether that future is discounted or not. There are no certainties in energy forecasting.

HISTORIC COSTS CALCULATIONS

Over the whole period of the development of commercial nuclear power in the UK until 1983 the two major utilities, the CEGB and the South of Scotland Electricity Board habitually made use of comparative cost figures based on historic rather than current costs, a calculation which took little or no account of inflation and which therefore distorted the figures in favour of nuclear power because of its high capital costs. This practice was discontinued in 1983 after it was roundly condemned in the *First Report of the Parliamentary Select Committee on Energy* (House of Commons 1981). Its effect can be illustrated by considering the case of the Dungeness B AGR power plant on the coast of southeast England, which was ordered in 1965 for commissioning in 1971, but which did not come on line until 1985. On 3 July 1978 Duncan Burn, in an article in the *Guardian* newspaper, wrote that a CEGB spokesman had said that the then ten year time lag 'was not a delay, and the plant will be cheap, costing £344 millions, whereas a plant ordered now would cost £600 millions'. Burn pointed out that:

> Many people in the CEGB would know this piece of misinformation to be nonsense. The first figure is the sum of payments in the prices current when they were made, and the second figure includes financing costs of £180 millions. The Dungeness B figure in 1976 prices plus financing costs, enormous for a 15-year construction period, would be £900 millions or more, and with other large extra costs to add the cost in resources used will be probably twice the current price.

It is worth noting that many of the earlier nuclear power plants were built under turnkey contracts. In the UK the five construction consortia involved in building the Magnox stations often incurred substantial financial losses. The UKAEA carried out basic design work on the Magnox stations, but a great deal of engineering development was left to the engineering firms. Associated Electrical Industries, for example, suffered large losses in building the Berkeley power station on the Severn estuary (*Economist,* 25 April 1959: 362). Similarly, during the 1960s in the USA, the private electrical utilities feared wide federal involvement in nuclear energy and thus accepted heavy financial losses in order to claim nuclear power as a field for private companies. Likewise, the manufacturers of reactor equipment were moved to a degree of self-sacrifice. The US General Electric Company was reported to have contributed tens of millions of dollars towards the 180 MWe Dresden nuclear power station built by the Commonwealth Edison Company southwest of Chicago (*Economist,* 23 June 1963: 1,251). On 7 November 1966 *Barron's Weekly* reported that

Babcock and Wilcox and Combustion Engineering were losing money in the USA on overall atomic activities.

Economic justifications for building a power station, or for choosing one kind rather than another, depend upon accuracy in answering two vital questions in the difficult areas of energy forecasting. What are the future trends going to be in the *relative* cost of different fuels in particular energy markets? How is the demand for electricity going to develop? As far as the choice between nuclear and fossil-fuelled electricity generation is concerned, much rests upon estimates of the advance of nuclear technology and construction timetables on the one hand and on the future movement in the cost of conventional fuels on the other. The cone of uncertainty in the calculations underpinning answers to these questions widens rapidly with increasing time from the present into the future, and the high level of uncertainty increases the opportunity for a committed and skilful anti-nuclear lobby to develop and deploy its economic arguments. The South of Scotland Electricity Board claimed that there was a 'robust' economic case for commissioning Torness at least four years before it could be justified by demand trends. In a critical report published in 1987 the Scottish Consumer Campaign (SCC) argued that the Board appeared to make unjustified cost assumptions. The SCC pointed out that the Board assumed that coal costs would rise at 5–6 per cent per annum in real terms until 1986–7 and 2 per cent per annum to the year 2000: in fact, between 1975 and 1981 coal costs to the Board increased in real terms at around 1 per cent per annum. SCC pointed out that in its economic justification for the power station the Board assumed that nuclear fuel costs would remain constant until 1995, whereas they had actually risen by 2.5 per cent per annum in real terms over the decade 1971–81 (Scottish Consumer Campaign 1981).

THE UK SIZEWELL SCENARIOS

One of the most detailed assessments of nuclear *vis-à-vis* other fuels for electricity generation provided for any energy market in recent years is that produced in 1985 by the CEGB for the Sizewell B Public Inquiry. Sizewell B nuclear power station, on the Suffolk coast, was seen by the CEGB and by nuclear power opponents as a key element in the development of the UK's nuclear power programme, not least because it was an application to build a PWR station of US origin for the first time in the UK. At an expensive public inquiry, lasting 340 days to 7 March 1985, the application to build the 1,200 MWe Sizewell B power station was justified by the CEGB on three grounds. First, it promised a better economic return than the construction of an AGR or a coal-fired station and would reduce the overall cost of generating electricity under a wide range of future circumstances. Second, it would make a contribution to achieving fuel diversity in a coal-dominated generating system by displacing the equivalent of 2.25

million tonnes a year of coal. Third, the extra capacity would be likely to be needed to replace plant reaching the end of its life and to meet any increase in electricity demand.

Five scenarios were postulated to examine the range of possibilities. Scenario A postulated a high rate of economic growth for the UK based on the service industries, with high regard for the environment and rapid progress with energy conservation, against a background of high world economic growth. Scenario B postulated a high rate of economic growth along traditional lines based on expansion of manufacturing industry against a background of medium world economic growth. Scenario C postulated medium economic growth both nationally and in the world as a whole, broadly continuing the trends underlying the previous set of electricity demand estimates. Scenario D envisaged stable low economic growth accompanying de-industrialization and a declining or stagnating level of economic activity, still with medium growth in the world as a whole, maintained living standards through a reduced need for investment and, in the longer term, through returns on overseas investment. Scenario E assumed an inability in the UK to overcome economic conflicts or to achieve stable policies combined with low world economic growth and a socially divided society, leading to low unstable economic growth.

These scenarios, summarized in Table 9.6, cover a wide range of future possibilities or 'futures'. For example, the gross domestic product (GDP) varies by a factor of two in the year 2000 and by a factor of over three in 2030, and the primary energy demand in 2000 varies by ± 25 per cent on the 1982 level. The corresponding maximum generating capacity varies from the 1982 level by +40 per cent to −20 per cent. Scenario C represented the most likely future, in the view of CEGB, but there were considerable uncertainties, as the CEGB itself readily acknowledged.

One such uncertainty, that of the size of the future generating capacity requirement, seems to have prompted heavy reliance on economic arguments in the case for Sizewell B. The declared net electricity generating capacity of the CEGB at the end of March 1985 was 51.1 GWe, comprised as follows: coal, 32.6 GWe; oil, 7.2 GWe; gas turbine 2.6 GWe; nuclear, 4.5 GWe; hydroelectricity and pumped storage, 2.2 GWe; dual-fired coal–oil, 2.0 GWe (CEGB Annual Report 1984–5). A further 6.4 GWe of capacity under construction in mid-1985 included 5 GWe of AGRs at Dungeness B, Hartlepool and Heysham (I and II). The CEGB *Statement of Case* for Sizewell B, produced in 1982, indicated an intention to avoid disconnections (but not voltage or frequency reductions) in all but four winters per century. On any sizeable generating system a generating capacity in excess of the maximum demand is customarily required. The CEGB laid out its capacity calculations to take account of the following:

(a) the average availability of generating plant, estimated at 85 per cent;
(b) the estimated variation in availability, corresponding to a standard deviation of 3.75 per cent;

Table 9.6 Some parameters of the Sizewell B scenarios

	A	*B*	*C*	*D*	*E*
GDP (1979–80 = 100)					
2000	171	171	124	91	91
2030	187	310	167	91	91
GDP growth (%/year)					
1980–2000	2.6	2.6	1.0	−0.4	−0.4
2000–30	0.3	2.0	1.0	0	0
UK fossil fuel, Mtce (1980 = 337)	375	385	355	325	305
Industrial fuel prices in 2000 (pence/therm at 1982 prices)					
Coal (1980 = 14)	32	37	28	–	27
Firm gas (1980 = 23)	82	93	72	–	41
Electricity (1980 = 91)	137	160	125	–	108
Fuel oil (1980 = 27)	67	80	67	–	36
Uranium (£/lb U_3O_5) (1980 = 15.4)					
2000	44	47	38	–	25
2030	173	94	94	–	71
Energy demand in 2000 by final users (thousand million therms)					
Electricity (1980 = 7.9)	10.0	12.3	9.1	–	6.6
Gas (1980 = 15.1)	16.7	19.7	17.9	–	16.8
Oil (1980 = 26.6)	19.8	24.4	21.9	–	18.0
Solid fuel (1980 = 8.8)	8.2	11.5	8.1	–	–
Solar, CHP etc.	0.3	0.3	0.2	–	–
Total (1980 = 58.4)	55.1	68.2	57.2	–	44.1
Primary energy in 2000 (1980 = 342 Mtce)	338	418	339	–	258
Electricity in 2000					
Generation (1980 = 225.8 TWh)	285.8	351.1	259.0	–	187.8
Max. demand (1980 = 44.1 GW)	51.3	61.7	46.9	–	35.1

Source: CEGB 1982; Layfield 1987

(c) variation of the weather from the 'average cold spell' conditions adopted in the forecasts, giving a standard deviation of 3.95 per cent;
(d) variations in peak demand from that forecast, at a standard deviation of 9 per cent.

These factors combined to require a margin of 28 per cent above the future forecast maximum demand, the so-called planning margin, to meet the intended security of supply. The *Statement of Case* pointed out that on this basis the 1982 peak demand of 42.6 GWe required generating capacity of 52 GWe, some 3.2 GWe less than the capacity actually available. For the future, using a 28 per cent planning margin, the additional capacity provided by Sizewell B would not be necessary under scenario E, but would be required for the higher growth scenarios A and B. The scenario C position became finely balanced; new capacity would not be needed until 1997, but 9 GWe would be needed by 2000. It was argued that to allow time to build up the capacity by 2000 a first order should be placed in 1986. But the CEGB also revealed at the Inquiry that its forecasts of electricity demand had been 26 per cent too high over the previous ten years.

However, it is the economic assessment that has been regarded by the CEGB as the main part of its case for Sizewell B. The economic analysis estimated the net effective cost (NEC) of adding a new station to the system. All the costs were discounted to the same date using a 5 per cent discount rate, and allowance was made for the effect on the whole generating system, including operating the power stations in merit order, and its relationship to load factor. The results for PWR, AGR and coal-fired stations for scenario C are given in Table 9.7.

Table 9.7 shows the Sizewell B PWR to have a lower NEC regardless of how the generating system is subsequently developed. Furthermore, the NEC is negative, showing that it would be economic on cost-saving grounds alone to build Sizewell B to replace existing fossil-fired plant with a positive net avoidable cost. A breakdown of the components of the NEC shows that the dominant contribution comes from the saving in fuel costs over lower merit order stations, far outweighing the capital costs of Sizewell B.

The *Statement of Case* seems a formidable riposte to cost arguments that have been advanced from time to time in the UK by various nuclear power critics. It systematically adds, brick by brick, telling additional arguments. Mathematical analysis is used to test the optimal proportion of nuclear plant in the future CEGB system. For scenario C, this exceeds 70 per cent, showing that a further tranche of nuclear capacity is expected to be economic. Later PWRs, commissioned in the year 2000, would still have a negative NEC. A sensitivity analysis showed that a 10 per cent variation in the capital cost and a 50 per cent variation in fossil fuel price increases

Table 9.7 Comparative net effective costs on scenario C (£/kW/year)

Type of station	*Future generating background*		
	No nuclear	*Medium nuclear*	*High nuclear*
Sizewell B PWR	−93	−69	−43
AGR	−55	−27	−10
Coal-fired	10	14	22

Source: As for Table 9.6

would not affect the overall conclusion. However, in a dominantly coal-burning utility, such as the CEGB, *if the price of coal were to remain constant in real terms*, the choice between new coal-fired and new nuclear capacity would be finely balanced. A fall in the price of power station coal in real terms, such as may be achieved by large new pits or very cheap imports, completely alters the picture.

The CEGB said that, at March 1982 prices, Sizewell B could be built for £1,172 million. It calculated at the Inquiry that the PWR would produce a net saving at March 1982 prices of more than £1.55 billion over its lifetime through reduced usage of conventional fossil-fuelled power stations. By December 1985 it had lowered its estimated saving to £1.32 billion, because of several factors, including the increasingly long delay in the planned commissioning date caused by the length of the Public Inquiry into the application to build the power station. The cost estimate was increased to £1,300 million. By July 1989 the figure was £1,700 million.

Despite its thoroughness, the Sizewell analysis has not gone unchallenged. One study has argued that the CEGB ignored the effects of the nuclear choice on employment, the UK balance of payments and national income (Fothergill *et al.* 1983). While accepting that Sizewell B offered the possibility of providing cheaper electricity after 1992 than a coal-fired plant, Fothergill *et al.* estimated that up to 8,000 jobs would be lost, with the coal industry hit hardest. Conservation bodies such as Friends of the Earth have argued against the Sizewell PWR (Cannell and Chudleigh 1983) and J.W. Jeffery has challenged CEGB's accounting methods (Jeffery 1988). At the Public Inquiry the National Coal Board (NCB) disputed the CEGB calculations and assumptions about trends in coal prices, arguing that the demand for coal was no higher in 1980 than in 1983 and was likely to remain low. The NCB maintained that new sources of supply from Poland, Colombia and South Africa would ensure only modest increases in the price of coal.

A common complaint about the CEGB has been the virtual impossibility of reconciling the different assumptions that it has made about coal price increases from one study to another (Hockley 1983), and there is no doubt that both coal and nuclear costs in the UK are very sensitive to trends in real sterling exchange rates. A problem that arises in the use of a discounting technique, as used at Sizewell by the CEGB, is that expenditures on waste management appear insignificant. However, these are costs that will appear some years into the future. It is important also that the CEGB did its calculations before the UK Government announced its plans to float the electricity industry on the Stock Exchange as part of its privatization programme. The impending privatization of the industry brought decommissioning costs into sharp focus. On 18 July 1989, the London brokers UBS Phillips and Drew reported that the UK's entire nuclear industry could have a negative value because of the high costs of decommissioning and waste disposal. The widely publicized report suggested that by the end of the century National Power, the larger of the two companies to be created from the CEGB, 'could face a decommissioning bill of £12 billion' (Wilson 1989). The government had previously pledged £2.5 billion to cover decommissioning costs after the flotation. At the same time, the Centre for Policy Studies produced calculations indicating that nuclear electricity in the UK was produced at 6.0 pence/kWh, twice the coal-fired figure, and at the Hinkley Point public inquiry for the UK's second PWR the CEGB moved away from justifications based on cost to those emphasizing the need for a diverse and secure supply.

On 24 July 1989 the Energy Secretary withdrew the Magnox reactors from the privatization programme, and later the other nuclear power stations were excluded too. Effectively the nuclear plants remain a public responsibility within a privatized electricity industry, organizationally grouped as one of the four successor companies to the CEGB. Plans for more PWRs after that under construction at Sizewell were frozen in 1989 until a review in 1994 of costs and output potential. It was rather a forlorn gesture for the Department of Energy to allocate some new sites in 1990, at Druridge Bay in Northumberland, Pembroke in Wales and Denver, near Downham Market, Norfolk, 'in case the review proves favourable'. Until the Government makes a firm commitment to indemnify institutions against any escalation in nuclear costs after privatization there is little hope that a PWR programme will be sponsored by the private sector.

THE ECONOMICS OF REPROCESSING SPENT FUEL

In the 1970s reprocessing plants were seen as a vital link in the chain under which first-generation nuclear power stations would progressively feed plutonium-burning fast breeder reactors which would in turn breed usable plutonium. Fuel reprocessing plants were seen, thereby, as opening the way to genuinely cheap and plentiful nuclear power. Fast breeders were also regarded as a way of allowing the West to make a considerable cut in its dependence on uranium imports. It is now clear that by 2005 there are likely to be very few commercial fast breeders in operation. Thus the main civil outlet for plutonium is no longer there and this has left projects such as the expansions at Sellafield (THORP) and Cap de l'Hague in an awkward position. In the FRG, Wackersdorf has been cancelled. At Sellafield, uncomfortably large backlogs of used fuel are building up in wet storage awaiting reprocessing (Lean and Leigh 1989). If the CEGB had remained in existence, it might have preferred to build dry-storage bunkers for some of its spent fuel, at Heysham for example, rather than continue to pay BNFL's steadily rising reprocessing charges, but it is also possible that the UK will attempt to reduce its plutonium stockpile and buttress continued reprocessing by burning plutonium as well as uranium in its new PWRs. Recent (1989) reports from the OECD/NEA indicate that economic differences between direct disposal and reprocessing, based on specified expectations of fuel stage prices, vary internationally but often are not large.

THE ECONOMIC GEOGRAPHY OF THE CASE: THE INFLUENCE OF COSTS UPON THE LOCATION OF NUCLEAR POWER STATIONS

The comparative cost, and hence the competitive position, of electricity produced by nuclear units *vis-à-vis* that produced by other means varies from one energy market to another, and energy markets themselves may be variously defined. Geographical definitions emphasize such things as the size and shape of the area to be served, the strength and density of the grid distribution system, the type and location of demand (load), the pattern of existing and possible future power station sites, and the availability and cost of particular fuels for electricity production. The scale of the market can vary from scattered isolated farmsteads or villages in a rural area, which may best be served by small diesel generating units, to massive urban–industrial complexes requiring large central power plants to provide base load. Similarly, the administrative and organizational arrangements adopted by utilities may range from large nationalized bodies such as Electricité de France (EdF) through regional supply organizations like Ontario Hydro in Canada, down to quite small concerns with one or two modest generating units serving a purely local electricity distribution system. It may be thought that scale economies would make electricity provided by very small utilities rather expensive, but this is not inevitably the case. For example, in the USA the Idaho Falls Electricity Utility is owned by the township of Idaho Falls, and has a very low electric utility rate because it purchases its electric power from the Federal Bonneville Power Administration, where the power plant capital costs are subsidized by the US government. To have real meaning, therefore, attempts to establish the competitive position of nuclear power in relation to other means of generating electricity has to be related to a particular energy market, an individual utility or administratively definable organization such as EdF or Ontario Hydro, or a country.

The task for decision-makers in choosing power station locations within such areas is no less complex an issue than the assessment of the factors influencing the costs themselves. An attempt is made here to suggest the influence of costs on location by considering selected cases on a variety of scales, but in a manner intended to be illustrative and suggestive rather than comprehensive.

INTERNATIONAL COMPARISONS

On *a priori* grounds the presence or absence of nuclear power stations in a country might be expected to depend to some degree upon the cost of

Table 9.8 National comparative cost calculations on a single-station levelized cost basis in constant money terms

	Cost per kWh	
	Nuclear	*Coal*
Canada ($\times 10^{-2}$ Can$ at 1 Jan 1981)	1.08	1.82
France ($\times 10^{-2}$ Fr francs at 1 Jan 1982)	19.3	31.4
FRG ($\times 10^{-2}$ ECU at 1 Jan 1981)	–	4.64
Netherlands ($\times 10^{-2}$ Gld at 1 Jan 1982)	7.9	10.7
UK ($\times 10^{-2}$ £ at 31 Mar 1982)	2.61	3.88
USA ($\times 10^{-3}$ US$ at 1980)		
New England	43.0	46.8
Rocky Mountains and Great Plains	42.3	31.0

Source: OECD/NEA 1983a
Notes: (a) Discount rate 5 per cent; 40 year life for nuclear plants and 35 years for coal.
(b) Only the US figures include taxes.

nuclear power *vis-à-vis* other means of electricity production. According to P.M.S. Jones (1987) the only satisfactory method for international comparison of base load plants is the single-station-lifetime levelized cost approach. This takes the total discounted cost in constant money terms of building, operating and ultimately dismantling the plant, including all spent fuel processing, waste management and disposal costs, and divides this by the discounted sum of the expected net electrical output over the plant's lifetime (Table 9.8). The resultant cost, in constant money terms per kilowatt hour, is the projected lifetime average cost of supply at the station busbar which, if recovered, would exactly meet all incurred expenditures and provide a real return or investment equal to the discount rate employed. This method takes all the money costs into account and allows for the different capacity, lifetime and expected load factors of any plants being compared.

Discounted levelized costs in real terms are widely accepted as a useful basis for international comparison or for inter-fuel comparison of plants operating under equivalent conditions. However, even when the levelized real cost method is used, the answers will depend on the assumptions adopted. Thus, reactor economic lifetimes, discount rates, and equilibrium load factors may vary a great deal, as shown in Table 9.9. The figures include all stages of the fuel cycle and all investment and long-term plant costs. The only useful comparison that can be made is between nuclear costs and those of electricity production by contemporary (1995–2000

Table 9.9 Summary of levelized discounted electricity generation costs, 1995–2000, for OECD NEA/IEA reference case

Country	*Nuclear (mills/kWh)*				*Coal (mills/kWh)*				*Ratio of coal to nuclear*
	Investment	*O & M*	*Fuel*	*Total*	*Investment*	*O & M*	*Fuel*	*Total*	
Belgium	15.6	5.4	8.1	29.1	13.9	9.2	29.1	52.2	1.79
Canada									
Central	13.2	2.4	4.0	19.6	8.2	1.9	15.9	26.0	1.33
E-n/E-c	15.4	7.6	3.1	26.2	9.1	3.7	15.1	27.9	1.06
E-n/W-c	15.4	7.6	3.1	26.2	10.3	3.3	7.9	21.5	0.82
France	12.9	5.4	9.1	27.4	11.3	4.8	23.5	39.6	1.45
FRG	21.4	7.4	11.1	39.9	10.1	8.5	38.1	56.7	1.42
	21.4	7.4	11.1	39.9	10.1	8.5	25.3[e]	43.9	1.10
Italy[a]	23.4	6.3	10.7	40.4	12.7	6.9	23.5/38.1	43.1/57.7	1.07/1.43
Japan	21.4	8.7	13.2	43.3	18.0	13.3	24.4	55.7	1.28
Netherlands	17.6	6.4	10.6	34.6	9.7	4.0	19.1	32.8	0.95
Spain[b]	25.4	8.7	8.5	42.6	12.5	6.0	22.8	41.3	0.97
Sweden	–	–	–	–	12.6	8.4	25.0	46.0	–
Turkey	22.5	3.7	6.0	32.2	11.6	2.8	19.4	33.8	1.05
UK[c]	22.6	6.6	6.6	35.8	13.1	6.9	18.1	38.1	1.06

Table 9.9 continued

USA[d]									
Midwest	21.7	11.4	5.6	38.7	15.0	6.0	14.4	35.4	0.91
West	21.4	11.4	5.6	38.4	15.1	4.1	11.9	31.1	0.81
East	21.7	11.4	5.6	38.7	16.3	4.7	20.5	41.5	1.07

Source: OECD/NEA/IEA 1989: 77

Notes: Jan. 1987 mills per kWh. O & M, operation and maintenance. E-n/E-c, Eastern located nuclear compared with eastern coal fuelled; W-c, Western coal fuelled.

The table includes the ratios of costs of generation in coal plants to the costs of generation in nuclear plants *on the basis of the utility projections for each individual country*. Thus, comparisons *across* countries are not particularly informative.

(a) Coal prices in high and low estimates.
(b) Coal data are for imported coal.
(c) Investment includes decommissioning, fuel includes cost of coal stocks.
(d) Nuclear O & M costs include decommissioning.
(e) Imported coal case.

The levelized cost in constant money value terms is the average cost per unit of electricity fed into the grid which, when discounted over the total lifetime output of the plant, is exactly equivalent to the present discounted value, per unit of electricity, of the capital costs of the plant including interest charges, plus operating, maintenance and fuel costs, and the costs of waste management and ultimate decommissioning of the plant.

commissioned) coal-fired power stations *in the same country.* These data are the levelized generating costs for mainly light water reactors and normal coal-fired stations with a lifetime levelized load factor of 72 per cent, a discount rate of 5 per cent, and an assumed lifetime of 30 years.

It is clear from Table 9.9 that nuclear power would have an economic advantage for these reference conditions in all the listed territories except for Spain, the Netherlands and the western and mid-western USA. Substantial changes adverse to nuclear power in major parameters such as discount rates or coal prices could change this position. If, for example, the discount rate were to double to 10 per cent, coal generation would be cheaper in most of the countries listed in the Table.

INFLUENCE OF COSTS ON NUCLEAR POWER PLANT LOCATION WITHIN COUNTRIES

A second proposition relating costs to location is that spatial variations in generating costs have influenced the geographical pattern of nuclear power plants. Within many countries, and within particular electricity markets, fossil fuel costs vary and this alone might be assumed to have had some influence on nuclear power plant location, not least through the greater mobility of nuclear power stations in relation to fuel supply.

In the UK, until 1967, nuclear power stations were located so as to meet growth in electricity demand in areas where power station coal was scarce or costly, e.g. the southwest, the southeast and the northwest (Mounfield 1961). The 1967 White Paper on Fuel Policy suggested that the Generating Boards should base their choice of fuel on an economic assessment of the method of generation that would enable them to supply elec-

Table 9.10 Electricity generating costs in France (French francs/kWh)

	Oil	*Coal*	*Nuclear*	*900 MWe PWRs*	*Average all plants*
Fuel costs	0.254	0.134	0.056	0.054	0.110
Other operating costs	0.029	0.041	0.021	0.016	0.029
Capital charges	0.055	0.039	0.071	0.068	0.072
Total generating cost	0.034	0.210	0.150	0.140	0.210
Sterling equivalent (pence/kWh)	3.1	2.0	1.4	1.3	2.0

Source: Electricité de France 1982

Note: Costs for individual stations have not been published but the introduction of the 1,300 MWe units from 1984 has not produced the economies of scale expected. In December 1988 *Nuclear Engineering International* ('Datafile: France': 61) indicated generating cost at 22.7 centimes/kWh for nuclear and 32.9 centimes/kWh for coal without desulphurization (9 per cent discount rate, 25 year lifetime, average availabilities).

tricity at the lowest cost. The CEGB's policy came to be one in which nuclear power stations were located in areas where the greatest savings in fuel costs could be made. The Hartlepool nuclear power station, for example, in the heart of a traditional coal-mining district in northeastern England, was justified on the grounds that cost estimates suggested a 25 per cent advantage for a nuclear plant compared with one that was coal fired.

After the OPEC oil price increases in 1973–4 some industrialized countries, short of indigenous supplies of low cost coal, decided that imported oil was becoming too expensive to burn freely in power stations. Japan was one such country and France another. Japan imports over 85 per cent of its energy requirements, including all its uranium, but according to the Japanese Institute of Energy Economics, since 1980, when 15,700 MWe of nuclear capacity was available, Japanese utilities have generated nuclear electricity for about half the cost of that produced by oil-burning power stations (Figure 9.4).

The comparative costs of oil-, coal- and nuclear-generated electricity in France, shown in Table 9.10, are based on the first 21,634 MWe of nuclear capacity for the twenty-three PWRs which EdF began building in the 1970s. The table suggests that France is generating nuclear electricity from its PWRs at about 70 per cent of the cost of coal-fired electricity, and well under half the cost of oil-burning stations. However, France is exceptional within Western Europe. Most other European countries for one reason or

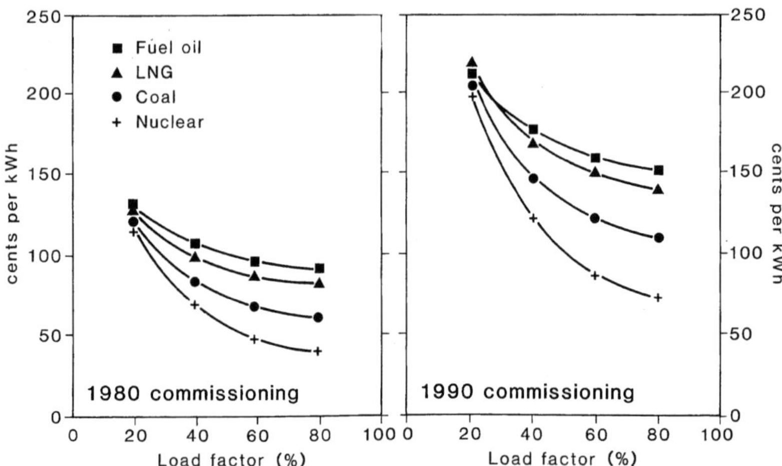

Figure 9.4 Comparative costs for electricity from Japanese power stations by fuel type
Source: adapted from Fishlock 1986
Notes: (a) Costs levelized during station lifetimes at 1986 prices.
(b) LNG, liquified natural gas.

another continue to rely heavily upon oil and gas or coal and hydroelectricity for their electricity requirements. Nevertheless, the pattern of new power station construction is steadily increasing its nuclear proportion (Figure 9.5).

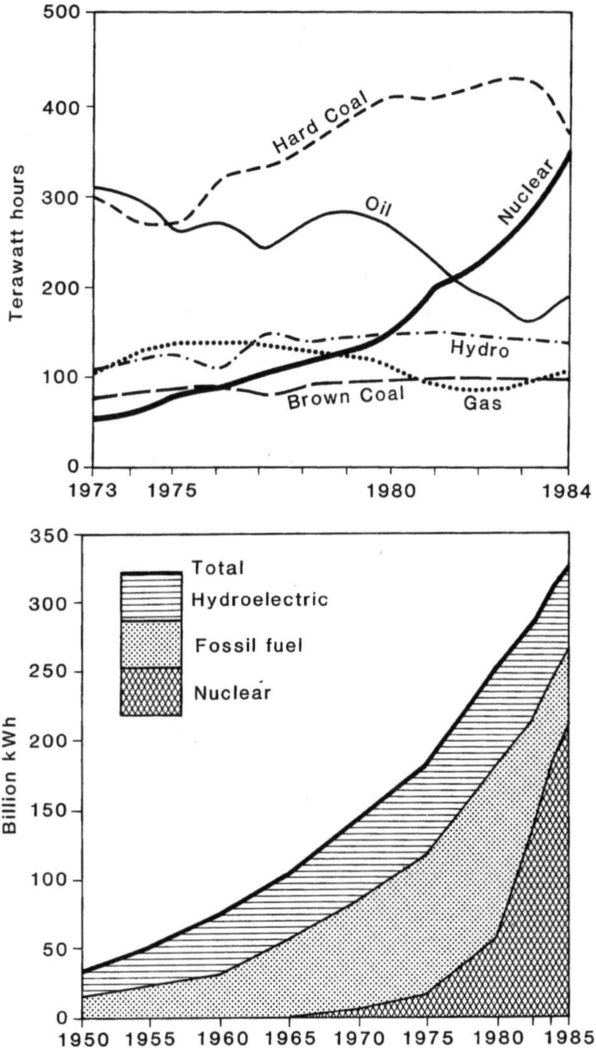

Figure 9.5 Trends in fuels used for (a) EC and (b) French electricity production
Sources: Annual Reports of Eurostat and of Electricité de France
Note: The 1984 figures in (a) are distorted by the National Union of Mineworkers' strike in the UK.

To take account of size differences, perhaps the USA ought to be compared with the whole of Western Europe rather than with just one European country. The aim in the early stages of US nuclear power development, shared by Congress and the USAEC, was to make nuclear energy competitive within ten years in areas of the USA where fuel costs were high (*Economist*, 3 May 1959: 429). Commercial development accelerated rapidly after the Oyster Creek order of 1964 but wide variations in fuel costs for electricity generation existed then in the USA, as they do now. Also, the US electricity market is fragmented. The grid system pattern indicates heavy concentration of demand in the East and the far West; the industry is run by a large number of utilities, and the pattern of interconnection between them has created regional power blocs (see Chapter 3, Figure 3.4).

Figure 9.6 provides an overall picture of regional variations in fuel costs in the Federal Power Commission (FPC) regions of the contiguous USA for 1966, a year within a crucial innovative period for commercial nuclear power. Coal dominated in four of the regions: East North Central, East South Central, South Atlantic and Middle Atlantic. In three others – West North Central, Mountain and New England – coal competed on reasonably level terms and in two areas – West South Central and Pacific – natural gas was dominant. The map also indicates that the high cost fuel areas were New England, Middle Atlantic, South Atlantic and Pacific. The fact that these four accounted for many of the first US commercial nuclear power stations built from the mid-1960s indicates a degree of economic rationality in the nationwide pattern of nuclear power station location that emerged in the following years. The low cost fossil fuel areas in 1954–64 were East North Central, West North Central, Mountain, East South Central and West South Central.

According to one industry source, by the beginning of the 1980s coal-fired capacity was cheaper than nuclear power for base load current in only one of the five regions, and for stations above 700 MWe at load factors above 50 per cent nuclear power generally offered the lower costs (Newsletter 1982a). However, in a later analysis carried out by an Expert Group for the OECD it was concluded that nuclear power held a clear economic advantage in Central and Atlantic Canada, and a 5–10 per cent advantage in the northeast and southeast parts of the USA (OECD/NEA 1983a). In the circumstances of the late 1980s and early 1990s a coal-fired plant on or close to one of the North American coalfields is likely to be the cheapest option for new plants even when they are equipped with desulphurization systems. The Expert Group argued that nuclear power prospects in the USA would improve only if capital costs could be stabilized and made more predictable by reducing construction times and hence interest charges, by adopting standardized designs and by co-siting several reactors on single sites with common services. By 1988, some commentators were arguing that the southwest USA was the only region where completion of

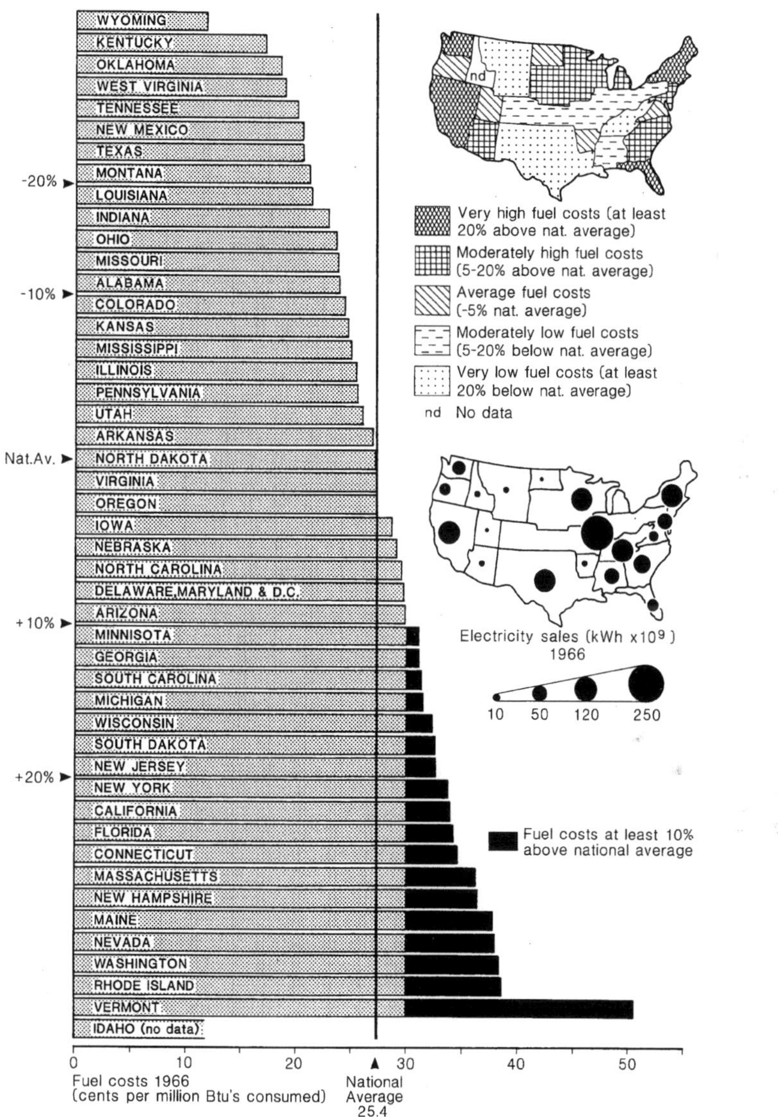

Figure 9.6 Spatial variations in fuel costs for electricity generation in the USA 1966
Source: Edison Electric Institute 1967

nuclear power stations under construction could be justified economically. It was argued that so little additional generating capacity would be needed before 2000 that finishing partly completed nuclear plants would throw many utilities into financial crisis (Feldman *et al.* 1988).

In the 1980s some individual US power utilities remained convinced that, despite its problems, nuclear power has been a financially profitable investment. The Commonwealth Edison Company (CECo) has been one of these. In 1982 it had 5,058 MWe of nuclear capacity in operation with 6,636 MWe under construction. It also had seventeen smaller coal- and oil-fired units, five large oil-fired units and six large coal-fired units. The large coal units and six of the nuclear units were built over a comparable time with similar load factors in mind. Coal and nuclear provide approximately equal shares of the total electricity generated in the system, but since 1970 the nuclear plants have outperformed the coal plants, partly because the latter were designed to burn Illinois coal with a high sulphur content but are burning low sulphur western coal for environmental anti-pollution reasons. It has been estimated that in 1980 CECo customers had to meet over US$500 million in 'environmental costs'. The nuclear plants have exhibited generating costs of about half the corresponding cost for the large coal-fired plants (Zimmerman 1982). It has been argued that the major advantage held by nuclear over coal and oil in this utility, as in many others that had nuclear plants in operation by the mid-1970s, is that fossil fuel costs have been particularly sensitive to inflation, while nuclear fuel costs have been less so. Between 1973 and 1980 CECo's coal costs grew at an average of 22.2 per cent per annum and oil costs increased at 34.6 per cent each year, while nuclear fuel costs grew at only 11.0 per cent per year, Edison's savings in fuel expenditure over the period as a result of having substantial nuclear generating capacity were enormous. Compared with low sulphur coal, the nuclear units saved more than US$2.4 billion over the period 1973–80, and more than US$6 billion when compared with residual oil.

The CECo case indicates that measures needed to reduce pollution hazards affect the economics of coal-fired plants, just as safety modifications imposed in nuclear power stations have pushed up their cost, especially when they have to be 'backfitted'. CECo, for example, is burning low sulphur western coal shipped in over long distances when it would prefer to burn high sulphur coal from Illinois mines. In the mid-1960s a cost advantage of around 10 per cent accrued to nuclear plants as fossil-fuelled power stations were forced to reduce sulphur dioxide by using fuels low in sulphur or by installing smoke controls at the chimney stacks (*Business Week*, 11 March 1967: 66). It was an advantage wiped out by the difficulties that overtook nuclear power in the late 1970s.

The heavy concentration of nuclear power stations in the Ukrainian region of the USSR is an outstanding example of the locational influence of energy costs in a centrally planned economy. The USSR is one of the few countries in the world which, to date, has not needed to import energy supplies, but it does have an energy crisis, arising from the cost of extracting and transporting fuels to the principal consuming areas in the USSR. Soviet energy planners feel that the country can no longer rely on supplies of oil and coal to fuel the electricity industry, and there is a shortage of both fuels in many East European countries. Thus, of the three major fossil fuels in the USSR, only gas has given cause for optimism over the past few years, but gas has recently become one of the more reliable means of obtaining hard currency from Western Europe.

The eastern Ukraine has long been one of the major industrial regions of the USSR, especially for the production of coal, steel and chemicals. The region's resource base for such activities has been intensively exploited. Iron ore, uranium and coal reserves remain in quantity, but stagnation in the Donetsk coalfield is an important reason for the widespread development of nuclear power in the Ukraine as a replacement for fossil fuels. The coalfield is not about to run out of coal – at current rates of extraction there is enough coal still underground to last for another hundred years – but most of that remaining is in relatively narrow seams less than 1.5 m thick. The seams often dip very steeply and the mines themselves are now very deep (1,200–1,600 m). Poor organization has led to a lack of proper equipment to cope with these conditions, and so in mining operations a great deal of waste material is brought out of the pits with the coal, increasing the ash content. Thus Ukraine (Donbass) coal costs twice as much to mine as Kuzbass–Kuznetsk coal and several times more than Ekibastuz coal. In the late 1980s Donbass coal was also costing the lives of more than two miners for every million tonnes extracted.

A political struggle has developed amongst those responsible for energy planning and utilization in the USSR between a group committed to reinvestment in the Donetsk coalfield and those who wish to expand output from the Siberian and other non-European USSR coalfields, especially the Kuznetsk coalfield. 'Given a choice between producing machinery capable of mining [Donetsk] seams or concentrating on strip mining in Siberia, Soviet officials are finding the latter increasingly attractive' (Marples 1987). Labour productivity at the opencast mines is three times higher than that of the Donetsk underground mines and the unit cost of coal is 50 per cent less. Thus the centre of gravity of coal production in the USSR is moving eastwards, but distance from the consuming areas is a major problem for the Siberian fields. In any year, much more than 5 million tonnes of coal equivalent is used in moving coal from these remote areas to their markets. Nevertheless, the Donetsk coalfield declined as a supplier of fuel to power stations from 64.8 million tonnes in 1975 to 59.7 million tonnes in 1980. At the same time the Ekibastuz coalfield of northern Kazakhstan supplanted Donetsk as the largest coal supplier, rising from 44.1 million tonnes in 1975 to 63.4 million tonnes in 1980.

Oil is a major alternative to coal within the USSR energy sector, but the

Table 9.11 USSR oil output

	Output (million tonnes)	*Annual average rate of increase (%)*
1960	147.9	
		23.8
1970	353.0	
		27.8
1975	490.8	
		24.6
1980	603.2	
		1.5
1984	612.7	

Source: Marples 1987

Soviet leadership is attempting to conserve oil reserves as an important future source of hard currency and as an energy source for CMEA partners. The annual rate of growth of the oil industry has slowed down markedly since 1980, as Table 9.11 shows.

West Siberia has been the main growth area as the country's more accessible oil resources have been used up, but in Siberia the industry has suffered from a lack of adequate housing for oil workers, and from poor equipment and bad management.

The problems of obtaining energy supplies from Siberia have had a direct effect on the Soviet Ukraine, which is a major energy market. At the same time, the Ukraine's location on the western border of the USSR makes it a very suitable area from which to forge closer economic links between the USSR and its East European neighbours. As Marples has put it:

> Even in the 1970s the Soviet authorities were seeking alternative supplies of energy, and moreover supplies that could be more or less guaranteed, that did not depend on a Donbass miner working every weekend in a month, or on the Soviet railway or supply system to a distant oilfield. The demand for electricity was considerable. A means to this end was sought in nuclear energy, which appeared to the Soviet authorities to be an industry that was both economical and reliable.
> (Marples 1987: 50)

Marples points out that nuclear power was an avenue that was technologically available to the Soviet authorities ever since the Obninsk nuclear power station went into operation in 1954, but it was an option that had not previously been explored as a major form of energy supply. In the 1970s it came to be seen as a way of plugging the energy gap in the Ukraine, and the export of electricity from nuclear power stations was regarded as a convenient way of reducing the demand for Soviet oil in Eastern Europe.

Thus the Ukraine came to assume a key role in the USSR's nuclear power programme, with two powerful (750 kV) interconnections between the East European grid, including one to the Chernobyl station. Of the five Ukrainian nuclear plants operating in 1986, three were exporting electricity to Romania, Bulgaria, Poland and Hungary.

SUMMARY

For a large electricity utility the ultimate economic justification for a nuclear power programme involving several nuclear generating units is that the costs of nuclear plant and fuel are low enough to improve the total system generating costs. It is generally agreed that historic cost figures provide a poor basis for future cost comparisons between nuclear and other fuels. Even very careful calculations for the future are faced with the facts that costs can change rapidly and that published information on costs varies significantly between reactor designs and for stations of the same reactor design projected to be built at different sites or to start operating at different dates.

Station capital costs and fuel costs are two of the main items used in calculating the operating cost (cost per kilowatt hour) of a particular power station. They have to be amortized over an assumed economic station lifetime. The total electricity generated in the economic lifetime depends on the load factor, which in turn depends upon a plant's capability and the use made of it within the generating system. Electricity costs published in different countries and the breakdown into capital, operation and fuel components are not, in general, directly comparable.

Within countries, or within large utilities, the comparative economic position of nuclear and fossil-fuelled power stations may be more readily compared, and there is sufficient evidence to indicate that in many countries the costs of other fuels for electricity generation have influenced the locational pattern of nuclear power stations. Ultimately, though, nuclear and fossil-fuelled station roles must be complementary to, rather than competitive with, one another within many electricity markets. At some point in time it will be most advantageous to add a new fossil-fuelled power station to the system, and at other times a new nuclear power station, in order to obtain the combination that will give the whole generating system improved performance and reduced overall costs (Bainbridge 1972). Within each 'family' or class of plants there will be good and poorer performers, with some overlaps between the classes, and generating costs quoted for fixed load factors can only be taken as indicative of relative economic merit. Thus, in the economic appraisal of nuclear plants, it is important to appreciate that generating costs cannot be regarded in isolation but must be related to the economic operation of the electricity supply system as a whole (Hinton *et al.* 1960).

Despite the complications, economic evaluations of nuclear versus fossil fuel and hydraulic plants will continue to be made. Moreover, the consensus view *in the industry* is that nuclear power has an economic advantage over coal for base load generation except in situations where cheap coal is readily available close to a load centre, as in many parts of the USA. The economic competitiveness of nuclear power, however, depends crucially on assumptions regarding the cost of the capital employed and the costs of decommissioning. In late 1988 the UK's CEGB presented evidence at a public inquiry into proposals to build a 1100 MWe PWR at Hinkley Point in Somerset. Their calculations suggested that the PWR would cost £1,500 million compared with £770 million for an equivalent coal-fired plant. The analysis showed that, with a real return on capital of 5 per cent, nuclear would have the advantage but that with an 8 per cent real return the two would be about equal. In 1989 there existed in the UK confusion and uncertainty regarding the cost of decommissioning the Magnox units and with respect to the economics of reprocessing fuel for the AGRs.

The US experience and that of several other countries such as Sweden and Austria indicates that final decisions on the type of power plant to be built are becoming less directly based on conventional economic evaluations. In many cases it is not the direct costs but indirect ones in the form of environmental impact and public concern over safety which have become key elements in the selection procedure. Furthermore, the debate on these issues is being extended to stages of the nuclear fuel cycle following the use of uranium in the reactor, especially the treatment and disposal of radioactive waste. The discussion turns now to these issues.

FURTHER READING

Chapman, P.F. (1974), 'Energy costs: a review of methods', *Energy Policy* 2 (2) 91–103.

House of Commons Energy Committee (1990), Fourth Report Session 1989–90, *The Cost of Nuclear Power*, vols 1 and 2, London: HMSO.

Komanoff, C. (1981) *Power Plant Cost Escalation: Nuclear and Coal Capital Costs, Regulation and Economics*, New York: Van Nostrand Reinhold.

OECD NEA (1989), *Plutonium Fuel An Assessment Report by an Expert Group*, Paris.

OECD NEA/IEA (1989) *Projected Costs of Generating Electricity from Power Stations for Commissioning in the period 1995–2000*, Paris: OECD.

Sweet, C. (1983) *The Price of Nuclear Power*, London: Heinemann.

Teitelbaum, P.D. (1963) *Energy Cost Comparisons*, Washington, DC: Resources for the Future Inc.; first published in *Papers Prepared for the United Nations Conference on the Application of Science and Technology for the Benefit of the Less Developed Areas*, vol. 1, Washington, DC: US Government Printing Office.

10 Safety issues in the siting of nuclear power plants

When a utility decides upon a nuclear power reactor site, it has to consider the exposure to possible harm of the public and of operating personnel in the event of a serious accident at the plant. The probability of such an accident is primarily a function of reactor design and operator competence; the risk in terms of exposure is mainly a function of location in relation to population, but the two are not independent and competent examination of the suitability of any proposed site requires analysis and evaluation of environmental features which may influence the safety of the power station to be built and operated on it. Site characteristics may dictate the inclusion of specific engineered safeguards, and a proposed design may in turn have a marked influence on the acceptability of a suggested site. In general, those reactors whose characteristics are well known tend to offer fewer siting problems than more advanced types, such as fast breeder reactors.

ACCIDENT CONDITIONS

The nuclear power industry has always maintained that there could be an explosion within a nuclear power reactor but that in no conceivable circumstances could the blast damage of such an explosion be comparable to the devastation arising from the detonation of a nuclear or atomic bomb. The worst possible explosive energy release in a power reactor could be of a magnitude sufficient to destroy the reactor and, possibly, the various containment structures in which it is housed, but direct damage from the *blast* of such an explosion would be unlikely to extend much beyond the boundaries of the exclusion area formed by the perimeter of the site itself.

A far greater hazard under such serious accident conditions is radiation exposure and contamination as the fission products stored in the reactor are released to the environment. The most volatile of the radioactive by-products produced during the nuclear fission process are xenon, krypton, bromine, iodine and strontium. Some of these have relatively short half-lives, but others, particularly strontium and iodine, have long ones and are amongst the radioactive substances most hazardous to man. As long as they are confined to the fuel, the reactor core or the reactor system, these fission products present a negligible hazard, but even before Chernobyl it was possible to conceive of an accident which could release them in the form of particles fine enough to be dispersed over a wide area. Death and genetic damage at distances of many miles from the plant, and injury and property damage for hundreds of families, could occur in such circumstances, and did occur at Chernobyl in 1986.

For these reasons, the best nuclear power stations are designed with a high safety margin to ensure that, under postulated accident conditions, the safety of the public is maximized. The safety margin takes the form of multiple lines of defence, often called 'defence in depth', against the release of the fission products. First, safety features are incorporated into the equipment essential to the everyday operation of the plant. Second, there are protective devices and equipment intended to limit the damage that could result from the failure of any part of the process equipment. Third, there are containment provisions to prevent the escape of radioactive matter in dangerous quantities from the building should both process equipment and protective devices fail. The first and second of these defence lines are usually called 'engineered safeguards'; the third is termed 'reactor containment'.

CONTAINMENT

Until the Chernobyl disaster, the phrase 'maximum credible accident' (MCA) was commonly used in the nuclear power industry to refer to extreme accident situations. Pre-Chernobyl conceptions of the MCA varied somewhat according to the type of reactor. In light-water-cooled reactors, whether PWRs or BWRs, analyses indicated that serious pipe rupture causing a loss of coolant and a meltdown of the reactor core would result in a release of fission products on a scale not likely to be exceeded by any other 'credible' accident. This situation, normally called 'core meltdown' was designated the MCA for these reactors. Comparable MCAs have been postulated in a similar way for gas-cooled and other reactor types.

The two major biological hazards associated with core meltdown and attendant fission product release are (i) direct radiation and (ii) inhalation and ingestion of radioactive material. Protection against radiation can be accomplished by providing sufficient distance between the reactor and the surrounding population, by providing an adequately engineered concrete biological shield or by both of these together. Figure 10.1 shows the effect of distance on direct radiation, and this diagram demonstrates that the dose rate at a given distance is approximately in inverse proportion to the square of the distance. The diagram also shows that a foot (30 cm) thickness of concrete shielding reduces the intensity of direct radiation at any point by about a factor of five which, in turn, approximately halves the distance to population that would be required if no shielding were provided.

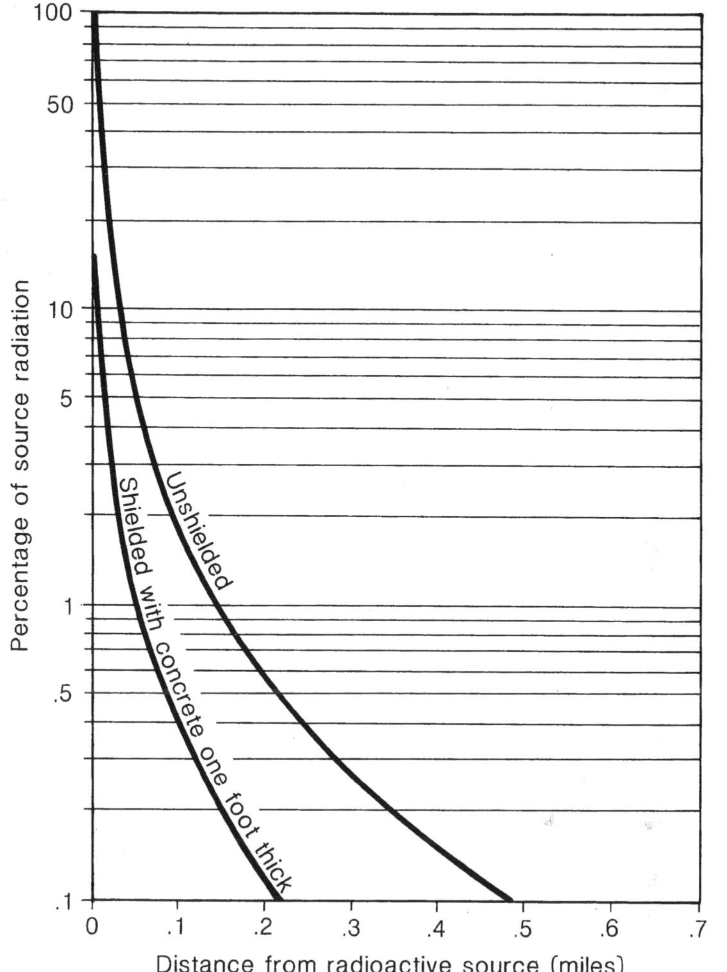

Figure 10.1 Influence of biological concrete shield on dose intensity
Source: Malay and Chave 1964

The inhalation and ingestion hazard, wherein gaseous and volatile fission products are assumed to be released from the reactor system, is more serious than direct radiation and it is generally the limiting hazard that plant designers have tried to guard against. Before Three Mile Island and Chernobyl it was accepted that under certain meteorological and accident conditions these fission products could be dispersed to the surrounding population where their inhalation would cause serious injury. The quantity of fission products necessary to cause such injury is small. Of all the volatile and gaseous fission products in the reactor core, the worst potential offender is generally iodine and its isotopes. Figure 10.2 shows the effects of distance on reducing dosage intensity from released radioactive gases. These curves are based on average atmospheric dilution conditions and do not include any additional dilution resulting from variations in wind speed and direction (see Chapter 1). With release occurring at ground level the only method of control is to limit the amount of gases released. By providing the discharge at an elevation, significant changes in the pattern of release can be achieved depending on the height of the chimney stack and the quantity of dilution air available.

Thus the principle of *compensating safeguards* applies in power reactor siting, i.e. where unfavourable environmental site characteristics exist at a proposed site, that site may nevertheless be found acceptable by licensing and regulatory bodies if the design of the power station includes appropriate and adequate compensating engineered safeguards.

However, beyond the application of good judgement to the sites available to an operator, and despite several attempts to devise a rigorous siting methodology, nowadays using computers, it is difficult to match site characteristics and reactor design in a precise way. Growing acceptance of this fact, combined with the economic attractiveness of sites near to the areas of

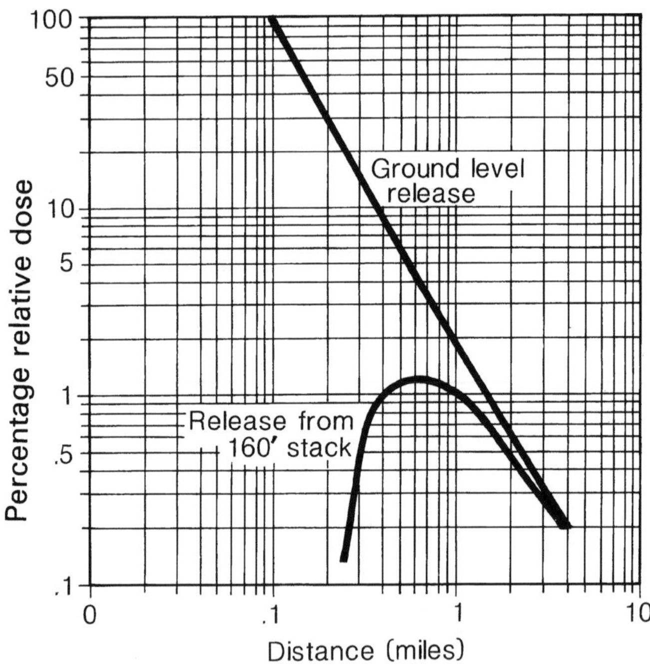

Figure 10.2 Influence of distance on dose intensity from released radioactive gases
Source: Malay and Chave 1964

demand, has caused shifts in siting policy over time from remote siting to sites close to load centres. In the mid-1960s it was argued in some quarters that there was a great deal of difference in reactor siting amongst the major users of nuclear power (Davis and Robb 1967). Subsequent examination revealed that siting practices were not very different (Hake 1969; Ergen 1970; Hake and Palmer 1970). Farmer in the UK has often pointed out that if only one-tenth of the risk represented by the fission-product inventory of a 500 MWe reactor were to be released, the health prospects of many thousands of people in a metropolis such as London could be affected even if the station were 100 km away (Farmer 1967a). Practically, sites do not differ much with respect to the population dose in the case of an accident and therefore it does not make sense to approve a reactor for one site and not for another (Ergen 1969). The small difference in population dose is a result of the assumption of large releases and of the fact that doses above the low level of around 150 mrem are considered; consequently the release is felt over large distances from the reactor. Similar arguments are advanced in the USA but here a segment of opinion holds that the important releases are the smaller ones; the 'significant' consequences are then confined to a relatively small area around the reactor.

In the early days of the nuclear power industry in the USA, Canada and the UK, it was thought that isolation of a site would give sufficient protection to people against the risk that a nuclear reactor represented (Figure 10.3). However, it is now clear that the risk can be limited substantially only if means other than or additional to isolation are used and that the hazard to public health can be controlled only to a limited degree through the choice of site. In the UK, Japan, the FRG and much of the rest of industrial Europe, the opportunity to use distance in this way is very restricted. In the USA and Canada it may be more feasible, though costly, to use distance to mitigate a large risk. The main advantage of distance from the reactor is nowadays considered to be for evacuation purposes.

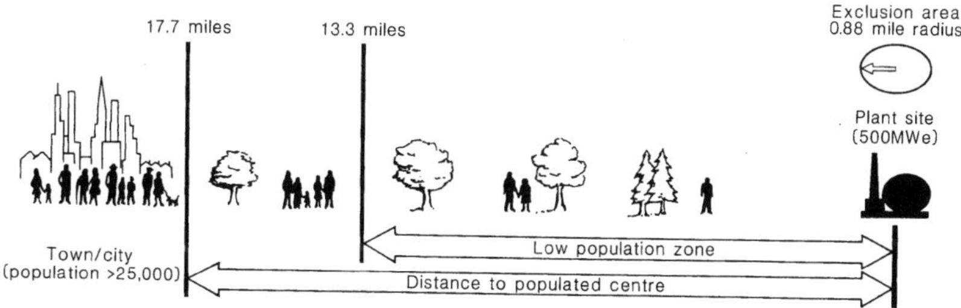

Figure 10.3 Distance as a safety factor in siting power reactors. The concepts of an area around the plant where all but plant personnel were excluded, of a surrounding low population zone and of a long distance to urban centres were incorporated in the siting of many, but not all, of the world's early commercial nuclear power plants but were not used in a uniform or rigorous fashion
Source: Burns and Roe Inc. 1966

PHASES IN THE SITING OF NUCLEAR REACTORS

Initial experimentation and military requirements

Nuclear reactors were built as a necessary part of the early stages of atomic energy research in the USA, the UK, Canada, the USSR and France. These reactors were very small, often not much more than 10–50 MWe, they were built to see how they would behave under various operating and technical conditions, the electricity they produced could be readily disposed of by means of modest power lines and security was the main locational constraint.

Soon, larger reactors of about 100 MWe were built to produce plutonium for atomic weapons, e.g. Calder Hall in the UK, Novotroitsk in the USSR, Hanford in the USA. Remoteness and secrecy remained dominant locational requirements. There was no formal licensing procedure through regulatory bodies before the 1950s, but even at the earliest stages reactor siting was given some thought. In 1944 G.C. Laurence provided one of the first statements on siting policy, in relation to the selection of the Chalk River site in Canada:

> Isolation was desirable to avoid supposed hazards to population from possible explosion or release of radioactive dust into the atmosphere, and to simplify control of secrecy. It was suggested that it should not be less than about ten miles from a town or village. Ample supply of soft and cooling water would be required for cooling the reactor, and a large river or lake near the plant would be required to dilute the effluence from the reactor which might contain a small amount of radioactive materials.
>
> (Eggleston 1965)

Although Laurence was referring to Canadian conditions, the same line of reasoning was applied in the UK, the USA and the USSR.

Commercial development

During the early commercial phase from the mid-1950s reactors were built and linked to power grids with the prime objective of providing electricity or proving the technology. In the USA some of the early reactors were ‘demonstration plants’ built to convince utilities of the feasibility and desirability of using nuclear power. In the UK and USSR this was less necessary, but in all three countries the size of reactors steadily increased to achieve scale economies. Licensing procedures through regulatory bodies were instituted and more attention was given to devising guiding principles for site selection. In general, during the early part of the phase of commercial development until about the mid-1960s, a remote siting policy dominated, as distance from population centres was regarded as an integral part of site selection (see Figure 10.3).

From about 1965 onwards, siting criteria in relation to population were gradually relaxed and remote siting policies were increasingly used for unfamiliar or experimental reactors. New large reactors were incorporated in power stations built much closer to large centres of population. The rationale for this change was threefold.

(a) It was sometimes argued that remote locations in rural areas were increasingly difficult to find outside areas of high landscape and amenity value (Faux and Stone 1963).
(b) New approaches to possible accidental happenings indicated such low probabilities of harm to the public that the cost of remote siting far exceeded any possible benefit. The nuclear industry was particularly concerned about the economic penalties of remote siting. Transmission, including construction and maintenance, cost US utilities in the neighbourhood of $250,000 a mile, while electricity losses for the line and increased system instability could incur similar additional costs (Rolph 1979).
(c) Experience in reactor operation indicated that containment and engineered safeguards provided the best form of protection against hazards, and that long high voltage transmission lines to remote sites were an unnecessary expense (Barry 1970; Gammon and Pedgrift 1983).

These shifts in power reactor siting philosophy, and the reasoning that underpinned them, can be illustrated by considering the evolution of siting policies in the UK and the USA.

THE UNITED KINGDOM

Until 1955, the siting of nuclear installations, all operated by the UK Atomic Energy Authority (UKAEA), was free from any formal licensing process (Haire and Shaw 1979). The UKAEA was empowered to use its own approval procedures and safety committees to deal with safety and siting issues; Calder Hall and Chapel Cross were therefore sites selected by the UKAEA and not by the Central Electricity Generating Board (CEGB). The first programme for commercial nuclear power in 1955 made provision for the installation of a total of 2,000 MWe, and at first it was proposed to build four gas-cooled Magnox nuclear power stations at Berkeley, Bradwell, Hinkley Point and Hunterston.

A Reactor Location Committee was set up in 1955 with a majority of members from the UKAEA, the only body with real experience, plus members from industry and the Central Electricity Authority (later the CEGB). This committee considered the choice of sites for proposed nuclear power stations until 1959 and the CEGB was for some years a pupil to the UKAEA in nuclear matters (*Economist*, 3 June 1961: 1,029).

Guiding principles for site selection which attempted to incorporate safety considerations in relation to population distribution in a systematic way were worked out by T.M. Fry in 1955, were published in part in a paper to the Geneva Conference in 1955 (Marley and Fry 1955) and were later summarized by Charlesworth and Gronow (1967). It was proposed that sites should be classified on a comparative scale from A to D, with class D sites being recommended for Magnox reactors in steel pressure vessels. It was suggested that sites should be selected such that the following conditions held.

(a) Very few people would be exposed to extreme risk. Plans should be made for them to be evacuated very rapidly in the event of an emergency.
(b) The possibility of protracted evacuation or severe restriction on normal living should be imposed only on very small centres of population.
(c) In the event of an accident, temporary evacuation or severe restriction on normal living would not be necessary for more than 10,000 people in any but exceptional weather conditions, and if an accident were to coincide with exceptional weather conditions, not more than 100,000 people should ultimately be affected.

The clear intention was for the first nuclear power stations in the UK to be kept away from towns (Mounfield 1961). An attempt to give effect to this objective was made through specifying population characteristics for the area around a possible site. These are shown for a class D site in Table 10.1.

The implication of designating such a site as class D is that other site classes, A, B, C and, perhaps, E, F, G, were also defined. This was in fact the case, and a change from one class of site to the next carried the implication that the site could accept a fission product release which was a factor of four greater (class E) or four less (class C) in geometric progression. If a site under consideration had a population distribution less favourable than the specified all-round limits, it was graded to the next more restrictive class.

In 1957 there was a serious accidental release of fission products at

Table 10.1 Population limits imposed for the first Magnox nuclear power stations in the UK, class D site

	Within 0.33 mile (0.50 km)	*1.5 miles (2.40 km)*	*5 miles (8.0 km)*	*10 miles (16.00 km)*
Max no. in any 10° sector	Few	500	10,000	100,000
Max no. all round site	Few	3,000	60,000	600,000

Source: Charlesworth and Gronow 1967: 145; Haire and Shaw 1979: 174

Windscale (now Sellafield) which underlined the importance of the radioactive isotopes of iodine in accidents involving inadvertent fission product release to the environment. Slight modifications of the site assessment procedure were made as a result, but a radical change was proposed by F.R. Farmer in 1962 when he was head of the Safeguards Division of the UKAEA. Farmer suggested that the population distribution around a possible site should be examined in relation to a system of weighting factors derived from the dispersion of iodine in stable air conditions in the downwind direction. The product of population numbers and weighting factors was summed out to a range of 12 miles (19.0 km) for various 30° sectors subtended from the proposed reactor site. These products enabled a relative rating for each sector to be calculated and the sector giving the highest product was designated as the rating factor for the site. Sites were then classified according to their site rating factors and a reactor type was designated as suitable or not for a given class of site (Farmer 1962; Charlesworth and Gronow 1967). The site classifications according to site rating factors are given in Table 10.2 and the site categories of the earlier system are indicated for comparison.

Class I sites, with a site rating of less than 750, were suggested as suitable for Magnox reactors with steel pressure vessels; the critical threshold was raised to 1,500 in 1963 for Magnox power stations with prestressed concrete pressure vessels.

This approach to site assessment proved difficult to implement. The application of an evaluation to a class I site would permit the edge of a town of 100,000 people to lie at a distance of 5 miles (8.0 km) from the site, assuming that very few people lived closer to the site in that 30° sector and that no other major town lay in that sector unless at a considerable distance. Relaxation from a class I to a class II site could bring the edge of the nearby town to 3 or 3.5 miles (4.8 or 5.16 km), still assuming the town to be constrained to a single 30° sector. In practice it would be impossible to constrain such a large town in a single sector; it would be certain to spill over into two or even three sectors, with the result that on strict application

Table 10.2 Modified UK system of site classification on the site rating scale

Class	*Site rating*	*Percentage of maximum UK urban population density in most populated 30° sectors*	*Equivalent category in earlier site assessment method*
I	Up to 750	6	D
II	Up to 1,500	12	D/E
III	Up to 3,000	25	E
IV	Up to 6,000	50	F

Source: Charlesworth and Gronow 1967: 145

of the sector method the population could be allowed to live even closer to the station. Rapid population growth, such as that represented by urban housing programmes, provided a major difficulty for both the original Fry and the new Farmer site evaluation techniques. The derivation of weighted factors in a radial direction was designed to discriminate against close-in populations, but left it as a matter of judgement whether developments in individual sectors were to be treated in the same way as a multisector development generated by a large city or conurbation. It was also possible to identify sites in a relatively low classification where, for a given radial distance, significantly fewer people were at risk than around a nominally more restrictive class of site (Charlesworth and Gronow 1967). Whether these things mattered very much is doubtful for, although it was recommended in 1963 that Magnox reactors and AGRs in prestressed concrete pressure vessels should be acceptable on class II sites because they were inherently safer than their predecessors, all the Magnox sites had been decided by 1960–1.

The 1963 recommendation was based on the evaluation of the relative safety features of reactor plant designs. The introduction of the prestressed concrete pressure vessel which enclosed the reactor core, the primary coolant circuit and the heat exchangers was considered to be a major improvement in the overall safety of the plant. This enabled a relaxation in siting restrictions to be permitted, so that this type of plant could be sited nearer to urban populations. Cost estimates made this desirable economically and the CEGB claimed to have found difficulties in obtaining sites under a remote siting policy (Mounfield 1967). The sites were in rural areas and frequently in areas of high landscape or amenity value. The adverse effect on amenity of the power stations and their transmission lines at such sites had come in for considerable criticism (Faux and Stone 1963). The desirability of a fresh siting policy was also underlined by the application of probability analysis to reactor safety. It began to be argued that the building and acceptance of nuclear power stations involved the acceptance of a finite degree of risk, and that, because no engineering plant or structure built to hold it could be entirely risk free, there was no logical way of distinguishing between 'credible' and 'incredible' accidents. The incredible, it was argued, is often made up of a combination of very ordinary events, and the credible may actually be exceedingly improbable (Farmer 1967a). The maximum credible accident approach came in for criticism in some quarters because it took no specific account of this low probability (Farmer 1967a; Barry 1970).

Accident probability as an approach to reactor siting

The probability approach to safety in reactor siting was first discussed in the mid-1950s (Siddall 1957), but was further expounded and elaborated by F.R. Farmer and others in the 1960s. The key characteristic of the approach is that accident probability is treated statistically: the consequences of an uncontrolled release of fission products are accepted *on the condition that the likelihood of such an event could be shown to be extremely low.*

An assumption of failure in a piece of equipment or operator error provides the starting point. A measure of risk can be obtained by estimating the probability of the failure or error and assessing the consequences. Any initiating event, e.g. failure of pipework, loss of coolant circulator power or operator error in the use of control systems, may set up an *incident sequence* which can follow many *pathways.* The notion of equifinality is that the same disaster can be reached along different paths, but the likelihood of following any one path depends upon the performance of many items of equipment and from knowledge of this performance a probability can be assigned to each pathway to produce an *event tree.* Essentially, an event tree maps the course of possible sequences of multiple equipment failure which may jeopardize some of the safety barriers interposed between the nuclear fuel in the reactor and the general public.

A simple event tree is shown in Figure 10.4. If it is assumed that the initiating event, for example, a loss of feedwater as happened at Three Mile Island, occurs with a frequency of once per year and that the failure probabilities P_1, P_2, P_3 are each 1 in 100 or 0.01, then the frequency of occurrence of each possible sequence can be calculated. For this set of figures full protection is provided for 97 per cent of the time, partial

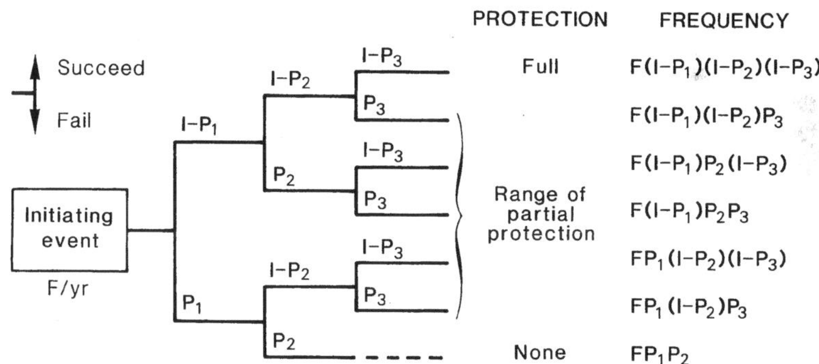

Figure 10.4 An event tree. The diagram shows the technique used to construct a fault tree from an initiating event. The frequency *F* of the event is calculated from engineering and other criteria and the cascade of failure possibilities which it might entrain are constructed in the form of a fault tree in which, at each step, the failure probabilities P_1, P_2, P_3 etc. are given a calculated value. The overall probability of each possible sequence can therefore be calculated, but relationship with reality in the event of failure depends on the accuracy with which real events can be simulated in the computer model
Source: Collier 1979

protection for 2.99 per cent of the time and no protection for 0.01 per cent of the time or just once in every 10,000 years of reactor operation.

A full safety evaluation would comprise a spectrum of events with associated probabilities and consequences. In the 1960s Farmer suggested the construction of a probability consequence diagram as shown in Figure 10.5 using as a convenient scale for probabilities the average time interval between events, i.e. reactor years, and for consequences the equivalent ground-level release of I-131. The aim of the Farmer criteria was to ensure an inverse log–log relationship between the frequency of accidents and their consequence as measured by the amount of radiation released. It is seen that in Figure 10.5 area A is one of low risk and area B is one of high risk and all parallel lines of equal slope −1 join points of equal risk. One such line, CD in Figure 10.5, might be used as a safety criterion by defining an upper boundary of permissible probability for all fault consequences. Although this line joins points of equal risk in curies per year, it may not represent an equal risk of casualties (a casualty is defined as one case of thyroid carcinoma). Furthermore, a relatively heavier penalty ought to be imposed against the possibility of a large release than a small release. This would lead to a boundary line of greater negative slope. One with a slope of −1.5 would reduce by three orders of magnitude the frequency of an event whose severity increases by two orders of magnitude.

A full probability safety evaluation of a reactor results in many culminating points for numerous event trees. Some of the culminating points will be within the safe zone, by several orders of magnitude in probability or severity, and so will make little contribution to the total risk. Some points may be near the boundary line and their summation may provide a basis for overall risk assessment.

Farmer (1967b: 318) calculated the risk level for a hypothetical reactor sited in a heavily populated urban area and concluded that it was very low indeed. He also argued that the final presentation of reactor risk by *any*

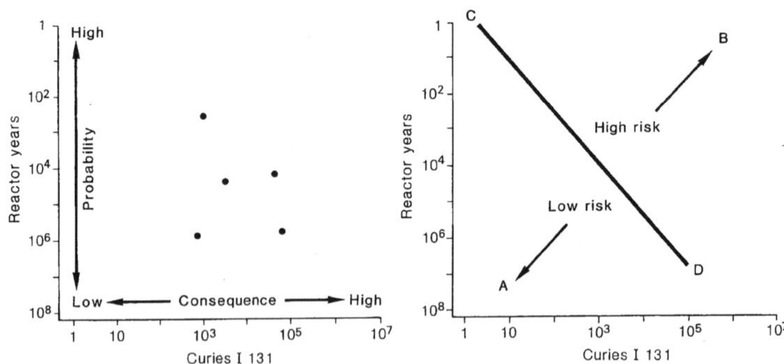

Figure 10.5 Probability consequence diagram
Source: Farmer 1967b

method was inexact because of lack of information on plant component behaviour. He maintained that sites in the UK near urban populations which might be used in a power network, as distinct from truly remote sites, varied by a factor of ten at most in population density. It was further suggested that the estimate of risk associated with a reactor carried an uncertainty of about one order of magnitude, and that the position of the boundary line (Figure 10.5) can also be placed by informed opinion within one order of magnitude. He therefore concluded that reactors could not be matched to sites with any precision:

> If reactors can only be graded within a factor of 10, is there any merit in a more refined sub-division of sites? There is obviously a development period in all countries permitting a growth of reactor technology and skills in operation and inspection; a maturing of siting policies and of public opinion. In this phase site selection is very much a matter of political wisdom.
>
> It is already clear in the UK that reactors must attain a standard of safety which will permit complete freedom in siting and a similar situation will be reached shortly in a number of countries. When this goal is reached, site categorisation will disappear and all reactors will need to meet a single high standard if they are to play a significant part in a power network. Less populated sites might be reserved for the development of new types.
>
> (Farmer 1967b)

Clearly the key to safety in siting was to be the safety and reliability of the reactor and associated equipment rather than the characteristics of particular sites.

Farmer's arguments met with an interested response in the nuclear power industry and in relevant government circles. On 6 February 1968 a revised siting policy, later to become known as 'the relaxed siting policy', was announced to Parliament by the Ministry of Power (*Hansard*, 6 February 1968). The announcement did not give any quantitative criteria for site selection but considered that gas-cooled reactors in prestressed concrete pressure vessels could be constructed and operated much nearer to built-up areas than was previously permitted. In a later statement on 12 March 1968 it was said that it was not yet contemplated that sites would be licensed within a mile or two of full urban population densities. The population within any 30° sector within 1–3 km of a reactor had to be capable of being evacuated within two hours; no other definitions were given for site requirements. However, it was considered that the population characteristics of the sites at Heysham and Hartlepool were examples of acceptable sites under the new policy.

Despite this 'relaxed siting policy', the Nuclear Installations Inspectorate (NII) considered it desirable to have some quantified siting criteria specifying acceptable population distributions as a guide to prospective licensees in the examination of potential sites, to set standards against which applications might be judged and to help in controlling housing developments and population growth around a site. The method proposed by Charlesworth and Gronow (1967) was revised and updated by Shaw and Palabrica (1974). Siting was considered to be of secondary importance to engineered design features of the plant, but was recognized as the only effective means of controlling the exposure of the population to radiation in the event of an incident leading to a release of radioactivity into the environment. The siting criteria developed to implement the policy included the use of sites in or near large centres of population, and the extent of provision for emergency action became an essential feature of siting policy. Sites were only accepted if it could be shown that effective emergency evacuation and possible medication could be taken for all persons within 0.67 mile (1 km) radius of the site, and that the population density and distribution, local communications, roads, transport, reception centres etc. enabled the emergency evacuation to be extended, if necessary, to all persons within a 2 mile (3 km) radius of the site. These characteristics had to be maintained for the life of the station and the local planning authorities were required to consult the NII before giving permission for certain categories of development within a radius of 2 miles (3 km). There were no restrictions on purely industrial development provided that it did not present a hazard to the safe operation of the nuclear plant or any incident arising from the nuclear plant presented any unusual situation at the industrial site, and that the workforce could be effectively included in the emergency procedures.

In assessing a site, consideration is also given to the population size and distribution in the 2–20 mile (3–30 km) radial zone. Development in this area is not subjected to control but it is unlikely that a licence would be granted for a site where development to full urban density was probable during the life of the station.

The present method of assessing a proposed site in the UK proceeds by first deriving a set of characteristic curves for the site and for the worst 30° sector. These curves provide a measure of the population distribution around the site and give an assessment of risk in terms of thyroid dose due to airborne I-131 in the event of an incident leading to a release of radioactivity. The curves are called site and sector risk curves (Haire and Shaw 1979). The region within a 20 mile radius of the site is divided into seven radial zones with zonal distances of 0–1, 1–1.5, 1.5–2, 2–3, 3–5, 5–10 and 10–20 miles. The population P_i in each zone is multiplied by a zone weighting factor W_i and the site characteristic curve is obtained by plotting

$$\sum_{i=1}^{n} P_i W_i$$

against the outer radius of the ith zone. The weighting factor for the zone represents the potential thyroid inhalation dose that a hypothetical 'standard person' located at the centre of the zone would receive from a unit ground-level release of I-131 under specified wind conditions. Once weighting factors have been derived they can be used to determine allowed population distributions by using limiting site and 30° sector risk curves. The choice of these limiting curves 'gives the licensing authority flexibility in its approach to site assessment' (Haire and Shaw 1979).

The limiting curves used for gas-cooled reactors in prestressed concrete vessels can be varied according to government policy at any time. These limiting curves, not surprisingly, encompass the population characteristics of the sites at Heysham and Hartlepool and do not render unacceptable any previously approved sites.

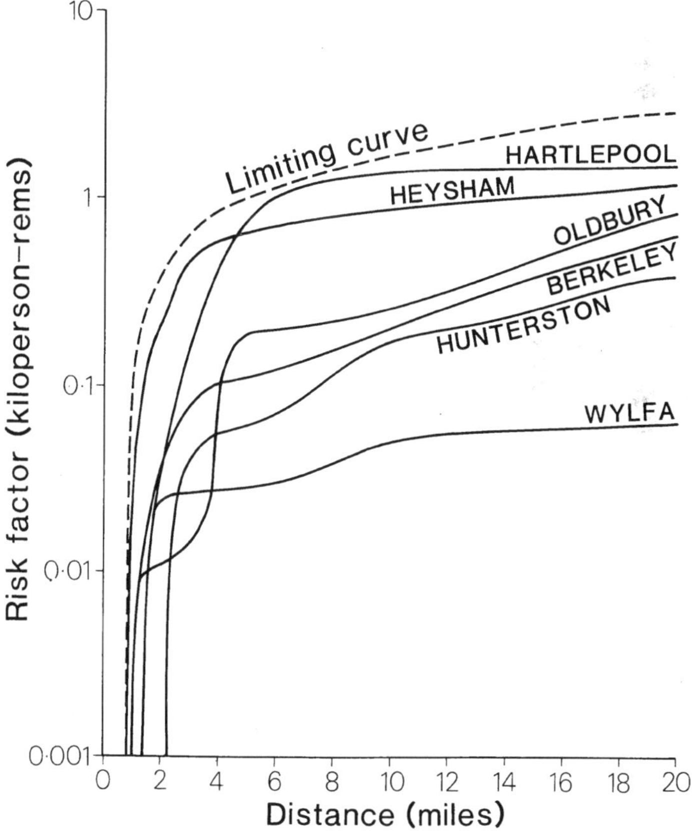

Figure 10.6 Theoretical site limiting curve and actual curves for a selection of nuclear power station sites in the UK
Source: Haire and Shaw 1979

Table 10.3 Population weighting factors and limiting populations for metric zonal distances

Zone no.	*Zonal distance (km)*	*Site weighting factor (mrem)*	*Sector weighting factor (mrem)*	*Site risk factor (kiloperson rem)*	*Sector risk factor (kiloperson rem)*	*Site population*		*30° sector population*	
						Zone	*Total*	*Zone*	*Total*
1	0–1.5	28.2	22.7	0.10	0.032	3,600	3,600	1,400	1,400
2	1.5–3	14.0	10.57	0.40	0.133	21,400	25,000	9,600	11,000
3	3–5	5.73	4.11	0.66	0.218	45,000	70,000	21,000	32,000
4	5–8	3.20	2.19	0.98	0.328	100,000	170,000	50,000	82,000
5	8–15	1.48	0.95	1.57	0.520	400,000	570,000	202,000	284,000
6	15–30	0.61	0.36	2.58	0.850	1,656,000	2,226,000	916,000	1,200,000

Source: Haire and Shaw 1979

The present site and 30° sector limiting curves are shown in Figure 10.6 (with distances shown in miles), and Table 10.3 shows the population weighting factors and limiting populations currently taken into account in the siting of power reactors in the UK.

If the actual site and 30° sector curves for a potential site are greater than the limiting curves at any point, the site is not considered to be acceptable and is given no further consideration. If the actual risk curves are lower than the limiting curves, then further consideration is given to the problems associated with the possible evacuation of persons situated within the 2 mile zone. Restrictions such as communications, transport, possible medication and provision of reception centres could mean that a proposed site is still not acceptable.

The UK-population-based computerized meteorological approaches to siting have been criticized on three counts (Openshaw 1982). First, their application is based on data that require fudging because the spherical zoning systems used are incompatible with the actual areas for which population data are available. Second, the location of the 10°, 30° or 60° sectors is arbitrary. Different results may be obtained if the origin is not at 0°. Third, the maximum distance limit of 32 km (20 miles) is inadequate when major reactor accidents could affect populations out to 600 km (310 miles) or more. The CEGB claim that suitable sites were in short supply in the 1960s has also been challenged by Openshaw, and he has published a set of four computer-generated maps which are the result of the interactions between the various siting criteria and the population geography of the UK (Figure 10.7). The dark areas are where population siting criteria would inhibit reactor siting. Openshaw claims that the maps show not only the widening of changing criteria for reactor location, but also that many more sites were available in the 1960s than indicated by the CEGB. Figure 10.8 shows the kind of maps that the CEGB was publishing in the mid-1960s to illustrate its view of the problem of obtaining remote sites. Even now *different* kinds of sites are being reserved, some remote for untried future reactors and some close to population centres for well-tried commercial power reactors. In effect this is the basis of the nationally organized siting strategy and it is not entirely certain how this will be influenced by the privatization of the electricity industry.

The exclusion area and zone of low population density

The first nuclear power plants in the UK were located in sparsely populated regions. In addition to reactor containment and other engineered safeguards, distance from densely populated areas was relied upon as a means of protecting the public from radiation in case of an accident. As experience in designing, building and operating reactors accumulated, and as some of the limitations of a population density siting strategy were realized, later nuclear power plants were sited nearer to large population

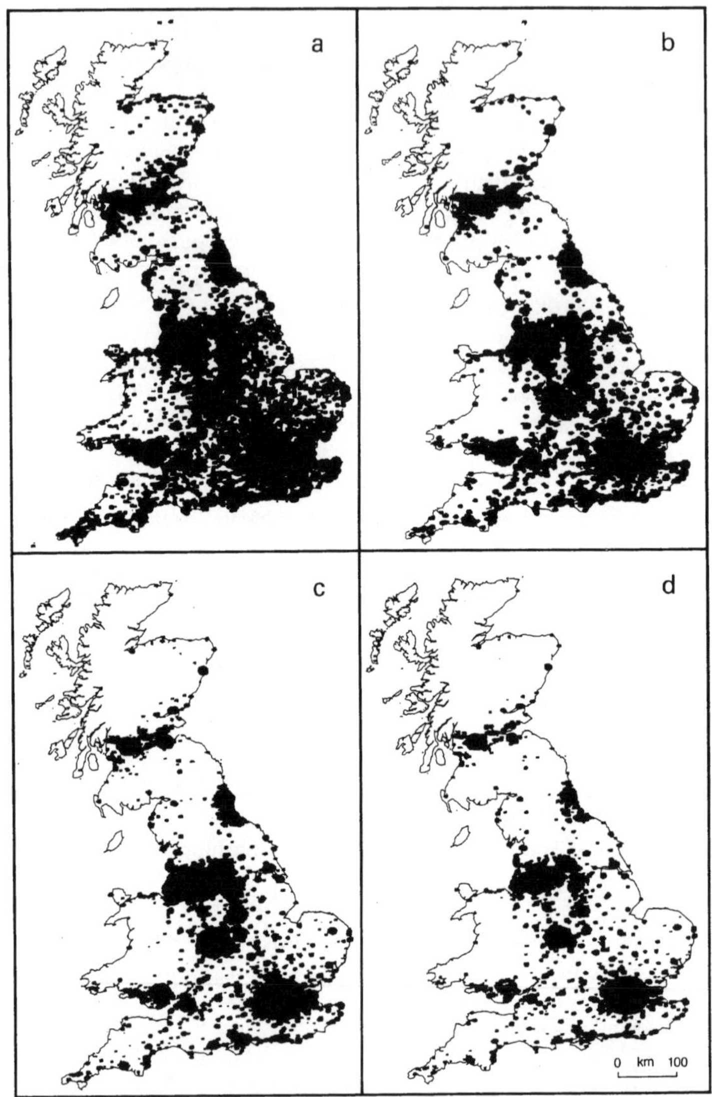

Figure 10.7 Unsuitable sites for nuclear power stations in the UK: (a) according to Fry (1955), (b) according to Farmer (1962), (c) according to Charlesworth and Gronow (1967) and (d) according to Shaw and Palabrica (1974)
Source: Openshaw 1982

centres (Mounfield 1967; Gammon 1979). The Hartlepool AGR, for example, is only 5.5 km (3.5 miles) from the town centre, and only 3 miles (4.8 km) separate the Heysham AGR from Morecambe on the Lancashire coast. Even so, distance from population centres and planning control over housing developments near nuclear power station sites have not been entirely abandoned as part of a safety-in-siting strategy. Indeed, as nuclear power sites have been chosen close to population centres, the exclusion area, with a perimeter fence, from which the public are excluded entirely and a zone of low population density around the exclusion area from which people could readily be evacuated have become even more important than before. But just how important?

In stable air conditions the atmospheric dispersion of any radioactive gases and particulate matter released in an accident gives a dilution factor of 10^2 from 100 yards (91 m) out to a distance of 1 mile (1.6 km), and a further factor of ten is attained at a distance of 8.0 km (5 miles). Thereafter, dispersion is slow and a further dispersion factor of ten is achieved only at 50 miles (80.5 km). Thus, for a large release, a significant public hazard would extend to several tens of miles. Thus, in a small and densely populated country such as the UK, remote siting alone can never be relied

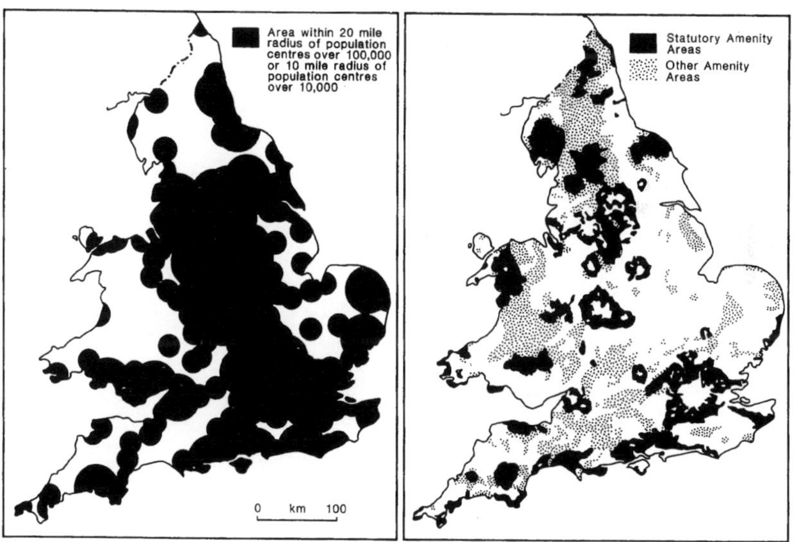

Figure 10.8 Influence of population centres and amenity areas on power reactor siting in the UK. Early attempts by the CEGB to indicate the difficulty of choosing nuclear power station sites in the UK. The black areas on the left-hand map were indicated to be unsuitable because of the presence of population centres; the black and shaded areas on the right-hand map were to be avoided as far as possible because of the landscape or amenity value
Source: Faux and Stone 1963

upon to safeguard the public. On the other hand, less severe accidents, which result in a relatively small release of gaseous and volatile fission products have happened, and in terms of this size of release, siting can mitigate the consequences to the public and differences between sites can be recognized. Reference to the release probability curve shown in Figure 10.9 helps to illustrate the significance of the total exclusion and low population zones. The diagram shows the probability of uncontrolled fission product releases per reactor year (Iansiti and Sennis 1967) and, since it is doses or integrated concentrations that are involved, it is apparent that the existence of an exclusion area shifts the curve towards the probability axis according to the ratio between the concentration at the exclusion area limit and the concentration in the immediate vicinity of the plant. *The main reason for the exclusion area is that the immediate vicinity of the plant cannot be represented with any reliability by the normal analytical models since it is particularly affected by local microclimatic conditions.* For example, atmospheric diffusion is strongly influenced by details of the release (concentrated or extended source, ground or elevated level), by the effect on air turbulence of the plant structures and by the actual wind speed and direction during an accidental release (see Chapter 1).

The low population zone around the plant makes it possible to plan and carry out evacuation in emergency situations. It has been suggested that the low population zone should extend to between 4,000 and 7,000 m (2.5–4.3 miles) (Iansiti and Sennis 1967), and that such a distance would again

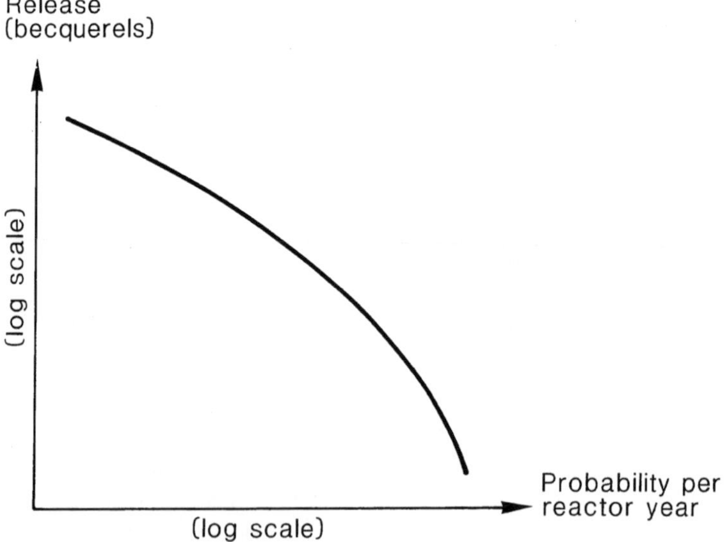

Figure 10.9 Release probability diagram
Source: Iansiti and Sennis 1967

mean a shift of the probability–concentration curve towards the probability axis by a factor of twenty to forty with respect to the concentrations that could occur at the limit of an exclusion area of about 600 m. The reality of any evacuation would be conditioned by the settlement pattern and the shape of the exclusion area: people in scattered single farmsteads would take much longer to gather together and move out than if they were in one or two compact nucleated villages.

It was on such grounds as these that on 23 October 1970 the Health and Safety Executive rejected an application by the CEGB to build an AGR power station just outside Stourport, Worcestershire. This was the first time that the UK electricity industry was refused a licence for a nuclear power station site. The 1,300 MWe power station would have been only 1.5 miles (2.4 km) from Stourport town centre, nearer to a population centre than ever before.

THE USA

The task of licensing nuclear power reactors in the USA rested with the US Atomic Energy Commission (USAEC) until January 1975, when it became the responsibility of its successor, the Nuclear Regulatory Commission (NRC). Over the years these bodies have set the standards for power reactor siting practice as part of a statutory responsibility under federal law, and it is federal law that no one may build or operate a nuclear power plant without first obtaining a construction permit and then an operating licence. However, a power utility has to go through preliminaries in the negotiating procedure before it even reaches the stage of applying for a construction permit. Application has to be made to the State Public Utilities Commission for its approval and all interested state and local authorities have to be notified of the proposed project, and over the years states have become increasingly involved in power reactor siting. If the proposed site does not have the appropriate planning zone classification, an application has to be made and granted for a zoning change. Alternative sites have to be considered, feasibility studies made and financial arrangements explored. A public announcement has to be made, usually through a news release, of the utility's intention to build the plant. Thus a nuclear power plant proposal accumulates a lengthy history before the licensing and regulatory machinery is put into motion.

USAEC staff reviews of the first commercial-size reactors were generally limited, and guidance to power utilities and construction firms was often based on direct communication. In the author's file of nuclear power station photographs is one of construction work in progress at the Tennessee Valley Authority's Brown's Ferry plant which was taken in 1967 some months *preceding* the granting of a formal construction permit. There is a very large hole in the ground, and a lot of construction equipment already in use. There is little doubt that in the early 1960s staff limitations at the USAEC meant that experts from outside that organization's Hazards Evaluation Branch frequently had to be called in for technical support and staff positions on specific design issues were often casually codified and documented (Komanoff 1981a). This state of affairs did not last long, however, and by early 1967, with twelve large reactors under construction and over a dozen more construction permits lodged, the USAEC began to expand its licensing division to cope with the growing case load. Word-of-mouth approval of design approaches was superseded by detailed examination requiring extensive documentation, and formal licence application procedures were established. Figure 10.10 shows the main steps in the USAEC application procedure that came into use. The first step was for the power company to prepare and submit a formal application. In addition to providing a detailed description of the proposed project and covering the applicant's technical qualifications and financial responsibility, the application analysed various accident possibilities, including the MCA, and described the safeguards to be provided against these accidents. The analysis included calculations of possible radiation exposure based on the design characteristics of the proposed plant and the meteorological, hydrological and other characteristics of the proposed site.

On 30 June 1966 the USAEC issued a *Guide for the Organization and Contents of Safety Analysis Reports*, reinforcing the message that satisfying the USAEC that a nuclear power plant would be safe was the chief prerequisite for obtaining a construction permit. However, at this stage the USAEC's regulatory authority was limited essentially to matters of radiological health and safety, defence and security. It had no control over zoning, cost and availability of land, aesthetics or environmental impact, including the thermal effects of cooling water discharges. However, taking into account the size and design of the proposed plant and the nature of the proposed site, there were many safety-related factors affecting the siting of nuclear reactors that were within the aegis of USAEC through the limits prescribed by its radiation protection regulations. These regulations had been set out in a document published on 11 February 1961 which, with subsequent modifications, was to guide reactor siting policy statements for many years: *Reactor Siting Criteria, Title 10, Code of Federal Regulations Part 100* (10 CFR 100). In this document were guidelines for assessing the potential hazards from a reactor assumed to have experienced core meltdown. Maximum acceptable limits were specifically defined for the following:

(a) the amount of fission products released from the reactor core to the containment;
(b) the amount of fission products available for leakage from plant containment to the environment;
(c) exposure dose to the public.

By using a demonstrable leak rate and conservative assumptions

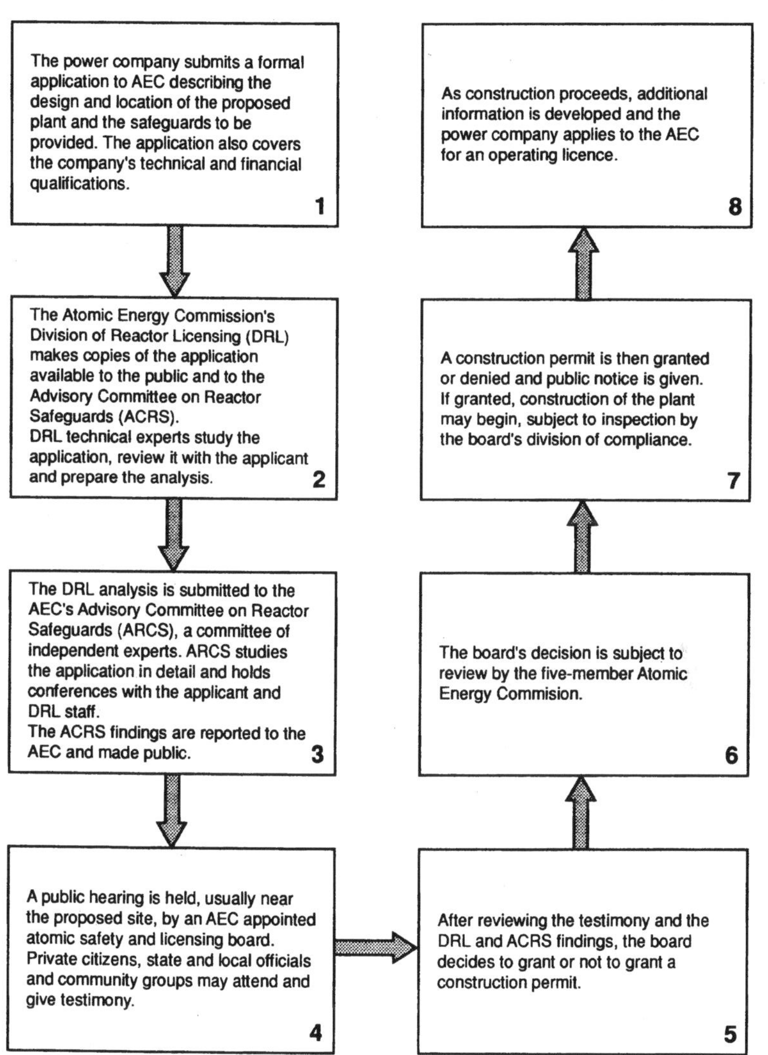

Figure 10.10 Steps in the USAEC application procedure for a nuclear power station construction permit in the 1960s

regarding meteorological conditions at a site, it proved possible to derive an exclusion area, a low population zone and a population centre distance which were defined as follows:

(a) A *population exclusion area* is an area of such a size that an individual located at any point on its boundary for two hours immediately following the onset of the postulated fission product release would not receive a total radiation dose to the whole body in excess of 25 rem or a total radiation dose in excess of 300 rem to the thyroid gland from iodine exposure. There should normally be no residents within the exclusion area, and the people inside the area should be plant personnel readily subject to emergency control by the reactor operators in case of accident.
(b) *A low population zone* is a zone of such size that an individual located at any point on its outer boundary who is exposed to the radioactive cloud resulting from the postulated fission product release, during the entire period of its passage, would not receive a total radiation dose to the whole body in excess of 25 rem or a total radiation dose in excess of 300 rem to the thyroid from iodine exposure.
(c) A population centre (over 25,000 residents) distance must be at least one and a third times the distance from the reactor to the outer boundary of the low population zone. In applying this guide, due consideration should be given to the population distribution within the population centre, and where very large cities are involved a greater distance may be necessary because of total integrated population dose consideration.

The intention was to guide nuclear power reactors to sites away from major concentrations of population where protective counter-measures in case of a serious release to the atmosphere would be difficult to implement. On 23 March 1963 the USAEC expanded on these guidelines in *Calculations of Distance Factors for Power and Test Reactor Sites* (TID 14844) by showing how the distances for these zones around a reactor site could be calculated for a PWR, given a number of assumptions concerning the nature of the accident and the meteorological conditions at the time. At the time, the assumptions seemed to be conservative; for example, atmospheric dispersion was assumed to occur under inversion type weather conditions. The results of the calculations performed for the inhalation (iodine) dose and the external gamma radiation dose produced estimated distances for the exclusion area and low population zone for reactors of different power levels. These are listed in Table 10.4.

Under the conditions assumed, the doses resulting from the inhalation of iodine isotopes provided the main control for the low population zone distance and the population centre distance. The population centre distance was equal to the low population zone distance increased by a factor of one-third (see Figure 10.3).

Table 10.4 Calculated radii for water-cooled reactors of various power levels in the USA

Power level of reactor (MWe)	*Exclusion area distance (miles)*	*Low population zone distance (miles)*	*Population centre distance (miles)*
500	0.88	13.3	17.7
400	0.77	11.5	15.3
300	0.63	9.4	12.5
200	0.48	7.2	9.6
100	0.31	4.5	6.0
30	0.25	2.2	2.9
3	0.13	0.5	0.7

Source: TID-14844: 31

A number of nuclear power plants were in operation or under construction when TID 14844 was published in March 1962. Comparison of actual and calculated distances for these reactors is provided in Table 10.5. It can be seen that in two cases (Pathfinder and PG & E) the actual population centre distance is less than the calculated distance, and that in four cases (Conn.Ed., Hallam, PG & E and Piqua) the actual distance for the exclusion area was smaller than the calculated distance.

To achieve scale economics, the trend in the USA was soon to build nuclear plants of ever-increasing size. By 1964 a closed-cycle water reactor was on offer to the power utilities which was of such a size that, without additional engineered safeguards, the USAEC guidelines indicated the population centre distance to be about 28 miles (45 km) from the plant site. In most cases such a site would have been regarded as economically unattractive because of the transmission costs (Malay and Chave, 1964), but the USAEC made it clear to the utilities that a site with unfavourable population characteristics could nevertheless be approved if appropriate compensating safeguards were supplied. It was indicated that where detailed documented safeguards were supplied to the regulatory agency, demonstrating sufficient effectiveness and reliability, reactor sites could be approved with some reduction in the distance factors that otherwise would be required. Many were approved. San Onofre I (440 MWe) went onto full power in 1968 at San Clemente, California. Calculated on the basis of TID 14844 it should have had a minimum exclusion radius of 4,050 feet, a low population zone out to 11 miles (18 km) and a population centre distance of 14.7 miles (23 km). The actual figures were 2,600 feet, 4 miles (6.5 km) and 10 miles (16 km). Malibu should have had an exclusion radius of 4,840 feet (1,475 m), a low population zone out to 15 miles (25 km) and a population centre distance of 20 miles (32 km). The actual figures were 1,700 feet (5.8 m), 5 miles (8 km) and 10 miles (16 km). Connecticut Yankee should have had an exclusion radius of 5,150 feet (1,570 m), a low

Table 10.5 Calculated and actual distances for selected US reactors in existence or under construction in 1962

Reactor	*Power level (MWe)*	*Exclusion area*		*Calculated low population area distance (miles)*	*Population centre distance*	
		Calculated distance (miles)	*Actual distance (miles)*		*Calculated distance (miles)*	*Actual distance (miles)*
Dresden	210	0.50	0.50	7.4	9.9	14.0
Conn. Ed.	188	0.48	0.30	7.0	9.4	17.0
Yankee	161	0.42	0.50	6.3	8.4	21.0
PRDC[a]	100	0.31	0.75	4.5	6.1	7.5
PWR	90	0.31	0.40	4.1	5.6	7.5
Consumers	80	0.30	0.50	3.9	5.2	135.0
Hallam[a]	80	0.30	0.25	3.9	5.2	17.0
Pathfinder	67	0.29	0.50	3.4	4.6	3.5
PG&E	67	0.29	0.25	3.4	4.6	3.0
Phil. Elec.[a]	38	0.26	0.57	2.4	3.2	21.0
NASA	20	0.22	0.50	1.6	2.1	3.0
CVTR	20	0.22	0.50	1.6	2.1	25.0
Elk River	19	0.22	0.23	1.5	2.0	20.0
VBWR	16	0.21	0.40	1.4	1.9	15.0
Piqua[a]	16	0.21	0.14	1.4	1.8	27.0

Note: [a]Not water-moderated reactors; included in the table for illustrative purposes only.

population zone out to 16 miles (25 km) and a population centre distance of 21.4 miles (34 km). The actual figures were 1,700 feet (518 m), 9.5 miles (15 km) and 9.5 miles (15 km) again.

Such examples could be multiplied from the plants approved and built during the 1970s. Nuclear power plants quickly became larger and were located ever closer to the load centres they were designed to serve. Because the economic attractiveness of such sites was clear they came to be regarded as safe enough for this to be justified. Weaknesses in this approach to safety in siting were not obvious to the nuclear industry until the Three Mile Island nuclear accident in Pennsylvania, from 28 March to 2 April 1979. As the accident at that plant assumed crisis proportions, it became clear that, in assessing whether or not evacuation of the public from areas around the plant should take place, fundamental questions could not be answered. The first concern was the measurement of the amounts of radiation. The extent of radiation was difficult to determine because experts were not certain if readings were one-time or steady releases, and frequent changes in wind direction alternating with a flat calm rendered predictions on the areas of impact unusable. Thus a decision was made to use a circular form of evacuation around the accident site. A second problem was the size of the area to be evacuated. The NRC had based its original recommendation on a low population zone, which at Three Mile Island was less than 5 miles (8 km) from the plant, while the State of Pennsylvania was prepared for a 5 mile evacuation. The Pennsylvania State Bureau of Radiological Protection recommended no evacuation, while the NRC recommended a 10 mile (16 km) evacuation. While plans were under way for a 10 mile evacuation the Governor of Pennsylvania ordered a partial evacuation of pregnant women and preschool children, ordered all others within 5 miles to stay indoors and requested a single NRC spokesman from the US President. This spokesman arrived with a recommendation for a 20 mile (32 km) evacuation, which included the state capital itself.

In the event, no total evacuation proved necessary, although some 144,000 people actually evacuated themselves from a radius of 15 miles from the plant, but Three Mile Island demonstrated that responses to accidents are a continuum all the way from minor incidents to total catastrophe and should be treated as an entire management system (Fischer 1981). Figure 10.11 indicates the existence of three phases in an accident: the pre-accident phase of prevention and contingency planning, the accident phase of actual responses directed toward on- and off-site accident control and the post-accident phase of recovery and incorporation of lessons learned back into the first phase. At Three Mile Island resources were overconcentrated in the prevention aspect of the accident management system but, even so, produced a management response that worked very badly.

In June 1979 the NRC began a formal reconsideration of the role of emergency planning to ensure the protection of the health and safety of

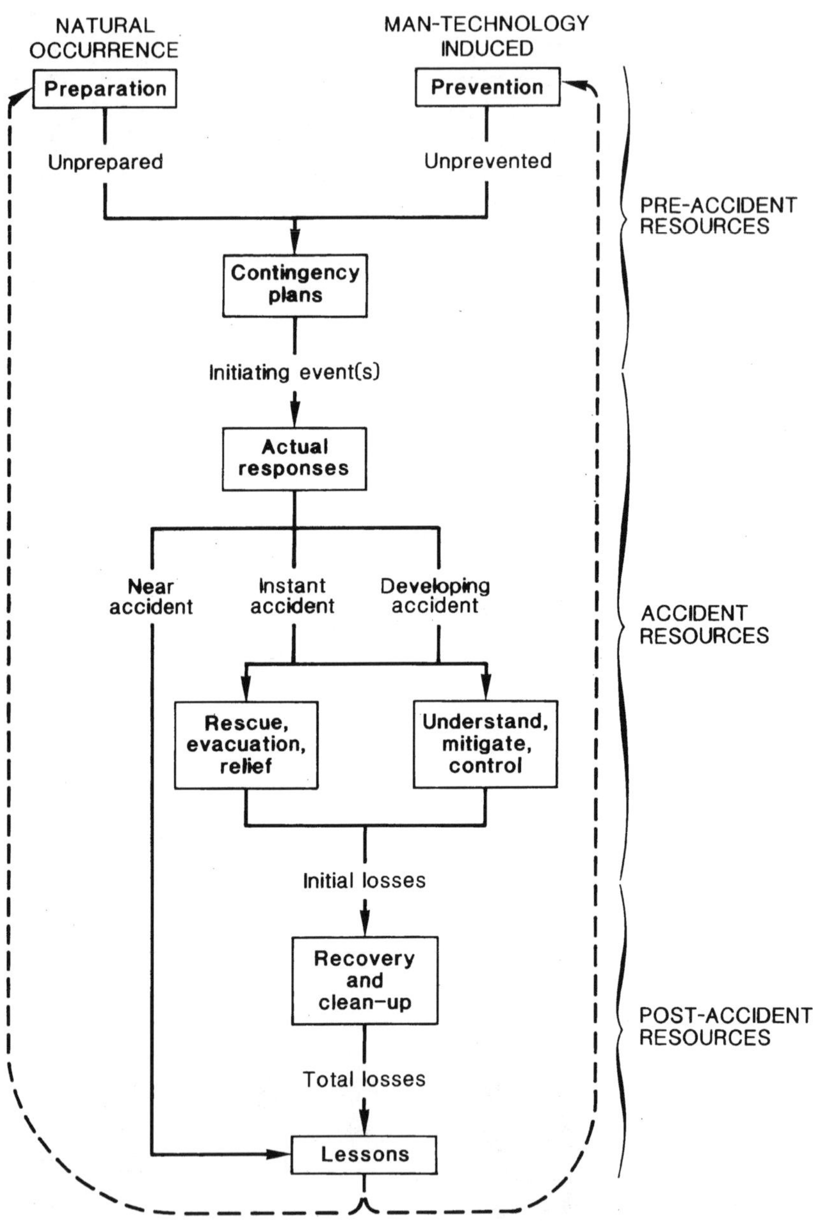

Figure 10.11 General accident management responses and resources
Source: Fischer 1981

people in areas around nuclear power stations. In a policy statement published on 23 October 1979 the NRC stated that two Emergency Planning Zones (EPZs) should be established around each nuclear power plant. For larger more recently built plants, and those under construction, the EPZ for airborne exposure was given a radius of 10 miles (16 km) based on plume pathway: the EPZ for contaminated food and water, based on ingestion pathways, was given a radius of 50 miles (80 km) (Figure 10.12). The small older power reactors of less than 250 MWe and the gas-cooled reactor at Fort St Vrain were allowed 5 and 30 mile zones (8 and 48 km), and it was decided to set zones for other existing reactors on a case-by-case basis. Nuclear power plant operators were required to submit their emergency plans for these areas to the NRC, together with the emergency response plans of state and local governments (Grimes *et al.* 1982). The NRC and the Federal Energy Management Agency (FEMA) reviewed the adequacy of these plans, and all power reactor licencees were required to submit their emergency plans by 2 January 1981. The exact size and shape of each EPZ were decided after consideration of specific conditions at each site.

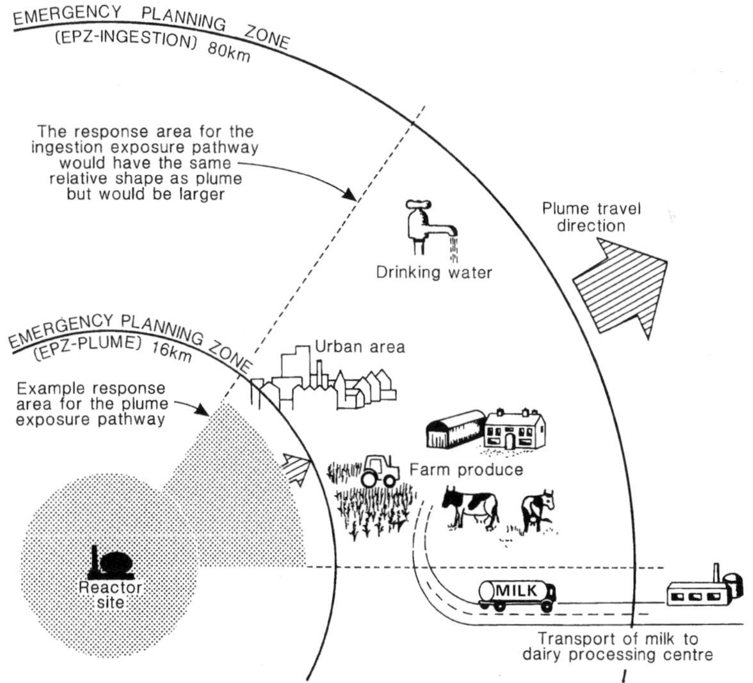

Figure 10.12 Concept of an emergency planning zone in the USA

EMERGENCY PROCEDURES IN THE UK

In the UK the design base accident (DBA) which could lead to an off-site release is called the 'reference accident'. Emergency plans have been prepared for each nuclear power station site in the event of a DBA. These provide for radioactivity monitoring up to 40 km (25 m) from the site and the declaration of a state of site emergency which could be followed by (i) sheltering (public advice to stay indoors and close doors and windows), (ii) issue of potassium iodate tablets to people who might be exposed during the passage of the plume to minimize the effects of uptake of radioactive iodine, (iii) evacuation of the public from the downwind sector up to a distance of 2–3 km from the site to reduce the risk of exposure and (iv) control of potentially contaminated food and water supplies. Detailed guidance is provided in each station emergency plan for the dose levels at which the Emergency Controller would recommend these actions (Health and Safety Executive 1982; Hill *et al.* 1988). In an emergency the power station's management becomes responsible for the site, taking advice from the NII. Each nuclear installation possesses an Operational Support Centre (OSC), located between 10 and 30 km from the installation, to act as a control room to deal with off-site consequences of nuclear accidents.

The advantage of sheltering is that it is a measure capable of being implemented rapidly, possibly preceding an evacuation, but opinions differ as to the degree of protection that would be provided from the gamma radiation of a cloud passing overhead (Atkinson *et al.* 1988). Administering stable iodine tablets would be a very effective measure to reduce the dose received provided that the tablet is taken before or very soon after the arrival of the radioactive plume. Once the cloud has passed the benefit dwindles rapidly because radioactive iodine has already entered the body and begun to concentrate in the thyroid. The administration of iodine is quicker, cheaper and easier than evacuation. Evacuation, though, is the most effective of all the emergency counter-measures as it removes people from the source of contamination and can thus reduce dramatically their total dose. Ideally people should be removed before the arrival of the cloud, but considerable benefits would still be derived from later evacuation to avoid the cumulative effects of radiation. The greatest problems with evacuation would be the provision of adequate transport and reception centres capable of providing adequate bathing, showering and laundering facilities for personal decontamination at the earliest possible moment, and road accidents arising from panic on overcrowded roads. It was an inadequate road system for evacuation that finally precipitated the decision not to commission the nuclear power station at Shoreham, Long Island, USA. The control of food and water is a longer-term rather than emergency counter-measure, as is the decontamination of property following the accident (Howarth and Sandalls 1987).

No emergency plan can be other than open-ended and incomplete, and all plans suffer from degradation with the passage of time (Sills *et al.* 1982). Nowhere is this clearer than in the UK where the reference accident has come to be accepted by the nuclear authorities as the limit of the credible, with the result that detailed consideration of the consequences of an event of greater magnitude is not considered necessary (Atkinson *et al.* 1988). In 1989 the CEGB believed that it could make any OSC operational within three to four hours, but emergency counter-measures are most effective if initiated before the arrival of a radioactive cloud. With light or moderate winds, the plume would be likely to reach the maximum downwind extent of current emergency planning zones in the UK in half an hour or so. It is difficult to envisage any way in which sheltering or distribution of stable iodine could be deployed effectively under the present OSC system, for an effective response within the first hour should be a minimum requirement out to 10 km from the source of the plume. An effective response also requires a much greater public awareness of the requirements of an emergency situation than exists in the UK.

Things are better arranged in the USA. Within the 10 mile EPZ stringent counter-measures have been designed and are now regularly exercised. Emphasis is placed upon public education and upon speed of notification within the EPZ in the event of an accident and FEMA has taken over responsibility for approving emergency plans as part of the site licensing arrangements and for monitoring exercises. The Three Mile Island accident stimulated the development of these arrangements, which as yet (1988) have no UK counterparts.

THE ACCIDENT RECORD

In any discussion of the nuclear power industry's safety record it is necessary to make three things clear at the outset. The first is that most industries are dangerous to some degree – quarrying and mining kill people, so does transportation. In 1988–9 far more people were killed in the UK in rail and air accidents than in all the power industries put together. Second, in the nuclear power field it is necessary to distinguish between incidents having no demonstrably deleterious effects on human health or the environment, even though contamination may have to be dealt with, and accidents of a scale such that actual or potential irradiation, injury and environmental degradation is involved. Third, it is necessary to specify exactly which part of the nuclear fuel cycle is being discussed; here it is the production of heat energy in the nuclear reactors and *not* fuel processing, waste disposal or uranium mining.

Between 1940 and 1980, ninety-eight separate accidents in which persons were injured by accidental exposure to ionizing radiation were recorded world-wide. Among these were fourteen criticality (reactor) accidents with five fatalities, all before 1965. By far the greatest number, fifty-six, were the result of mishandling industrial sealed isotope sources or the inadvertent exposure to X-rays used for quality control in industry (Sagan and Fry 1980). Of the 562 individuals involved in the ninety-eight reported accidents, sixteen died as a direct result of the acute effects of their accidental exposure, five of the deaths occurring in the USA.

Some of the fourteen reactor accidents occurred at experimental facilities designed to operate at essentially zero power. Such an accident occurred in 1958 at an unshielded experimental reactor at the Boris Kidrich Institute at Vinca, Yugoslavia, resulting in one death and several radiation injuries. The accident was the result of improper operating procedure, and happened when the facility was being prepared for an experiment and safety devices were not in service. In 1961 a criticality accident occurred in a small prototype BWR (SL-1) designed to meet specialized military requirements and located at the National Reactor Testing Station in Idaho. The accident happened during a shutdown period when the reactor was undergoing maintenance and was due to the sudden manual withdrawal of the central control rod by a technician attempting to reconnect the rod to its drive mechanism in preparation for the resumption of operation. This created a super-critical condition in the reactor, and the sudden generation of heat that followed melted the core assembly and produced a pressure surge sufficient to cause serious damage to the reactor installation. Three technicians at work on top of the reactor at the time of the accident were killed and the shed housing the reactor was heavily contaminated with radioactivity.

A partial review reveals other criticality accidents in experimental and power reactors, and a number of other adverse incidents such as partial fuel meltdowns due to improper coolant flow. The Fermi FBR near Detroit, Michigan, got into trouble in October 1966 when two fuel channels were blocked by dislodged metal plates. In October 1969 there was a partial fuel meltdown at St Laurent 1, a 480 MWe gas-cooled French power reactor near Orleans. The accident occurred through the error of an operator when he by-passed several interlocks of the automatically controlled fuelling machine and introduced a graphite piece which restricted coolant flow. Five elements sustained melting and the reactor was put out of action for a year. On 21 October 1973 an accident occurred at the San Onofre 1 plant, at San Clemente, California. Damage was caused to the reactor's emergency cooling system by a turbine throwing a blade. This set off vibrators which upset the flow of water and steam. Temperatures rose in the reactor and activated another back-up system separate from the emergency core cooling system. Pumps sent water gushing through pipes to the reactor core, but since nothing was wrong with the primary cooling system, this extra water slammed into valves leading to the core. Trapped air caused an air hammer effect which damaged pipe restraints and motor mountings on a power-operated valve. Noting the drop in pressure, an operator tried to restore it by pulling out one bank of control rods, thus effectively pressing the accelerator instead of the brake.

The subsequent USAEC report described this act as 'injudicious'. In March 1975 an accident which has entered the folklore of nuclear power occurred at TVA's 2,100 MWe Brown's Ferry nuclear power station in Decatur, Alabama. An electrician used a naked candle flame to test the efficiency of seals where control and instrument cables entered the reactor containment by seeing if any air flow caused the flame to flicker. The open flame ignited the polyurethane foam sealant; the fire spread to and put out of action a number of control cables, including those needed to operate the emergency core cooling system, and took several hours to put out. One reactor was out of use for weeks and the other for months, and the cost to TVA ran into tens of millions of dollars. In July–August 1976 the 2400 MWe Biblis power station developed cracks in a water container of the cooling system. Almost simultaneously, routine inspection revealed twenty loose screws in the heart of the reactor. Apparently loosened from a coolant pump by vibration, they had been taken along with the coolant to the reactor's pressure vessel. Their removal from the radioactive area, involving the use of remote control devices, took much longer than expected, and as a precautionary measure some operators were excluded for a while from work in the radiation area because their occupational radiation dose levels were approaching permissible maxima (Buschschluter 1976). The 687 MWe Wolsung 1 CANDU reactor, near Ulsan in South Korea, was put out of action for more than a month after 24 tonnes of radioactive cooling water seeped out of a steam generator pipe during a safety check on 25 November 1984. In March 1987 Philadelphia Electric Company's two nuclear units at Peach Bottom were shut down by the NRC after operators were found asleep or playing video games. After operator retraining the units were reopened in June 1989.

As far as can be ascertained from published evidence, none of these occurrences resulted in physical injury to plant workers or exposed the general public to radiation levels above those defined by then current radiation protection standards. Although they provided both experience for plant operators and publicity material for anti-nuclear power groups, they were at a level of incident occasionally experienced in many branches of heavy industry and were far less significant in their immediate effects than accidents that have taken place in chemical plants, such as the Flixborough disaster in the UK on 1 June 1974, when twenty-eight people were killed, and the world's worst industrial disaster at Union Carbide's Bhopal plant on 3 December 1984 which caused over 2,000 deaths. However, there have been three reactor accidents which have gone well beyond the level of those described so far. One was at Windscale in the UK, another at Harrisburg (Three Mile Island) in Pennsylvania, USA, and the third at Chernobyl in the USSR.

Windscale

On 10 October 1957 a fire occurred in a production reactor at the UKAEA's Windscale (now Sellafield) works. The reactor was fuelled with metallic uranium, moderated with graphite and cooled with a once-through flow of air, i.e. the cooling air discharged directly to the environment rather than being recirculated in a closed equipment loop. In such a reactor it is necessary to raise the temperature of the graphite periodically to rid it of stored energy. In this instance too much heat was applied, causing some of the fuel cladding to fail. The metallic uranium thereby exposed to the hot air oxidized rapidly and caught fire. The fire spread to a substantial section of the core before it was extinguished by flooding the pile with 5 million litres (1.1 million gallons) of water. Quantities of radioactive material (of the order of 20,000 Ci of I-131) escaped from the fuel into the air and caused contamination of an area downwind from the reactor. Activity was identified across the UK and into Europe. Fourteen workers at the plant were exposed to serious radiation by this release and there was sufficient effect on milk supplies through the selective intake of radio-iodine by dairy cattle to warrant temporary suspension of milk production and financial compensation to farmers within the most affected locality. The NRPB has estimated that about 30 deaths may have resulted over subsequent years from the fallout from the Windscale fire, about half of them due to polonium-210 which was being produced by the irradiation of bismuth in one of the number of side-channels in the pile at the time (Pearce 1988).

Three Mile Island

The TMI nuclear power plant at Harrisburg, Pennsylvania, has 2.6 million people living within a 50 mile (80 km) radius. On 28 March 1979 a series of events at Unit 2, shown in Table 10.6, formed part of an accident that captured world-wide attention and had enormous repercussions on the US nuclear industry. This table summarizes the first four hours, but the TMI accident consisted of three stages spread over a week and a fourth stage lasting much longer:

(a) the loss of cooling capability and lack of operator recognition of the initiating event leading to inappropriate responses which contributed to a misunderstanding of the nature of the accident and led to its severity;
(b) an unplanned single release of radioactivity into the atmosphere;
(c) the formation of a hydrogen bubble in the cooling system which had not been foreseen and which was thought to be potentially explosive;
(d) the cooling of the reactor core and securing of the plant to a safe condition.

The first stage was an on-site contained emergency that escalated into an accident with potential adverse off-site consequences. The second stage

Table 10.6 Sequence of events in the Three Mile Island accident

4.00 a.m.	
Zero	Feed-water pump failed Pressure rise in primary circuit Relief valves opened
8 s	Reactor shut down
13 s	Relief valves failed to close but indicated closure in the control room
13 s	Extra cooling water (normal) injected by the operator
1 min 45 s	Steam generators dried out
2 min	Pressure in primary circuit fell and emergency cooling water injection started
3 min 30 s	Indication of high level of water in the pressurizer caused operator to shut off one emergency cooling pump and reduce flow in the other
5 min 30 s	Steam generated in the core Operators drained off small volumes of primary water in response to rising level in pressurizer
11 min	Alarm indication of high level of water in the containment building sump Sump water water being pumped to auxiliary building
39 min	Sump pumps stopped by operator
60 min	Primary coolant pumps vibrating owing to presence of steam
1 h 14 min	Two primary pumps stopped by operator
1 h 45 min	Two remaining primary pumps stopped by operator
1 h 50 min	Temperature in outlet ducts high enough to show presence of superheated steam – significance not recognized
2 h 22 min	Block valves on the relief valve lines closed by operators
3 h	Site emergency declared
3 h 20 min	Indication of 800 rem/h in containment building
3 h 24 min	General emergency declared
4 h	Containment building isolated automatically but operators continued to transfer primary let-down water to tanks in auxiliary building
4 h 25 min	Local radio carried an item based on a statement from Metropolitan Edison's headquarters that a general emergency was a red-tape sort of thing and there was no danger off site

Source: Dunster 1980

was the public knowledge that a greater off-site emergency existed because of an initial high reading of radiation on an overflight of the plant which, however, did not continue. The third stage involved the growing hydrogen bubble which not only blocked coolant from reaching the reactor but was viewed as a source of hydrogen explosion potentially affecting thousands of people. The fourth stage was the gradual cooling of the plant to a safe shut-down and subsequent beginning of decontamination procedures. Such procedures were continuing in 1989, when 34 tonnes of fuel remained in the reactor, 8 tonnes in the form of a hard and brittle mass at the bottom of the reactor vessel.

The initial danger subsided on 2 April, but in the intervening period degassing of primary coolant water and leakage out of the waste-gas-handling system allowed radioactive noble gases to enter and pass through the building air filters to the atmosphere. The resulting off-site radiation levels at surrounding locations were much greater than those due to routine operation (Figure 10.13). The dosage received by the population was mainly due to xenon-133. The health consequences of the releases were subsequently analysed and found to be minimal, and the most important health effect of the accident on surrounding populations seems to have been mental stress suffered during the emergency (Fabrikant 1981). Three site workers received doses slightly exceeding the NRC's maximum permissible quarterly dose.

TMI had a number of consequences, including some relevant to siting. It cast doubt on the relevance of the MCA approach, for it was pointed out that, prior to TMI, technological prevention of a total or massive accident occupied so much attention that little was known or guidance available about intermediate accidents and their appropriate responses (Dunster 1980; Fischer 1981). It emphasized the fact that siting is more closely linked with the case of making effective emergency plans than with the amelioration of the consequences of accidents. Choice of site changes the number of people at risk without changing the level of individual protection. Except for carefully chosen combinations of size of accident and distribution of close-in population, siting in a practical European context, and maybe in some parts of the USA also, can influence the consequences of an accident by a factor of no more than ten. This is certainly not insignificant but it is not large compared with the factors available from protective measures at the reactor itself.

This suggests that the complete spectrum of design, construction, operation and maintenance features of a reactor should be extended to cover the emergency plans and siting considerations as part of the process of granting a licence. It follows that siting policy would not then be based on numerical rules, although quantitative assessments would play a major role (Dunster 1980). It has already been shown in this chapter that US emergency arrangements for nuclear sites were radically restructured after TMI.

Chernobyl

The town of Chernobyl, an ancient twelfth-century foundation, is located at the junction of the Prypiat and Uzh rivers, 133 km (83 miles) north of Kiev. Twenty kilometres (12.5 m) to the north is an RBMK four-reactor nuclear power station at which on 26 April 1986 occurred an accident described by the UK's CEGB Chairman, Lord Marshall of Goring, as 'the most traumatic in the entire history of civil nuclear power'. By Soviet

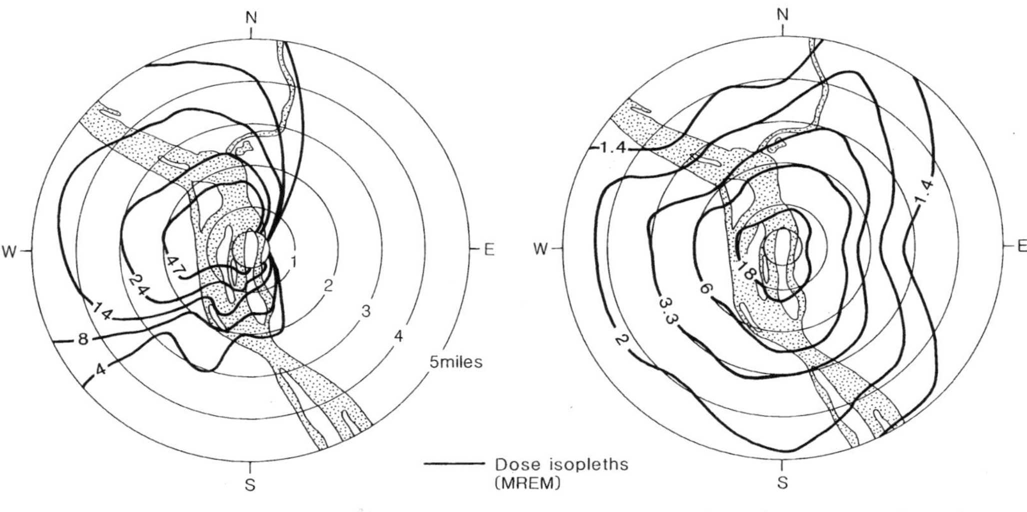

Figure 10.13 The Three Mile Island accident. The diagrams show dose isopleths in millirems out to 5 miles from the TMI nuclear power plant. The innermost circle represents the plant exclusion boundary
Source: Pasciak *et al.* 1981

standards the power station is located quite close to population centres at Kiev, Chernigov, Gomel and the new town of Prypiat, but it is in an area of sandy podzolic soils, marshlands and peatbogs, devoted mainly to dairy farming and constituting one of the more sparsely populated parts of the Ukraine. The population of the 2,000 km^2 Chernobyl *raion* in the late 1970s was around 47,000, with 19,000 in Chernobyl itself. Although this is not a major grain-growing *raion*, the local network of rivers connects it to the Ukraine's grain-growing regions.

The Chernobyl accident was triggered by a turbogenerator experiment with reactor number 4 when the reactor core contained water at just below the boiling point but little steam (USSR State Committee 1986; Gittus *et*

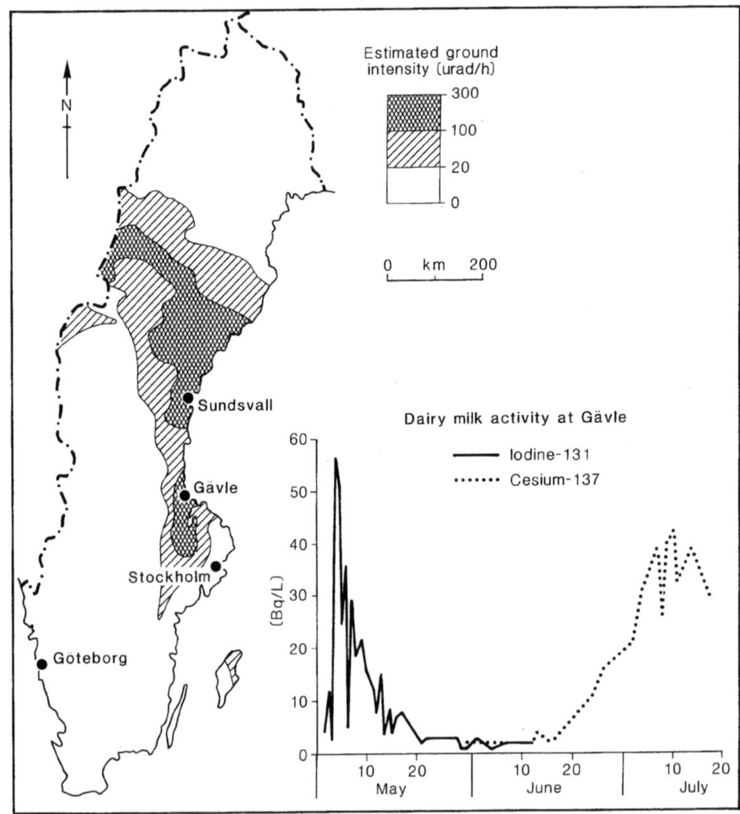

Figure 10.14 Post-Chernobyl contamination in Sweden. The map shows contamination levels for 9 May 1986 based on aerial surveys. The graph shows the radioactivity in dairy milk in becquerels per litre at Gälve, Sweden, from two radioisotopes (I-131 and Cs-137) following the accident on 26 April
Source: Reisch (1987) 'The Chernobyl accident – its impact on Sweden,' *Nuclear Safety* 28 (1):31, 33

al. 1987). The experiment was to test a voltage-regulating scheme on the turbogenerator and, when it began, half the main coolant pumps were slowed down; the flow reduction caused the water in the core to start boiling vigorously. The bubbles of steam that formed absorbed neutrons less strongly than the water they displaced and the number of neutrons in the core began to rise. This increased the power of the reactor; more steam

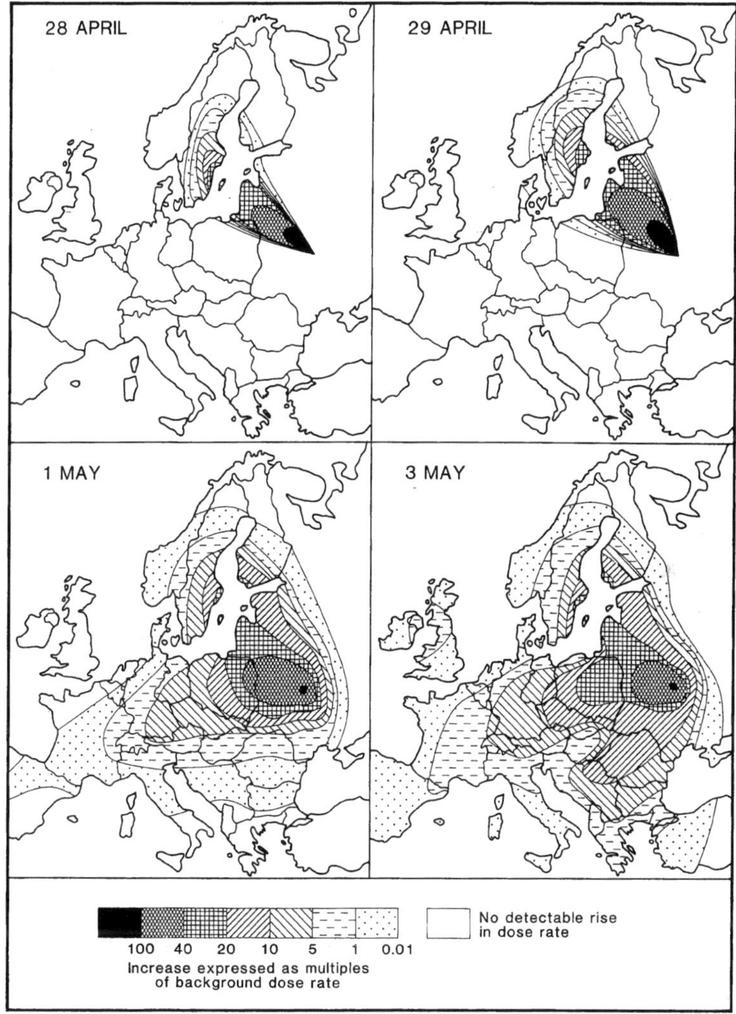

Figure 10.15 Increases in radioactivity dose rates in Europe between 28 April and 3 May 1986 consequent upon the Chernobyl reactor accident
Source: Gittus *et al.* 1987

was produced and even fewer neutrons were absorbed, a phenomenon known as 'positive feedback'. The reduced neutron absorption led to a power surge which caused the fuel to heat up, melt and disintegrate. Fragments of fuel were ejected into the surrounding water, causing steam explosions that ruptured fuel channels and led to the cap of the reactor being blown off – the MCA had occurred. Radionuclides escaped into the atmosphere where the wind carried them to distant countries (Figures 10.14 and 10.15).

The cause of the accident was operator error. Station engineers violated their operating instructions in attempts to perform an injudicious experiment and the reactor had design characteristics that exacerbated the errors. At low power it was intrinsically unstable, the shutdown system was too slow in operation and there were no physical controls to prevent staff from operating the reactor in its unstable regime or with safety systems seriously disabled.

Atmospheric dispersion in Europe of radionuclides from Chernobyl

Increased activity levels were first reported on 28 April from environmental monitoring stations in Finland and Sweden, where external dose rates in certain locations exceeded normal background levels by a factor of ten or more (Figure 10.15). On succeeding days increased radioactivity concentrations were detected throughout Europe until almost compete coverage had been achieved by 3 May. The progression of this pattern with time is illustrated in Figure 10.15. By 29 April the contamination had spread further across the Scandinavian countries and on 30 April had reached central Europe, reflecting a trajectory initially northwesterly, but subsequently veering westwards in the vicinity of the Baltic Sea. From 1 to 3 May the plume spread to the west, north and south, essentially covering Europe. This further spreading was influenced by an anticyclone moving eastwards across central Europe. The contamination of the UK resulted from air convected northwards behind the area of high pressure; the higher contamination in the north (see Chapter 8, Figure 8.6) was due to higher rainfall during the passage of the plume (F.B. Smith 1986).

After the Chernobyl accident thirty-one people, mainly those engaged in fighting fires at the stricken reactor, died from the effects of very high levels of radiation exposure. Around 200 others were diagnosed as suffering from acute radiation effects. Agricultural products in a 30 km (19 mile) zone around the ractor were irradiated and could not be harvested. Over ten days 135,000 people in the worst affected surrounding areas were evacuated pending decontamination of an area out to about 30 km; subsequently 'clean patches' were found within this zone counterbalanced by serious contamination outside it, especially to the north in Gomel province. Late in 1988 buildings around Chernobyl regarded as not safely habitable within their expected lifetimes were razed to the ground.

Outside the USSR, feeding grounds for Lapland reindeer and pastures for sheep farmers in North Wales and Cumbria in the UK remained contaminated to the extent that affected livestock had to continue to be destroyed two years and more after the event (see Chapter 8, Figure 8.6). The total collective dose, summed over all countries in both eastern and western Europe, has been estimated to be approximately 1.8×10^5 man Sv. The average dose in the eastern European countries is estimated to be three to four times that in western Europe (Gittus 1987). In the fifty years from 1986 to 2036, 1,000 extra deaths from cancers of all types are anticipated in the EC, including some 100 deaths from thyroid cancer, from an estimated 2,000 extra cases. It must be recognized, however, that the patchy nature of the deposition from the plume of radionuclides due to wide variation in rainfall as it passed over various areas gave rise to wide geographic variations in the dose increment (Smith and Clark 1986).

Even in terms of long-term casualties, however, Chernobyl will not be the world's worst nuclear power accident. Nor does it prove that the nuclear industry generally, throughout the world, is inherently unsafe. It simply earns a place in any inventory of large industrial disasters that have caused loss of life and environmental damage in the last half century. It is a human disaster comparable with that on 3 December 1984 at Bhopal, India, when at least 2,000 people were killed and 180,000 were treated for ailments when poisonous gas escaped from a tank containing 45 tonnes of methyl isocyanate, or to that on 27 March 1977 when two fully laden Jumbo jets collided on the runway at Teneriffe airport, killing 583 people, or to that on 21 October 1966 when 116 schoolchildren and 21 adults were killed at Aberfan, Glamorgan, South Wales, when torrential rain caused a coal mining waste heap to slide onto and engulf a school, or to the Piper Alpha oil platform disaster on 6 July 1988 in which 167 men died. None of these events has brought about a call for cessation of production in the relevant industry and none can match the 55,000 people killed by the earthquake in Armenia in 1988.

Disaster in the Urals

Chernobyl may not have been the USSR's first major accident in the area of nuclear power technology. In 1976 it was suggested by Zhores Medvedev that a vast explosion of nuclear waste buried in the vicinity of Kyshtym at the end of 1957 or the beginning of 1958 had spread radioactive contamination over a huge area in the southern Urals between the two cities of Sverdlovsk and Chelyabinsk (Figure 10.16), resulting in many deaths and eradication of plant and animal life (Medvedev 1976; Roberts and Medvedev 1977). There were eyewitness accounts of a dead, defoliated and lifeless landscape, with ruined and empty buildings destroyed so as to make the dangerous zone uninhabitable, but without official Soviet confirmation views regarding the causes have been varied. An analysis for the Los Alamos National Laboratory in the USA stated that there was no evidence of an explosion, but that the damage had been caused over several years by a general disregard for environmental pollution as a result of military pressure to maintain plutonium production. The Los Alamos report (LA 9219 MS, 1982) confirmed the military nuclear complex near Kyshtym between the major industrial cities of Chelyabinsk and Sverdlovsk in the eastern foothills of the Urals as the source of the contamination. The first plutonium production reactor at Kyshtym, designated Unit O, was graphite moderated with an open-circuit light water cooling system fed from Lake Kysyltash. A large artificial lake was dug before August 1950 to take the radioactive discharge water and by 1952 cooling water from two further reactors, designated 301 and 701, was also discharged to this lake. The lake itself discharged to the River Techa, and at least as early as 1953 this river provided a chronic source of radioactive contamination. A second source of environmental pollution from Kyshtym was the plutonium processing plant, similar to that at Hanford in the USA. It has been reported that after this plant was commissioned its chimney emitted a yellowish mixture of steam and acids for many years, yellowing and killing trees and grasses for a distance of 15–20 km. The third source of pollution identified was high level radioactive waste stored in a lake 5 km southeast of the Kyshtym complex with natural drainage routes to the Techa river, providing further contamination of the river valley.

The Los Alamos report states that serious health problems developed among the sparse rural population of the valley and their livestock. Evacuation of the region took place and destruction of the dwellings discouraged attempts to return. Later, closed-loop cooling was installed to obviate further contamination from the reactors, but it is suggested that the high level waste storage pond presented considerable problems. This waste, with ammonium nitrate and hexone present, may have caused an explosion, scattering dust throughout the region. Alternatively, the storage pond may have dried out, so that the strong intermittent winds, which blew at up to 100 km/h in February and March, may have started a dust storm. The Los Alamos researchers believe that, at some time in the 1960s, long-sentence prisoners were used to dump soil over the lake bed. The high radiation doses involved gave rise to the description ‘death squad’, and this area, about 8 km in diameter, has been used as a radiological training area.

SUMMARY

The approaches to safety in power reactor siting and design described here have been mainly those adopted in the USA and UK. While international guidelines exist for setting safety standards, their implementation by regulatory agencies shows some variation between countries. The USA perceived a need to improve reliability and safety systems following the accident at Three Mile Island, with the main emphasis on the reduction of vulnerability

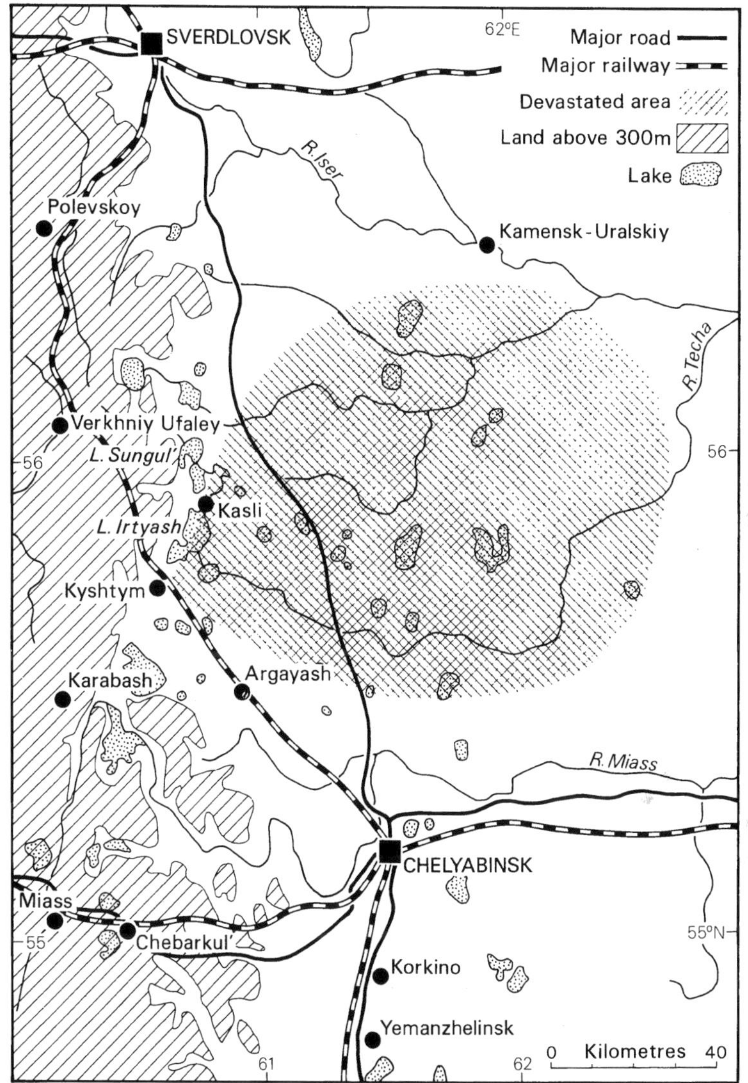

Figure 10.16 Approximate area in the Southern Urals suggested to have been contaminated in 1957–8. On 17 June 1989, in the course of an inquiry to build a civilian nuclear power station in the same area, Soviet authorities admitted this accident, stating it to have been caused by the explosion of a tank containing radioactive waste at an atomic defence establishment at Kasli. It was stated that nobody was killed in the accident

to severe accidents. In Japan, high reliability is emphasized. France defines safety objectives very simply. As a general objective, French design engineers are instructed to aim at a probability of unacceptable accident consequences of less than 10^{-6} per reactor per year; an unacceptable consequence is defined as the necessity for immediate evacuation around the site. France undertakes extensive evaluation of reactor operating experience to detect precursor incidents that may warn of an impending major accident. Prevention of major accidents by monitoring the sequence of more minor incidents is also emphasized in the FRG. The FRG demands nuclear power stations that are able to resist substantial external impact, such as may result from an aeroplane crash. Therefore enhanced vital service systems, such as emergency power and feed-water supplies, are provided. In the FRG the issue of severe accidents has not been addressed in terms of quantitative frequencies and consequences. The emphasis has been on the need to maintain core cooling under adverse circumstances. Core cooling and containment are also emphasized in Japan.

Thus, within the context of broadly similar responses to international guidelines, national approaches to design safety vary, but in all the term 'site selection' applied to nuclear power reactors has come to refer almost exclusively to their planned siting for public safety. There is also a common and continuing interest amongst the utilities and on the design and construction side in seeking ways to reduce the cost of generating electricity without increasing risks to the population. The reduced capital costs and lower transmission costs of short transmission lines have been attractive enough to provide an incentive to locate nuclear power plants close to the areas of demand, i.e. close to the towns and cities that they are built to serve.

The first stage of commercial nuclear power station siting in both the UK and the USA was developed under a strategy aimed at limiting possible injury to the public through a combination of engineered safeguards, reactor containment and remote siting. The most significant subsequent change in this strategy was the adoption of an approach which treated accidents in terms of statistical probability, and which accepted the consequences of an uncontrolled release of fission products provided that the probability that this would happen was extremely low. This approach was prompted in part by the difficulty experienced of matching site characteristics and reactor design in any statistically meaningful way.

The probability approach can handle component failures that coincide with or follow one another, and variables such as human error, or wind direction and wind speed following a theoretical accident, can also be built into the relevant models. However, it is a type of analysis that can spell out the *likelihood* of a particular type of incident but it cannot say *when* the event may occur. Moreover, it has been argued that certain combinations of events, for example loss of feed-water and relief valve failure, such as happened at Three Mile Island, and their consequences are inadequately modelled in safety tree procedures (Etemad 1979).

Nevertheless, in the USA, the UK and some other countries, such procedures became part of a 'compensating safeguards' philosophy which allowed favourable safety design features to offset the unfavourable site characteristics of proximity to population despite the fact that site and plant characteristics could still not be matched except in the crudest terms. This approach allowed the previous remote siting strategy to be relaxed and permitted large power reactors to be sited much closer than before to urban populations, with 'remote' sites being retained only for reactors with a large experimental element, such as fast breeders. Early guidelines which had attempted to relate siting to population density were frequently breached, and the accidents at Three Mile Island and Chernobyl should have reinforced the importance, for evacuation purposes, of retaining a zone of low population around a nuclear power site. Since the impact on surrounding populations of a release of fission products depends upon the size of the release, its height, the length of time for which it lasts, the contents and shape of the plume, meteorological factors such as wind speed and direction, inversion conditions and the physiography of the surrounding areas, to name just a few variables, the practical value of an exclusion area measured only in hundreds of metres or yards and of a population evacuation zone with a radius of only a few miles or kilometres might seem problematic, but there can be little doubt that evacuation procedures are more difficult and time consuming in densely populated areas and that population control zones have real value for dealing with an accidental release.

FURTHER READING

Barry, P.J. (1970) 'The siting and safety of civilian nuclear power plants', *Critical Reviews in Environmental Control* 1 (2): June.

Haire, T.P. and Shaw, J. (1979) 'Nuclear power plant licensing procedures in the United Kingdom', *Progress in Nuclear Energy* 4: 161–82.

Openshaw, S. (1986) *Nuclear Power: Siting and Safety*, London: Routledge & Kegan Paul.

Marples, D.R. (1988) *The Social Impact of the Chernobyl Disaster*, London: Macmillan.

Smith, F.B. (1989) 'Debris from the Chernobyl nuclear disaster: how it came to the UK, and its consequences to agriculture', *Journal of the Institute of Energy* 62 (450): 3–13.

11 Siting in relation to exceptional environmental events: earthquakes and faults, tornadoes, tsunamis and floods

The judgement of whether or not a particular site is suitable for a nuclear power station or radwaste depository involves not only the normal day-to-day conditions under which the plant is expected to operate but also the likelihood of an exceptional natural event at the site, such as an earthquake or tornado. Seismological and meteorological evaluations and plant design have to include such extreme events, which are more common in some parts of the world than in others. Risk of destructive earthquakes is higher in Japan, California and Italy than in France or the UK, while nuclear power stations in the USA may be more at risk from tornado damage than those in the FRG. Similarly, 62 per cent of the known important tidal waves (tsunamis) have occurred in the Pacific Ocean, 20 per cent in the Indian Ocean, 9 per cent in the Mediterranean and 9 per cent in the North Atlantic Ocean, while they are virtually unknown in the South Atlantic (Newmark and Rosenblueth 1971).

SEISMIC HAZARDS: EARTHQUAKES AND FAULTS

Earthquakes pose an exceptional hazard to power plants because they can affect all portions of the plant simultaneously. Thus an earthquake can both damage the plant and disrupt protective plant safety mechanisms, and this means that more stringent design requirements are required for nuclear than for conventional power stations.

It was shown in Chapter 1 that a nuclear power plant is a fairly complex engineered system consisting of a containment structure, an internal primary support that also functions as a biological shield, the reactor vessel and its internals, the reactor coolant loops, consisting of several large vessels, piping and other equipment such as pumps and valves, and ancillary equipment such as control rod drive mechanisms, overhead cranes and fuel transfer equipment. Conventional parts include the turbine, the condenser, switchgear, emergency power systems and electrical distribution equipment. The possible modes of failure in these areas, which may occur simultaneously under earthquake conditions, are shown in Table 11.1.

Many phenomena may give rise to earthquakes, including volcanic

Table 11.1 Principal nuclear power station systems and potential failure modes

System	*Potential failure mode*
Containment building	Loss of containment (not a major structural failure, but rather a local failure, probably near a support or penetration for pipework etc.)
Reactor vessel	Dislocation of vessel, fuel damage, instrumentation damage, reactivity effects and control-rod interference
Primary coolant loop	Cracked welds, coolant leaks, damage at nozzle, pump or steam-generator connections, steam-generator tube failures, pump damage due to shaft flexure or bearing damage, dislocation of large equipment items
Turbine generator	Bearing or blade damage
Auxiliary systems	
Emergency power	Dislocation, severed cables and ducts, failure to start or operate properly
Electrical distribution equipment	Mechanical damage, broken welds, collapse, overturning, failure to operate
Control and instrumentation	Failure to operate properly or intermittent operation

Source: C.B. Smith 1973:588

activity, explosions and collapse of the roofs of large underground caverns, but from an engineering viewpoint by far the most important and frequent earthquakes are those of tectonic origin (i.e. those associated with large-scale strains in the crust of the earth) which release the most energy and affect extensive areas. The *magnitude* of an earthquake is a measure of the maximum amount of ground motion caused by an earthquake and is directly related to the amount of energy released. It is measured logarithmically on the Richter scale, from zero to 8.9. A 6 on the Richter scale is ten times more powerful than a 5 and a hundred times more powerful than a 4. The *intensity* of an earthquake is a measure of its local destructiveness; all intensity scales are subjective and similar to the Modified Mercalli scale which ranges from 1 to 12 (as shown in Table 11.2). For example, scale 1 on the Modified Mercalli is one where 'hanging objects swing; vibration is felt like the passing of a heavy lorry or truck, or the sensation of a jolt like a heavy ball striking the walls; standing motor cars rock; windows, dishes, doors rattle; glasses clink; crockery clashes. In the upper range of 4, wooden walls and frames crack.' Thus, any one earthquake will be associated with a single magnitude, while its intensity will vary from place to place.

Earthquakes occur offshore as well as under land, sometimes producing

Table 11.2 Modified Mercalli intensity scale (abridged)

I	Not felt except by a very few under especially favourable circumstances
II	Felt only by a few persons at rest, especially on upper floors of buildings. Delicately suspended objects may swing
III	Felt quite noticeably indoors, especially on upper floors of buildings, but many people do not recognize it as a earthquake. Standing motor cars may rock slightly. Vibration like passing of truck. Duration estimated
IV	During the day felt indoors by many, outdoors by few. At night some awakened. Dishes, windows, doors disturbed; walls make cracking sound. Sensation like heavy truck striking building; standing motor cars rocked noticeably
V	Felt by nearly everyone; many awakened. Some dishes, windows etc. broken; a few instances of cracked plaster; unstable objects overturned. Disturbance of trees, poles and other tall objects sometimes noticed. Pendulum clocks may stop
VI	Felt by all; many frightened and run outdoors. Some heavy furniture moved; a few instances of fallen plaster or damaged chimneys. Damage slight
VII	Everybody runs outdoors. Damage negligible in buildings of good design and construction; slight to moderate in well-built ordinary structures; considerable in poorly built or badly designed structures; some chimneys broken. Noticed by persons driving motor cars
VIII	Damage slight in specially designed structures; considerable in ordinary substantial buildings with partial collapse; great in poorly built structures. Panel walls thrown out of frame structures. Fall of chimneys, factory stacks, columns, monuments, walls. Heavy furniture overturned. Sand and mud ejected in small amounts. Changes in well water. Disturbs persons driving motor cars
IX	Damage considerable in specially designed structures; well-designed frame structures thrown out of plumb; great in substantial buildings, with partial collapse. Buildings shifted off foundations. Ground cracked conspicuously. Underground pipes broken
X	Some well-built wooden structures destroyed; most masonry and frame structures with foundations destroyed; ground badly cracked. Rails bent. Landslides considerable from river banks and steep slopes. Shifted sand and mud. Water splashed (slopped) over banks
XI	Few, if any (masonry), structures remain standing. Bridges destroyed. Broad fissures in ground. Underground pipe lines completely out of service. Earth slumps and land slips in soft ground. Rails bent greatly
XII	Damage total. Waves seen on ground surfaces. Lines of sight and level distorted. Objects thrown upward into the air

Source: Wood and Neuman 1931 by US Geological Survey, *Earthquake Information Bulletin* 6 (5):28, 1974, in Keller, E.A., *Environmental Geology*, 3rd edn, Columbus, OH: Merrill

huge tidal waves. In 1976 the town of Mindanao, in the southern Philippines, was inundated and all but destroyed and thousands of lives were lost when an undersea earthquake at the mouth of the crescent-shaped Morro Bay caused a tidal wave 5 m high that engulfed the shores of the bay. Such an ocean earthquake off San Francisco would devastate the Bay area, and could inundate large parts of the Sacramento and San Joaquin Valleys.

Paradoxically, designers of nuclear power stations demand precision and accuracy in forecasts required from the geologist, when it is virtually impossible to predict precisely when, where and to what extent earthquakes and surface faulting may occur. From information on past seismic activity it is possible to estimate the frequency and magnitude of earthquakes in a

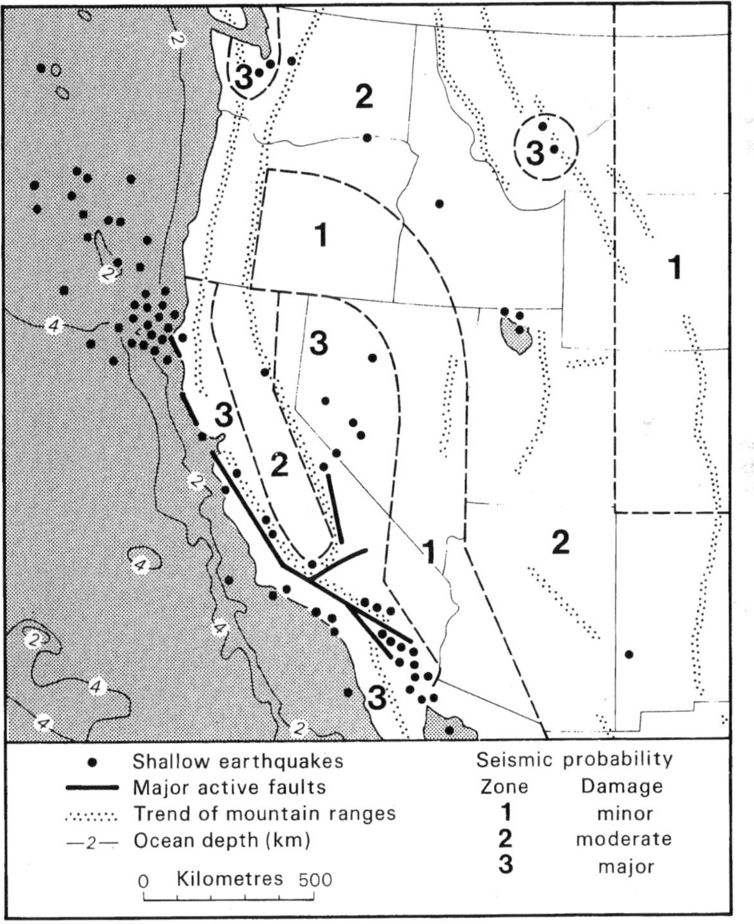

Figure 11.1 Seismic probability and earthquake record in the western USA

given region which may then be used to provide some measure of the seismicity of a particular site. However, it is necessary to recognize that these are only approximations and that, from a practical standpoint, there is no absolute upper limit to earthquake intensity, that a margin of error exists in the application of regional data to the relatively small area of land that may be occupied by a nuclear power station and that, no matter how conservative a building design may be, there is always a finite probability of structural failure in any finite interval of time. Thus a map such as Figure 11.1 showing the location of known epicentres and least-risk zones is only a crude guide.

The simplest grading of earthquakes by means of intensity or local destructiveness on the Modified Mercalli scale is too crude a basis for most nuclear engineering purposes. Procedures for evaluating a possible nuclear power station or waste depository site may involve three main steps. First, the maximum ground motions likely to occur at a site are determined. Second, the response of the proposed structures to the ground motions are estimated. Third, the ability of the nuclear power station or store to respond safely is calculated. For any particular site, the first two objectives can be achieved by producing a graph in which the amplitude of an anticipated vibratory ground motion is plotted against the periodicity or frequency of the components of the motion (a ground motion spectrum). For design purposes an 'envelope' defined by the maximum ground acceleration, velocity and displacement can be used as the spectral envelope of maximum ground motion. The response of the structures to such ground motions is then calculated by determining the extent to which they may *amplify* the movement of the ground (much in the way that a jelly can be made to amplify the movement of a plate on which it stands). Figure 11.2 shows a typical ground motion spectral envelope and illustrates how a building amplifies the ground motion.

Use of this technique aids the design of containment buildings capable of responding elastically to anticipated ground motions. These tend to be low broad structures, partially buried in the soil. In addition, seismic qualification of safety-related equipment is required so that in the event of an earthquake the reactor can be safely shut down.

Careful seismic design alone, however, cannot compensate for adverse discrepancies between the design desiderata specified for particular pieces of equipment and their actual performance under seismic stress. Moreover, while geologists understand tectonic processes sufficiently well that the relative possibility of the occurrence and magnitude of vibratory motion can be determined to some extent, surface faulting and differential ground displacement are less well understood parts of the tectonic process. Surface faults may be located on land in such a way that the direction of future movement can be predicted with a relative probability of occurrence, but the magnitude of displacement cannot be predicted with reasonable assurance. In addition, offshore geological information for possible coastally

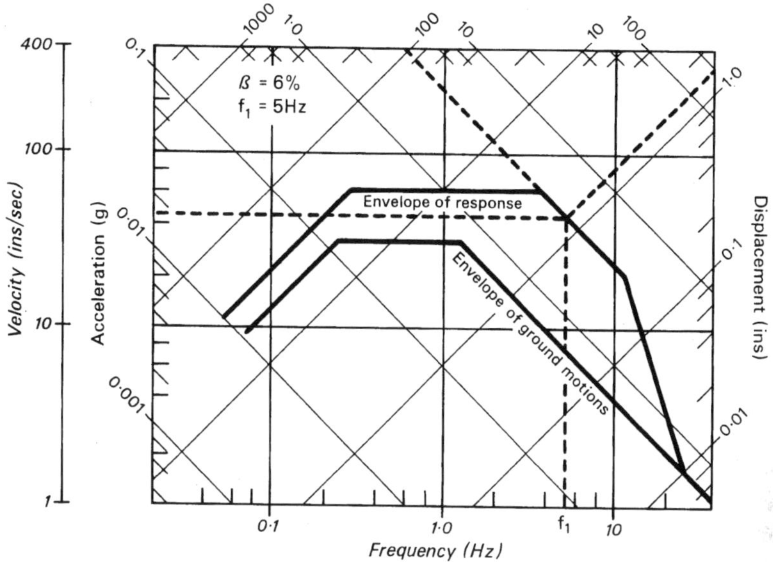

Figure 11.2 A typical ground motion spectral envelope. Earthquakes produce random ground motions which are characterized by simultaneous but statistically independent horizontal and vertical components. The horizontal and vertical ground motions may be filtered *and* amplified by the buildings in which equipment is housed. Knowledge of the fundamental frequency of the building and its damping value allows an estimate of its response to be made. In this example the damping is 6 per cent, the fundamental frequency *f* is 5 Hz and the maximum response, as read from the graph, is 1.4*g*, about 40 in/s and 12 in displacement
Source: Smith (1973) 'Power plant safety and earthquakes', *Nuclear Safety* 14 (b):591

located nuclear power station sites is often obtainable only by expensive surveying and measuring techniques. In such circumstances the best that the geologist can do is to advise the power company to choose a reactor location at a reasonable distance from known faults considered to be capable of significant surface movement.

Many of these generalizations can be illustrated by the experience of nuclear power station siting in California and Japan. California lies in a significant area of weakness in the earth's crust where major structural units or 'plates' adjoin one another. Most of California lies on the North American plate, but the section to the west of the San Andreas Fault lies on the North Pacific plate. Although the North Pacific plate appears to be rotating slightly, its movement in California is to the northwest and periodically, when the pressure between the plates becomes too much, it eases this pressure by jerking northwest along the 9 mile (14.5 km) deep, 650 miles (1,046 km) long San Andreas fault. Lasers beamed from Californian mountain tops measure the average movement at 2 inches (50 mm) yearly, but the extent of dislocation along this right-slip fault, so called because if one stands on either side of it after an earthquake the other side appears to have moved to the right, is crucial in relation to the magnitude of the shock. Occasionally there may be no lateral slip at all. Sometimes it may be only a foot or two, but a lateral slip of 40 feet (12 m) is very possible because the San Andreas fault is strained like an overwound watch spring, ready to snap. If an earthquake equivalent to that of 1906, which measured 7.9 on the Richter scale, were to hit San Francisco in the rush hour, federal studies predict that there would be of the order of 10,000 deaths, 50,000 injuries and 100,000 rendered homeless.

Design of nuclear power stations now involves the selection of a 'maximum design earthquake' (SSE or safe-shutdown earthquake) representing the most severe ground motion to be expected for a particular site. An 'operating-basis earthquake' (OBE) has also been defined as one that is expected to occur once or more during the life of the plant and for which it is required that the plant remain operational. If ground motions exceeding those specified by the OBE are likely to be experienced, it is necessary to demonstrate that the plant is still capable of safe operation, or shutdown may be required. From the viewpoint of public safety during an accident, it is not judged necessary to consider this smaller earthquake, but reliability of the plant during moderate earthquakes is important from the viewpoint of economic power generation, for a loss of power to institutions such as hospitals during a moderate earthquake could be very important.

From the beginning, some standards were set for the design of Californian nuclear power stations, but the fact remains that the application of geological knowledge to the siting and design of nuclear power plants involves a *historical* not an experimental science, which has an incomplete record of evidence and which requires the interpretive judgement of geological and seismic information. Moreover, the regulatory agencies did not at first take earthquakes fully into account in site and design assessment. One nuclear power plant in California was closed down in 1979 because of earthquake problems, and the importance of the geological evidence in deciding whether or not a nuclear power plant be built at all is emphasized by the case of the proposal made in 1961 to build such a plant at Bodega Bay, near San Francisco. The fact that the interpretation of the geological evidence can take different forms and lead to different conclusions is very clearly illustrated by tracing the course of the proposal made in 1963 to build a nuclear power station at Malibu, near Los Angeles. The case histories of both proposals are sufficiently important to warrant detailed description.

THE BODEGA BAY CASE

In the 1960s, the selection of a nuclear power reactor site in California was guided by normal USAEC rules, but the State Public Utilities Code gave the Californian Public Utilities Commission cognizance over investor-owned utilities such as the Pacific Gas and Electricity Company (PG&E). Publicly owned utilities, such as the Los Angeles Department of Water and Power were not regulated by the Commission (Peter E. Mitchell, formerly Public Utilities Commissioner, State of California, personal communication). The California Department of Public Health reviewed environmental factors under the Health and Safety Code and the views of the State Water Quality Control Boards, the State Resources Agency, including the Division of Fish and Game, and other state agencies working in collaboration with the State Co-ordinator of Atomic Energy Development and Radiation Protection were sought and considered. Under section 1001 of the California Public Utilities Code, a public utility wishing to construct a nuclear power plant had to file with the Public Utilities Commission an application for a certificate of public convenience and necessity. In other words, the utility had to prove that the project be both necessary and convenient to the public. A public hearing could be arranged by the Public Utilities Commission and the views of interested parties expressed at the hearing.

PG&E filed its application with the Public Utilities Commission on 4 October 1961, requesting a certificate of public convenience and necessity to construct a 325 MWe BWR power station at Bodega Bay, Sonoma County, approximately 50 miles (80 km) north of San Francisco. The actual site suggested for the plant was at Campbell Cove, on the harbour side of Bodega Head. PG&E indicated that the primary reason for selecting the site was its proximity to load centres, but it also offered excellent dispersion of the warm water discharge, a solid granitic type of rock providing good foundations, isolation from population centres and a harbour making possible water transportation of the heavy components during construction and for shipping the irradiated fuel elements for processing. Support for the project came from a variety of state and county organizations who claimed that it would result in great economic benefits to the local community through increased employment, tax revenues and business.

Immediately following the announcement of the proposed plant vigorous protests were made to the granting of the application, and these continued throughout the entire period of licensing effort. Initially, objections were made on the basis of recreation and conservation, spearheaded by the Sierra Club. The Club's objection to the plant was not that it was a technically inappropriate use of Bodega Head but that it was not the highest and best use; alternative uses were considered to be exceptional and would be rendered impossible if the plant were built.

On 2 November 1962, the California Public Utilities Commission granted PG&E an interim certificate of public convenience and necessity for the power station. This decision withstood a number of protests and requests to reopen the hearings, and ultimately had to be sustained by a decision of the California Supreme Court, but the fight was not yet over.

PG&E took initial action to obtain a USAEC construction permit in July 1962 and filed a formal application in December 1962. The application was terminated by PG&E on 30 October 1964, after submission of nine amendments to the application and a number of meetings with the USAEC staff and the Advisory Committee on Reactor Safeguards.

Initial objections to the plant had been based on conservation issues, but as time went on the emphasis shifted to concern for safety because of proximity to the San Andreas fault zone which lay approximately a quarter of a mile to the east of the reactor building. At the hearing before the State Public Utilities Commission in November 1962 concern was expressed by a number of witnesses as to the serious consequences that might arise if high level radioactive materials were released as a result of earthquake damage to the plant. This concern was increased when it was pointed out during the hearings that Bulletin 118 of the California Division of Mines and Geology, published in 1943, contained a map indicating a trace of a fault running directly through Bodega Head in a northwesterly direction. A geologist from the Division of Mines testified that this same line indicating a fault on Bodega Head appeared on the preliminary compilation of map sheets forming the basis of a new set of geological maps to be issued by the Division of Mines, and that he could see no reason for removing the line from the new maps. This witness further stated that his knowledge was not sufficient to allow him to testify as to whether or not the fault as indicated by the line was active.

When the Bodega Bay site was acquired by PG&E, it was done with full knowledge of the proximity of the San Andreas fault zone because the site was felt to be safe and, in the opinion of PG&E and its consultants, the best available power plant site between San Francisco and Humboldt Bay. However, during the two year period of effort to obtain a construction licence, a substantial difference of opinion developed between PG&E and its consultants on the one hand and the US Coast and Geodetic Survey and the US Geological Survey on the other. The differences appeared with respect to both the maximum credible earthquake to be taken as the design basis for the Bodega Bay plant and the magnitude of a possible tsunami to be considered in plant design. To meet these differing viewpoints, PG&E agreed to adjust its proposed design to meet higher seismic criteria, and reasoned that this design would also provide for safe plant shutdown in the unlikely event of a maximum tsunami.

After extensive study, the Advisory Committee on Reactor Safeguards in a letter of 20 October 1964 to the USAEC expressed the opinion that the power reactor facility as proposed could be constructed at the Bodega Bay site with reasonable assurance that it could be operated without undue hazard to the health and safety of the public. However, the Regulatory Staff of the Division of Reactor Licensing in a separate report concluded that:

> Bodega Bay is not a suitable location for the proposed nuclear power plant at the present state of our knowledge ... because of the magnitude of possible consequences of a major rupture in the reactor containment, accompanied by failure of emergency equipment, we do not believe that a large nuclear power reactor should be the subject of a pioneer construction effort based on unverified engineering principles, however sound they may appear to be.

On the basis that a reasonable doubt as to the ultimate safety of the plant at the Bodega Bay site did exist in the mind of the regulatory staff, on 30 October 1964 PG&E withdrew its application to construct the nuclear power station without proceeding to a public hearing.

THE MALIBU CASE

In November 1963 the Department of Water and Power of the City of Los Angeles (LADW&P) filed an application to construct and operate a 490 MWe PWR at Corral Canyon, Malibu, California.

Corral Canyon is on the south flank of the central Santa Monica Mountains, approximately 29 miles (47 km) west of Los Angeles City Hall, in a region where the dominant relief features are south-trending ridges alternating with steep-walled canyons draining southward to the Pacific Ocean. Along the coastline is a narrow beach, merging on the seaward side into a shallow coastal shelf sloping to a water depth of approximately 300 feet (91 m) at a distance of 3 miles (4.8 km) from the beach. In 1963, there were approximately 9,500 residents within a 5 mile (8.0 km) radius of the site, and 3.1 million people within a 30 mile (48 km) radius.

There was very little organized local resistance to the power station project at the start, but towards the end of 1963 a group of interested individuals forming the Malibu Citizens Group argued that the nuclear power station would change the residential character of Malibu, depress land and property values, and be hazardous to the health and safety of people in the surrounding area. At a public hearing on 4 November 1963 the group opposed LADW&P's application to the Los Angeles Regional Planning Commission, but the Planning Commission approved zoning exemption for the site, part of which previously had been reserved for light industry. Subsequently, the opposition group appeared twice before the Board of Water and Power Commissioners, requesting postponement or relocation of the project. The Board eventually approved the project. The Malibu Citizens Group then appealed to the County Board of Supervisors, which temporarily reversed the Regional Planning Commission's decision.

After the formal application for a construction licence by LADW&P, the USAEC issued a Notice of Public Hearing on 9 February 1965 and forty days of hearings ensued, held in intermittent sessions throughout 1965. There were five principal intervenors in the proceedings, three of whom actively opposed the application and participated throughout the hearings by presenting evidence, proposals for findings, and conclusions and briefs. These were the Marblehead Land Company, the Malibu Citizens for Conservation Inc. and a private individual. An additional Intervenor, the State of California, participated by the presentation of evidence.

The whole of the Californian coast, together with much of the inland part of the state, lies in an area designated seismic zone 3, a zone where major damage is possible from seismic activity (see Figure 11.1). Thus the major part of the Malibu Hearings was devoted to consideration of the geology and seismicity pertinent to the reactor site, especially ground acceleration or shaking, and permanent ground displacement or rupture.

The Malibu site is on the southern edge of the Transverse Range province. Structurally, this area is rather different from other areas in California for it trends generally from east to west, whilst the more common structural trend is from northwest to southeast.

The Malibu Coast fault separates two quite different kinds of basement rock, suggesting that it is one of California's major faults. It extends certainly for 25 miles (40 km) and maybe between 13 and 75 miles (21–120 km) further if offshore geological structures are taken into account. Horizontal thrusting and crustal shortening along and across the fault have produced a band of greatly deformed rock about 1 mile (1.6 km) wide, and included in this zone of deformation was the site proposed for the nuclear power station.

Geological investigations produced a wide range of disagreement between the various interested parties with regard to the nature and significance of faulting at the site. Contrasts in the mapping by different geologists consulted by the parties are shown in Figure 11.3. Each of the faults in the diagrams is associated with a band of sheared or fractured rock of varying widths, but it is worth noting that faults F and X were discovered after LADW&P had submitted all the evidence that it initially proposed to offer on geology and seismicity. The Atomic Safety and Licensing Board requested extra information that could be provided only by further trenches, and it was in the course of these excavations that faults X and F were discovered. The USAEC staff and Marblehead both stated that the geological evidence firmly established the existence of faults F and X. However, LADW&P did not recognize fault F as a fault, asserting that the character of the rocks at this location reflected a sedimentary contact zone between strata, representing soft sediment deformation without tectonic significance. Fault A, recognized by all the parties, ran through the heart of the site.

In view of the disagreement concerning the faults it is necessary to point out that the youngest known displacements at the proposed reactor site probably took place more than 10,000 years ago and possibly as much as 180,000 years ago. There is no known evidence for movement of the rock

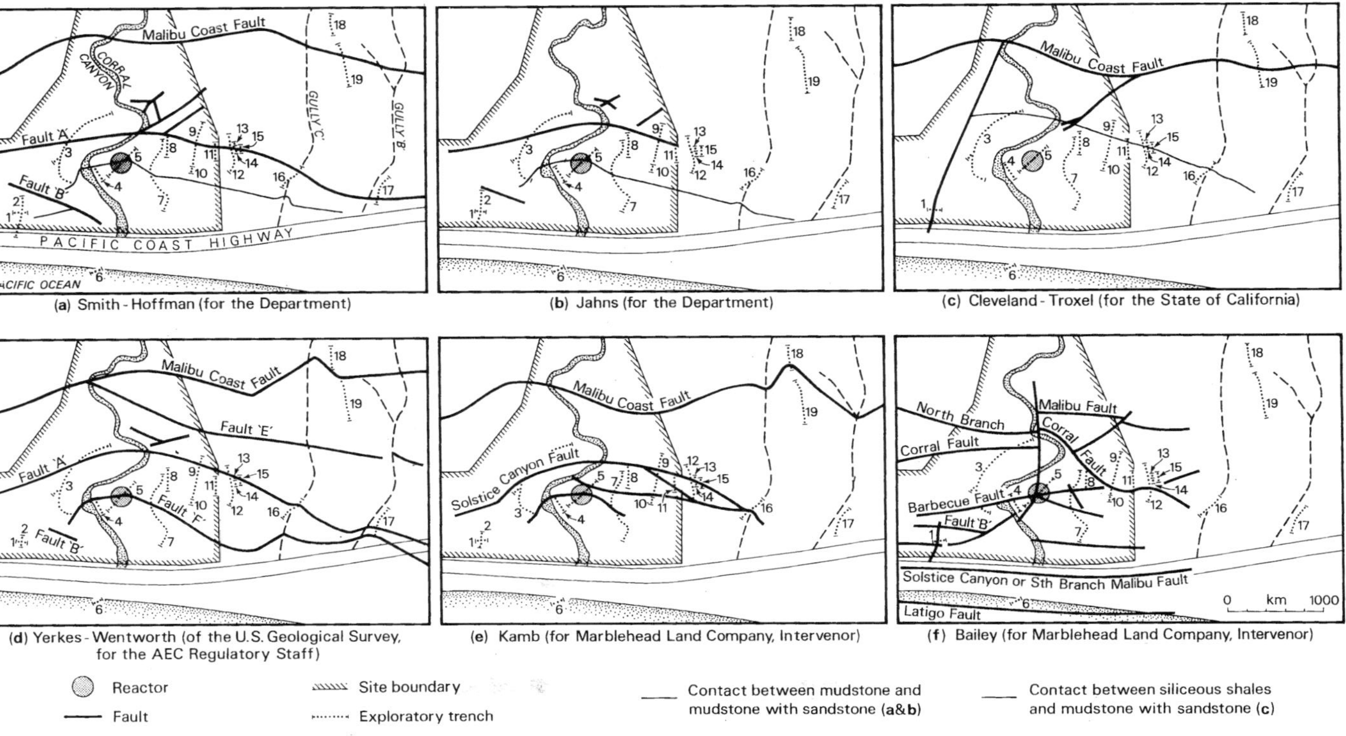

Figure 11.3 Alternative versions of the position of faults at and near the proposed Malibu nuclear power station in California. Trenches dug by LADW&P are shown by the numbered designations. It is interesting to compare (a) and (b), prepared by geologists for LADW&P, with (e) and (f), prepared by geological consultants for objectors, and (d) prepared by the US Geological Survey for the federal regulatory body.
Source: Redrawn from Figure 1d of the LADW&P's rebuttal testimony of 19 October 1965

on the faults of the Malibu Coast zone in recent geological times (less than 10,000 years). The alluvial deposits at the bottom of Corral Canyon are more than 3,000 years old. On the other hand, it is not known for what period of time a fault can lie quiescent and still be active, and the faults lie within a system that is tectonically active at depth.

Three alternative viewpoints regarding the likelihood of ground displacement at the reactor site were advanced during the hearing. The 'recurrence interval theory', which assumes a mechanism that leads to periodicity in the occurrence of earthquakes, was used by LADW&P to support the conclusion that the longer the time since the last displacement, the lower the probability of displacement in a given short time interval and that, if the last movement of a given fault occurred, for example, 180,000 years ago, the recurrence interval is likely to be of the same length of time or more. The USAEC staff took the view that there could be no reasonable assurance that occurrences are periodic, that the degree of regularity of the intervals, if they exist, was not known and that there could be no guarantee that the next occurrence would not occur tomorrow. The LADW&P also suggested that the 'uncocked gun' theory could apply to faulting in the area of the reactor site. This hypothesis suggests that all strain and stress has been relieved by past events and that there was no evidence to suggest a current accumulation of pressures that would cause a ground displacement. The third point of view advanced was that no valid statement could be made regarding the likely recurrence, intensity and magnitude of further faulting because of a lack of evidence. From this last premise, Marblehead and the USAEC staff arrived at different conclusions. Marblehead asserted that the LADW&P proposals required the reactor containment structure to remain airtight in the event of maximum credible accident and that the structure must therefore be able to withstand any credible forces that would rupture it. Marblehead further contended that such forces could result from permanent ground displacement under the containment structure, and that this displacement was a credible event as a result of faulting during the life of the reactor. Marblehead's conclusion in the light of these premises was that it would be an undue risk to the health and safety of the public to permit construction and operation of the Malibu power station. The USAEC regulatory staff concluded that, despite all the uncertainties and the inability to make a positive finding as to the time for a recurrence of faulting, and its intensity and magnitude, the probability of faulting was so low as to be negligible. Furthermore, the USAEC staff stated that the intervals of 10,000–180,000 years without evidence of ground displacement were long enough to justify the risk of neglecting permanent ground displacement in the design. However, the opinion of the USAEC differed from that of its regulatory staff on this point. The staff argued that:

> to deny Corral Canyon as a location for the proposed nuclear power plant because of inability to predict the next occurrence of an event

which has not happened for at least 500 generations would, in our judgement, represent an unwarranted, extreme viewpoint which would not be consistent with the standards applied in other areas of nuclear power plant design.

The USAEC on the other hand, found that none of the presentations defining the faults as major and minor, active or inactive, were convincing. The USAEC pointed out that

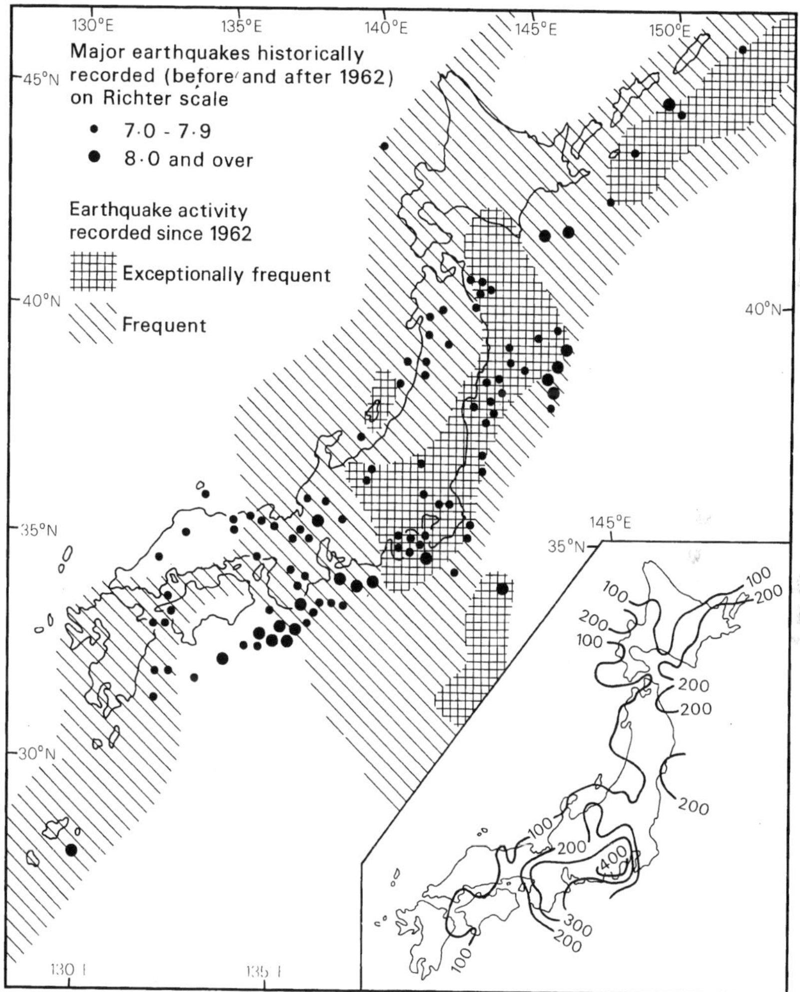

Figure 11.4 Frequency of earthquake activity in Japan. The main diagram shows the frequency of earthquake activity in Japan since 1962 and the distribution of *major* historical earthquakes before and after 1962. The inset shows the expectancy of maximum acceleration of earthquakes in 100 years (in gals)

> we are endeavouring to ascertain if an undue risk exists that the containment building will rupture and result in a release of radioactive effluents to cause an over-exposure of the surrounding population.... We find that the structural character of the plant site portends a potentiality of disruptive earthquake forces. We find ... that the probability of faulting and permanent ground displacement is high enough so that we cannot conclude that there is reasonable assurance that no undue risk is involved until an analysis and determination has been made of the resistance of the containment structure to permanent ground displacement.

However, the USAEC was also of the opinion that design criteria could be applied to proposals for the Malibu plant which would permit the containment building and nuclear reactor to withstand permanent ground displacement and that, with the requisite design features built into the plant, it could be constructed and operated without undue risk to the health and safety of the public. Thus the final order of the USAEC authorized the LADW&P to construct the power station, provided that the design criteria were modified to include adequate provision for ground displacement at the reactor site from earthquake activity, and, with this proviso, a construction permit was granted by the Division of Licensing and Regulation on 14 July 1966. However, sufficient doubts had been generated during the hearing for the utility to decide not to proceed and in the end the plant was not built.

JAPAN

The island chain that is Japan is part of the circum-Pacific orogenic zone and has been subjected to frequent tectonic movements of high magnitude and intensity since Palaeozoic times. Most of the country's major present geological structures were formed by post-Mesozoic and pre-Tertiary earth movements, but Japan is still in orogenic movement. The Japanese islands lie on the subduction zone along the edge of the Pacific Plate. The occurrence of earthquakes in Japan has been carefully recorded by one means or another since 599 AD. Figure 11.4 shows the distribution of the epicentres of the major earthquakes that have occurred (the *focus* of an earthquake is the point in the earth's crust where the seismic waves originate, and the *epicentre* is the vertical projection of the focus onto the earth's surface), and the inset to the figure shows the expectancy of maximum ground acceleration over a 100 year period. It is clear that the epicentres of the earthquakes are distributed along the axis of the islands, that on the Pacific side of the islands epicentres of earthquakes of magnitude 8 on the Richter scale occur in fairly clear lines, and that localized earthquakes of magnitude 7 on the Richter scale happen very frequently on the side of the Sea of Japan.

The word tsunami is of Japanese origin and refers to transient sequences of water waves which may generate potentially destructive pressures. The causes of these so-called tidal waves have nothing to do with tides; most commonly they result from a local change of elevation of the ocean floor, in turn associated with an earthquake or with coastal and submarine landslides which may be triggered by earthquakes. They can cause serious loss of life and widespread property damage. One 90 foot (27 m) tsunami which hit part of Japan on 15 June 1896 killed 27,000 people and destroyed 10,000 houses. Coastlines directly facing the epicentre of the related earthquake usually suffer the highest run-ups, as do steeply sloping coasts. Offshore coral reefs afford some protection. Tsunamis have importance not only because they are relatively frequent phenomena in Japan but also because all the Japanese nuclear power stations now built or under construction are at coastal sites. The catchment areas of the Japanese rivers are rather small, and their seasonal fluctuation tends to be large, with low dry season flows; inland lakes and reservoirs are also of comparatively small capacity, and so it is considered in Japan that the only source of sufficient supplies of condenser cooling water is the sea (Inouye 1973).

In a tectonically very unstable environment the Japanese authorities have recognized the need to find nuclear power station sites that are geologically as stable as possible, and they have also taken both earthquakes and tsunamis into account in engineering design criteria.

HIGH WINDS AND TORNADOES

A tornado is a vortex of air of great intensity extending downwards from a thundercloud. It is usually visible as a funnel-shaped cloud with a broad base at the cumulo-nimbus and a narrow extension to the ground. The diameter may vary from a few metres in the case of a dust devil to several hundred metres, but most destructive tornadoes have a large diameter. The funnel of the tornado is visible because it is the region where winds are so high that the barometric pressure is low enough to condense the water vapour in the air. This zone of high rotational winds forms a wall that prevents air from moving in or out across it. The upward helical movement of air inside the funnel is compensated for by air rushing into its lower tip creating a suction effect much like that of a vacuum cleaner on a giant scale (Figure 11.5). In theory, tangential and vertical wind velocities within a tornado could reach 500–600 miles/hour, but winds of 200 miles/hour would represent most destructive tornadoes. The speed of the funnel relative to the ground can vary between 20 and 80 miles/hour.

Much of the damage caused by tornadoes is attributable to the external pressure drop caused by the pressure field, but whilst the pressure force is prominent in causing damage to the roof and leeward side of a building, direct wind force also causes damage on the windward surface. Structures must also be able to withstand wind-induced oscillations and the impact of heavy objects projected at the building.

In the relevant technical literature the maximum wind speed is usually regarded as a random variable for a particular site. Therefore, on the basis

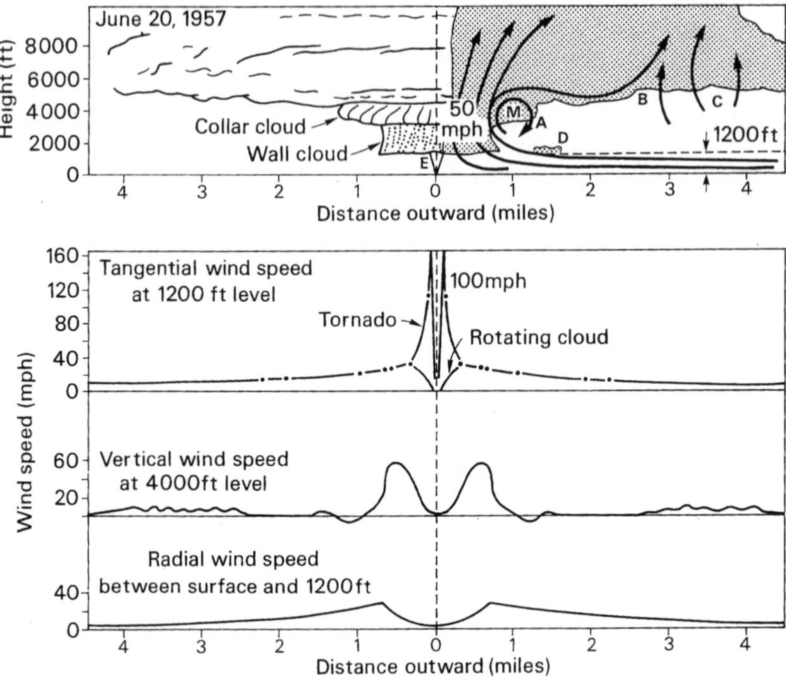

Figure 11.5 A sketch of the wind fields as drawn by Fujita in his study of the Fargo tornado on 20 June 1957
Source: Doan (1970) 'Tornado considerations for nuclear power plant structures', *Nuclear Safety* 11 (4):297

of the values recorded at meteorological stations, methods are used to evaluate the maximum wind speed as a function of its probability. Figure 11.6(b) shows the distribution of the maximum annual wind speed averaged for five minute intervals for some US meteorological stations of different heights above sea level as a function of the frequency on linear probabilistic charts. It is clear that the points of the same station are aligned on a Gaussian curve, but that values of different stations are aligned on different Gaussians whose parameters are functions of the site. It is also evident from Figure 11.6(a) that in the USA certain areas lie along trajectory pathways of tornadoes and are therefore subjected to strong winds more frequently. Thus a local study of tornado probability at a site has importance.

The threshold speed for some wind damage to normally constructed buildings is usually taken to be 75 miles/hour, and so the damage path width of a tornado can be taken as that part of the diameter where winds of 75 miles/hour exist. The expectation value for a tornado path area in the USA has been calculated at 2.821 square miles (7.25 km^2), and on this

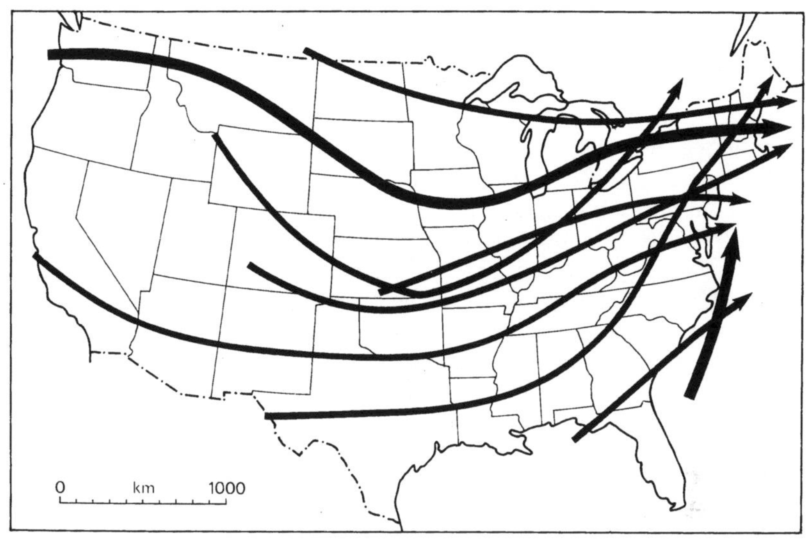

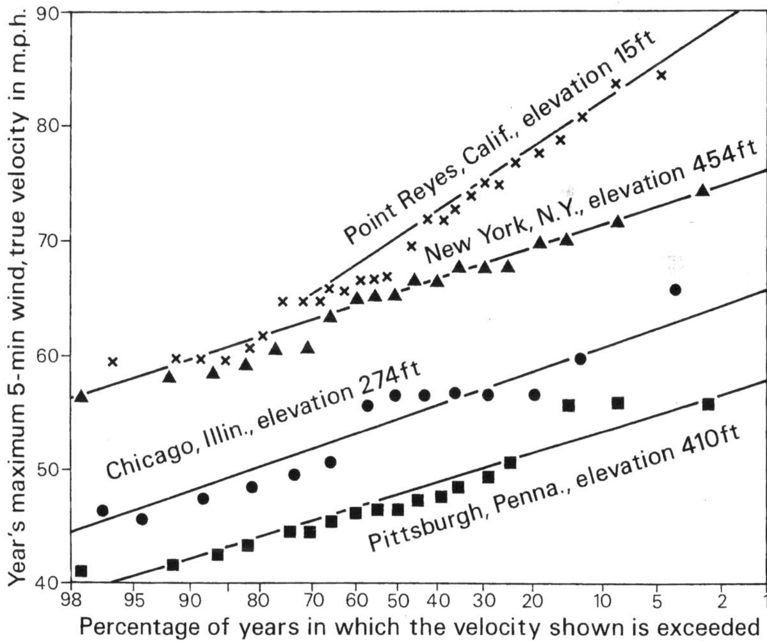

Figure 11.6 (a) Average tracks of cyclones in the USA; (b) Distribution of maximum yearly 5 min winds at particular localities in the USA
Source: Wing 1932 (b only)

basis Doan (1970) has calculated the probability of a destructive tornado striking a nuclear power station to be of the order of 10^{-3} per year for areas of high tornado activity such as the US Midwest to 10^{-4} per year for areas of lower tornado activity such as the East Coast. The lifetime probability of a tornado strike for a US nuclear power plant is of the order of 10^{-2}, but tornadoes of design proportions have a probability of 10^{-4} (Table 11.3).

In most instances the 'design tornado' taken into account has a 300 miles/hour horizontal wind with an assumed pressure drop of 2–3 lb/in^2 in a period shorter than 3 s. This is a very violent tornado, and the probability of occurrence is of the order of 1 in 10,000 over the lifetime of a nuclear plant. Primary containment vessels and underground equipment can be designed to resist such a tornado, but the storage pools for spent fuel elements could suffer loss through having several feet of their water blown off. Exposed pipework and electrical switchgear equipment would be exposed to damage, as would service water systems.

Table 11.3 Tornado probabilities for selected nuclear plant sites in the USA

Latitude band	*Plant site*	*Probability of a strike per year (x 10^{-4})*	*Tornado interval (years)*
26°30′	Turkey Point, FL	9.94	1,010
28°	Crystal River, FL	6.05	1,650
34°	Browns Ferry, AL	10.00	1,000
34°	Brunswick, NC	5.72	1,750
35°	Oconce, NC	3.62	2,760
35°	H.B. Robinson, NC	5.79	1,730
37°	Surry, VA	3.72	2,690
38°	Calvert Cliffs, MD	3.77	2,660
40°	Cooper, NB	18.60	537
40°	Fort St Vrain, CO	5.43	1,840
41°	Fort Calhoun, NB	15.70	635
41°	Three Mile Island, PA	5.51	1,810
41°	Peach Bottom, PA	6.30	1,590
41°30′	Susquehanna, PA	3.18	3,150
41°30′	Quad Cities, IL	10.30	970
42°	Dresden, IL	16.00	623
42°	Indian Point, NY	4.82	2,080
42°30′	Connecticut Yankee, CT	8.07	1,240
42°30′	D.C. Cook, MI	9.68	1,030
42°30′	Palisades, MI	4.04	2,480
43°	Point Beach, WI	6.50	1,540
43°	Nine Mile Point, NY	2.24	4,110
44°30′	Maine Yankee, ME	4.16	2,400
35°	Tornado 'worst site' (latitude 35°, longitude 98°: Oklahoma)	36.20	276

Source: Doan 1970

Hurricanes and tornadoes were first considered in the USA in the construction permit review for the Turkey Point plant in Florida in 1966 and subsequently were included in the reviews of all East Coast and Gulf Coast sites, and many flood-prone inland sites (Okrent 1981). Similarly, tornado-protection requirements were first applied to new reactors in the late 1960s after the first review for a reactor in a high tornado area established that tornadoes occurred sufficiently frequently in most parts of the country to warrant uniform defences in design.

FLOODS

Most of those nuclear power stations not sited at the coast occupy riverside sites, but a few, such as those on the Great Lakes or that at Trawsfynydd in North Wales, depend for their cooling water supply on an inland lake. Methods of calculating the probability of floods are well established and the flood hazard at any given site can usually be calculated more readily than the earthquake probability.

There are two established methods of flood evaluation. In the first, the flood probability is calculated using an empirical formula in which the main variables are the hydrological characteristics of the drainage basin and its rivers. The second method is based on statistical analysis of the maximum floods of the river under examination, with some corrections derived from the hydraulic characteristics. Floods of defined magnitude can then be related to defined levels of damage at the plant producing specified releases of radioactive materials, and the adequacy of the plant design and acceptability of a site can be determined from the flood adequacy curve and the calculated levels of damage-related release.

Many of the methods of hydraulic evaluation are based on statistical extrapolation, and it seems important to take into account changes that may occur over time in the nature of the river and river basin characteristics to avoid dangerous extrapolation. Nevertheless, flood damage of such severity as to cause any release of radioactivity seems one of the less likely eventualities at nuclear power stations.

SUMMARY

Guatemala, Soviet Central Asia, Tangshan in northern China, the Philippines, Iran, Mexico, Greece and Turkey have all recently suffered much more from earthquakes than California or Japan. Moreover, only four major earthquakes are known to have occurred in the past in California, although two of them, Owens Valley in 1872 and San Francisco in 1906, were very high on the Richter scale. No major earthquake has ever been recorded in the central area of the San Andreas fault system, and the last major earthquake in the southern San Andreas section was in 1857. It is also true that physical damage to buildings under earthquake conditions rarely occurs when they have hard rock foundations; it is filled-up land, marshy land, drained or undrained, and alluvial soil which spell the most danger. In parts of Italy, a country in which seismic risk is relatively high (Chapter 4, Figure 4.4), there are towns and cities that have been standing for hundreds of years longer than the likely economic or technological lifetime of any power reactor. The implication is that good design and careful geological survey can be combined to reduce the risks that may be associated with power reactor siting in seismically active areas. Once again, however, there is a balance to be struck between costs, benefits and risks. It is generally stated that in the event of an earthquake a reactor must be capable of being safely shut down and maintained in a safely shutdown condition almost indefinitely (the SSE situation). Worldwide, a substantial proportion of the engineering effort that goes into nuclear power station design is developed against environmental hazards. Japan's first nuclear power station, a Magnox plant built by the General Electric Company of the UK incurred an increase of civil engineering costs of between 30 and 40 per cent because of additional anti-earthquake design features (*Economist*, 11 April 1959).

With the passage of time power utilities have experienced decreasing freedom in their choice of sites for nuclear power stations. Locational decisions have become increasingly influenced by public attitudes, particularly in the USA where a sufficiently aroused public has exerted extreme political pressure on policy. In California, opponents of nuclear power have used fears over earthquake liability to reinforce the case for a moratorium on nuclear power. Prior to the Bodega Bay and Malibu cases in the early 1960s the USAEC had not considered seismic phenomena in licensing and had no familiarity with them (Komanoff 1981a). Two early Californian reactors licensed in the 1950s, the Vallecitos test reactor and the commercial Humboldt Bay plant, received no detailed seismic review at that time and subsequently shut down in the late 1970s rather than accept the high cost of upgrading to seismic safety standards. Shortly after, the USAEC commissioned seismological and geological research which indicated considerable seismic potential in parts of the eastern USA also. Subsequently, eight *Regulatory Guides* spelling out methods of calculating earthquake forces and specifying the instrumentation, structural reinforcement and component reliability necessary to reduce susceptibility to damage and accidents were drawn up.

Some power stations then under construction were caught by this new framework. Diabolo Canyon was started in 1978. It had to be completely redesigned and strengthened four years later to take account of an offshore fault.

To the layman there may seem to be little commonsense justification for building nuclear power stations in parts of the world notorious for their earthquake activity or tornado liability. Engineered safeguards in the design and construction of plant buildings can cope with tornadoes and

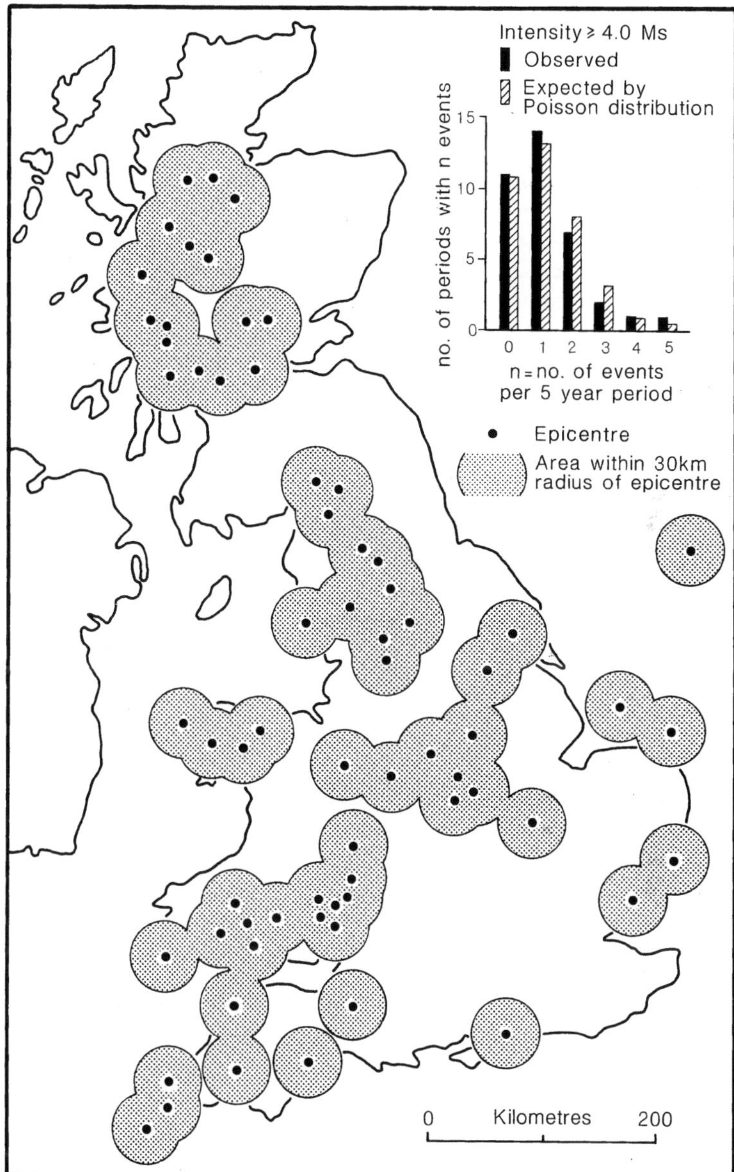

Figure 11.7 Seismic events in England, Wales and Scotland recorded from 1800 to 1982 (magnitudes 3.5–6.0 on the Richter scale). Although the British Isles are seismically stable compared with areas such as Italy, Japan or California, this map shows that it would be wrong to regard them as free from seismic events significant for nuclear power station design. Early nuclear power stations, however, were not seismically qualified in the way that the components of more recent units such as Heysham have been

Source: Institution of Mechanical Engineers 1984: 23

floods, but earthquakes, it is argued, are a different matter. In tectonically unstable areas it is impracticable to provide a structure totally capable of resisting *large-scale* ground failure. Any rational attempt at earthquake-resistant design must rest on a probabilistic analysis of the variables concerned, and among these the characteristics of future earthquakes involve by far the greatest uncertainties. So far, nuclear power plants in the vicinity of an earthquake have stood up well to the ground motions, but such a situation may not last for ever. Some older units are particularly vulnerable to severe damage simply because they were developed and built at a time when possible earthquake damage was not taken into account. This applies to early reactors in the UK, a country that lies well away from the edges of the world's plates but in which earth tremors have shaken buildings in Cornwall, South Yorkshire and Stoke-on-Trent in recent years (Figure 11.7).

FURTHER READING

Maclean, A. (1978) *Goodbye California*, London: Collins.
Newmark, N.M. and Rosenblueth, E. (1971) *Fundamentals of Earthquake Engineering*, London: Prentice-Hall.
Smith, C.B. (1973) 'Power plant safety and earthquakes', *Nuclear Safety* 14 (6): 586–96.

12 The back-end of the nuclear fuel cycle: storing and transporting radioactive waste

Radioactive waste (radwaste) is produced in the extraction or subsequent use of natural or artificial radionuclides and can be said to be discharged when the user disposes of it without intention of using it further. Growing interest in environmental pollution has heightened public awareness of the issues involved in the disposal of such waste. All stages of the nuclear fuel cycle produce radwastes in the form of gases, liquids, sludges and solids which, because of their inherent toxity, must either be treated to remove most of the contaminants before being released to the environment or must be so diluted that contaminants are below permitted levels. However, it is at the so-called back-end of the nuclear fuel cycle – the stages following removal of used fuel from the reactors – where radwaste management comes into sharpest focus. The relevant stages are shown in Figure 12.1. At the time of its discharge from the reactor, the spent fuel is intensely radioactive and generates significant heat. One tonne of spent fuel from an LWR typically generates about 2000 kW of thermal decay heat at the time of discharge. This drops to only 10 KW after one year and 1 kW after ten years. For this reason the fuel is first stored under water in cooling ponds on the power station site for periods ranging typically from several months to several years.

Not all countries have adopted fuel reprocessing as part of the nuclear fuel cycle. In the *once-through cycle* the fuel, a solid ceramic material, can itself serve as a form suitable for long-term disposal. In the countries that use reprocessing, the used fuel elements are transported to a reprocessing plant where high level liquid waste arises from the process of separating reusable uranium and plutonium. This waste contains nearly all the radioactive material which is left after the separation operation. Chemically it contains 99.9 per cent of the non-volatile fission products, about 0.5 per cent of the uranium and plutonium, and many of the other transuranic elements formed in the fuel during its burnup.

The management process from the reprocessing stage onwards involves control and capture of toxic gaseous effluents, interim storage of high level liquid wastes and their conditioning and conversion into solid form, storage, from a few years to perhaps half a century, to allow further decay

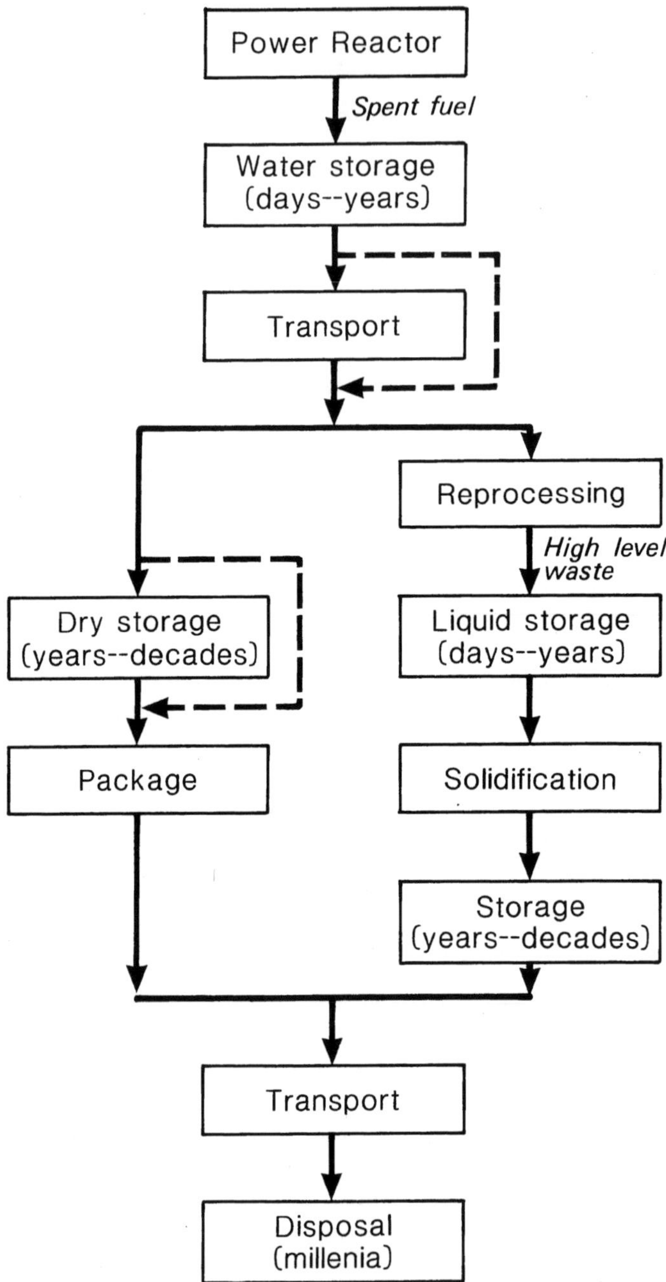

Figure 12.1 High level waste management
Source: Blomeke and Johnson 1985

of radioactivity and, finally, transport and ultimate disposal. In a once-through fuel cycle, the spent fuel elements themselves become the main radwaste stream.

Two basic concepts are inherent in radwaste disposal practice: the first is 'dilute and disperse to the environment', and the second is 'concentrate, contain and isolate'. The first approach may be used only for waste that is low in radioactivity; the second involves the provision of proper containment or storage of highly radioactive waste to exclude as far as possible the likelihood of radioactivity returning to man by chemical or biological pathways. In some cases a combination of the two methods, 'delay and decay', can be used, with the wastes being stored for a time to allow cooling down by decay before storage or disposal. Thus it may be possible to store relatively short-lived wastes in a retrievable manner in shallow burial trenches for about 300 years, after which the total concentration of radionuclide in these low level wastes would be low enough for dispersal to be acceptable.

An important measure of successful waste disposal policies is that what is done today should not leave unsolved problems for tomorrow's generations. It must be acknowledged, therefore, that provision of the ultimate safe disposal of wastes should be the responsibility of the generation producing them, and temporary or interim storage should not be used as an expedient to pass on the problems to future generations. Long-term interim storage of high activity waste with no intention of ultimate disposal is widely recognized as an ethically untenable option.

The implications for public health of the disposal of some radwastes need to be examined at local or regional levels while others, such as the gaseous effluent krypton-85, demand a world-wide scale of attention (Table 12.1). There are also international differences in the degree of commitment to development and implementation of means of geological disposal for highly radioactive waste categories. Attitudes seem to be influenced by the size of national waste totals and attitudes to fuel reprocessing (Orlowski 1986). For example, Spain has a 'no reprocessing' policy and consequently has not been very active in the search for ultimate disposal sites.

Table 12.1 Approximate content in 1 tonne of used LWR fuel after a normal burnup of 33 GWd/tonne

	Half-life (years)	*Approximate content (g/t)*	*Activity (Ci/t)*
Krypton-85	10.80	29	11,000
Iodine-129	16.16	170	0.03
Carbon-14	5,730	0.6	2.2
Tritium	12.3	0.06	570

Source: Hebel *et al.* 1986

TYPES OF RADIOACTIVE WASTE

Radwaste is produced at all stages of the nuclear fuel cycle and as a result of the decommissioning of nuclear power stations. This waste can be categorized according to the volume produced, the amount of activity within it and its form (liquid, solid or gaseous).

High volume low activity solid wastes result from mining and uranium ore processing at the front-end of the fuel cycle, from reactor operations and from final plant dismantling (decommissioning). Generally speaking, low activity wastes are characterized by radionuclides with short half-lives and could also be called short-lived wastes.

Each gigawatt of electrical generating capacity (GWe) at a nuclear power plant requires around 150 tonnes of uranium per year which, for an average ore content of 0.2 per cent, produces 25,000 tonnes of mining waste (the figures decrease with fuel recycling). In the OECD countries this amounts to a waste accumulation of 15 million tonnes per annum. In the early days of the uranium industry mining activities were conducted without thought about, and without much knowledge of, any possible adverse environmental effects from radioactive ore tailings (see Chapter 7). A large programme of remedial action on tailings heaps (Uranium Mill Tailings Remedial Action (UMTRA)) began in the USA in 1983 as a result of federal legislation, and in 1987 clean-up and stabilization operations were being carried out at twenty-four sites (White and Miller 1983). In addition, the Grand Junction Remedial Action Programme was attempting to decontaminate 740 contaminated structures on private properties in and around Grand Junction, Colorado, in whose construction tailings material had been used as building aggregate. In 1985 similar decontamination activities were taking place at Canonsburg, Pennsylvania, as a result of uranium processing by a local firm which closed in 1980.

In normal reactor operations low activity solid waste is produced in the form of protective clothing, paper and equipment originating from the active area of the plant. Commonly, this type of waste is buried in designated shallow trenches, and much of it is almost contamination free, but at the upper level of contamination are items with low beta and gamma activity accompanied by alpha contamination. Such wastes also stem from fuel enrichment and fuel fabrication activities. In the past sea disposal has been used to remove much of this waste.

The disposal of *low activity liquid waste* from nuclear power plants and fuel reprocessing factories depends upon the siting of the works. Those with a coastal location, or on a large river or lake, remove sufficient radionuclides from liquid streams by distillation or floc precipitation to produce effluents of 'acceptable' purity prior to discharge into the adjacent water body. Some of the inland nuclear fuel reprocessing plants in the USA and USSR have discharged wastes into seepage ponds.

Medium volume medium activity wastes are produced by reactor operation

Table 12.2 General characteristics of radioactive waste categories relevant to waste disposal

Waste category	*Important characteristics*[a]
I High level long lived	High β–γ Significant α High radiotoxicity High heat output
II Intermediate level long lived	Intermediate β–γ Significant α Intermediate toxicity Low heat output
III Low level long lived	Low β–γ Significant α Low – intermediate radiotoxicity Insignificant heat output
IV Intermediate level short lived	Intermediate β–γ insignificant α Intermediate radiotoxicity Low heat output
V Low level short lived	Low β–γ Significant α Low radiotoxicity Insignificant heat output

Source: IAEA 1981

Notes: [a]The characteristics are qualitative and can vary in some cases. 'Insignificant' indicates that the characteristics can generally be ignored for disposal purposes. An alpha emitter (α) is a radionuclide which emits alpha particles. An alpha particle consists of two protons and two neutrons and has a net charge of +2. Alpha emission is a high linear energy transfer radiation. A beta emitter (β) is a radionuclide which emits beta particles and has a mass and charge equal to an electron (−1). Beta emission is a low linear energy transfer radiation. Gamma rays (γ) are protons emitted from the nucleus of a radionuclide during radioactive decay.

and fuel reprocessing, e.g. ion exchange resins, sludges and precipitates, and may include some plutonium-contaminated material (PCM). These liquid wastes are produced at a rate of about 60 m^3 per tonne of fuel reprocessed. Some of these liquid wastes are concentrated by evaporation and the concentrate then stored for a number of years to allow the relatively short-lived activity to decay; others may be treated by floc precipitation enabling the sludge to be stored and the decontaminated liquid to be treated as low activity waste.

Solid low volume high activity waste comprises mainly fuel element cladding and solidified material from reprocessing. The volume of contaminated fuel cladding lies in the range 0.25 – to 0.6 m^3 per tonne of fuel processed. At present, the common practice is to store cladding waste pending the discovery of a satisfactory means of ultimate disposal. About 0.98 m^3 of solidified radioactive waste is produced each year by every 1,000 MWe of continuous thermal operation of nuclear power stations.

High activity liquid waste is produced entirely in fuel reprocessing operations, at a rate of approximately 5 m^3 per tonne of fuel reprocessed, although in the case of high burnup oxide fuel it can be reconcentrated to about 0.25 m^3. This material is stored in double-walled tanks, a practice that may be acceptable for tens of years but not for longer.

There are operational definitions for the different radwaste categories; in the UK, for example, the division between low level waste and medium or intermediate level is set at 4 GBq/t for alpha activity and 12 GBq/t for beta–gamma activity, but the thresholds tend to vary internationally (Hill and White 1982). The classifications of waste are essentially 'shorthand' references, and arguments about whether a particular waste is 'high', 'intermediate' or 'low' level are much less relevant than having consistent standards and approaches used in disposing of each type of waste (Chapman *et al.* 1987). However, the IAEA has produced a fivefold classification of radwastes on the basis of their radioactivity and heat output which provides a useful frame of reference (Table 12.2).

Disposal of high activity wastes currently presents problems because of their potential as biological hazards. Not only are they highly active, but they also contain some very long-lived activity. Fission products resulting from the splitting of the fuel are generally relatively short lived but they include strontium-90 and caesium-137, which both have half-lives of about thirty years. The fuel itself is one of the chemical group of elements known as the actinides, as are the products formed by the activation of uranium. These include americium-241 (half-life 458 years) and plutonium-239 (half-life 24,000 years). Even when the plutonium-239 has reduced to a lower level (after, say, half a million years) there remain significant amounts of longer-lived nuclides such as plutonium-242 (3.8×10^5 years) and neptunium-237 (2.2×10^6 years). An additional problem is that of dissipating the heat caused by radioactive decay during the first few centuries and, if wastes are in liquid form, to prevent the liquids from boiling (Beale 1982).

It has been widely argued that the long half-lives of many of the elements associated with the high activity wastes require their isolation from the earth's biosphere over geological time-scales. The contrary case has also been made that after 600 years the potential risk is no greater than that from the ores from which the uranium was obtained (Pigford 1974; Eisenbud 1981). This difference of opinion is important. Any conclusion that high level wastes can adequately be isolated from the biosphere requires agreement on the length of time for which isolation is necessary. Storage times of a few thousand years are within the range of human experience, e.g. the tombs of ancient Egypt. In the relevant technical literature the notion of storage of high level radwaste for around 1,000 years seems to be increasingly common. There is general agreement that such repositories must be remote from man's environment, deep on land or at sea, and metallurgists are confident that casks capable of maintaining their integrity

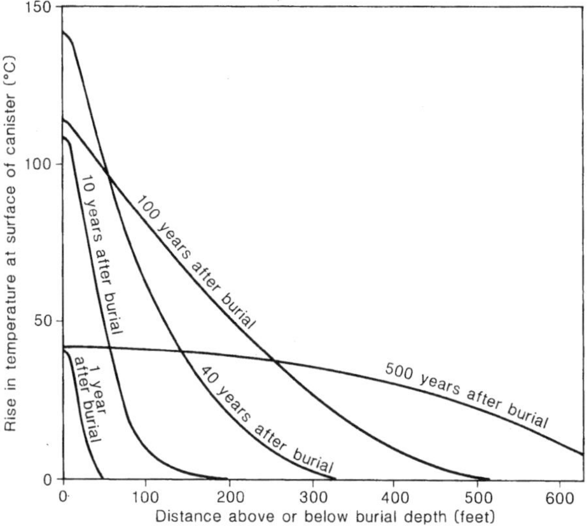
150
100
50
0
0
100
200
300
400
500
600
Rise in temperature at surface of canister (°C)
Distance above or below burial depth (feet)
1 year after burial
10 years after burial
100 years after burial
40 years after burial
500 years after burial

(a)

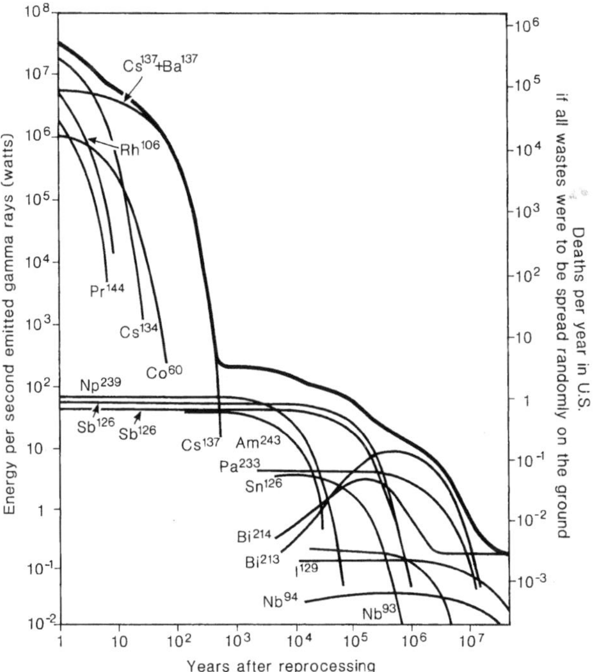
Energy per second emitted gamma rays (watts)
Years after reprocessing
Deaths per year in U.S. if all wastes were to be spread randomly on the ground
Cs137+Ba137
Rh106
Pr144
Cs134
Co60
Np239
Sb126
Sb126
Cs137
Am243
Pa233
Sn126
Bi214
Bi213
I129
Nb94
Nb93

(b)

for 500 – 1,000 years can be produced economically. Figure 12.2 gives an indication of the time-scales involved in high level waste storage from the viewpoint of reductions in temperature and gamma radiation levels.

DISPOSAL ON LAND

Three types of land disposal are either in use, at an experimental stage or under design.

Shallow burial in trenches, tumuli, tunnels, concrete bunkers or caissons

Shallow burial has been used for many years to deal with the large volumes of low level solid wastes produced at nuclear establishments. At Los Alamos over a quarter of a million cubic metres of contaminated materials placed in steel drums or wooden containers now lie buried under the Idaho Desert. In the UK at Drigg, in Cumbria, low level radwaste has been buried in shallow soil trenches since the early days of the industry, and these trenches will remain in use until at least the year 2000.

Shallow trench disposal may be temporarily safe, but much of it takes place in ground above the water-table and a hazard to be guarded against is the leaching of radionuclides from the buried waste by rainfall or soil water movements. Drigg has been criticized in this respect (House of Commons 1986), but it is noteworthy that, although there are around 30,000 people living within 50 km of the site, the Savannah River disposal site in the USA has a surrounding population of over 500,000 within the same radius. Rainfall figures also vary between disposal sites. Los Alamos has an annual average rainfall of only 46 cm and a deep water-table (300 m), whereas the rainfall figures for Drigg and Savannah River are 100 cm and 120 cm respectively. At Savannah River the waste material has been buried in soil

Figure 12.2 Temperature and gamma radiation curves for high level radiation wastes. (a) This diagram shows that the heat produced by high level radwaste can best be dealt with by delaying burial for 10 years after reprocessing. In rock of average thermal conductivity the maximum average temperature of the rock just above and below the burial depth would be reached 40 years after burial (140 °C). Burial in salt, which has greater thermal conductivity, would mean that the rise in temperature at burial depth would be much less (85 °C). (b) Gamma radiation emitted by decaying nuclei presents the most direct health hazard represented by high level radwaste. The biological damage done by a gamma ray is roughly proportional to its energy in most situations. Thus in this graph Cohen has plotted on the left the gamma ray energies emitted per second by various radioactive isotopes in the wastes resulting from a full year of an *assumed* all-nuclear US electric power system (400 × 100 MWe power plants). The bold black line shows that between 8 and 400 years after reprocessing the total gamma ray hazard falls by more than four orders of magnitude. The scale on the right indicates the total number of fatal cancers expected per year if the source of this amount of gamma radiation were spread at random over the entire land surface of the USA
Source: Cohen 1977: 174, 175

that has high infiltration, poor run-off characteristics and low absorptive capacity because of the presence of kaolinite as the clay fraction.

At some sites the possibility of migration through leaching has been anticipated by building trenches with concrete linings, e.g. at the Centre de la Manche in France, or by using buried concrete or steel storage containers. These allow the use of the shielding properties of the soil without permitting immediate migration but are more expensive than simply excavated backfilled trenches of the Drigg and Los Alamos type. At the Hanford Reservation in the USA concrete pipes with concrete bases sunk vertically in the soil have been used for some low volume intermediate level wastes, and this method of storage has been used at Tokai, Japan, and by Euratom at Ispra. At Trombay, India, sunken pipes have been used because ground conditions do not permit direct burial. At Trombay 2 m of clay-gravel topsoil are underlain by 10 m of weathered basalt that crumbles when dry. Normally the water-table is 8 m below ground level, but during the monsoon it rises to within half a metre of the ground surface and the danger of overflowing and surface contamination excludes the use of simple trenches.

Deep injection of liquid wastes into porous strata, with or without added cement, by means of boreholes drilled from the surface

This technique was used in the late 1960s at the Oak Ridge Nuclear Laboratory (ORNL) in the USA for disposal of some intermediate level wastes, but deep injection has now been concluded to be unacceptable. It is difficult to carry out, and predicting the fate of the waste is very complex.

Deep burial in an existing mine adapted for the purpose, or in a specially built deep vault or 'geological repository', or in a deep borehole

Some European countries do not allow shallow trenches. In several, disused mines have been adapted to dispose of waste at greater depth. In Czechoslovakia, for example, a former limestone mine that was used as an underground factory during the Second World War has been used for the purpose. Mines are also used in Spain and the FRG.

Shallow burial is widely practised for the disposal of high volume low level radioactive waste not only from the nuclear power industry but from hospitals, research laboratories and industry. The second and third techniques are either at the experimental development stage or have gone through the process of conceptual design or modelling. At present, only Sweden is building anything approaching the category of a long-term depository. As a result, waste with higher levels of activity is becoming an increasing embarrassment for many governments as delays occur in the siting and building of repositories. Nowhere is this clearer than in the way spent fuel is accumulating at many nuclear power station sites. Conventional spent fuel storage facilities at nuclear power plants in the form of water-filled cooling plants were designed in the expectation that the fuel would be transported offsite for reprocessing within a year of removal from the reactor. However, because of problems in finding sites for long-term repositories and because the development of a fuel reprocessing industry has been slower than originally anticipated, there is now a widespread need to store spent fuel for longer periods than initially planned.

The design of long-term repositories

Table 12.3 lists considerations relevant to the design, modelling, building and operation of a geological repository, but the major natural process to be guarded against is the leaching of the waste and transport by ground water. Thus, in designing such facilities, the concept of multiple barriers to these processes must be employed with each barrier resting inside the other and being brought into operation only when those in advance of it have failed to prevent the ingress of water. The behaviour of each barrier is quantifiable and verifiable. The complete process of selecting a nuclear waste disposal site, constructing a deep repository, filling it with waste and then closing and abandoning it has not yet been carried through anywhere. However, over the past decade a great deal of the necessary research and design work has been undertaken and the technical expertise needed to build such a repository is widely available (Chapman *et al.* 1987). In this work, the barriers are conceived to be as follows:

(a) the matrix in which the waste is set (e.g. borosilicate glass);
(b) the waste container or canister;
(c) the backfill in the hole or walling of the vault to provide a 'buffer';
(d) the host rock and the geological environment in which the repository is located which isolate the waste from the surface and control water movement and the ground-water geochemistry.

For practical purposes it is convenient to break up the system of repository barriers into two zones identified as the *near field* and the *far field*. The near field is the zone in the host strata which is significantly altered by the presence of the repository (e.g. by thermal activity) while the far field is the surrounding undisturbed natural geological system. With the insertion of a repository the near field becomes 'a zone of great chemical complexity in which waste matrix, containers, backfill and rock may all interact with groundwater and with each other' (Chapman *et al.* 1987). What happens within, and to, this region controls any release from the repository, and the near field is calculated to extend for some tens of metres from the location of the radwaste itself. The far field is assumed to remain in a steady state. It controls the rate at which water can enter the near field and it would retard transport of activity away from the repository when all the barriers are

Table 12.3 Site selection factors for deep geological repositories

1 Topography
2 Tectonics and seismicity
3 Subsurface conditions
    Depth of disposal zone
    Formation configuration: thickness and extent, consistency, uniformity, homogeneity and purity of strata
    Nature and extent of overlying, underlying and flanking beds
4 Geological structure
    Dip or inclination
    Faults and joints
    Diapirism
5 Physical and chemical properties of host rock
    Permeability, porosity, solubility and dispersivity
    Inclusions of gases and liquids
    Mechanical and plastic behaviour of rock
    Thermal gradient: regional and local
    Thermomechanical and thermohydraulic responses
    Thermal conductivity and specific heat
    Sorption capacity
    Mineral content of water
    Radiation effects
6 Hydrology and hydrogeology
    Surface waters: occurrence, form, volumes
    Ground-waters: occurrence, volumes, chemistry
7 Future natural events
    Hydrological changes
    Uplifts and subsidence
    Seismicity
    Intrusions and faulting
    Climatological changes
    Topographical changes
8 General geological and engineering conditions
    Site area and buffer zone
    Pre-existing boreholes
    Exploration boreholes
    Spoil disposal
    Waste transport
    Engineering and construction of repository
    Operational safety and stability of repository
9 Societal considerations
    Resource potential
    Land value and use
    Population distribution
    Jurisdiction and rights of the land
    Accessibility and services
    Other environmental impacts
    Public attitudes

Source: After IAEA (1982) *Site Investigation for Repositories for Solid Radioactive Wastes in Deep Continental Geological Formations*, Technical Report Series 215, Vienna

breached, and such activity as is released would be diluted in the far field before entering into systems in the biosphere.

For the normal-case model it is accepted that ground water will eventually cause corrosion of the waste canisters, but that the process of release will be gradual. An event such as an earthquake might cause simultaneous failure to many barriers causing much more sudden release. Thus repositories must be sited away from tectonically active or marginal areas.

The normal-case model shown in Figure 12.3 is constructed around the credible mechanism by which components of the waste could be transported back to the surface in ground water. Free water is present in all rocks and it moves through the pores and fissures in response to differences in hydraulic head or pressure. Even where no flow occurs in the body of water so contained, the pores of the rock can allow diffusion of radionuclides at very slow rates.

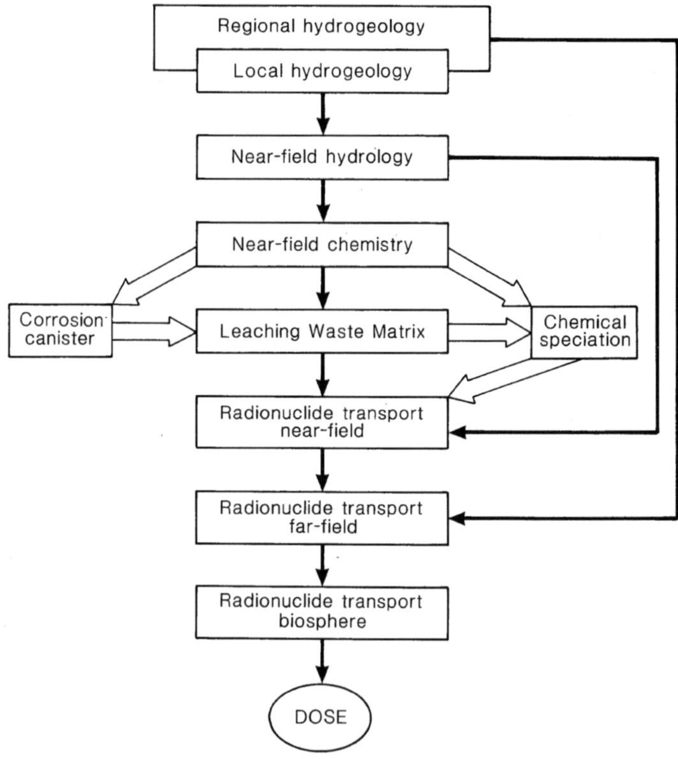

Figure 12.3 A typical normal-case model chain for safety analysis purposes
Source: From NAGRA, Project Gewähr 1985, NGB 8509, Switzerland

However strong and long lived the repository, the corroding and leaching action of the water will eventually mobilize the waste in solution or suspension. Therefore the normal-case model assumes that progressive geological processes exist which must inevitably cause the return of some nuclear waste to the biosphere. It begins with the assumption that water gains access to the near field of a depository, penetrates the backfill, corrodes the package, dissolves the waste from the matrix, mobilizes the waste elements in chemical forms and transports them into the biosphere. It is not a worst-case scenario because the process of breakdown is assumed to be very gradual, and this would not necessarily be the case with an earthquake. Thus analysis of the far field becomes important in assessing the likely routes along which water may move, how long it takes to travel those routes, and how far and in what form and concentration radioactivity is likely to be carried. These are fundamental questions to be asked, capable of being properly answered in different geological environments, before realistic safety assessments can be made for any proposed repository site. Clearly, however, the nature of the far field needs to be treated seriously in site assessment and depends very much upon the characteristics of different rock types. Particular rocks offer different capacities to retard the rate of movement of individual radionuclides (OECD/NEA 1984a). Three principal rock types are believed to be best able to provide adequately the requirements for siting a repository for the disposal of high level radwaste. These are (i) argillaceous or clay-rich rocks, (ii) hard crystalline igneous and metamorphic rocks and (iii) evaporites (salts, either bedded or in salt domes).

Argillaceous rocks

The main advantage of clay formations is their ability to absorb active species that may be leached from stored waste by flowing ground water. Although their water content may be relatively high, the rate of water movement by ground water through these formations is relatively low, adding to their capacity to absorb leached activity. The main difficulty is that argillaceous series offer properties whose incompatibility raises problems for siting of a depository. The softer plastic clays have low permeability, respond to stress without fracturing and are good at absorbing leached radionuclides, but they occur at relatively shallow depths, often in association with permeable rocks, and they have low thermal stability. Mining is difficult in clays, a fact important in designing tunnel supports. Nevertheless, Belgium has gone a considerable distance in exploring the potential of clay deposits, followed by Italy, the UK and the USA.

Crystalline rocks

Hard massive igneous or metamorphic rocks such as granite, gabbro and basalt have very low total porosity. However, they can be extensively fissured and fractured, and the pattern of fissuring which does much to control water flow, sometimes over many miles, introduces an element of variability. Civil engineering experience of such rocks from hydroelectric power station construction is considerable, but the thermal conductivity of hard rock is relatively low, so that heat dissipation at a repository would be slow.

In the past, work on underground storage in the UK has tended to concentrate on hard crystalline rocks as part of EC programmes of research into underground disposal of high level waste. Such programmes have tended to be divided into components according to types of geological formation, with the UK and France studying mainly crystalline rocks such as granite, the FRG and France concentrating on salt deposits, and Italy and Belgium considering clay formations. During 1988–9 the behaviour of granite under specific site conditions was investigated at the Swedish project Stripa, 15 km north of Lindesberg. Hard rock sites that have been suggested as possibilities from time to time in the UK include Loch Doon, near Dalmellington in Dumfriesshire, Morven, near Aberdeen, North Harris on Lewis, one of the largest of the Scottish Western Isles, the island of Pabbay between Lewis and North Uist, and the island of Taransay off the southwest coast of Lewis.

Evaporites

Evaporites are salt deposits formed in the geological past by the evaporation of shallow lagoons and lakes.

Since 1955, considerable attention has been given to the possibility of storing highly radioactive wastes within deep salt formations. Interest in salt for this purpose has been particularly strong in the USA and the FRG, countries in which underground salt formations are common. Salt offers a number of advantages. Bedded salt deposits are relatively free of circulating water and have a low water content, and the properties of salt are such that any fractures that might develop are healed by plastic deformation and recrystallization. Thus it is argued that materials deposited in salt formations are unlikely to come into contact with leaching solutions over very extended periods of time. Bedded salt formations are preferable to more geologically disturbed formations, such as salt domes, because they make it easier to predict future stability, but dome salts are thermally more stable than bedded salts. Salt deposits can be mined with ease, but salt itself has good structural properties, with a comprehensive strength and radiation shielding characteristics similar to those of concrete. Ideally, the depth of a suitable salt formation should not exceed 600 m because of the difficulty and extra cost of operating at greater depths; many of the known salt beds in the USA occur at usefully accessible depths, in areas of low seismicity. One disadvantage is that salt is soluble in fresh water if sufficiently fractured.

Table 12.4 Status of repository development in NEA countries with nuclear power programmes as at 1984

	Country	*Generic studies*	*Survey of sites*	*Surface investigations*	*Test boreholes*	*Access shaft or tunnel sunk*	*Test facility*	*Test facilities at depth (location)*
Crystalline rocks								
Granite	USA	×	×					
	Switzerland	×	×	×	×			
	Sweden	×	×	×	×			
	Finland	×	×	×				
	Canada	×	×	×	×	×		Whiteshell
	France	×	×	×	×			
	Japan	×	×					
	Spain	×	×					
	Sweden	×	×	×	×		×	Stripa
	Switzerland	×	×	×	×	×	×	Grimsel Pass
	UK	×	×	×	×			
	USA	×	×	×	×	×	×	Climax
Other								
Gabbro	Sweden			×	×			
	Canada	×		×	×			
Diabase	Japan			×	×			
Evaporites								
Salt diapirs	Denmark	×	×	×	×			
	FRG	×	×	×	×	×		Gorleben
	FRG	×	×	×	×	×	×	Asse
	Netherlands	×	×	×	·			
	USA	×	×	×	×		×	Avery Island

Table 12.4 continued

Bedded salt	Spain	×	×					
	USA	×	×	×	×		×	Lyons
	USA					×	×	WIPP
Anhydrite	Switzerland	×					×	Felsenan
	USA	×	×					
Other sedimentary rocks								
Clay	Belgium	×	×	×	×	×	×	Mol
	Italy	×	×	×			×	
	Switzerland	×						
	UK	×	×	×	×			
Shale	USA	×			×			
	Japan	×	×					
	Spain	×		×	×			
Other								
Mixed marine sedimentary sequence	FRG			×	×			Konrad
Basalt	USA	×		×	×		×	NSTF Hanford
Tuff	USA	×	×	×	×			NTS
	USA	×	×					

Source: OECD/NEA 1984a

Table 12.5 Major past or present underground research laboratories

	Rock	*Laboratory name*	*Country*
Salt	Bedded	Salt Vault (Kansas)	USA
	Dome	Avery Island (Louisiana)	USA
	Dome	Asse	FRG
	Bedded	WIPP (New Mexico)	USA
	Dome	Hope	FRG
Crystalline rock	Granite	Stripa	Sweden
	Granite	Grimsel	Switzerland
	Granite	Edgar Mine (Colorado)	USA
	Granite	URL (Manitoba)	Canada
	Granite	Climax Mine (Nevada)	USA
	Granite	Fanay Angeres	France
	Granite	Akenobe Mine	Japan
	Basalt	NSTF (Washington)	USA
	Tuff	G-tunnel (Nevada)	USA
Argillaceous rock	Plastic clay	Mol	Belgium
	Clay marl	Pasquasia	Italy
	Mixed sediments	Konrad Mine	FRG

Source: Chapman and McKinley 1987:155

Another is that overlying sedimentary deposits are frequently aquifers offering the potential of flooding, and brine could result in a very corrosive environment for any waste repository.

In the USA salt formations underlie 193,000 km^2 (74,520 square miles) in portions of twenty-four states. They were first recommended for the disposal of radwastes by the National Academy of Sciences in the early 1960s when the Lyons mine of the Carey Salt Company in Rice County, central Kansas, 65 miles (105 km) northwest of Wichita, was identified as an available accessible non-producing mine. The location was chosen with some care; road and rail access were needed, but it was felt that such a facility could not be sited near a large town, and from 1965 to 1968 radwaste materials were stored in the mine in an attempt to demonstrate the feasibility of the concept for the long-term disposal of highly active waste. The results were so encouraging that at one stage it seemed almost certain that Project Salt Vault, as it was known, would become a reality as a repository capable of storing all the high level wastes to be produced in the USA until the year 2000. Towards the end of 1971, however, it was decided to drop the project, for several reasons. In the first place, concern developed regarding the possibility of water getting into the near field. One possible mode of invasion could have been through an unplugged or improperly plugged drill hole or via a naturally produced fissure in the salt

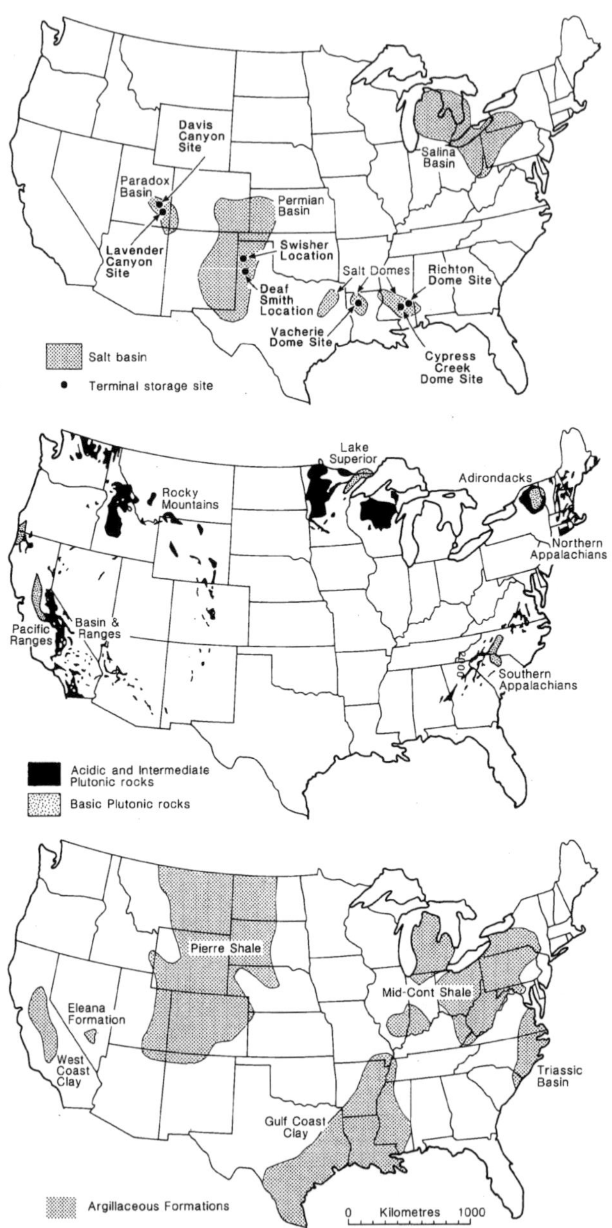

Figure 12.4 Geological environments for radwaste depositories that have been considered in the USA: (a) salt basins; (b) crystalline rocks; (c) argillaceous formations

bed. Twenty-nine abandoned gas and oil wells were known to exist in nearby salt deposits. Twenty-six of these holes could have been successfully plugged to prevent water penetration, but the remaining three would have been very difficult to deal with. Concern also developed about the possibility of progressive natural erosion of the eastern edge of the salt bed. The third mode of water invasion that came to be feared was that from brine-filled cavities into the canister storage holes. Although the Carey mine was defunct in the 1960s, the solution mining activities of the neighbouring American Salt Company reached to within 560 m of the repository. Solution mining removes all the salt from a bed, leaving no supporting pillars to prevent subsequent collapse of the mined out areas. A combination of events became conceivable which might have permitted water to invade the waste storage site and provide an exit for water contaminated with radwaste to escape from the salt. Thought along these lines may have been influenced by the fact that in one solution-mining operation the adjacent American Salt Company pumped 800,000 litres of water into its salt deposit and the water simply disappeared. The company was unable to determine whether the water broke through to geological formations below the salt or whether it had moved laterally through the salt bed.

Thus, at the end of 1971, site work at Lyons stopped, plans to use the mine as a large-scale radwaste depository were abandoned, and it became doubtful whether geological disposal of this kind would find ready acceptance in the USA. It seemed that the environmental stability of evaporites may have been upset by the past drilling of large numbers of unrecorded boreholes in the USA, unwittingly providing potential water routes into dry salt beds. There was a long pause in activity, but by 1988 the USA had again become concerned with exploring the possibilities of storing high activity radwaste in evaporites through work at the Waste Isolation Pilot Plant Project (WIPP) in New Mexico, the Avery Island experiments in Louisiana (Gulf Coast) and the conceptual design for a depository in Deaf Smith County, Texas (Tables 12.4 and 12.5 and Figure 12.4).

In the FRG a salt mine at Asse, southeast of Hanover, has been used since April 1967 for the disposal of low activity wastes and has been the scene of a broad research and development programme aimed at establishing its suitability for the ultimate storage of solidified high activity waste (Figure 12.5). There is confidence in the FRG that this approach will prove satisfactory for the long-term storage of high level radwaste, but Asse has proved to be rather expensive for low level waste disposal because of the transport costs incurred in bringing the wastes to the mine. In co-operation with Euratom a second repository in salt is being developed for the disposal of high and intermediate level salt beneath Gorleben, in a salient of the political border with the GDR.

A question that is being explored in the development of these facilities is that of the 'Wigner effect'. The high activity radwastes will bombard the surrounding salt walls with intense radiation, causing energy to be stored in

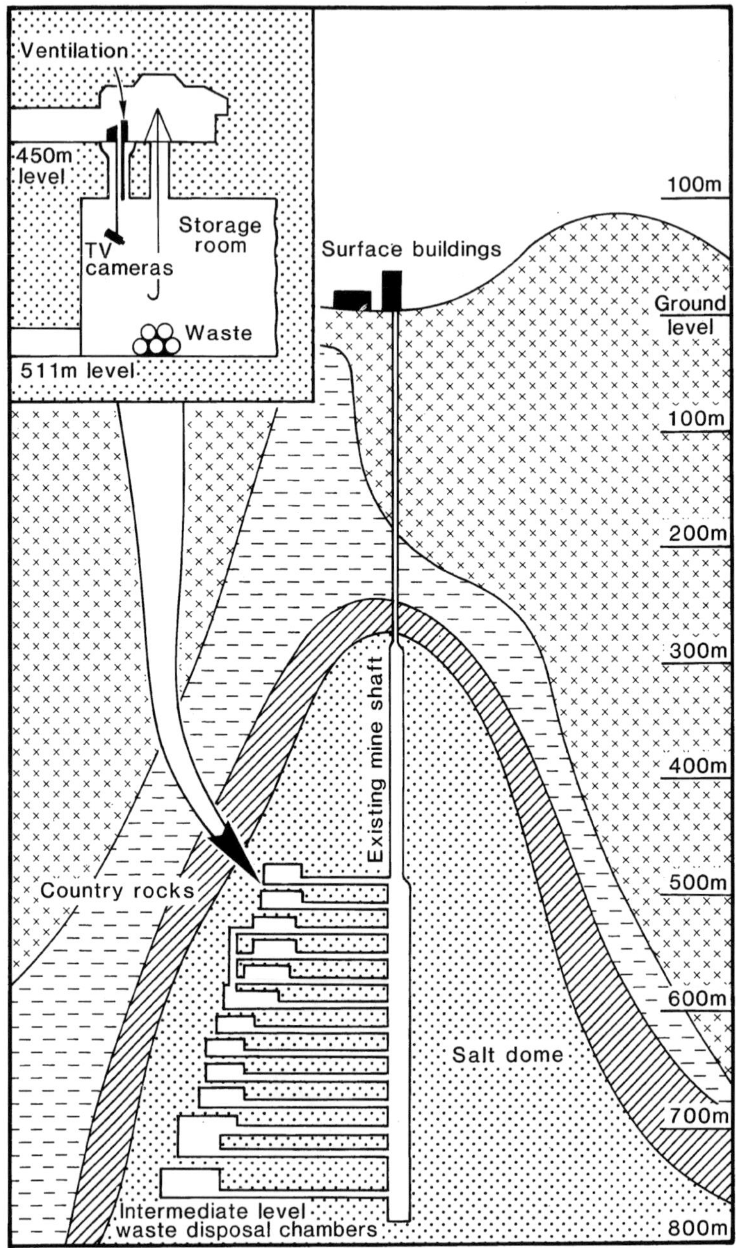

Figure 12.5 Diagrammatic representation of nuclear waste disposal facilities at the Asse salt mine in the FRG
Source: House of Commons 1986: xlivi

the salt, and it is theoretically possible to postulate conditions under which this energy could be released in a sudden burst of heat. An extreme event under such conditions might be explosive upheaval which could release radioactivity to the world above.

The status of programmes for waste disposal

Conceptual designs for high activity waste repositories are numerous; actual depositories are much harder to find. The scientific community recognizes that deep geological disposal in a repository is the final target for any high level waste strategy (Draper and Stanford 1982). Most of the work necessary for demonstration projects either has been done or has been carried through to an advanced stage. The available technology includes processes and techniques for the following processes:

(a) Solidification of high level liquid waste and encapsulation of this waste, or spent fuel, into canisters believed to be suitable for geological disposal. The most thoroughly tested solid matrix for high level waste is borosilicate glass. France has built a successful borosilicate demonstration plant known as AVM (Atelier de Vitrification de Marcoule) and several countries including the UK have adopted the process under licence. In the USA present NRC regulations require that the canister and waste package have a life of 300 – 1,000 years and similar parameters operate in the UK (Marsh 1982). In Sweden a packaging specification for the disposal of spent fuel using thick high purity copper has produced canisters with a life expectancy of hundreds of thousands of years, but the Swedish proposals would use 7,200 tonnes of copper and 4,700 tonnes of lead to dispose of 6,000 tonnes of spent fuel.

(b) Interim storage of the waste in either solid or liquid form. The storage of spent reactor fuel in water pools has been used world-wide for over twenty-five years and so far has proved to be an economical means of short-term storage. A number of the leaks at Sellafield, however, would not have happened if dry depositories had been used rather than placing spent fuel rods in water-filled containers.

(c) Transport of packaged solidified waste and spent fuel. Up to 1985 there had been no shipment of solidified high level waste but many thousands of individual movements of spent fuel had taken place. On the basis of this experience it is claimed by the nuclear industry that canisters of waste can be shipped safely (Blomeke and Johnson 1985).

(d) Design and construction of geological repositories and the emplacement and retrieval of waste before final closure of a repository. Here, full understanding of the hydrological characteristics of rock fracture systems is an area needing further study.

Despite this state of technological near-readiness, industrial-scale applications to implement the technology are not widespread. In the EC local storage facilities at nuclear power plants are becoming saturated and there is a need for centralized storage facilities to cope with the problem (Orlowski 1986). Centralized interim storage of vitrified waste from the reprocessing of gas-cooled reactor fuel has been practised in France, at Marcoule and under Comega, for over ten years (Chometon and Cantin 1983). Buffer storage facilities are expected to enter into operation near the La Hague reprocessing plant around 1995 and later on will feed into a future underground repository. Other storage facilities of similar kind will enter into operation near the Sellafield plant in the UK after 1990, and vitrified waste will be stored there for at least fifty years. These are all *interim stores,* however, and in many countries the process of selecting sites for long-term repositories, even for simpler facilities for low level waste, have run into the 'not in my backyard' (NIMBY) problem of public acceptance. Nowhere is this clearer than in the USA and the UK.

Impasse in the US nuclear waste disposal programme

Spent fuel has been accumulating at reactor sites in the USA in some quantity over the last decade. In an attempt to solve the problem the US Congress passed the Nuclear Waste Policy Act in 1982. The Act was designed to enable construction of two high activity waste depositories, one in the West and one in the East. For the first time the Act delineated guidelines, schedules, actions, procedures, funding and modes of state, tribal, and local participation in the decision process leading to the establishment of repositories for high level nuclear waste, but implementation has proved difficult. Geographical screening processes resulted in the identification of three favoured sites, one in Nevada, in volcanic tuff, one in Texas, in salt, and one in Washington State, in basalt. These sites were announced for further investigation in May 1986 by the US Department of Energy (USDoE) and the White House. At the same time the USDoE called off further screening for the second repository which was to have been located in the eastern part of the USA. Possible sites for the second repository had been identified in the upper Midwest, in New England and in the Southeast. Westerners immediately viewed the indefinite postponement of the second repository as an act of political expediency which disregarded the attempt made by Congress in framing the Nuclear Waste Policy Act to obtain regional balance and fairness (Carter 1987b). Members of Congress from the West took the lead in cutting spending for the waste programme and in bringing about a one-year moratorium on the excavation of deep exploratory shafts at the three nominated Western sites. Nor was the screening process happily conducted. The Texas site sits beneath an aquifer that is the regional ground-water supply, the site in Washington is next to the Columbia River, and only the site in Nevada at Yucca Mountain, remote in the desert 150 km northwest of Las Vegas, is relatively free from such adverse environmental considerations. It is in a corner of the Nevada

Test Site, where nuclear weapons have been tested since the early 1950s.

The 1982 Act gave state governors a strong role in the consultation process including a power of veto capable of being overidden only by a vote of both House and Senate in the US Congress. Fierce resistance to the development of a repository at Yucca Mountain came from Governor Richard H. Bryan of Nevada who said that he would not allow the state to take 'the stigma of being the country's nuclear wasteland'. Public opinion polls indicated that the majority of Nevadans shared their Governor's position. Yet, a 1984 study by the US Geological Survey indicated that numerous potential repository sites are to be found in the Basin and Range Province, of which Nevada and the Great Basin are an important part. Moreover, 87 per cent of Nevada is federal land, much of it is public domain land not yet withdrawn for specific purposes and it is an arid state with a low population density.

Similar pressures surfaced in Tennessee. The USDoE proposed to build a monitored retrievable-storage surface facility at Oak Ridge, but the State of Tennessee fiercely opposed the project. Tennesseans worry that a spent fuel store at Oak Ridge may turn out to be not temporary but permanent if agreement is not reached on the siting and construction of a deep repository. Meanwhile growing amounts of spent fuel are being stored at nuclear power station sites. The reprocessing industry in the USA has experienced a number of problems which have held back its growth and, while the Carter administration offered to provide away-from-reactor (AFR) spent fuel storage similar to those that have been built in Sweden and the FRG, this was not endorsed by Congress. The Reagan administration considered spent fuel storage to be primarily a utility responsibility, but the industry believes that the federal government must make provision for AFR spent fuel storage prior to the construction of one or two high activity waste repositories.

The disposal of low level waste is also causing problems in the USA. Of the initial six regionally dispersed low level waste disposal sites in the USA, only three remained open by 1982, one in the southeast, the other two in the far west. All these sites have imposed restrictions on the waste that they accept. There is obviously a need for additional sites, particularly in the northeast and central regions, but whereas it is generally agreed that the federal government has responsibility for high level waste disposal, in 1980 Congress assigned responsibility for low level waste disposal to the state in which it is produced.

The United Kingdom: public suspicion and opposition

The political decision-making process for dealing with radioactive waste in the UK has unfolded in a somewhat haphazard way and has been described as essentially a process of in-house site selection by the responsible agencies (Kemp and O'Riordan 1988). Responsibility has moved around from one government agency to another (Chicken 1982). Until the Sixth Report of the Royal Commission on Environmental Pollution (the Flowers Report) responsibility lay with the UKAEA, acting for the Department of Energy. The Flowers Report recommended that no large-scale programme of nuclear power be adopted until it had been demonstrated that a method exists to ensure the safe containment of long-lived highly-radioactive waste beyond reasonable doubts for the indefinite future. Following this recommendation, responsibility for selecting a site or sites was transferred to the Secretary of State for the Environment together with the Secretaries of State for Scotland and Wales. On their behalf the National Environment Research Council (NERC) was asked to arrange geological investigations. Then, in 1982, the Nuclear Industry Radioactive Waste Executive (NIREX) was formed and given the task of site selection.

Moving the responsibility around from one agency to another has heightened rather than reduced public anxiety. It is an anxiety which expresses itself by distrust of the organizations responsible for disposal plans and dissatisfaction with the decision-making processes employed. The UK has no repository for high activity long-life waste. Investigations at two possible sites in northern Britain, one at Rothbury in Northumberland and the other in Carrick Forest, were abandoned in 1981 after intense public opposition at two public inquiries.

After 1981 attention was switched to categories of waste that might be sent for immediate disposal. Sea dumping of low level and some intermediate level wastes in the northeast Atlantic ceased effectively in 1983 and officially in 1985, and the shallow trenches at Drigg in Cumbria, with only fifteen years life left, remain the country's major disposal site for these waste categories, especially from those produced at nearby Sellafield. Drigg has been used for a quarter of a century and has accumulated over 500,000 m^3 of waste material (Gregory 1984).

In October 1983 NIREX announced proposals to dispose of long-lived medium level waste in a disused anhydrite mine at Billingham in Cleveland, and short-lived medium waste and low level waste in shallow trenches to be excavated in clay at a CEGB site at Elstow in Bedfordshire (Figure 12.6). Again vehement local public opposition came to the forefront, in the course of which ICI, the owner of the Billingham mine, withdrew its support for the proposal and Elstow was left alone. Consequently, three other shallow burial sites were added to the list as possibilities: Fulbeck in Lincolnshire, Bradwell in Essex and South Killingholme on Humberside. Opposition emerged again from many quarters, including the Conservative Members of Parliament in whose constituencies the four sites occurred (one of them was the Government Chief Whip). Despite this widespread antagonism NIREX began investigative work at the four sites in July 1986. However, less than two weeks before the dissolution of Parliament prior to the General Election that year, the Secretary of State for the Environment announced the abandonment of work at the sites. The ostensible reason

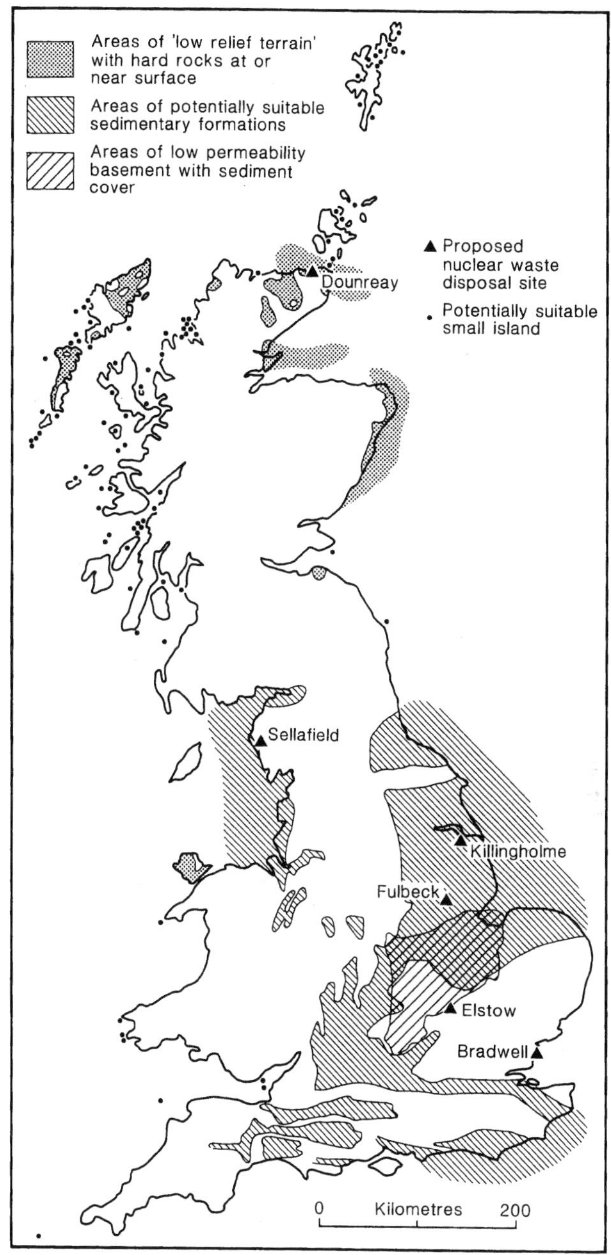

Figure 12.6 Geological environments considered suitable for the deep disposal of intermediate level radwaste in the UK
Source: adapted from Chapman *et al.* 1987; NIREX 1987

was that new cost estimates indicated little difference if low level waste were placed into a deep repository along with all the medium level waste.

In 1988 NIREX was in the process of assessing three different deep-disposal locations and was undertaking site selection in conjunction with the British Geological Survey (BGS). The geological environments examined by BGS included inland basins, seaward dipping and offshore sediments, low permeability basement rocks under a sedimentary cover, hard rocks in areas of low relief and small offshore islands (see Figure 12.6). Once more, however, the agencies came under criticism for not having incorporated an effective means of public consultation in the process and for not explaining their actions clearly (Roberts 1987; Kemp and O'Riordan 1988). The disposal of radioactive waste has become the most intractable of the public relations problems besetting the nuclear industry in the UK in recent years, and the current position is a long way from being satisfactory. Much time, money and skill has been spent without any significant outcome in terms of disposal facilities. A new mechanism is needed to enable the political strategy to be worked out before implementation of an environmental policy is attempted. It may also be needed in Eastern Europe. In June 1988, villagers from Ofaln, Feked, Vemend and Mecseknadasd were fighting to stop a hilltop near Ofaln in the Mecsek Hills becoming Hungary's main nuclear burial ground.

Progress in Sweden

While the USA and the UK have got into convoluted decision-making positions regarding deep repositories, Sweden has proceeded to build one for intermediate and low level wastes. Swedish nuclear waste is deposited at a central interim storage facility for spent fuel (CLAB) adjacent to the nuclear power station at Oskarshamn. A final depository for intermediate and low level wastes has been engineered at the nuclear power station at Forsmark on the east coast north of Stockholm (Figure 12.7). This is a submarine facility in crystalline rock. It is situated immediately offshore from the power plant to which it is connected by tunnel from a small island linked to the mainland by a causeway. The project includes both caverns and silos for different types of waste. It may well be the precursor for a deep high level waste repository at Stripa, near Lindesbeg (Ahlstrom *et al.* 1981).

SEA DISPOSAL

Countries with nuclear power stations have commonly disposed of highly diluted liquid radioactive waste into the sea; fewer have used sea dumping to dispose of low and intermediate level solid radwaste.

Coastally located fuel reprocessing plants such as those at Sellafield and Cap de la Hague discharge liquid radwaste to the sea, as do some nuclear

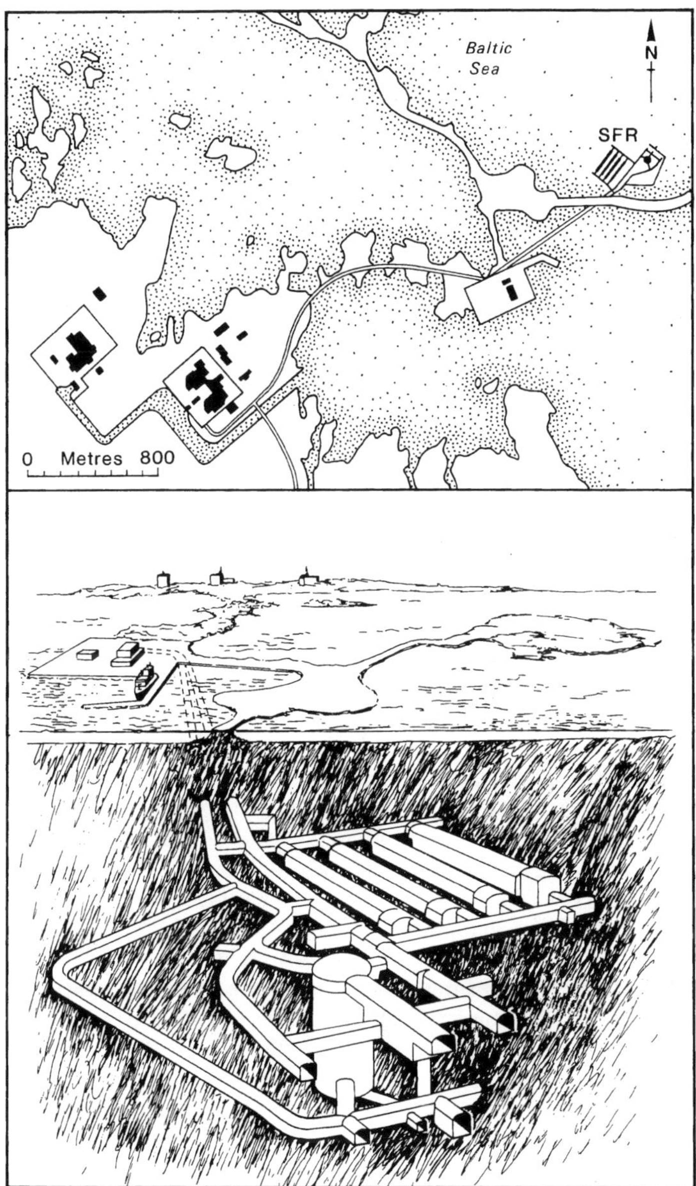

Figure 12.7 The Swedish final repository (SFR) for reactor wastes at Forsmark. A site 60 m below the seabed was chosen to ensure minimum ground-water flow. The first waste was placed in the respository in 1988 but ultimately it is envisaged that the special roll-on-roll-off ship Sigyn will transport waste from all Sweden's nuclear power stations, which are coastally located, to this repository

power stations and research establishments. At Lulworth in Dorset, for example, low level low activity liquid radwaste has been discharged from the Winfrith Atomic Energy Establishment through a pipeline running out to sea for 3 km from Arish Mell. The BNFL uranium enrichment plant at Capenhurst, near Chester, has fed fluid containing technetium-99 and traces of uranium, mixed with sewage, into Liverpool Bay off Meols on the Wirral, Merseyside.

Solid low level waste and some intermediate level waste have been sea dumped in the past by the USA and some West European nations. Ocean dumping was used by the USA from 1946 to 1970. Three sites took 90 per cent of the waste but fifty different locations were used in the Pacific and Atlantic Oceans and in the Gulf of Mexico (Holcombe 1982). European nations, led by the UK used ocean dumping at various locations in the northeast Atlantic from 1949 to 1983 (Figure 12.8). The OECD/NEA supervised these operations from 1967. In 1985 a moratorium was imposed on sea dumping at the Atlantic sites as a concession to political pressure from environmental groups.

The philosophy of marine disposal of both solid and liquid radwaste rests on the notion that there are quantitative limits that can be used to guide the amount, location and timing of disposal in the marine environment. Theoretically, such disposal of radwaste could result in some changes in that environment, but it has been argued that it is important to distinguish between change and damage. Parallels have been drawn with the provision of land defences against seawater inundation, dredging and harbour construction, all of which can change the marine environment at particular places without necessarily damaging it, and the disposal of radwaste which 'may similarly change the ocean, but only in a trivial sense' (Dunster 1959). It has also been accepted that large amounts of radwaste could damage parts of the marine environment and limit human use of ocean resources. Between these extremes it is assumed that there exist limits that can be used to guide policies for the satisfactory marine disposal of radwaste, and the guidelines are those provided by the ICRP (see Chapter 8). However, there have been differences between the leading nuclear nations in the interpretation put upon the ICRP guidelines and one result has been profound disagreement as to the permissibility of discharging wastes to the oceans.

The practical difficulty of working with the notion of 'acceptable limits' is that these will vary markedly from one part of the marine environment to another. Thus there is now general acceptance that the potential impact of radwaste releases on local ecosystems and assessment of the concentrations of different radioactive elements that may be discharged with impunity at any locality depends upon the critical path approach, identifying potential routes of radioactivity back to man, discussed in Chapter 8.

Because sea water is not drinkable, and any desalination technique will also remove radioactivity, the potential hazard to man from radioactive sea

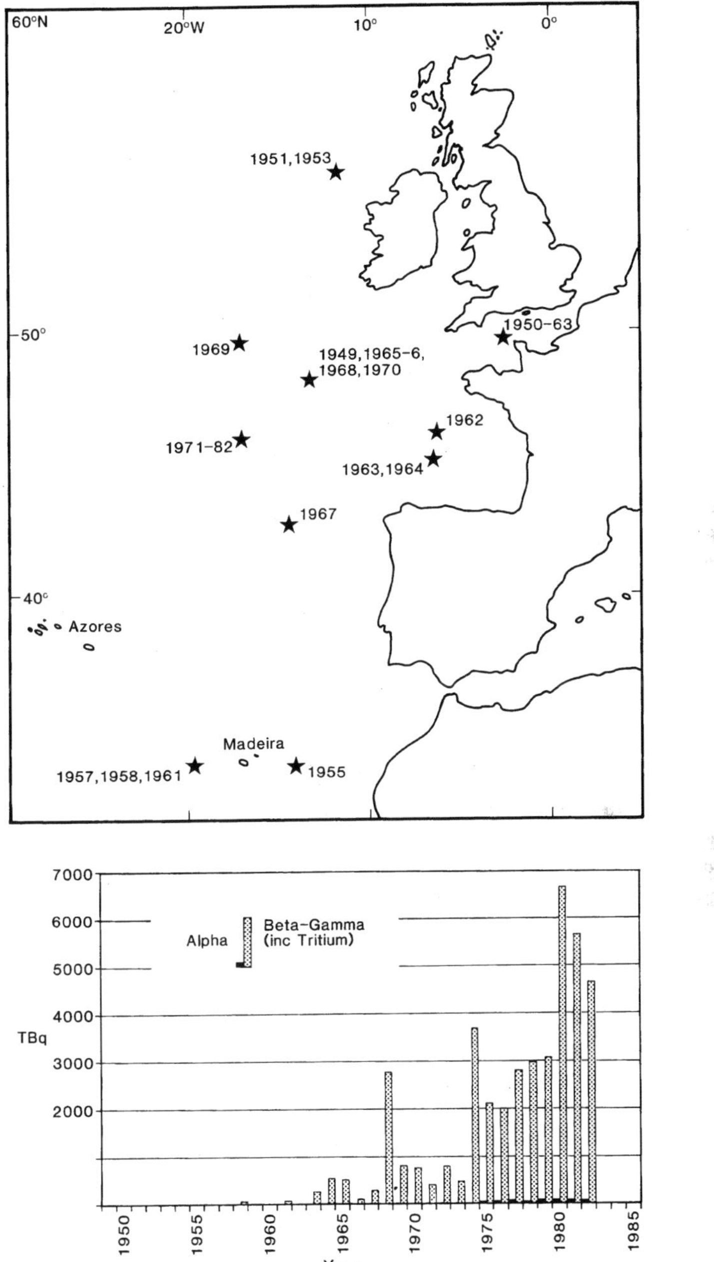

Figure 12.8 Locations of dumping areas and quantities of radwaste dumped in the northeast Atlantic 1949–84
Source: OECD/NEA 1985b: 434, 435

water is through the consumption of seafood which may have selectively reconcentrated radionuclides previously diluted by good waste management. Thus, in the UK it is the biological food-chains that have received most attention, for it is possible to measure concentration factors by the edible organisms and thereby determine permissible discharges (see Chapter 8, Table 8.3). However, the importance of particular food-chains varies from one site to another. At Sellafield in Cumbria, for example, effluents from the reprocessing of fuel elements are discharged to the Irish Sea through two pipelines. Surveys in the locality in the late 1950s pinpointed two important biological food-chains, one through the edible seaweed *Porphyra umbilicalis* and the other through fish. Three different critical groups of people were identified. One was a group of local salmon fishermen, the second consisted of a hundred fishing families who ate an average of about 250 g per day of locally caught fish and the third was another group of about a hundred people who ate more than 80 g of seaweed per day. The members of this last group live in South Wales where *Porphyra* collected from the Cumbria coast was sent for processing into laverbread, a delicacy eaten regularly by perhaps 25,000 people. In 1959 the intake by those who ate 80 g per day of seaweed in laverbread was as high as 74 per cent of the then current ICRP permissible concentrations for the general population, and it was this food-chain, together with that of the two other groups, which had to be taken into account in restricting the radioactivity in the Sellafield (then Windscale) effluent. A combination of changes in the harvesting of *Porphyra* has rendered this pathway dormant since 1973 (Pentreath 1982). A different critical pathway applied to the discharge of effluents from Bradwell nuclear power station, built at the mouth of the Blackwater estuary in Essex. The proximity of an oyster hatchery and the ability of oysters to concentrate such elements as zinc from sea water had to be taken into account. At Berkeley, on the Severn estuary, the external dose to fishermen from concentrations of radionuclides on silt was the most restrictive limitation on discharges, whereas at Hinkley Point it was the external radiation from the fishing gear and sediment.

As with all point sources of release, the hazard to man generally decreases with distance from the point of release as a result of various dispersal factors in the water and consequently lower levels of concentration in the biota, but there are some food-chains which are only initiated further downstream and hence can present a greater specific hazard at that point; the reconcentration of zinc-65 by oysters is one illustration.

Given the existence of these kinds of pathways from the environment back to man, careful attention must be given to the question of safely dispersing the discharges of liquid radwaste to the oceans and the discharges into river systems that eventually reach the oceans or inland water bodies such as the Great Lakes. On a world scale there have been three outstandingly important sources. Sellafield is one, Cap de la Hague is another and Hanford, through its discharges into the Columbia River and eventually into the Pacific Ocean, is the third. The external doses to individuals in and near the Columbia River prior to the time that the water reaches the ocean have been greater than doses received in or near the ocean; in effect, the river has channelled the liquid radwaste and prevented its immediate widespread dilution. Sellafield feeds its waste into what is almost an inland sea rather than open ocean (Figure 12.9) and with careful measuring techniques radwaste discharges from the plant can be detected at distances of up to 100 km, although at that distance they are not readily differentiated from the larger background of natural radioactivity. Discharges from Cap de la Hague have been of concern to the population of the Channel Islands.

The main risks due to the radwastes from the nuclear industry arise from the liquid effluents which run into the Irish Sea from Sellafield. At one stage the collective radiation dose to the British public reached 12,000 man rems per year. By 1986 this was down to 7,000 man rems, statistically equivalent to two cancer deaths per year of operation. It is anticipated that after the THORP facility is opened at Sellafield in 1991 the rate should drop to one cancer death in five years (Fremlin 1986).

The solid waste that has been subject to ocean dumping has been low in beta and gamma activity but has sometimes had alpha contamination from plutonium traces. This waste comprised gloves, paper and items of equipment which either had had direct contact with radioactive materials or were known to be contaminated. The waste materials came not only from operations in the nuclear fuel cycle but also from hospitals, research laboratories and industry. The waste was packaged in steel drums or concrete containers and then dumped at sites in the northeastern Atlantic Ocean at depths averaging 4,000 m (see Figure 12.8).

The dumping sites in the northeast Atlantic were not selected at random by the seven nations involved. Essential considerations in choosing the sites were as follows.

(a) There must be no chance of recovering the wastes by processes such as trawling. The area should have a depth of at least 2,000 m and be clear of the continental shelf.
(b) The area must be free from known undersea cables.
(c) The area must be suited for the convenient conduct of the dumping operations and must be chosen to avoid unreasonable financial penalties due to long steaming distances, the likelihood of bad weather conditions or undue navigational difficulties.
(d) The possibility of turbidity currents should be taken into account.

Since 1977 the OECD/NEA has assessed the suitability of sea dumping sites proposed or used by the various national authorities every five years. It was the conclusion of the Expert Group in its 1985 Report (OECD/NEA/1985b) that the North Atlantic was suitable for continued dumping

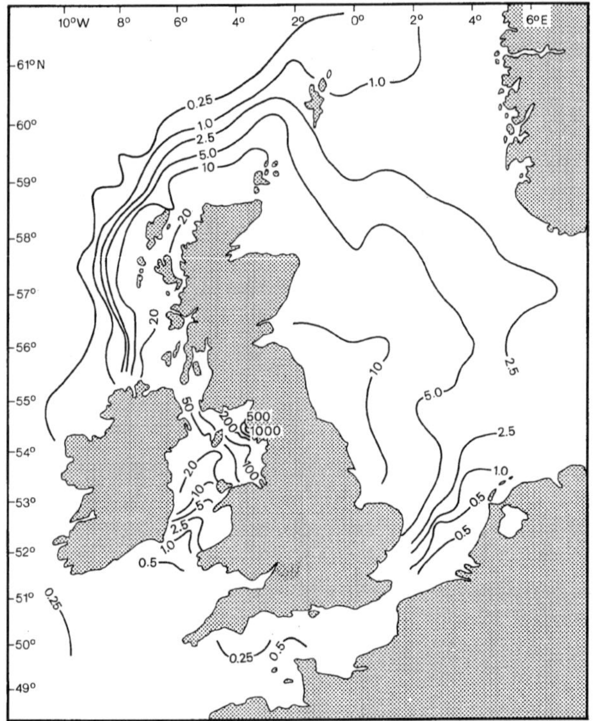

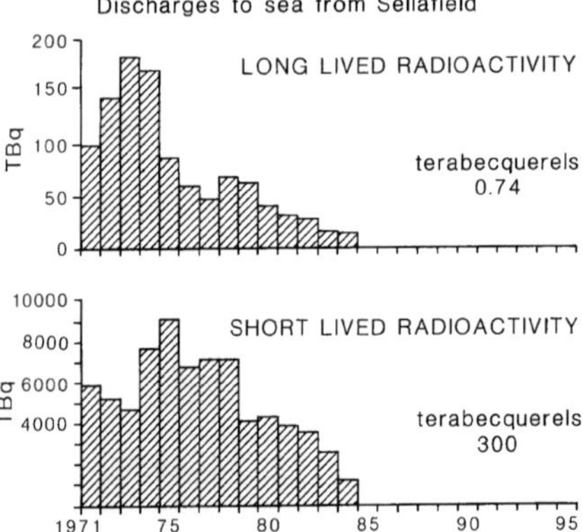

Figure 12.9 Caesium-137 concentrations in surface waters around the British Isles, March–June 1978. The heavy concentration around Sellafield has received much publicity and has prompted pressure for better management practices and new processing equipment at the Sellafield Works. The graphs show that the situation has improved but the fact that years are shown from 1985 onwards could be misleading for the graphs were published in 1985 on the basis of data up to and including 1984

Sources: (graphs) BNFL; (map) © Crown copyright, 1981. Reproduced from Lee and Ramster 1981

at sea at given rates. However, pressure from a variety of quarters, including environmental groups, grew to such an extent that the 1985 moratorium brought an end to the practice, for a time at least.

The surface of the world is seven-eighths ocean, the mere size of which has tempted many organizations with waste to dispose of to treat it as a limitless refuse pit. A major step towards controlling the pollution of the sea was taken in November 1972 when representatives of seventy-eight governments and various international organizations, including the EC Commission, the IAEA, the International Labour Organization and UNESCO, met in London and formulated a Convention to deal with aspects of waste dumping at sea. The Articles of the Convention included some designed to control and regulate the disposal of radioactive materials and to record the nature and quantities of material dumped, something not always done previously. The basic approach embodied in the Convention was the safeguarding of man against the damaging effects of radiation based on ICRP recommendations. The London Dumping Convention of 1972, although recognizing that control over marine dumping remained with the national authorities, gave the IAEA responsibility for defining high level wastes considered unsuitable for sea disposal. In 1977 the OECD established a Multilateral Consultation and Surveillance Mechanism for Sea Dumping of Radioactive Waste by which dumping in the northeast Atlantic was monitored. The 1982 Convention on the Law of the Sea did not directly refer to any sea-bed burial of nuclear wastes, but Articles dealing with marine environmental protection and regulation of the international sea bed are relevant.

To the extent that the Conventions introduced a semblance of order to a previous *ad hoc* situation, they must be welcome, but the essential problem remains that the dynamics of the ocean are imperfectly understood. Knowledge of the oceans and their currents has grown rapidly over the past two decades and new measurements have led to new theories of ocean circulation. The notion of a *two-layer ocean* is now widely accepted by oceanographers. The pattern of surface currents has been well known for centuries but there is still uncertainty about the degree to which surface and deep layers, below about 2,000 m, mix with each other, and about the speed and direction of movements in the deep layers. Figure 12.10 is a diagram published by the OECD/NEA in 1985. It is an idealized map of the patterns of deep water flow (black lines) and surface water flow (grey lines). There is sinking of North Atlantic deep water in the Norwegian Sea and recooling of water along the perimeter of the Antarctic Continent, with compensatory upwelling to balance the deep water generation. Figure 12.11 is an earlier diagram by the oceanographer Stommel. It suggests that sources of deep water, in the form of sinking cooled water, are to be found in the North and South Atlantic, that there are relatively narrow streams of water moving at several kilometres a day in the deep and bottom layers and that the deep water rises to the surface in many other parts of the ocean

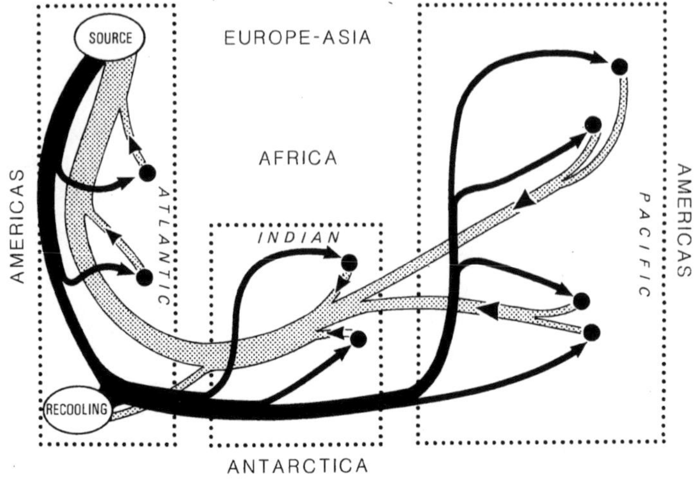

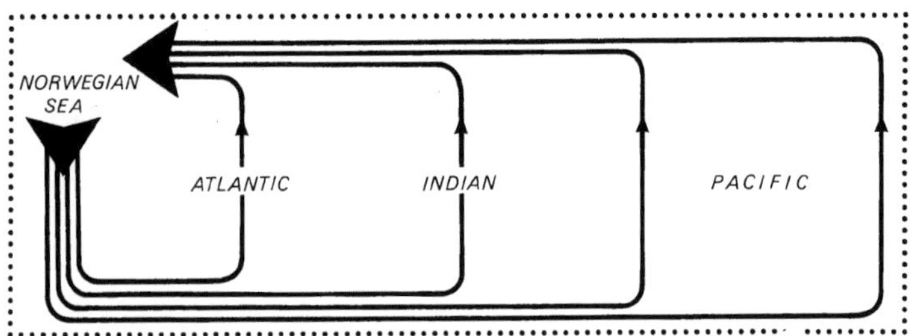

Figure 12.10 Global pattern of ocean deep water flow (black lines) and surface water flow (grey lines). The lower diagram is an idealized section running from the North Atlantic to the North Pacific showing the main advective flow pattern
Source: OECD/NEA 1985: 191

with a slow vertical velocity. There is clear indication of a strong southward current running at a third of a knot some 3.2 km below the Gulf Stream, and water with properties typical of Antarctic conditions can be traced far into the northern hemisphere.

Arguments that have occurred in the case of the Atlantic Deeps make it clear that even well-established factual evidence can be open to significantly different interpretations. The capacity of the Atlantic Deeps to accept low level radwaste was assessed by the NRPB in the early 1970s. The vertical structure of the ocean from the surface to the sea bed in the northeast Atlantic contains a horizontal layer of water at a depth of about 1,000 m which has high salinity and greatly restricts the mixing of deep and surface layers. The layer originates from the Mediterranean and is known to oceanographers as the Mediterranean Barrier. The NRPB report argued that the presence of the barrier would reduce the concentration of radioactivity as it diffuses slowly upwards from corroded or breached canisters on the ocean bed at a depth of 5,000 m or more, and that this provided an additional measure of safety for man's use of surface waters, such as fishing. On the other hand, it was suggested by Anthony Tucker in *The Guardian* of 17 August 1973 that if the diffusion and hence dilution were limited by such a barrier in one part of the Atlantic then the concentration in the remainder, below the barrier, must ultimately be higher than under conditions of uniform mixing.

Such large quantities of water lie below the Mediterranean Barrier that Tucker's argument in this case seems a shade tendentious. Nevertheless,

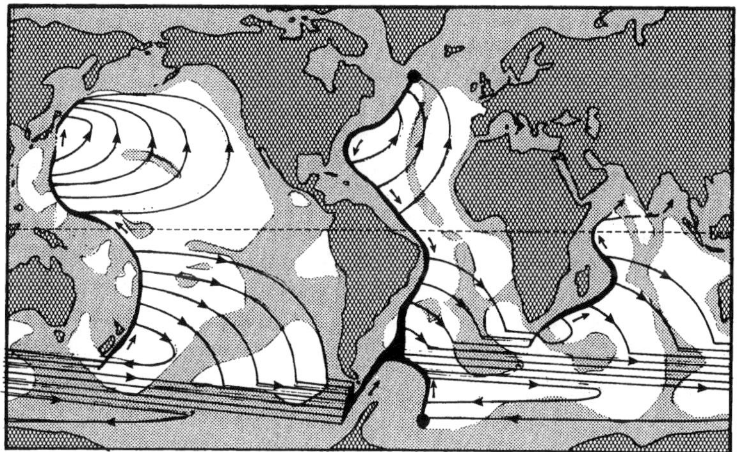

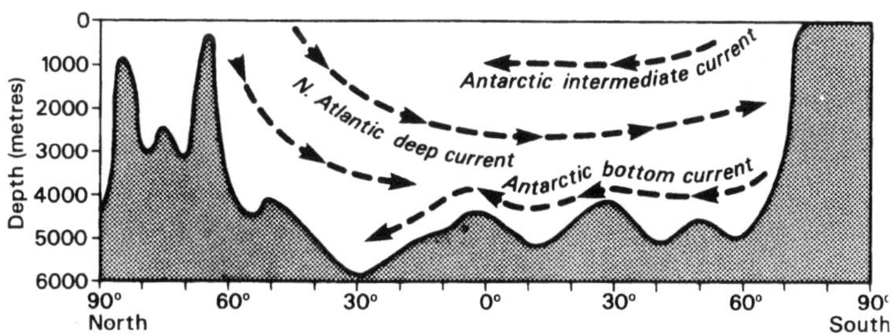

Figure 12.11 Deep circulation of the ocean basins as suggested by Stommel. The vertical section through the Atlantic Ocean from north to south shows the distribution of salinity in parts per thousand and indicates the principal features of the water circulation.
Source: Crease 1959

there are considerable uncertainties in assessing the amount of radioactive material that can be tolerated in the sea. It is necessary to emphasize this because the general consensus view in scientific papers published since 1973 seems to be that the burial of high level nuclear waste within the deep sea clays of the oceanic basins in conjunction with a (yet to be perfected) multibarrier containment concept could prove technically and environmentally feasible (e.g. Eaker 1983; Murray *et al.* 1986). In the conceptual designs for such depositories the sea-bed sediment becomes the primary barrier. Candidate 'study areas' have been identified in the Atlantic (Great Meteor East and Nares Abyssal Plain) and the Pacific by the OECD/NEA Sea bed Working Group (OECD/NEA/1984b). The current US view is that the ocean should not be perceived as a medium for waste dilution but as a disposal location that will ensure undisturbed isolation and containment. The thought remains that even carefully controlled dumping of high level long-lived radwaste in the ocean deeps may be some way from being an ideal form of ultimate disposal, not least because it is bad practice to put long-lived radwaste where it cannot be recovered until it is absolutely certain that there is no need to do so. Moreover, there are difficulties involved in modelling possible detriment, especially that of identifying a critical path. Within the water the density of marine organisms declines quite steeply from about 200 m to around 25,000 m, and from this depth downwards there is little biological presence until about 100 m above the sea bed. No fish are caught commercially below 1,500 m, and even the deep-sea fish that are caught commercially do not appear to have food-chain links with the fauna at about 4,000 m and below (Pentreath 1982). Thus the only pathway back to man that realistically can be envisaged is one involving dissolution of the waste into the water of the deep ocean which, by physical processes alone, subsequently contaminates the shallow waters of the continental shelf in which commercial fishing is carried out. This would seem to make deep sea disposal a safe option, but paradoxically the lack of measurable critical pathways makes it difficult to say how safe.

GASEOUS RADIOACTIVE WASTE FROM NUCLEAR FUEL REPROCESSING PLANTS

When spent fuel elements arrive at reprocessing plants they are fed into automatic decanning machines which remove the nose and tail pieces, split and peel off the cans and then chop them into pieces for storage. The decanned uranium fuel is then dissolved in nitric acid to form the nitrates of uranium, plutonium and the fission product elements. Gaseous effluents containing radionuclides or radioactive matter arise when the fuel is dissolved and from the ventilation of cells and vessels. Most prominent are krypton-85, iodine-129, carbon-14 and tritium. Their approximate content in 1 tonne of used LWR fuel after a normal burnup of 33 GW days/tonne is shown in Table 12.1. Aerosols are also produced at the stage of fuel

Table 12.6 Typical annual discharges to atmosphere from CEGB nuclear power stations (UK)

Activity	*Annual discharge Ci*
Argon 41[a]	up to 100,000
Argon 41[b]	up to 5,000
Sulphur–35	up to 5
Iodine–131	less than 0.1
Aerosols	0.01

Source: F.H. Passant 1985
Notes: [a]Magnox steel pressure vessel
[b] Magnox concrete pressure vessel (Oldbury, Wylfa, AGRs)

processing. All these effluents are strongly diluted by air moving through the nitric acid and dissolving fuel. High level waste vitrification and low or intermediate level waste incineration at fuel reprocessing plants are other sources of airborne activity (e.g. caesium-137 and ruthenium-106). It is clear, therefore, that arguments for nuclear power on the grounds that it contributes less pollution to the atmosphere than coal- or oil-fired plants must be qualified by the recognition that it has its own contribution to make to environmental pollution.

In managing the radioactive gaseous wastes the requirement is an effective means of capturing those that are particularly toxic or long lived so that they can go into long-term repositories, combined with controlled release of the others to the atmosphere in highly diluted form (Table 12.6). Capture is possible with carbon-14 but not yet with Tritium. Iodine-129 has a half-life of 16 million years and the best that can be done is to control the time, place and manner of dispersion into the biosphere (Figure 12.12). Krypton, which is a noble gas, is being released to the atmosphere in small amounts, but as reprocessing capacities grow the releases from individual plants might have to be restricted, not only to keep the radiation exposure down as far as reasonably retrievable in the local environment but also to avoid an increased accumulation of krypton-85 in the global atmosphere. Techniques to capture krypton are under development at Mol, Belgium, but the task is not easy and in 1988 they were only in the experimental stage.

DECOMMISSIONING WASTE

Recently the nuclear power industry and regulatory bodies have been providing and discussing scenarios for the decommissioning of nuclear power stations. By the mid-1990s over twenty nuclear power plants built to produce electricity during the late 1950s and 1960s will have been shut

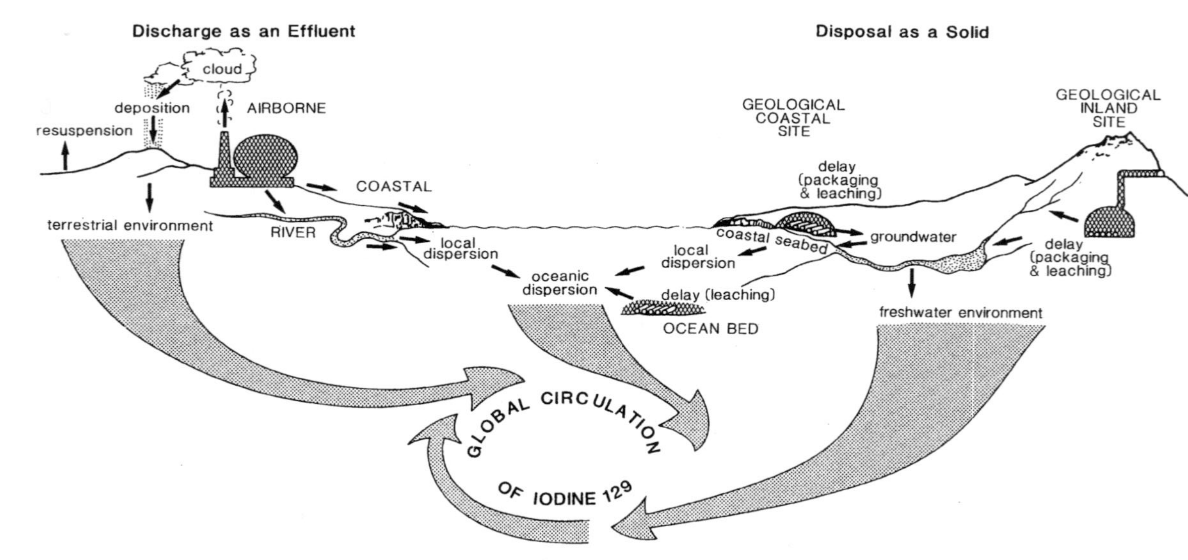

Figure 12.12 Options for the discharge and disposal of iodine-129 showing simplified environmental pathways

down and decommissioned. The management of this process will draw upon experience gained from small experimental and power demonstration units already closed down and dismantled, especially in the USA. OMRE, an organically moderated 12 MWt research reactor at the Idaho National Engineering Laboratory, was shut down in 1963, decommissioned between 1977 and 1979 and then dismantled. SRE (Sodium Reactor Experiment) at Santa Susana Field Laboratory near Los Angeles, a 20 MWt unit, was shut down in 1964 and later completely dismantled. The Elk River reactor in Minnesota (58 MWt BWR) was shut down in 1968 and dismantled. Hallam, a 254 MWt sodium-cooled graphite-moderated unit at Hallam, Nebraska, was entombed following a failure of fuel element cladding. At the Hanford Reservation, eight old graphite-moderated water-cooled single-phase plutonium-producing reactors are now in 'safe storage'. Shippingport (236 MWt PWR) was closed in 1962 and almost immediately dismantled. In Quebec, Gentilly I, a prototype 250 MWe CANDU BWR on the same site as Gentilly 2 (600 MWe PHWR) was closed in 1979 and decommissioned between 1984 and 1986. In the FRG Niederaichbach nuclear power station was shut down in 1974 and has since been in a state of 'safe enclosure' prior to decommissioning. In Sweden the small (80 MWt) Agesta reactor was shut down in 1974 and has been defuelled, drained and put into a state of 'storage with surveillance'.

Decommissioning is the term used to describe the whole process which follows final reactor shutdown. In includes a number of major events, some of which may happen sequentially and others in parallel:

Table 12.7 Estimate quantities of decommissioning wastes from UK Magnox power stations

Power station	*Postulated shutdown date*	*Power output (GWe)*	*Active waste arisings (m^3)*
Berkeley	1992[a]	0.27	5,400
Bradwell	1992	0.24	5,400
Hunterston A	1994	0.30	6,000
Hinkley A	1995	0.43	8,600
Trawsfynydd	1995	0.37	7,400
Dungeness A	1995	0.39	7,800
Sizewell A	1996	0.39	7,800
Oldbury	1998	0.33	6,600
Wylfa	2001	0.62	12,400
		3.34	67,400

Source: Knowles 1985
Notes: There is no doubt that decommissioning of these power stations will be expensive. In 1988 the CEGB set aside £161 million for a decommissioning fund that stood at £723 million.
[a] Actual 1989

(a) defuelling and shipment off the site of all spent fuels, a major task that could take several years for a gas-cooled reactor;
(b) decontamination (chemical cleaning) of accessible pipework and, if necessary, of parts of the site;
(c) packaging and removal of operational solid radwastes;
(d) dismantling and removal of all buildings and structures outside the biological shield, including contaminated boilers, circulatory systems and fuel ponds;
(e) 'mothballing' or 'entombment' of the reactor for a period of time (determined partly by the decay rate of cobalt-60);
(f) dismantling and removal of the reactor itself and the biological shield;
(g) site decontamination to enable free entry and reuse.

It is calculated that for a power unit of around 200 MWe the sequence up to but excluding reactor removal may take between twelve and seventeen years. Removal of the reactor and the biological shield would be unlikely to begin before ten years from shutdown, but after ten years 90 per cent of the radioactivity associated with these items would have disappeared and the half-lives of those remaining would be so long that there would be little to gain from further delay (Gregory 1984).

It is clear that decommissioning will make significant contribution to the quantity of radwaste material that will need disposal. By the end of the century almost 2 million tonnes of decommissioning waste will have been produced in the EC countries alone (Chapman *et al.*, 1987). More than 80 per cent will be low level waste, with higher levels of activity being concentrated in reactor waste. A necessary but not sufficient condition for dismantling is the availability of properly engineered carefully chosen waste disposal sites.

Table 12.7 provides an indication of the volume of radwaste that may be produced by decommissioning the UK's earliest commercial nuclear power plants. This table only hints at the eventual scale of waste from decommissioning. The small Magnox power stations constitute a tiny fraction of the total number of the 418 nuclear power reactors operating in the world in 1987 and of the 130 units then under construction. At the back of these are hundreds of supporting fuel cycle facilities. In the USA alone, counting both state and NRC licences, there were more than 20,000 nuclear material activities in the later 1970s (Stello 1978). All these nuclear installations, from the largest reactor to the smallest laboratory, will have radioactive residues. They will have to be decommissioned after their life-cycle and the residues safely dealt with.

At present there are major uncertainties relating to the issue of dealing with an irradiated reactor and to the length of time that sites will be subject to control and surveillance after final reactor shutdown. In the USA decommissioning of the production reactors at the Hanford Reservation is at an advanced state, but the ultimate dismantling, packaging and disposal of each reactor block after mothballing presents an interesting challenge.

The core of each reactor block is 8.4 m thick, 10.8 m high and 10.8 m wide. It consists of graphite blocks surrounded by a cast iron thermal shield and a 1.25 m thick biological shield of steel and masonite. The whole structure has absorbed heat and neutrons during its operational lifetime.

Eight of the nine reactors at the Hanford Reservation have been mothballed and this is likely to be the pattern at most commercial nuclear power stations; mothballing can significantly reduce the level of occupational exposure for construction workers as they take the reactors apart. In dismantling Elk River a total of less than 1 Sv of exposure was experienced by 100 people. It is anticipated that the radiation dose to workers dismantling a large reactor will be around 14 man Sv over a three year period.

Cobalt-60, which has a half-life of about five years, is the predominant source of the shorter-lived radionuclides, and so reductions in occupational exposure to be gained by delaying decommissioning can be directly related to the half-life of cobalt-60.[1] A twenty-year delay produces a reduction by a factor of 16, and a fifty year delay produces a reduction by a factor of 1000. It has been argued, therefore, that nuclear power station sites will need to be subject to control and 24 hour surveillance for about fifty years after shutdown, as an optimal period for a balance to be struck between economic and social considerations and occupational exposure (Crofford 1980). However, total radioactivity does not decline quickly after the first thirty years from shutdown (OECD/NEA 1985e).

The question of how soon after shutdown unrestricted use of a nuclear power station site may be allowed is an important one. At the time of retirement radiation levels in the plant will be quite high because of the neutron activation products. Since many of these have lifetimes of less than six years, benefits may be gained from delaying some stages of decommissioning for that period, but fifty years seems an undesirably long time. It promises the transfer of the problem from one generation to the next, and the possibility of institutional failures causing a lack of effective control and surveillance cannot be ignored. In the FRG, however, the working assumption is that a power reactor has a full-capacity operating life of thirty years which, at an average availability of 75 per cent, amounts to a total useful lifetime of forty years. At present the view is that the time frame for 'storage with surveillance' should also be about forty years. During this phase surveillance and restricted entry would take place for twenty four hours a day (OECD/NEA/1985e). Thus, in the German model, site control from start to finish would last for eighty years.

In the UK it has been the CEGB's declared policy to *replant* existing sites with new nuclear power units after obsolete plants are taken out of service whenever they are suitable for redevelopment. The advantage is that mothballed reactors would remain under surveillance and control until they are removed and the site released for alternative use (Gregory 1984).

TRANSPORT CONSIDERATIONS

Figures 0.1 (Introduction) and 1.2 (Chapter 1) might indicate the existence of four areas in the transportation of radioactive materials that could provide possible environmental hazards in the nuclear fuel cycle: (i) the shipment of uranium oxide from uranium mills to fuel fabrication plants; (ii) the transport of new fuel element assemblies to nuclear power stations from the fuel fabrication plants; (iii) the movement of spent fuel elements from the power stations to intermediate storage points or reprocessing plants; (iv) the transfer of solid wastes from reprocessing plants or intermediate stores to 'permanent' storage locations. Because of the non-existence of ultimate repositories, high level waste in liquid or solid form is stored at the reprocessing sites and has not entered into transport flows to date. Thus, in reality, it is only the third of these that has been of real consequence. Uranium metal is a naturally radioactive substance, but most of its radiation is alpha radiation which has virtually no penetrative power. Thus neither the oxide nor the fresh fuel elements are particularly dangerous or difficult to handle. Used fuel elements are a totally different proposition. They are intensely radioactive, and the gamma radiation given off is highly penetrative and biologically harmful. They also give off heat, which has to be removed to prevent the fuel cladding from melting. Thus, from the time when a spent fuel element is lifted from the reactor until the time when the fission products are removed, remote handling techniques are required.

Depending upon the particular reactor design, between a third and a quarter of the fuel in the reactor has to be replaced each year; this means the removal of several dozen fuel elements during the refuelling operation and eventually these must be transported either to intermediate storage points, in the case of a once-through fuel cycle, or to reprocessing plants. Immediately after removal from the reactor the spent fuel elements are placed in water in a pool or vault for a minimum period of four months to permit the radioactivity associated with short-lived fission products to subside. The pool is a substantial concrete chamber filled with water which is continuously circulated and purified by ion exchange. The spent fuel elements are suspended in a storage rack in the chamber with sufficient space between them to avoid any chance of their going critical. The water acts both as a radiation shield (affording the same protection against gamma rays as a 0.3 m thickness of lead or 1.2 m of concrete) and as a coolant. After the end of the storage period, the fuel elements are remotely handled into their shipping casks, ribbed steel vessels about 5 m long, and then transported to a reprocessing plant or off-site interim storage facility. Because there are so few reprocessing plants, some irradiated fuel elements travel long distances by sea and land, e.g. from Italy and Japan to the UK. One result of the lack of development of reprocessing and of the non-existence of high level radwaste depositories in the USA has been a pile up of irradiated fuel elements at reactor sites. Some utilities have replaced their original racks with new ones designed for closer spacing, some have gone through such a re-racking operation twice and others have installed storage racks in portions of pools previously planned as under water work areas (Draper and Stanford 1982).

A major accident that has killed, injured or apparently over exposed individuals or populations as a result of the transportation of radioactive spent fuel has not yet been recorded, but there have been a number of transportation accidents involving radioactive materials in some of which there has been measurable release of radioactivity. As the number of shipments and the amount of radioactive materials carried increase in the future, it seems inevitable that there will be an increase in risk.

There is a shortage of the kind of data needed to enable confident prediction of the detriment that may be suffered in the future by radioactive material packages in various types of accident in different transportation modes but, except for the USSR where nuclear fuel is moved mainly in armoured trucks with escort vehicles to front and rear, shipments move in routine traffic flows using modified conventional transportation equipment. In the USA, for example, commercial trucking firms handle the shipments which are therefore subject to normal transportation environments just like non-radioactive cargo. In the USA the national rate of 2.5 accidents per million vehicle miles for road transport of explosives and other dangerous materials is not significantly different from that for all cargoes (Brobst 1973).

In such circumstances the shipper has no effective control over the *likelihood* of an accident involving his vehicle, but he does have control over the *consequences* of an accident by means of the package design and contents of shipments. Control of the contents is important, because the criticality of the fuel assembly elements is avoided by limiting the number of subassemblies per cask and by inserting neutron absorbers into the basket structure that holds the individual assemblies in place. Loss of gamma ray shielding might result if a cask were to be cracked in an accident, but the construction of the casks is such that the probability of such an event is minimal. Sometimes safety in the transportation of irradiated fuel elements can also be increased by care in choosing a route. On 6 January 1966, for example, the first large-scale long-distance international movement of highly irradiated power reactor fuel took place from the Latina nuclear power station in Italy to the UK for reprocessing. The fuel elements were carried in the *Stream Fisher*, a charter vessel modified for the purpose by being fitted with a cooling tank and refrigeration and venting equipment. The chosen route was by road from Latina to Anzio, a journey of 17.7 km taking two hours under escort, followed by a sea journey of 4186 km from Anzio to Barrow-in-Furness. Before this route was chosen full consideration was given to transport overland across Europe (Kavanagh and Gualtier 1966). It is possible that when it is built the Channel Tunnel may be considered too risky for inclusion in such a journey,.

The *Stream Fisher* example indicates that shippers have a choice in the way in which irradiated fuel is transported. Rail transportation offers the advantage of larger payload capacities; loaded casks may weigh up to about 8 tonnes each with a 5 – 6 per cent payload (4 – 5 tonnes of fuel). Transportation by road is limited to casks weighing less than about 2.25 tonnes with a 2 per cent payload, or about 0.5 tonnes of fuel. Thus, while transportation by road is faster than by rail in most cases, the number of shipments needed to complete the movement of a batch of fuel to the reprocessing plant is larger because of the smaller capacity of the road vehicle.

Whilst the shipper has a choice of transport mode and route, the choice of cask is regulated. The IAEA has produced model specifications for the safe transport of radioactive materials. First drawn up in 1961, these have been revised on occasion since and they provide both the framework within which radioactive materials may be acceptably packaged for movement between countries and a model for national regulatory agencies. When laying down regulations for the movement of radioactive materials within their own territories, improvement in packaging standards is commonly apparent. In the UK, for example, there was no statutory control over the movement of radioactive material by road until 1971. This state of affairs was brought to an end by the Radioactive Substances (Carriage by Road) Great Britain Regulation of 1971 (S.I. 1970 No. 1826) and the Radioactive Substances, Road Transport Workers (Great Britain) Regulations of 1970 (S.I. 1970 No. 1827). The Department of Transport regulates the transport of radioactive materials in the UK, and for the purpose of the UK regulations radioactive material is defined as that having a specific activity greater than 70 Bq/kg (2 nCi/g). In the USA, until July 1966, the basic regulations concerning the shipment of radioactive materials were issued by the Interstate Commerce Commission (ICC) but the only hazards guarded against were radiation and contamination. This was sufficient when the movements were limited to low level products of the weapons programme, but in the mid-1960s fuel from commercial nuclear power stations began to emerge in significant quantities for reprocessing. The penetrating radiation from such fuel consists almost exclusively of gamma rays. Thus spent fuel shipping casks designed for this fuel use an extremely dense material, usually lead, for shielding. In 1966 changes were made in the US regulations to include precautions to safeguard against criticality (Bailey *et al.* 1973).

The IAEA regulations suggest specific limitations for the control of external radiation, heat emission, external contamination, criticality and toxicity, including the strength of packaging and limitation of radioactive content. Packages must generally comply with one or two standards. The first type A, must be able to withstand normal transport conditions including minor accidents, and the amount of radioactive material carried in it is limited to ensure that if the package were damaged in a severe accident, the amount of material likely to be released would be so small that it would be virtually harmless. The second, type B, must comply with specific tests to ensure that there will be no loss or dispersal of radioactive contents, nor 'excessive' radiation, under very severe accident conditions. If the transport package contains material of sufficient radioactivity the consignment is designated a 'large source' and additional operational controls including emergency procedures are required which take into account the proposed routes and modes of transport, and the whole movement must be authorized by an appropriate competent authority. In these terms a transport flask containing irradiated fuel constitutes a very large radioactive source.

Legislation designed to implement these standards varies in detail from country to country, but generally standards are set to ensure the integrity of a cask if it is dropped, hit by other objects, compressed, subjected to heat or immersed in water. The standards, however, are design desiderata, and design assessment is not always accompanied by physical testing of samples or prototypes; the alternative of calculative methods backed by materials control is sometimes adopted. Moreover, because of the rapid growth and development of the nuclear industry, and the cost of hiring casks, the tendency is for licensees to minimize shipping costs by using shipping packages as fully as possible. This, coupled with the changing nature of the materials to be transported (e.g. recycled plutonium and fuels irradiated to higher burnups) is causing designers to match actual operating conditions as closely as possible in the design of casks, and safety margins are becoming closer. Nevertheless, in one of the few efforts that have been made to quantify the degree of risk arising from transportation accidents involving irradiated fuel elements, working with conservative assumptions it has been calculated that in California the population receiving a limiting dose annually from an accident would amount to 2×10^{-3} in 1990 rising to 6×10^{-3} in the year 2000. The limiting dose in the calculations was taken to be 3 Gy to the thyroid, an *average* population density was assumed (80 persons per square kilometre) and the range of zero to one person affected varied according to meteorological conditions at the time of the accident (Yadigaroglu *et al.* 1972).

Regulation of the movements of nuclear material has become a matter of concern to authorities other than central government departments and regulatory authorities. Some local authorities in the UK have tried to exercise control over the movements of radwaste consignments through their areas by rail and road. This was a focus of attention for the Greater London Council (GLC) before its dissolution in 1984. The GLC example led some other local authorities in the UK to declare themselves 'nuclear-free zones'.

In the USA spent fuel from nuclear power stations is rarely reprocessed and is most commonly stored on site. In the UK, however, there are on average about 350 rail journeys a year involving nuclear waste, covering a total of 100,000 miles. British Rail's accident rate for freight trains averages 3.27 per million miles, although most of these are minor derailments which nuclear waste casks would readily survive. In 1984 CEGB organized a crash at British Rail's test track near Melton Mowbray, Leicestershire, when a flask survived intact after a diesel locomotive crashed into it at 100 miles/hour.

The waste container casks or flasks are normally taken by low-loader lorries from nuclear power stations to the nearest railway siding and transferred by crane to six-axled railway flat-cars. Special trains take the flasks to marshalling yards, including the Stratford depot in east London and Willesden in north London (Figure 12.13). There they are coupled to normal north-bound freight trains. Waste from Dungeness (Kent), Sizewell (Suffolk) and Bradwell (Essex) passes through London, usually on Thursday or Friday nights, and there are three or four nuclear trains travelling through London each week. Other cities and towns through which the radwaste passes on its way to Sellafield include Bristol, Gloucester, Cheltenham, Stafford, Chester, Preston, Lancaster (about six nuclear trains a week), Sunderland, Newcastle, Carlisle, Leeds, Bradford and Edinburgh, together with hundreds of smaller places lying on the relevant parts of the UK railway network.

Assessments of the significance of a derailment involving sufficient damage to a flask to cause a caesium release have produced conflicting results. One study of the health hazards of such a release at Willesden Junction, affecting an area defined by the broken line on the inset to Figure 12.13, concluded that even a severe accident would be of no great radiological concern (Surrey 1984). Another has concluded that a large release in fresh winds could cause over 2,000 deaths and require the evacuation of Kentish Town and Tufnell Park (Wakstein 1987). In the USA there has been continuing litigation between power companies and the railways since the mid-1970s, primarily focusing on rates, but indicating the railway's general unwillingness to carry spent fuel. The railways have embarrassed the federal government since 1974 by testifying that nuclear casks have not been designed to survive actual crash conditions by rail, but the US Government has conducted studies of possible routes of shipments by rail so as to avoid highly populated cities. Several state authorities have acted on their own to prevent spent fuel shipments from travelling through New York, Boston and other densely populated cities, but most shipments made to date have gone by road rather than rail.

A study by the National Academy of Sciences published in 1984 found the US regulatory programme for spent fuel transportation to be 'underdeveloped' (National Academy of Sciences 1984). The study included the road transport effects by the year 2004 of a postulated southern Nevada waste storage or disposal site, with massive waste corridors from the power reactors in the east and midwest (Figure 12.14(a)), of waste truck routes that would converge upon a high level waste storage facility at Hanford,

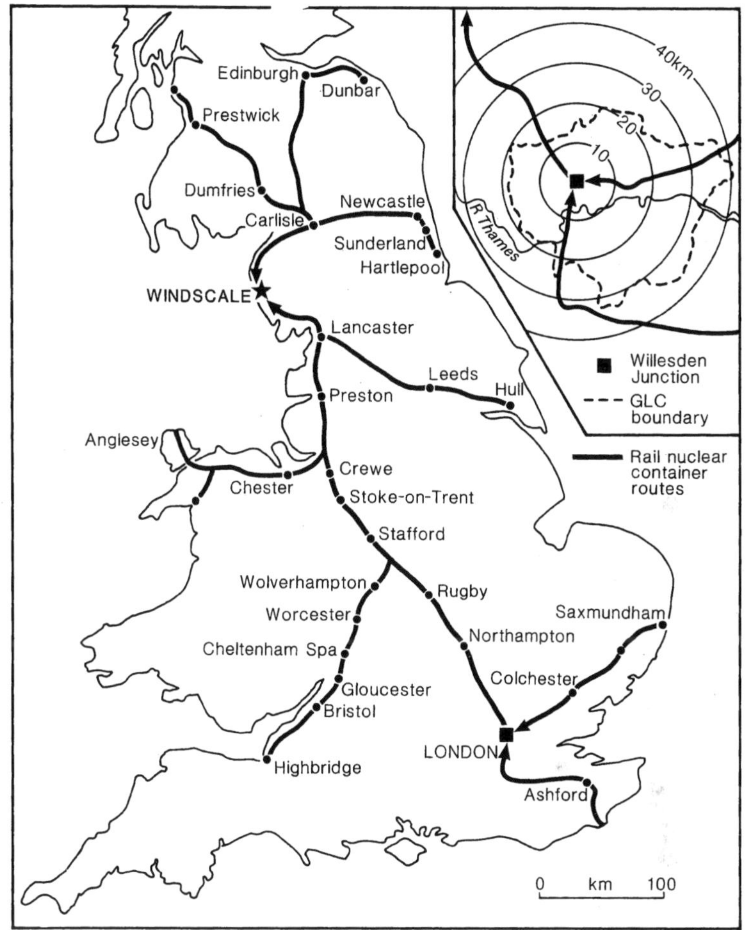

Figure 12.13 Rail nuclear container routes in the UK
Source: *Guardian*, 3 July 1979; Surrey 1984: 180 (inset)

Washington (Figure 12.14(b)), and of nuclear truck cargo routes that would converge on a proposed waste site located near Moab, Utah (Figure 12.14(c)). There were other maps for other localities also.

These route maps produced considerable media interest across the USA, and in ensuing discussions it became clear that a vocal part of the US public would not accept the waste shipments without being convinced of the sureness of safety regulations. There are also an increasing number of restrictions imposed in the USA by state and local jurisdictions. To provide a measure of uniformity nationwide and in response to the increasingly restrictive state and local actions, the Department of Transportation issued

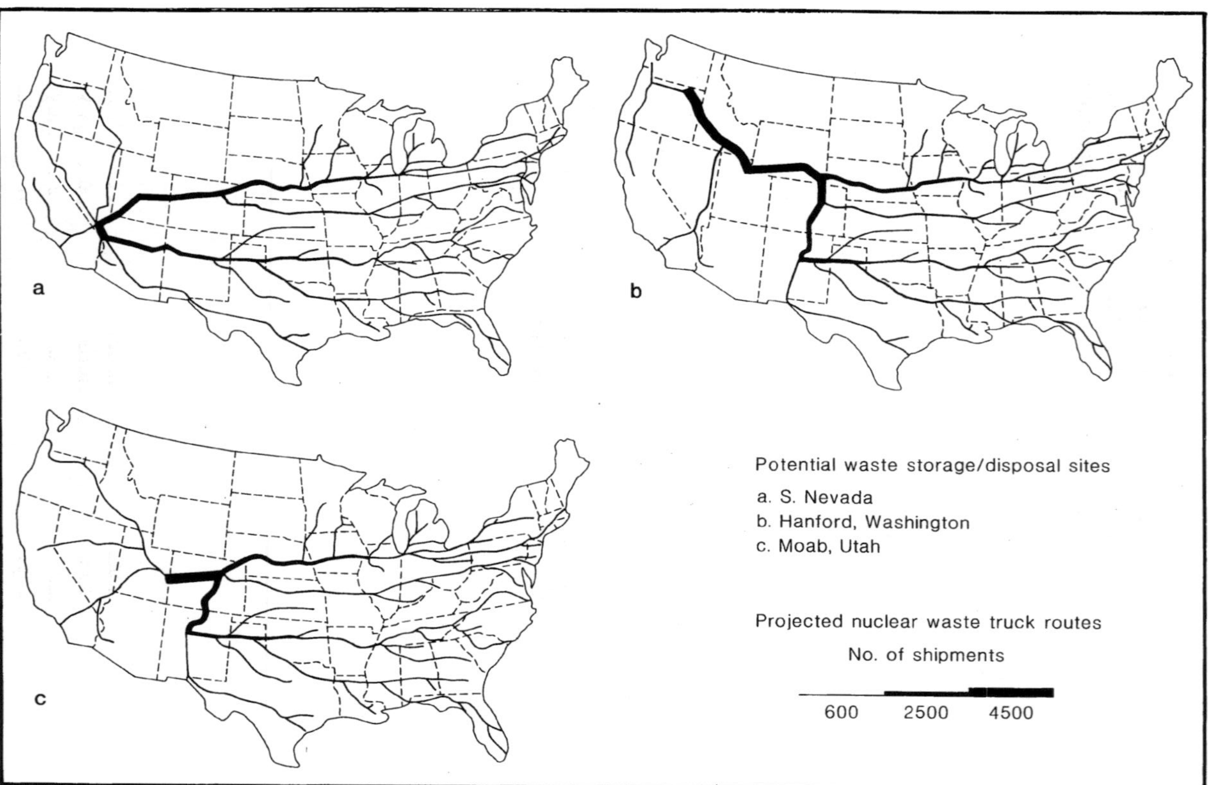

Figure 12.14 Postulated annual spent fuel shipments to a western storage site in the USA in the year 2004 based on road transport from all reactors.
Source: Kasperson *et al.* 1984

a highway routing rule in January 1981 that was intended to preclude state or local interference with the transport of spent fuel or low level waste by truck on interstate highways. These rules, however, have not pre-empted state and local rights to have pre-notification from shippers of movements, and the pre-notification requirements, often conflicting or overlapping, are an impediment to road shipments and an irritant for trucking firms.

Risks arising from theft

As the use of nuclear power grows, so does the number of consignments of fresh and spent fuel. By the end of the century, the USA expects something of the order of 10,000 movements of used fuel in a year. In the UK thousands of tons of used fuel have been transported to the Sellafield processing plant over the past quarter century without any hijack attempts on containers which usually travel without armed guards. Since 1973 some nuclear materials, such as plutonium nitrate solution, have been provided with armed security guards in the USA, in part because 2 kg would be sufficient to trigger a nuclear device.

In some quarters, not exclusively those opposed in principle to nuclear power, the loads of enriched and irradiated fuel are seen as a potentially tempting target for terrorists and political extremists who have shown by aircraft hijackings and forcible hostage-taking a readiness to take risks well beyond those that would be acceptable to normal people. The fear is that of theft of either radioactive ash, which might be used to poison water supplies, or certain nuclear fuels, which could be made into rather inefficient but nevertheless deadly bombs.

It is not always easy to establish whether uranium has been stolen, or whether it is simply still 'in the system'. On one occasion, the US Office of Nuclear Safeguards admitted to a Congressional House energy subcommittee that perhaps 0.5 per cent of the nuclear fuel in the USA was not accounted for (Zito 1976). Moreover, there have been several known apparent or attempted thefts of nuclear fuel. In 1969 'a few kilogrammes' were found to be missing from an experimental reactor at Strickler, Arkansas. In November, 1966, twenty fuel rods containing natural uranium were diverted from the nuclear power station at Bradwell, Essex, but the thieves and the rods were intercepted by the police in London. In the late 1960s a survey of the fuel inventory at the Nuclear Material Enrichment Corporation plant in Apollo, Pennysylvania, revealed losses of up to 6 per cent of material over a period of six years, losses which have never been satisfactorily explained. More recently, five unused fuel rods disappeared from the Wylfa nuclear power station in North Wales and were not recovered.

Natural uranium of the kind involved in the Bradwell and Wylfa thefts is of little monetary value to a thief and of no significance at all for weapons production. It contains only 0.7 per cent of U-235, and the cost and technical difficulties of separating the U-235 at the high purity needed for weapons are so enormous as to be beyond the capability of any terrorist organization.

The theft of plutonium is a more worrying possibility. The USA has introduced plutonium into conventional reactors and inevitably this has meant an increase in the carriage of the material; civil plutonium stores in the USA are likely to rise to about 600 tonnes by the end of the century. A few kilograms of weapons-grade plutonium, combined with a reasonable level of scientific competence, are all that are needed to make a simple fission bomb. However, the plutonium resulting from the civilian nuclear power programmes is not weapons-grade material. It consists of pellets of a compound of uranium and plutonium oxides, sometimes as a ceramic. The plutonium resulting as a decay product in the fuel elements used in conventional reactors comprises the complete spectrum of plutonium isotopes. To make a bomb, plutonium (Pu) that is *predominantly* Pu-239 is needed, and, whilst this could be produced in a conventional reactor, it would only be by means of a specially designed and highly uneconomic fuel cycle.

Thus anyone stealing plutonium with an eye to weapons production would be faced with chemical difficulties in separating the plutonium from the uranium and with the greater problem of then separating Pu-239 from isotopes of very similar mass – Pu-240, Pu-241 and Pu-237. Once degraded by the presence of these isotopes, plutonium is useless for weapons and cannot be upgraded again.

Nevertheless, plutonium is an extremely toxic material; the ICRP has set a limit of one-fiftieth of a millionth of an ounce as the maximum permissible body burden. Plutonium metal is highly inflammable, and if a few kilograms were to be burnt on top of a high building in a city centre under certain meteorological conditions it could spread over a wide area and a large population a smoke plume of such deadly toxicity that bomb-making might not be necessary. This could constitute a real terrorist threat. As far as bomb-making is concerned, the theft danger may lie not so much with the commerical-grade material as with some kinds of research fuels. The alloys used for such fuels, often involving a light metal such as aluminium, would be relatively easy to separate into their components and as little as 5 kg of the uranium, perhaps 93 per cent U-235, would be needed in a bomb. Fortunately very little of this kind of fuel is manufactured and it is transported infrequently and by irregular and unpredictable routes. It is also true that the 'degree of surveillance needed to detect and watch terrorists at any given time depends on the prevalence of terrorism as well as on the availability of uranium' (House of Commons 1977: 38).

SUMMARY

It has been clear for several years that the management of radioactive waste, particularly high level waste, poses problems for the nuclear power industry, and that waste disposal rather than reactor safety may turn out to be the industry's real Achilles' heel. The essence of the problem is that all the means currently used for the storage of high-level waste in engineered structures have to be regarded as interim methods. Despite many years spent on conceptual designs not one store for the ultimate long-term disposal of high level waste is yet available. Technically there may be no urgency; it may well be quite safe to continue to store increasingly large numbers of used fuel elements at nuclear power stations in the USA and Japan, and to keep high activity radwaste in dry storage at fuel reprocessing plants. But public anxiety regarding radioactive waste makes an early and convincing engineered solution to the disposal problem highly desirable. Liquids can escape more easily than solids, and solids that are massive and not easily leached are less likely to release activity to the environment. Thus a substantial improvement could be made in the containment of high activity wastes if they were all converted as soon as possible into a solid form such as borosilicate glass.

Even vitrification, however, is simply a safer method of storage; it is not a form of ultimate disposal. Moreover, while borosilicate glass enables each radwaste element to be chemically bonded into the atomic structure of an inert and stable solid, it also becomes unstable in the presence of temperatures approaching and over 200°C.

Thus the optimal sequence of events in the management of high level radwaste becomes clear.

(a) Storage of fuel elements in ponds for months to several years.
(b) Storage of highly active liquor produced in reprocessing fuel for not more than two decades.
(c) Solidification into borosilicate glass, after which the glass blocks will be artificially cooled, for between ten and twenty years.
(d) The encapsulation of the blocks and their emplacement in a final repository. Temperature limits in a final deep repository will be set by the thermal stress that can be tolerated by the surrounding rocks and it is unlikely that the blocks could be placed in a final repository until at least a decade after their manufacture. Even then the heat output will be significant, falling to 20°C above the surrounding rock only after 1,000 years.

The sequence outlined here requires three major types of AFR reactor storage; double-walled cooled tanks (which have been used for many years), intermediate stores and ultimate disposal facilities, deep in the ground or in the sea bed. Intermediate stores may be air or water cooled and situated either above or below ground, at the reprocessing plant, at the final repository or elsewhere. The construction of such a plant is well within the reach of the technologies currently used in the construction industry. In the autumn of 1979, the American authorities suggested a store for spent but unprocessed fuel elements on a Pacific island; Wake, Midway and

Palmyra were suggested as possibilities. The idea was for spent fuel pellets still contained in the sealed fuel elements taken from the reactors to be encased in 100 ton concrete monoliths, 7 m high and 4 m in diameter, standing on concrete pads above ground. The chosen island would become an 'interim spent fuel storage facility' for an entire region, not just for American reactors. The idea has not been implemented, but implicit in the notion is the thought that disposal facilities should be an international rather than a national responsibility, a concept that may need to be developed further if the option of sea-bed disposal gains favour.

The design, construction and successful operation of a final deep geological land-based repository seems feasible with existing technology. Geological disposal strategies can be designed and tested by well-established modelling methods; they appear to satisfy defined performance objectives set in terms of responsible levels of risk or radiation dose. If significant doubts remain after these attempts at model verification, the way is open to deliberate overdesign.

The size of such a repository would depend upon the rate at which high activity waste is produced in the region it is designed to serve. In general, each 1,000 MWe year of nuclear power will give rise to 3 – 4 m^3 of solidified highly active waste containing 15 per cent by weight of fission product oxides.

If the construction of a geological disposal facility of adequate size is well within reach, its location is not. Public opposition to the siting of such a facility is dominated by the NIMBY factor, and experience so far in several countries indicates that public acceptance is the cornerstone of decision-making for all types of radwaste disposal sites. It cannot be the safest option, however, to have radwaste piling up in facilities which were not initially designed for its reception and in arrangements which can only be regarded as short term. By far the safest public health option would be for the industry to be allowed to proceed with a small number of well-sited deep depositories as demonstration projects.

Ultimate storage seems likely to be based on the principle of multiple containment in which a number of barriers are inserted between the radwaste and man. The barriers most likely to be involved are the stability and durability of the solidified waste form itself, the nature of the containers surrounding the waste, limitations on the access of ground water to the repository and its contents through a buffer area and the ability of surrounding geological strata, the near field, to absorb or trap any released radioactivity. The choice of a site or sites of such a nature that *transport* of radioactive species via ground water or surface waters is slow seems of great importance, for one way in which radioactivity could find a pathway through to man from a repository would be by transport in ground water following corrosion or breaching of a container and leaching of the solid waste, or as a result of a catastrophic natural or man-made event. Thus it is important that the hydrogeological conditions in and around a repository are clear and simple, and that deep ground water movement is negligible.

The integrity and safety of solidified high activity radwaste in underground repositories would probably be assured for many decades and, perhaps, for several centuries given a stable political and geological context.

The Swedish Corrosion Institute, which has evaluated the service life of copper canisters, is of the opinion that it is realistic to expect that they will last for hundreds of thousands of years. The safety analysis for a central high life waste disposal facility therefore assumes that the first canisters will break down after 100,000 years and then continue to break down at a uniform rate for another 400,000 years. Thus deep geological disposal in a cavern or borehole seems to be the likeliest option, but discussion has ranged widely over alternative means of ultimate disposal. Methods involving nuclear incineration, extraterrestrial disposal and burial in an ice sheet have all been suggested, together with the incorporation of highly active waste in molten silicate rock. Incineration of the higher actinides in thermal power reactors, it is claimed, would reduce their effective half-life to between ten and one hundred years but would be very costly with existing technology. Extraterrestrial disposal may be feasible, but expensive, and has been much less discussed as an option after the tragic malfunction of the US shuttlecraft *Challenger* in January 1986. One suggestion has been to place 'hot' containers of high activity waste on the ice surface of selected regions in Antarctica and to let them melt their own emplacement shafts down to the land surface 2 km below. It has been claimed that this would isolate the waste for 250,000 years. The idea has been discredited; the next Ice Age could arrive in 10,000 – 50,000 years and this would cause the ice accumulations to become unstable. The molten rock system is no more than a conceptual notion.

Very few of the more speculative schemes seem likely to progress beyond the drawing-board stage for many decades, and the radwaste scenario most likely to be used for high activity materials between now and the year 2000 would seem to consist of storage in liquid form, solidification and then storage of the solidified waste in engineered repositories. The ultimate disposal of such wastes implies a high degree of confidence that there will be either no further release of radioactivity to the environment or that, if it does occur, it will be sufficiently slight and gradual to cause no problem.

In the short term, the main danger seems to be that if the difficulties of finding acceptable sites for final repositories prove insurmountable, intermediate or interim storage methods may come to be regarded as viable substitutes. Public anxiety regarding the possible consequences of mismanagement of radioactive wastes makes an early solution to the siting problem highly desirable. Full public participation in achieving that solution is a necessary if not sufficient ingredient for achieving acceptance.

World-wide, the current situation regarding the disposal of radioactive waste produced by the nuclear fuel cycle is becoming unsatisfactory. Elderly unlined shallow landfill trenches, overlong on-site storage and cessation of ocean dumping without proper land-based alternatives add up to a problematic picture. Public concern is compounded by media reports of serious violations of safety procedures in the handling of nuclear waste by reprocessing plants in the EC (*Guardian*, 23 February 1988).

It is at the back end more than anywhere else in the nuclear fuel cycle that central and local government, the nuclear power industry and the public need to co-operate to produce a coherent and widely acceptable management strategy. Otherwise, problems that are soluble now will be handed on to following generations in a form that makes them less easily solved then. A sense of the need for urgency in policy and practice is heightened by the impending quantities of waste to be added to existing streams as increasing numbers of nuclear power plants are decommissioned. When the waste to be disposed of is spent fuel, the very definition of waste attempted at the beginning of this chapter becomes problematical. Spent fuel is a potential mine; it can be reprocessed to obtain more fuel for power stations or material for weapons. However, the capacity of the few reprocessing plants that exist in the world today is sufficient for only a small portion of the spent fuel discharged from nuclear power stations to be reprocessed.

NOTES

1. The half-life is the time taken for the activity of a radionuclide to decay to half its original value.

FURTHER READING

Carter, L.J. (1987) *Nuclear Imperatives and Public Trust: Dealing with Radioactive Waste*, Washington, DC: Resources for the Future Inc.

Chapman, N.A. and McKinley, I.G., with Hill, M. (1987) *The Geological Disposal of Nuclear Waste*, Chichester: Wiley.

National Academy of Sciences (1984) *Social and Economic Aspects of Radioactive Waste Disposal: Considerations for Institutional Management*, Washington, DC: National Academy Press.

Organization of Economic Co-operation and Development/Nuclear Energy Agency (1984). *Geological Disposal of Radioactive Waste: an Overview of the Current Status of Understanding and Development*, Paris: OECD.

OECD/NEA (1989) *Interim Oceanographic Description of the North-East Atlantic Site for the Disposal of Low-level Radioactive Waste*, Paris: OECD.

13 Prospects for the future

He who rides a tiger cannot readily dismount

(Indian Proverb)

The desire for safety stands against every great and noble enterprise

(Tacitus, *Annals*, xv, ciio)

The future of the nuclear power industry depends crucially upon winning public acceptance. In many energy markets, nuclear power has suffered from slower growth rates in electricity demand since the mid-1970s. In some parts of the world, including the UK, it has experienced a reversal of the cost advantage over coal-fired power generation that seemed to exist from the mid-1960s. In the USA in particular the industry has had to cope with the growth of a regulatory framework of Byzantine complexity. The sheer speed of the West's take up of nuclear power in the decade from 1965 brought industrial difficulties which caused mistakes, failures, bad design and accidents. Most significant of all, however, has been the rise of public anti-nuclear power sentiment. In the UK, for example, polls indicate that 90 per cent of the respondents in representative samples of the population are against increasing the country's reliance on nuclear power to meet future energy needs (Young 1987).

The safety issue is now paramount in shaping public attitudes. At Shoreham, Long Island, USA, and at Zwentendorf in Austria lie two nuclear power stations, completed but unlikely ever to be used. Because of public resistance, proposed plants were not built at Bodega Head in California nor at Plogoff in France. In Armenia, the nuclear power station at Yerevan has been closed down in response to widespread public anxiety about another earthquake disaster like that in 1988 which, fortunately, did not damage the plant. Sweden has voted to phase out nuclear power gradually and no new nuclear power station has been ordered in the USA since 1978. Fast reactor programmes in the UK and the USA have been cut back substantially.

The 1,860 MWe Lemoniz power station, near Bilbao, contracted for Spain's largest private electrical utility Iberduero, is the most obvious example of a nuclear power station that has been used by a group with wider political ends. In February 1981 the chief engineer at the plant, Sr Jose Mari Ryan, was assassinated by ETA, the militant Basque separatist organization, which had vowed to prevent the plant functioning, and the Director of the plant, Sr Angel Pascual Mugica, died in the same way in May 1982.

At first, opposition to the building of nuclear power stations was confined to minority groups and was limited to areas close to proposed sites. The opposition has grown to national and international proportions in which well-organized pressure groups, acting as a federation of anti-nuclear power interests, have made use of the media, mass protests at sites and regulatory frameworks to obtain publicity and impose delays on construction and licensing.

A variety of themes has been evident in the anti-nuclear opposition, ranging from concern over the environment, through fear of terrorism, to anxiety that the spread of nuclear power will inevitably facilitate the diffusion of nuclear weapons manufacture. Fear has been the common denominator. The risks associated with radiation have become a source of widespread generalized public concern.

In addition, geographical research has shown that power reactor siting policies may have been flawed (Openshaw 1986), and the potential for subversion of civil liberties has been highlighted (Jungk 1979). In the UK, under the Atomic Energy Authority (Special Constables) Act 1976, the UKAEA police are given very wide powers without any direct political control. If plutonium were to be stolen from an atomic plant, a UKAEA constable could carry arms in pursuit and if he reasonably believed the person he is chasing to have stolen nuclear material he may lawfully follow and pursue the individual into a house, cinema, theatre, restaurant or other premises. The constable can legally carry a machine gun or semi-automatic weapon and the chief constable is accountable only to the UKAEA which is not an elected body.

As Alvin Weinberg has put it, 'the price of a near-inexhaustible energy source (nuclear power) is eternal vigilance against accidents, sabotage and even a loss of civil liberties'. Do people think the price too high, or are they satisfied with the terms of the bargain?

The evidence from a variety of sources is that many people are not satisfied and that they may think the price too high. These sources include attitude surveys and opinion polls, sociological profiles, motivation analyses, perceived risk research, case studies of community conflict over reactor siting and a smaller number of studies which have documented the nature of the public response over a substantial period of time (Pijawka 1982). The social and psychological background to public acceptance, or lack of it, has been investigated by individual academics, university research groups and by research organizations such as the Battelle Institute, the Rand Corporation, Decision Resources of Eugene, Oregon, the US National Research Council and others.

In the USA the accident at Three Mile Island (TMI) has had a significant impact on public acceptance of nuclear power, increasing opposition and decreasing support for the construction of new nuclear power plants. The trend towards increasing opposition was established earlier, however, between 1975 and 1978. Opinion polls have shown how, from a position of majority support (approximately 55 per cent in favour in 1971–5) and minority opposition (averaging 28 per cent from 1971 to 1975) there was a change to equivalent levels of support (47 per cent) and opposition (43 per cent) by 1978. By 1981 there was a majority of opposition (over 60 per cent) and only minority support (30 per cent) (Melber 1982). Some writers have argued that public concern before and after TMI was episodic – a single incident pushing concern to a high level from which it subsided over time (Mitchell 1980). Others have suggested that incidents of sufficient magnitude, such as TMI and Chernobyl, have raised the level of public anxiety to a new and critically higher level (Nealy *et al.* 1983).

Several surveys have shown that, both before and after the TMI and Chernobyl accidents, support for nuclear energy was stronger in communities located close to nuclear plants than in the population at large (Whyte 1977; Firebaugh 1981; Manning 1982; Ester *et al.* 1983; Lee *et al.* 1984). In Sweden, in the nuclear power referendum of 23 March 1980, 73 per cent of the voters around the Barsebäck nuclear station favoured nuclear power.

It may be erroneous to conclude from this that people living close to a reactor are better informed about nuclear power safeguards and thus perceive them to represent a lower degree of risk. Local residents would be in a classic situation of cognitive dissonance if they perceived the risk to be high *and* continued living near the station, and so their perceptions should be evaluated accordingly (Whyte 1977). They would also have vested interests in the jobs and incomes directly and indirectly provided by the plant for the local economy.

The nuclear industry has reacted rather sluggishly to the growing weight of opposition and has tried to counter it mainly through information programmes. In the USA, for example, nuclear power interests and some utilities combined after TMI to form the US Committee for Energy Awareness (CEA) with the object of emphasizing the importance of electricity in general and nuclear power in particular. In the UK, BNFL has spent a great deal of time and money to attract visitors to Sellafield. Such tactics might be effective if it has not been the case that, as has been suggested, the very intensity of the efforts to reduce risk without explicit and intensive communication of their achievements has actually increased the perceived risk and proved to be a cause of fear (Fremlin and Wilson 1982). Some writers have also suggested that crucial links may have been missing in the past in the vital chain of information from the scientific community to the public (Rothman and Lichter 1982; Cottrell 1985). Therefore more information, and above all more precise and accurate information, may help, but the situation may be too complex to be solved so simply.

People's images and understanding of alternative courses of action in the field of power supply, as in other areas of human activity, are not necessarily those of a scientific expert nor of the social scientist who is studying them (Whyte 1977). In the UK, for instance, research by members of the Psychology Department at the University of Surrey has shown that people are not appraising 'scientific facts' about radioactive waste as though they existed in a value-free vacuum (Lee *et al.* 1984). Lee *et al.* point out that:

> It is evident that beliefs about waste are inseparable from confidence in managements, worries about 'possible' harm in the future, and the relation of technological advance to established value systems. It is unlikely that the uncertainty surrounding waste management options now or in the future can be countered by scientific argument.

In their research programme the Surrey University group investigated the attitudes to nuclear power amongst 1,354 respondents in the southwest of England in 1982–3. Three attitude groups were identified: pro-nuclear anti-nuclear and uncommitted. The Surrey researchers used a computer-based discrimination technique to distribute key questions relating to knowledge about nuclear power, pro- and anti-nuclear beliefs and possible forms of action into a two-dimensional space, each being represented by a point. The programme arranges the points so that proximity is equivalent to similarity. The result is shown in Figure 13.1.

The Surrey team drew attention to a number of implications to be drawn from the diagram (Lee *et al.* 1984):

(a) The pro-nuclear group hold anti-nuclear as well as pro-nuclear beliefs, whereas the anti-nuclear attitude is associated with no pro-nuclear beliefs at all.
(b) The uncommitted are closer in the knowledge – belief – action space to the pro-nuclear group than to the anti-nuclear group.
(c) A belief that nuclear power creates jobs is highly influential with the pro-nuclear group and comes nearer than any other belief to the uncommitted position. A change in this belief would move the uncommitted position closer to the pro-nuclear position. The same applies, to a smaller extent, to the beliefs that renewable sources of energy are inexhaustible, that stations are vulnerable targets and that stations are a source of harmful radiation.
(d) Conversely, if the beliefs that less electricity is needed, that waste cannot be disposed of safely and that nuclear power threatens jobs were strengthened, the uncommitted position would move towards the anti-nuclear stance.

The polarization of anti- and pro-nuclear stances is exacerbated by the fact that the industry's expectation of the likelihood of occurrence of a serious accident differs from and is more optimistic than that of the public

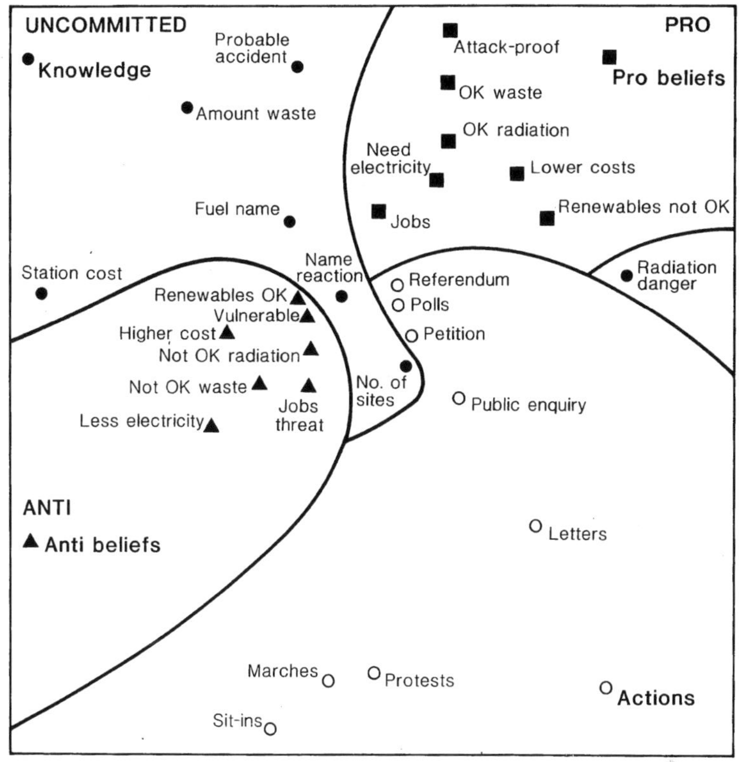

Figure 13.1 Guttman Lingoes smallest space analysis of knowledge, beliefs, action disposition and attitudes towards nuclear power
Source: Lee *et al.* 1984

(Whyte 1977; Young 1987). Here, the power utilities, the nuclear industry and the regulatory agencies cannot easily avoid the accusation of complacency. When risk studies gave a low probability of an accident's causing off-site damage, as in the Rasmussen Report (USNRC 1975), the working assumption in the industry too often became one that such an accident would not occur. The events at TMI, Chernobyl, and Windscale have reminded everyone that low probability estimates do not eliminate actual accidents. Thus, there is justification for attempts to work out the costs of such accidents on the assumption that they *will* happen. Otherwise it is impossible to perform proper meaningful cost-benefit calculations for nuclear power.

The extent to which nuclear capacity for electricity generation expands beyond the total afforded by plants now in operation and under construction depends upon the fulfilment of seven conditions. First, nuclear power stations must be very safe in reality and must be seen to be so by the general public. Second, effective emergency evacuation procedures for surrounding populations must be in place, and should be seen to be in place. Third, honest costings in particular electricity markets must show nuclear power to be economically competitive with available alternative means of electricity production. Fourth, a technically, politically and socially viable solution to the nuclear waste problem must be found. Fifth, people must be prepared to accept nuclear power stations and radwaste stores as neighbours. Sixth, nuclear energy must come to be regarded by the public as a legitimate part of an effective answer to the world's energy problem and to the greenhouse effect; it must clearly be seen that the benefits do outweigh the costs. Seventh, it has to be recognized that, because of questions concerning environmental quality and priority of competing land uses, notwithstanding those of public safety, the processes of site selection for nuclear power stations and waste disposal facilities, together with those involved in designing evacuation zones, will continue to pose problems, the resolution of which requires professional geographical expertise. Site selection by stealth or compelling force has never been an acceptable procedure in democratic society and neither has bribery by payment to populations living in proximity to nuclear installations.

Glossary

AERE
Atomic Energy Research Establishment, Harwell. The main research centre of the Research Group of the UKAEA.

AGR
Advanced gas-cooled reactor. An improvement on the Magnox type of reactor, and operates at higher temperature to give a greater burnup. The fuel consists of slightly enriched uranium oxide pellets canned in stainless steel.

Alpha particle
A positively charged particle emitted in the radioactive decay of some heavy nuclei, e.g. uranium and radium.

Atom
The atom is the smallest amount of an element which has the chemical properties of that element. The atom consist of a comparatively massive central nucleus carrying a positive electric charge, around which electrons move in orbits at relatively great distances away. The nucleus is made up of protons and neutrons. The number of protons present is the atomic number of the element and determines the charge on the nucleus, and hence its chemical properties. The sum of the number of protons and neutrons is called the mass number and determines the mass of the nucleus. The number of neutrons in an atom of a given element can vary, resulting in nuclei that have the same atomic number but different mass numbers; these variants are called isotopes of the element. The number of electrons in a neutral atom is equal to the number of protons in the nucleus and their charges balance the equal and opposite charge of the nucleus. All atoms of the same atomic number (i.e. same number of protons) are atoms of the same element, irrespective of the number of neutrons present.

Background (radiation)
Refers to those undesired counts or currents that are registered with a radioactivity detector even though no radioactive sample is present. This effect is due to the response of the detectors to cosmic rays, local contaminating radioactivity, insulation leekage, amplifier noise, power-line fluctuations etc.

Becquerel (Bq)
Has replaced the Curie (Ci) as the unit used for measuring quantity in terms of the number of radioactive disintegrations taking place in a material and is defined simply as the disintegration of one atom per second in any quantity of any radioactive substance.

Beta particle
An electron, positive or negative, emitted from the nucleus in certain types of radioactive disintegration.

BeV
American for GeV (1,000 million eV).

Blanket
Fertile material put around a reactor core to breed new fuel, e.g. uranium becomes plutonium by absorption of spare neutrons.

BNFL
British Nuclear Fuels Ltd (now British Nuclear Fuels plc). The publicly-owned company which on 1 April 1971 took over the work of the UKAEA's Production Group in fuel manufacture, reprocessing and marketing of fuel for nuclear reactors in the UK and abroad. It has three main sites, at Springfields, Capenhurst and Windscale. It also operates two nuclear power stations: Calder Hall and Chapelcross.

Boiling water reactor (BWR)
A reactor in which water is used as both coolant and moderator and allowed to boil in the core. Steam is produced directly in the reactor vessel, under pressure, and in this stage can be supplied to a turbine.

Bone seeker (bone-seeking element)
An element which tends to be deposited in the bones of the body (e.g. strontium, radium or plutonium) because either it is chemically similar to calcium or it takes part in bone-forming processes.

Boron
A non-metallic element, obtained from borax or boric acid, which readily absorbs the slow neutrons essential to the uranium fission process in a thermal reactor. It is used for control rods in such factors, often in the form of an alloy with steel.

Breeder reactor
A nuclear reactor which produces more fissile atoms than it 'burns'.

Breeding (see breeder reactor)
The process of generating nuclear fuel, e.g. plutonium from uranium, by absorption of neutrons.

Burnup
The fraction or percentage of atoms in a reactor fuel which has undergone fission; also the total amount of heat released per unit mass of fuel (usually expressed in megawatt days per tonne).

Caesium
A rare soft silvery white metal, atomic number 55, similar in chemical and physical properties to sodium. A long-lived (30 year half-life) gamma-emitting radioactive isotope (caesium-137) is found in fission products, the 'ash' of nuclear reactors.

Cascade process
A process, such as gaseous diffusion, which involves the repetition of the same process many times. Each stage involves some form of separation of phases; one of these passes up the plant and the other down. Consequently the feed at each stage consists of a mixture of two phases fed in from opposite directions.

Chain reaction (nuclear)
A process in which one nuclear transformation sets up conditions which permit a similar nuclear transformation to take place in another atom. Thus, when fission

occurs in uranium atoms, neutrons are released which in turn produce fission in neighbouring uranium atoms.

Collective dose or effective dose equivalent
The sum of the products of the individual dose equivalents and the number of individuals in each group in an exposed population. The collective dose commitment is the dose commitment multiplied by the number of individuals in the population.

Collective effective dose equivalent
This is a composite figure in which all the doses from different types of radiation to different body organs are combined together according to their capacity to cause fatal cancers. This quantity is often simply called 'dose'.

Control rods
Rods, plates or tubes of steel or aluminium-containing boron, cadmium or some other strong absorber of neutrons. They are used to hold a reactor at a given power level.

Coolant
A fluid which is circulated through or about the core of a reactor to maintain a low temperature and prevent the fuel from overheating.

Cosmic radiation (cosmic rays)
Very penetrating ionizing radiation which reaches the earth mainly from unidentified sources in outer space and occasionally from the sun.

Critical
The term used to describe the condition in which a chain reaction is being maintained at a constant rate, i.e. it is just self-sustaining.

Curie (Ci)
The unit of radioactivity. It is the quantity of a radioactive isotope which disintegrates at a rate of 37,000 million disintegrations per second. The activity of a gram of radium is approximately equal to 1 Ci. The weight of a curie of any particular substance depends largely on its half-life. Whereas 1 Ci of strontium-90 weighs less than a thousandth of an ounce, 1 Ci of natural uranium weighs more than a ton (see also *Specific activity* and *Bequerel*).

Daughter product
The product formed in the radioactive decay of a nucleus (called the parent). Synonym for decay product.

Decay
When a radioactive atom disintegrates it is said to decay. What remains is a different element. Thus an atom of polonium decays to form lead, ejecting an alpha particle in the process. In a mass of a particular radioisotope a number of atoms will disintegrate or decay every second, and this number is characteristic of the isotope concerned (see also *Radioactive decay*).

Depleted uranium
Uranium having less than the natural content, namely 0.7 per cent, of the easily fissionable uranium U-235, e.g. the residue from a diffusion plant or a reactor.

Derived working limit (DWL)
A limit derived from the basic recommendations of the ICRP for maximum permissible doses and dose limits in such a way that compliance with the DWL implies virtually certain compliance with the relevant ICRP recommendations.

Deuterium
The isotope of hydrogen of mass 2, often called heavy hydrogen (see also *Heavy water*).

Diffusion
(1) A method of isotope separation (see *Gaseous diffusion*).
(2) In nuclear physics, the passage of particles through matter in such a way that the probability of scattering is large compared with that of capture.

Dose
For any ionizing radiation it is the energy, measured in grays or rads, which is imparted to matter by the ionizing particles per unit mass of irradiated material (see *Committed dose and effective dose*).

Dose limits (see ICRP)
The recommended dose limits for controlling the exposure of members of the public to ionizing radiations.

Electron
The negatively charged particle (mass $m = 9 \times 10^{-28}$ g) which is a common constituent of all atoms. Its positively charged counterpart, of equal mass, is the positron.

Electronvolt (eV)
A measure of the energy and so, indirectly, of the velocity of a particle. A particle carrying a charge equal to that of the electron, when accelerated by a potential of 1 volt, acquires a kinetic energy of 1 electronvolt.

Enriched fuel
Nuclear fuel which has been enriched in the fissile component, e.g. uranium containing more than 0.7 per cent U-235.

Fall-out
Radioactive dust and other matter falling back to the earth's surface from the atmosphere after a nuclear explosion.

Fast reactor
A nuclear reactor in which most of the fissions are caused by neutrons moving at high speeds. Such reactors contain little or no moderator.

Fissile
Capable of undergoing fission; sometimes used to mean capable of fissioning when hit by a slow neutron, e.g. the isotopes U-233, U-235, Pu-239 and Pu-241 are fissile.

Fission
The splitting of a heavy nucleus into two (or very rarely more) approximately equal fragments – the fission products. Fission is accompanied by the emission of neutrons and the release of energy. It can be spontaneous, or it can be caused by the impact of a neutron, a fast charged particle or a photon.

Fission products
Atoms formed as a result of nuclear fission.

Fuel element
A unit of nuclear fuel for use in a reactor – generally uranium, as metal or oxide, enclosed in a can which may have an extended surface area, e.g. fins, to assist heat transfer.

Fusion
The process of building up more complex nuclei by the combination, or fusion, of simpler ones. The formation is usually accompanied by the release of energy.

Gamma ray
Electromagnetic radiation emitted by the nuclei of radioactive substances during decay, similar in nature to X-rays.

Gas centrifuge process
A method of isotopic separation in which heavy gaseous atoms or molecules are separated from lighter ones by centrifugal force.

Gaseous diffusion
A method of separating isotopes, for example those of uranium, by causing a gaseous compound to diffuse through a porous membrane; the lighter molecules diffuse faster than the heavy ones and consequently their concentration increases on passing through the membrane, for the initial portions of gas. In practice one allow a certain amount to diffuse through a cell membrane, and the rest, somewhat depleted in the light component, to flow past.

Graphite
A form of carbon used as a moderator in nuclear reactors. It is made from purified petroleum coke compressed into bricks and heated to high temperatures.

Gray
1 gray (Gy) equals 1 joule of energy absorbed per kilogram of tissue and the gray is now the official SI unit for measuring the amount of ionization produced in tissues by the absorption of radiation energy. It has replaced the rad for this purpose.

Half-life
The time taken for the activity of a radioactive substance to decay to half its original value, i.e. for half the atoms present to disintegrate. Half-lives may vary from less than a millionth of a second to millions of years, according to the isotope and element concerned.

Heavy water
Water consisting of molecules in which the hydrogen is replaced by deuterium, or heavy hydrogen. It is present in ordinary water as about 1 part in 5,000. It is used as moderator because it has a low neutron absorption cross-section.

Helium
The second lightest element and the least dense inert gas. It is a good heat transfer medium and absorbs very few neutrons.

Hex
An abbreviation for uranium hexafluoride, the gaseous uranium compound used in diffusion and in centrifugal uranium enrichment plants.

IAEA
International Atomic Energy Agency – an independent United Nations organization, of which the UK is a member, which promotes the peaceful uses of atomic energy and establishes international standards of safety.

ICRP
The International Commission on Radiological Protection. A body set up to define maximum permissible levels for external and internal radiation especially with regard to human beings.

International nuclear event scale
In an attempt to communicate the significance of a nuclear accident promptly and simply to the general public, the international nuclear event scale was devised in 1990 by experts from the IAEA and OECD/NEA. The first of the following tables shows the underlying logic of the scale, the criteria given in the matrix being broad indicators. The second table is the event scale itself. Nuclear incidents/accidents are assigned numerical levels from 0 (no safety significance) to 7 (major nuclear accident). The underlying objective of the scale involves taking into account off-site and on-site effects of the accident and the extent to which the defence-in-depth provisions were degraded. An odd feature of the scheme is that the Windscale and Three Mile Island (TMI) accidents are both assigned to level 5, despite TMI being less serious in orders of magnitude than the largely uncontained Windscale accident. The reason for this is the inclusion of core damage in the on-site impact criteria. The journal *Nuclear Energy* (1990, 29 (5), Oct.) argues that it might seem reasonable to omit this consideration and rely instead on the degree of radiation exposure to workers. That would then put TMI in level 3.

Underlying logic of the international nuclear event scale

Level	*Off-site impact*	*On-site impact*	*Defence-in-depth degradation*
7 (major accident)	Major release: widespread health and environmental effects		
6 (serious accident)	Significant release: full implementation of local emergency plans		
5 (accident with off-site risks)	Limited release: partial implementation of local emergency plans	Severe core damage	
4 (accident mainly in installation)	Minor release: public exposure of the order of prescribed limits	Partial core damage Acute health effects to workers	
3 (serious incident)	Very small release: public exposure at a fraction of prescribed limits	Major contamination Overexposure of workers	Near accident Loss of defence-in-depth provisions
2 (incident)			Incidents with potential safety consequences
1 (anomaly)			Deviations from authorized functional domains
0 (below scale)			No safety significance

The international nuclear event scale

Level	Criteria	Examples
Accidents		
7 (major accident)	External release of a large fraction of the reactor core inventory typically involving a mixture of short and long-lived radioactive fission products (in quantities radiologically equivalent to more than tens of thousands terabecquerels of iodine-131). Possibility of acute health effects. Delayed health effects over a wide area, possibly involving more than one country. Long-term environmental consequences.	Chernobyl, USSR 1986
6 (serious accident)	External release of fission products (in quantities radiologically equivalent to the order of thousands to tens of thousands of terabecquerels of iodine-131). Full implementation of local emergency plans most likely needed to limit serious health effects.	
5 (accident with off-site risks)	External release of fission products (in quantities radiologically equivalent to the order of hundreds to thousands of terabecquerels of iodine-131). Partial implementation of emergency plans (e.g. local sheltering and/or evacuation) required in some cases to lessen the likelihood of health effects.	Windscale, UK 1957
	Severe damage to large fraction of the core due to mechanical effects and/or melting.	Three Mile Island, USA. 1979
4 (accident mainly in installation)	External release of radioactivity resulting in a dose to the most exposed individual off-site of the order of a few millisieverts.* Need for off-site protection actions generally unlikely except possibly for local food control.	
	Some damage to reactor core due to mechanical effects and/or melting.	Saint Laurent, 1980
	Worker doses that can lead to acute health effects (of the order of 1 sievert).**	
Incidents		
3 (serious incident)	External release of radioactivity above authorized limits, resulting in a dose to the most exposed individual off-site of the order of tenths of a millisievert.* Off-site protective measures not needed.	
	High radiation levels and/or contamination on-site due to equipment failures or operational incidents. Overexposure of workers (individual doses exceeding 50 millisieverts).**	
	Incidents in which a further failure of safety systems could lead to accident conditions, or a situation in which safety systems would be unable to prevent an accident if certain initiators were to occur.	Vandellos, Spain 1989
2 (incident)	Technical incidents or anomalies which, although not directly or immediately affecting plant safety, are liable to lead to subsequent re-evaluation of safety provisions.	
1 (anomaly)	Functional or operational anomalies which do not pose a risk but which indicate a lack of safety provisions. This may be due to equipment failure, human error or procedural inadequacies. (Such anomalies should be distinguished from situations where operational limits and conditions are not exceeded and which are properly managed in accordance with adequate procedures. These are typically 'below scale').	

Below scale/zero

No safety significance

---

* The doses are expressed in terms of effective dose equivalent (whole body dose). Those criteria, where appropriate, also can be expressed in terms of corresponding annual effluent discharge limits authorized by national authorities.

** These doses also are expressed, for simplicity, in terms of effective dose equivalents (sieverts), although the doses in the range involving acute health effects should be expressed in terms of absorbed dose (grays).

Iodine
A volatile element, atomic number 53, which is vital to human life (e.g. the thyroid system). Iodine-131 is an important, but short-lived (8 day half-life), fission product.

Ion
A charged atom or molecule, i.e. one which has lost or gained one or more electrons.

Ionizing radiation
Radiation which knocks electrons from atoms during its passage, thereby leaving ions in its path. Electrons and alpha particles are much more ionizing than neutrons or gamma rays.

Irradiation
The exposure of materials to radiation. In nuclear research, and in the production of isotopes, materials are often exposed to neutrons in reactors. Intense irradiation can alter the physical and chemical properties of solids – in some cases weakening them (e.g. fuel elements and graphite), but in others hardening them (e.g. some types of plastics and rubbers).

Isotope (see *Atom*)
Two atoms are said to be isotopes if they are of the same chemical element but have different masses. This means that isotopic nuclei contain the same number of protons but different numbers of neutrons.

Krypton
A gas with a half-life of 10.7 years.

Load factor (see *Availability*)
The load factor of a generating plant reflects its availability and usage. It is the ratio of the units (kWh) actually generated to those which could have been if the plant had worked continuously at full power.

Magnox
The magnesium alloy used for sheathing the uranium in Calder Hall type fuel elements. Nuclear power stations using such fuel elements are often referred to as Magnox stations.

Maximum permissible doses (see *ICRP*)
The recommended upper limits for controlling the doses of ionizing radiation for individuals who are exposed to radiation in the course of their work.

Mega (M)
Prefix meaning a million.

Megawatt (MW)
A million watts or a thousand kilowatts, where a watt is the unit of power. In MWe or MWt the e signifies 'electrical' and the t means 'thermal power' or heat output.

Megawatt days per tonne
A unit used for expressing the heat output per tonne of fuel in a reactor, and hence the burnup. Thus in a reactor with 100 tonnes of fuel, operating at 150 MW for a year, the fuel is irradiated at an average of

$$\frac{365 \times 150}{100} = 547.5 \text{ MW days/tonne}$$

MeV
A million electron volts.

Micro (μ)
Prefix meaning a millionth part of.

Milli (m)
Prefix meaning a thousandth part of.

Moderator
The material in a reactor used to reduce the energy, and hence speed, of fast neutrons, as far as possible without capturing them. Slow neutrons are much more likely to cause fission in a U-235 nucleus than to be captured in a U-238 nucleus (see *Natural uranium*).

Natural uranium
Natural uranium contains both the heavier uranium isotope U-238, which is not readily fissile material and is the parent material from which plutonium is created, and the lighter isotope uranium U-235, which is the fission material or fuel of most reactors. In 140 parts of natural uranium, 139 parts are of U-238, and one part only is U-235.

NEA
The Nuclear Energy Agency of the Organization for Economic Co-operation and Development (OECD), comprises countries including the UK whose objective is the orderly development of the peaceful uses of atomic energy.

Neutron
A nuclear particle having no electric charge and the approximate mass of a hydrogen nucleus. It is found in the nuclei of atoms and plays a vital part in nuclear fission. Outside a nucleus a neutron is radioactive, decaying with a half-life of about 12 min to give a proton and an electron.

Non-proliferation treaty
A treaty with the aim of preventing the diversion of nuclear materials from peaceful uses to nuclear weapons or other nuclear explosive devices. It came into force on 5 March 1970, and was ratified by forty-seven countries.

NRPB
The National Radiological Protection Board (UK). Established in 1970 as the national point of authoritative reference in radiological protection. Its main duties are to advance the acquisition of knowledge on protecting mankind from radiation hazards and to provide information, advice and services relating to radiological protection.

Nuclear Installations Inspectorate (UK)
Regulates safety of nuclear power stations, nuclear fuel processing plants, research

reactors and isotope manufacturing plants licensed under the Nuclear Installations Act 1965 and 1969.

Nuclear reactor
A structure in which a fission chain reaction can be maintained and controlled. It usually contains a fuel, coolant, moderator and control absorbers and is most often surrounded by a concrete biological shield to absorb neutron and gamma ray emission.

Nuclide
An atomic species of a single atomic number and a single mass number.

Pico
Prefix meaning a millionth of a millionth (10^{-12}) part of.

Pitchblende
A rich uranium ore, consisting largely of uranium oxide, which has been mined in Canada and the Congo in commercially economic amounts.

Plasma
Very hot gas consisting mostly of positive ions and electrons in nearly equal concentrations. It is almost electrically neutral and is highly conducting.

Plutonium
The element, atomic number 94, produced by neutron irradiation of U-238. The isotope Pu-239 is an important fissile material and is usually made in reactors. It is used as a nuclear fuel, usually as plutonium oxide.

Poison (nuclear or reactor)
A material which absorbs neutrons and hence tends to stop a reactor working. Frequently a fission product, e.g. xenon.

Pressurized water reactor (PWR)
Reactor in which the water coolant and moderator is kept at a high pressure to prevent it readily boiling and hence to keep it liquid. This type requires enriched fuel. The water is taken out to a heat exchanger to generate steam which can run through a turbo-alternator to give electricity.

Proton
The nucleus of the hydrogen atom. It carries unit positive charge and has unit mass.

Rad (see *Roentgen*)
The unit of ionizing radiation absorbed dose: 1 rad is equal to an energy absorption of 0.01 joules of energy per kilogram of tissue (see *Gray*).

Radiation
A term which embraces electromagnetic waves, in particular are X-rays and gamma rays as well as streams of fast-moving charged particles (electrons, protons, mesons etc.) and neutrons of all velocities, i.e. all the ways in which energy is given off by an atom.

Radiation burns
If radiation is sufficiently intense or prolonged it causes surface burns on the skin which reddens and blisters, not unlike heat burns or severe sunburn. These are called radiation burns. Patients subject to intense irradiation, e.g. in cancer therapy by X-rays or gamma rays sometimes suffer surface or skin radiation burns.

Radiation risk
The risk to health from exposure to ionizing radiation.

Radiation sickness
The symptoms, such as nausea etc., induced by acute whole-body over-exposure to radiation, i.e. sudden large doses received over the whole body.

Radioactive
Possessing or pertaining to radioactivity. The term 'radio' is used as an abbreviation of radioactive.

Radioactive family (or series)
A series of radioactive elements, each except the first being the daughter product of the previous one; the final member for the natural elements is usually lead, which, although stable, is included in the family.

Radioactivity, radioactive decay
The property possessed by some atoms of disintegrating spontaneously with the emission of a charged particle and/or gamma radiation. The rate of radioactive decay is not affected by any normal change of temperature, electric or magnetic fields or chemistry.

Radioisotope
An isotope which is radioactive. Most natural isotopes of mass below 208 are not radioactive.

Reactivity
A measure of the amount of the possible departure of a reactor from the critical condition, where the reaction is just self-supporting. At any steady state of operation the reactivity is zero. Addition of positive reactivity causes divergence; addition of negative reactivity causes the reaction to die down.

Reactor core
The central portion of a nuclear reactor containing the nuclear fuel, such as uranium or plutonium, and the moderator, if any.

Reactor vessel
The container of a reactor and its moderator, if any, and coolant.

Rem (roentgen equivalent man)
The unit of dose equivalent. The dose in rems is equal to the absorbed dose in rads multiplied by appropriate modifying factors aimed at expressing different types of radiation and different distributions to absorbed dose on a common scale related to the possible long-term radiation risks. For many radiations, including X-, gamma and beta radiation, and for most dose distributions, the modifying factor is unity and rems and rads are numerically equal (see *Sievert*).

Reprocessing
The procedure of removing fission products from fuel before reusing it. One main aim is to remove poisons which would absorb and waste neutrons; another is to remove mechanical stresses due to irradiation especially in the case of metallic fuels.

Roentgen
The unit of exposure to X- or gamma radiation based upon the capacity of the radiation to produce ionization in air. For a wide range of radiation energies, 1 roentgen will result in an absorbed dose in soft tissue of approximately 1 rad.

Safety rod
A neutron-absorbing rod which, in an emergency, can be put into and shut down a nuclear reactor in a fraction of a second. It is normally operated by gravity so as to be independent of power supply.

Separative work
A measure of the work done in enriching material (e.g. natural uranium) from the initial concentration to the desired final enrichment. The term is used in costing enriched fuel supplies.

SGHWR
Steam generating heavy water reactor: this reactor uses boiling light water to produce steam and heavy water as moderator. The UKAEA prototype 100 MWe station began operating at Winfrith in December 1967. The reactor provides an important irradiation facility for water reactor fuels, as well as producing power for the national grid.

Shield – biological
A mass of absorbing material used to shield operating staff by reducing radiation to permissible levels. It is often of dense concrete, e.g. the biological shield round power station reactors can be a 7 or 8 foot thickness of concrete.

Shield – thermal
A metallic shield several centimetres thick placed on the inside of the biological shield to prevent the latter from becoming overheated.

Sievert (Sv)
The SI equivalent and replacement for the rem as the dose equivalent unit; 1 Sv = 100 rem.

Sodium
A chemically reactive metal, atomic number 11, which is liquid over a wide range of temperatures; it is sometimes used, alone or mixed with potassium, as a reactor coolant, e.g. for fast reactors.

Source
Concentrated radioactive matter used as a source of radiation.

Specific activity
The radioactivity of unit mass of a radioactive substance, usually expressed in curies per gram.

Stretch (or stretch capacity)
A capability initially designed into plant which in the future permits the attainment of greater output capacity without major plant design changes.

Strontium
A metal, atomic number 38, similar in chemical properties to calcium. A long-lived radioactive isotope of strontium (Sr-90) is produced in the fission of uranium and plutonium, and hence is present in fall-out.

Thermal reactor
A nuclear reactor which includes a moderator and therefore uses slow or thermal neutrons for fission of its fuel.

Thorium
A naturally radioactive metal, atomic number 90, the mineral sources of which are widely spread over the earth's surface, particularly in monazite beach sands. It can be converted to uranium-233, an excellent nuclear fuel, by neutron absorption.

Tokamak
A toroidal plasma confinement system formed by a combination of magnetic fields, one of which is produced by current flowing through the plasma.

Tonne
1,000 kilograms or 2,200 lb.

Transuranic elements
The artificial elements, atomic numbers 93 and higher, which have heavier and more complex nuclei than uranium. They can be made by neutron bombardment of uranium.

Trip
The sudden shutdown of a nuclear reactor, normally by rapid insertion of the safety rods initiated by an emergency or by a deviation from normal reactor operation via automatic controls or the reactor operator.

Tritium
The isotope of hydrogen of mass 3. It is very rare and is naturally radioactive. It can be made by neutron absorption in lithium and in deuterium or heavy water, and is present in fall out.

UKAEA
United Kingdom Atomic Energy Authority. It is controlled by a Chairman and a Board of Members, with headquarters in London.

Uranium
A heavy metal, atomic number 92. U-235 is the only naturally occurring readily fissile isotope, U-238 is a fissile material and U-233 is a fissile material that can be produced by the neutron irradiation of thorium-232. Natural uranium contains 1 part in 140 of U-235.

Uranium hexafluoride
A gaseous compound of uranium with fluorine used in the gaseous diffusion process for separating the uranium isotopes (commonly called hex).

Xenon (Xe)
An inert gas, atomic number 54, produced as a fission product in reactors. Xe-135 is one of the most important poisons in a reactor.

Zircaloy
The zirconium alloy used for canning fuel in water reactors because it is relatively non-absorbent of thermal neutrons.

The following are suggested as reference texts for energy terms and definitions:

Counihan, M. (1981) *A Dictionary of Energy*, London: Routledge & Kegan Paul.
Slessor, M., Bennet, D.J., Maver, T., Twidell, J., Gibb, W., Common, M., Howell, P. and Lewis, C. (1985) *Macmillan Dictionary of Energy*, London: Macmillan.

In producing this glossary reference has been made to *Quantities, Units and Symbols* (published by the Royal Society); *Glossary of Nuclear Power Terms* (published by the Personnel Department, UK Central Electricity Generating Board), *Glossary of Atomic Terms* (published by the UK Atomic Energy Authority), *Glossary of Nuclear Power Station Terms Used in the Design and Construction Industry* (Commercial and Economic Intelligence Department, British Nuclear Design and Construction Ltd, Whetstone, Leicestershire).

Bibliography

Abel, E. (1977) 'In the USA: Atomic energy on trial', *Nuclear Law* 13: 56–8.

Adams, C.A. and Stone, C.N. (1967) 'Safety and siting of nuclear power stations in the United Kingdom', in *Containment and Siting of Nuclear Power Plants*, Vienna: IAEA, pp. 129–42.

Addinall, E. and Ellington, H. (1982) *Nuclear Power in Perspective*, London: Kogan Page.

Adkins, B. (1973) 'Public understanding and acceptance of nuclear energy in Japan', *Nuclear Engineering International* July: 562–66.

Ahlstrom, P.E., Löfveberg, S., Nilsson, L.B. and Papp, T. (1981) 'Safe handling and storage of high level radioactive waste', *Radioactive Waste Management* 1 (1): 57–103.

Ahmed, S.B. (1979) *Nuclear Fuel and Energy Policy*, Lexington, MA: D.C. Heath.

Alesso, H.P. (1981) 'Proven commercial reactor types: an introduction to their principal advantages and disadvantages', *Energy* 6: 543–54.

Allardice, C. and Trapnell, E.R. (1972) 'The first pile: An historical account of the plutonium project', *Journal of the Institute of Nuclear Engineers* 13 (6): 163–8.

Allday, C. (1982) 'Nuclear fuels: development, processing and disposal', *Energy World* 75: 12–15.

Allibone, T.E. (1961) *The Release and Use of Atomic Energy*, London: Chapman and Hall.

American Assembly (1957) *Atoms for Power: United States Policy in Atomic Energy Development*, New York: Columbia University.

Angelini, A.M. (1965) 'Nuclear power stations in Italy', *Proceedings of the 3rd International Conference on the Peaceful Uses of Atomic Energy, Geneva, 1964* vol. 5, Geneva: United Nations Organization, pp. 240–7.

Anthony, L.J. (1966) *Sources of Information on Atomic Energy*, Oxford: Pergamon.

Appleyard, C. (1978) 'The East German atomic dilemma', *Guardian*, 9 January.

ApSimon, H.M., Wilson, J.J.N., Guirguis, S. and Stott, P.A. (1987) 'Assessment of the Chernobyl release in the immediate aftermath of the accident', *Nuclear Energy* 26 (5) 295–301.

Archer, V.E. and Wagoner, J.K. (1973) 'Lung cancer among uranium miners in the United States', *Health Physics* 25: 351–70.

Armstead, H.C.D. and Tester, J.W. (1987) *Heat Mining*, London: Spon.

Armstrong, J. (1985) *The Sizewell Report: A New Approach for Major Public Inquiries*, London: Town & Country Planning Association.

Atkinson, A., Davis, M. and Fergusson, M. (1988) *Nuclear Facilities and Emergency Planning in the United Kingdom*, A report to the National Steering Committee of the Nuclear Free Zone Local Authorities, London: Earth Resources Research Ltd.

Atomic Energy Control Act, RSC 1952, c11, Queen's Printer, Ottawa, amended by c47 1953–4.

Atomic Industrial Forum (1964) 'The Jersey Central Report', *Forum Memo* March: 3–7.

Attewell, P.B. and Taylor, R.K. (1984) *Ground Movements and Their Effects on Structures*, Guildford: University of Surrey Press (distributed by Chapman and Hall).

Atwood, G. (1975) 'The strip mining of western coal', *Scientific American* 233 (6): 23–9.

ATW Report. (1989) 'Die kraftwerke in Europa 1989', *Atomwirtschaft Atomtechnik* 6: 287–302.

Avery, D.G. and Kehoe, R.B. (1970) 'The uranium enrichment industry', *Journal of the British Nuclear Energy Society* 9: 163–72.

Bachert, R.P. (1973) 'Fuel financing', *Transactions of the American Nuclear Society* 16: 165–6.

Bailey, H.S., Evatt, R.N., Gyorey, G.L. and Ruiz, C.P. (1973) 'Neutron shielding problems in the shipping of high burnup thermal reactor fuel', *Nuclear Technology* 17: 217–24.

Bailey, R. (1982) 'Impact of the Euro-Soviet gas pipeline', *National Westminster Bank Review* August: 16–27.

Bainbridge, G.R. (1972) 'A future in nuclear power: the costs', *Journal of the British Nuclear Energy Society* 11 (3): 259–62.

Barry, P.J. (1965) 'A measurement of the occurrence of minimum dilution rates for stack effluents in the atmosphere', *Proceedings of the 3rd International Conference on the Peaceful Uses of Atomic Energy, Geneva, 1964*, vol. 14, Geneva: United Nations Organization, pp. 22–5.

Barry, P.J. (1970) 'The siting and safety of civilian nuclear power plants', *Critical Reviews in Environmental Control* 1 (2): 193–220.

Battey, G.C. and Hardy, C.J. (1981) 'Uranium resources, exploration and production in Australia', *Transactions of the American Nuclear Society* 36: 6–14.

Beale, H. (1982) 'Storage of high-level radioactive waste', *Nuclear Energy* 21 (4): 245–52.

Beaton, L. (1966) *Must the Bomb Spread?*, Harmondsworth: Penguin.

Beattie, J.R. and Bryant, P. (1970) 'Assessment of environmental hazards from reactor fission product releases', Report AHSB(S) R135, UKAEA.

Beavis, S. (1987) 'Where Britain's nuclear dustbin radiates quiet optimism', *Guardian*, 17 September.

Beck, C.K. (1967) Statement to the Joint Committee on Atomic Energy, 5 April, in *Licensing and Regulation of Nuclear Reactors*, 90th USA Congress, First Session, p. 68.

Beir, V. (1990) *Health Effects of Exposure to Low Levels of Ionizing Radiation*, Washington DC: National Academy Press.

Bell, G.D. and Charlesworth, F.R. (1963) 'The evaluation of power reactor sites', *Siting of Reactors and Nuclear Research Centres, Conference Proceedings*, Vienna: IAEA, pp. 317–29.

Beninson, D. and Lindell, B. (1985) 'Bases and trends in radiation protection policy', in *Interface Questions in Nuclear Health and Safety, Proceedings of NEA Seminar, 16–18 April 1985*, Paris: OECD, pp. 18–31.

Berrie, T.W. (1983) *Power System Economics*, Stevenage: Peter Peregrinus.

Berry, D.J. (1966) 'The Dungeness B decision – some economic and technical factors affecting reactor choice', *Atom*, 112: 37–43.

Bezdek, R.H. and Cone, B.W. (1980) 'Federal incentives for energy development', *Energy* 5 (5): 389–406.

Bhabha, H.J. and Dyal, M. (1965) 'World energy requirements and the economics of nuclear power with special reference to underdeveloped countries', *Proceedings of the 3rd International Conference on the Peaceful Uses of Atomic Energy, Geneva, 1964*, vol. 1, Geneva: United Nations Organization, pp. 41–52.

Birkhofer, A., Denton, H.R., Sato, K. and Tanguy, P.Y. (1985) 'The approaches to reactor safety and their rationale in OECD/NEA, in *Interface Questions in Nuclear Health and Safety, Proceedings of NEA Seminar, 16–18 April, 1985* Paris: OECD.

Black Report (1984) *Department of Health and Social Security Investigation of the Possible Increased Incidence of Cancer in West Cumbria, Report of the Independent Advisory Group*, Chairman, Sir Douglas Black, London: HMSO.

Blanchard, R.L., Fowler, T.W., Horton, T.R. and Smith, J.M. (1982) 'Potential health effects of radioactive emissions from active surface and underground uranium mines', *Nuclear Safety* 23 (4): 439–50.

Blank, J. (1966) 'Europe switches on to atomic power', *Newsletter* 63: 7, CEGB.

Blokhintsev, D.I. and Nikolayev, N.A. (1955) 'Atomic energy in the USSR', *1st International Conference on the Peaceful Uses of Atomic Energy, 1954*, Geneva: United Nations Organization.

Blomeke, J.O. and Harrington, F.E. (1968) 'Waste management at nuclear power stations', *Nuclear Safety* 9 (3): 239–47.

Blomeke, J.O. and Johnson, K.D.B. (1985) *The Management of High-Level Radioactive Waste: A Survey of Demonstration Activities*, Paris: NEA/OECD.

Blomeke, J.O. and Nichols, J.P. (1973) 'Commercial high level waste projections', Report ORNL-TM-4224, Oak Ridge National Laboratory.

Bonnell, J.A. (1983) 'Radiation risks and radiation protection', *Nuclear Energy* 22 (1): 33–6.

Bowie, S.H.U. (1974) 'Natural resources of nuclear fuel', *Transactions of the Royal Philosophical Society of London, Series A*, 276: 495–501.

Bowie, S.H.U. (1979) 'Theoretical and practical aspects of uranium geology', *Transactions of the Royal Philosophical Society of London, Series A*, 291: 255–420.

Boxer, L.W. (1973) 'A possibility for diversifying uranium demand', *Journal of the British Nuclear Energy Society* 12 (2): 153–7.

Boyd, F.C. (1974) 'Nuclear power in Canada: a different approach', *Energy Policy* June: 126–35.

Boyle, M.J. and Robinson, M.E. (1981) 'French nuclear energy policy', *Geography* 66 (4): 300–3.

Brayne, M. (1978) 'East Germany shuns caution in energy race', *Guardian*, 2 June.

Breach, I. (1978) *Windscale Fallout*, Harmondsworth: Penguin.

Brinck, J. (1967) 'Calculating the world's uranium resources', *Euratom* 6 (4): 109–14.

Brobst, W.A. (1973) 'Transportation accidents: how probable', *Nuclear News* 16 (7): 48–54.

Brodsky, A. (1987) *Radiation Measurement and Protection*, Boca Raton, FL: CRC Press.

Brown, G. (1984) 'Japan's nuclear power programme', *Atom* 335: 10–20.

Brown, G., Plant, J. and Simpson, P. (1975) *The Earth's Physical Resources*, Milton Keynes: Open University.

Brown, J.M. and White, H.M. (1987) 'The public's understanding of radiation and nuclear waste', *Journal of the Society for Radiological Protection*, 7 (2): 61–70.

Brown, L.C. (1967) 'Elliot Lake – the world's uranium capital', *Canadian Geographical Journal* 75 (4): 120–33.

Brown, M.J. and Crouch, E. (1982) 'Extreme scenarios for nuclear waste repositories', *Health Physics* 43 (3): 345–54.

Bryan, R.H., Nichols, B.L. and Ramsey, J.N. (1972) *Summary of Recent Legislative and Regulatory Activities Affecting the Environmental Quality of Nuclear*

Facilities, Springfield, VA: National Technical Information Service, US Department of Commerce.

Buchan, D. (1982) 'Czechoslovakia's ambitious nuclear plans', *Financial Times*, 8 December.

Budnitz, R.J. (1974) 'Radon-222 and its daughters, a review of instrumentation for occupational and environmental monitoring', *Health Physics* 26: 145–63.

Budnitz, R.J. (1980) 'How reactor safety is assured in the United States', in R. Wilson (ed.), *Energy for the Year 2000*, New York: Plenum.

Bunyard, P. (1988) 'The myth of France's cheap nuclear electricity', *Ecologist* 18 (1): 4–8.

Burda, T.J., Mazzola, C.A., Van Helvoirt, G.T. and Lyons, W.A. (1982) 'Simple lake breeze front position technique for offsite dose assessment', *Transactions of the American Nuclear Society* 43: 83–4.

Burkett, J. (1959) 'Nuclear energy developments in Asia', *Nuclear Energy Engineer* July: 365–7.

Burn, D. (1967) *The Political Economy of Nuclear Energy*, London: Institute of Economic Affairs.

Burns and Roe Inc. (1966) *Nuclear Power Plant Siting*, presented at Bonneville Power Administration Offices, Portland, OR, 21 June.

Burton, I., Kates, R.W. and White, G.F. (1978) *The Environment as Hazard*, Oxford: Oxford University Press.

Burton, I., Whyte, A., Hohenemser, C., Kates, R.W. and Guy, K. (1977) 'Risks and rare events: a three-nation study of disaster prevention in nuclear energy programs', Final Report for Ford Foundation.

Burwell, C.C. (1981) 'An existing site policy for the US nuclear energy system', *Nuclear Safety* 22, (2): 156–62.

Buschschluter, S. (1976) 'Leaks at the biggest nuclear plant', *Guardian*, 9 August.

California Public Utilities Commission (1962) Decisions 60 Cal.PUC 335–385.

Cannell, W. and Chudleigh, R. (1983) *The PWR Decision: How Not to Buy a Nuclear Reactor*, London: Friends of the Earth.

Carter, F.W. (1986) 'Nuclear power production in Czechoslovakia, *Geography* 71 (2): 136–9.

Carter, L.J. (1987a) *Nuclear Imperatives and Public Trust: Dealing with Radioactive Waste*, Washington, DC: Resources for the Future Inc.

Carter, L.J. (1987b) 'U.S. nuclear waste programme at an impasse', *Resources for the Future* No. 88: 1–4.

Carter, M.W., Moghissi, A.A. and Kahn, B. (1979) *Management of Low-level Radioactive Wastes*, vols 1, 2, Oxford: Pergamon.

Carter, T.J. (1982) 'Radioactive waste management practices at a large Canadian electrical utility', *Radioactive Waste Management* 2 (4): 381–412.

Catholic Institute for International Relations (1983) *A Future for Namibia: Mines and Independence*, London: CIIR.

Caufield, C. (1989) *Multiple Exposures: Chronicles of the Radiation Age*, London: Secker & Warburg.

Cawse, P.A. (1980) 'Caesium-137 and plutonium in soils in Cumbria and the Isle of Man', Studies of Environmental Radioactivity in Cumbria, part 4, Report AERE-R9851, UKAEA.

Cawse, P.A. (1988) 'Environmental radioactivity in Caithness and Sutherland, Part 1, Radionuclides in soils, peat and crops in 1979', *Nuclear Energy* 27 (3): 193–213.

Cawse, P.A. (1988) 'Environmental radioactivity in Caithness and Sutherland, Part 2, Radionuclides in arable soils and crops in 1980', *Nuclear Energy* 27 (5): 311–20.

CEGB (Central Electricity Generating Board) (1982) *Sizewell B Statement of Case*, vols 1–5, London: CEGB.

Chapman, J.D. (1989) *Geography and Energy: Commercial Energy Systems and National Policies*, Harlow: Longman Scientific and Technical.

Chapman, N.A. and McKinley, I.G. with Hill, M. (1987) *The Geological Disposal of Nuclear Waste*, Chichester: Wiley.

Chapman, N.A., Black, J.H., Bath, A.H., Hooker, P.J. and McEwan, T.J. (1987) 'Site selection and characterisation for deep radioactive waste repositories in Britain: issues and research trends into the 1990s', in *Radioactive Waste Management and the Nuclear Fuel Cycle*, UK National Issue, Chichester: Wiley.

Chappell, H. (1983) 'Sizewell and the people: what the community thinks', *New Society* 63 (1051): 7–10.

Charlesworth, F.R. and Griffiths, T. (1962) 'Licensing and inspection of nuclear installations in the UK', in *Reactor Safety and Hazards Evaluation Techniques*, vol. 2, Vienna: IAEA.

Charlesworth, F.R. and Gronow, W.S. (1967) 'A summary of experience in the practical application of siting policy in the United Kingdom', *Symposium on the Containment and Siting of Nuclear Power Plants*, Vienna: IAEA, pp. 143–70.

Chayes, A. and Lewis, W.B. (1977) *International Arrangements for Nuclear Fuel Reprocessing*, Cambridge, MA: Ballinger.

Chern, W.S. and Just, R.E. (1980) 'Regional analysis of electricity demand growth', *Energy* 5: 35–46.

Chicken, J.C. (1982) *Nuclear Power Hazard Control*, Oxford: Pergamon.

Chometon, P.L. and Cantin, P. (1983) 'Centralized disposal of spent fuel elements in France', *Nuclear Europe* 3 (2): 12–14.

CIA (Central Intelligence Agency) (1985) *USSR Energy Atlas*, Washington, DC: CIA.

Clark, G. and Addington-Lee, F. (1989) 'Enrichment overcapacity leaves buyers in the driving seat', *Nuclear Engineering International* 34 (419): 42–4.

Clarke, R.H. (1985) 'Radiological protection aspects of exemption levels in the nuclear fuel cycle', in *Interface Questions in Nuclear Health and Safety, Proceedings of NEA Seminar, 16–18 April 1985*, Paris: OECD, pp. 234–45.

Clarke, R.H. and Southwood, T.R.E. (1989) 'Risks from ionizing radiation', *Nature, London* 338 (6212): 197–8.

Clarke, R.H. and Utting, R.E. (1970) 'Radiation hazards', Report RD/B/N1762, CEGB.

Clarke, R.W. and Thompson, I.M.G. (1978) 'Radiation dosimetry and calibration-BNL sets the standards;, *CEGB Research* May: 22–31.

Clausen, A.W. (1983) 'Third world debt and global recovery', *Energy World*, May (103): 7–14.

CMND 884 (1959) *The Disposal of Radioactive Waste*, White Paper presented to Parliament and given effect in Radioactive Substances Act (1960), 8 & 9 Eliz. 2, Ch. 34, London: HMSO.

CMND 6618 (1977) *Nuclear Power and the Environment: The Government's Response to the Sixth Report of the Royal Commission on Environmental Pollution*, House of Commons Sessional Papers, London: HMSO, May.

Cochran, T.B. (1974) *The Liquid Metal Fast Breeder Reactor: An Environmental and Economic Critique*, Washington DC: Resources for the Future Inc.

Cockburn, F. (1986) 'Soviet radiation hot spots found', *Financial Times*, 5 June.

Coffin, B. (1984) *Nuclear Power Plants in the United States: Current Status and Statistical History*, New York: Union of Concerned Scientists.

Cohen, B.L. (1977) 'The disposal of radioactive wastes from fission reactors', in *Scientific American, Energy and Environment*, San Francisco, CA: W.H. Freeman.

Cohen, B.L. (1980) 'The cancer risk from low-level radiation', *Health Physics* 39: 659–78.

Cohen, B.L. (1982a) 'Radon daughter exposure to uranium miners', *Health Physics* 42 (4): 449–57.

Cohen, B.L. (1982b) 'Health effects of radon emissions from uranium mill tailings', *Health Physics* 42 (5): 695–702.

Cohen, B.L. (1983) *Before It's Too Late: A Scientist's Case for Nuclear Energy,* New York: Plenum.

Cole, H.A. (1988) *Understanding Nuclear Power: A Technical Guide to the Industry and its Processes,* Aldershot: Gower Technical.

Cole, J.P. (1984) *Geography of the Soviet Union,* London: Butterworths.

Colglazier, E.W. (1982) *The Politics of Nuclear Waste,* Oxford: Pergamon.

Collier, J.G. (1979) Calculating the unthinkable, *Guardian,* 8 November.

Collins, J.C. (1961) 'Environmental dispersion of radioactive waste', *Nuclear Energy* August 322–9.

Conference on Nuclear Power Plant Siting, Portland, OR, 25–28 August 1974, *Transactions of the American Nuclear Society* 19 (suppl. 1): 1–53.

Conrad, J. (1980) *Society, Technology and Risk Assessment,* London: Academic Press.

Cook, J. (1986) *Red Alert: The Worldwide Dangers of Nuclear Power,* London: New English Library.

Cooper, J.R., McColl, N.P. and Hill, M.D. (1988) 'The radiological impact of the UK nuclear fuel cycle', *Nuclear Energy* 27 (6): 377–84.

Corbett, J. (1987) 'The safety of nuclear power in a risky world', *Atom* 368: 20–3.

Cottrell, Sir Alan (1985) *Public Attitudes to Energy,* London: CEGB.

Covello, V.T., Flamm, W.G. Rodricks, J.V. and Tardiff, R.G. (1983) *The Analysis of Actual Versus Perceived Risks,* New York: Plenum.

Crease, J. (1959) 'Ocean currents', *New Scientist,* 10 September 402–4.

Crofford, W.N. (1980) 'Decommissioning standards', in M.M. Osterhout (ed.), *Decontamination and Decommissioning of Nuclear Facilities,* New York: Plenum, pp. 17–23.

Crosbie, W.A. and Gittus, J.H. (1989) *Conference on Medical Response to Effects of Ionising Radiation, 28–30 June 1989,* Paper from Mr J. Evans, Harwell: UKAEA.

Curie, E. (1938) *Madame Curie,* London: Heinemann.

Curtis, R. and Hogan, E. (1980) *Nuclear Lessons: an Examination of Nuclear Power's Safety, Economic and Political Record,* Wellingborough: Turnstone Press.

Dancy, D. (1986) 'Thermal reactors in perspective', *Atom* 362: 10–13.

Davidson, C. (1955) 'The raw materials of atomic power', *Discovery* 16 (6): 234–6.

Davis, J.P. (1974) 'The new federal water pollution control act and its impact on nuclear power plants, Part 1: Application of the FWPCA and related legislation to individual discharges through permit programmes', *Nuclear Safety* 15 (3): 263–74.

Davis, W.K. and Robb, J.E. (1967) 'Nuclear plant siting in the USA', in *Containment and Siting of Nuclear Power Plants, Proceedings of a Symposium,* Vienna: IAEA, pp. 3–15.

Dawes, F.A. and Mathies, J.D. (1971) 'A new approach to the construction and siting of nuclear power stations', *Journal of the Institute of Nuclear Engineers* 12: 129–31.

Deese, D.A. (1978) *Nuclear Power and Radioactive Waste: a Sub-Seabed Disposal Option,* Lexington, MA: D.C. Heath.

Del Sesto, S.L. (1979) *Science, Politics and Controversy: Civilian Nuclear Power in the United States 1946–74,* Boulder, Co: Westview.

Department of Energy, Mines and Resources (1983) *Uranium in Canada, Report EP 8333,* Ottawa: Government Printer.

Department of the Environment (1982) *Programme of Research into the Disposal of Radioactive Waste into Geological Formations*, Harwell: Department of the Environment.

Department of the Environment (1984) *An Incident leading to Contamination of the Beaches near to British Nuclear Fuels Ltd, Windscale and Calder Works, Sellafield, November 1983*, London: Radiochemical Inspectorate.

Department of the Environment (1986) *The Assessment of Best Practicable Environmental Option (BPEOS) for the Management of Low and Intermediate Level Solid Radioactive Wastes*, London: HMSO.

Department of the Environment (1987) *Radioactive Waste Management and Radioactivity in the Environment*, Report of Research Commissioned by the Department of the Environment 1984–6, London: HMSO.

De Vergie, P.C. (1983) 'Some aspects of US uranium production costs', in *Economics of Uranium Ore Processing Operations, Proceedings of a Workshop, 25–26 April*, Paris and Vienna: NEA/OECD with IAEA, pp. 155–76.

Dienes, L. and Shabad, T. (1979) *The Soviet Energy System, Resource Use and Policies*, Washington, DC: Wiley.

Doan, P.L. (1970) 'Tornado considerations for nuclear power plant structures', *Nuclear Safety* 11 (4): 296–306.

Döderlein, J.M. (1976) 'Nuclear power, public interest and the professional', *Nature, London* 264: 202–3.

Dodsworth, T. (1984) 'US nuclear plant problems pile up', *Financial Times*, 25 April.

Doll, Sir Richard (1989) 'The epidemiology of childhood leukaemia', *Journal of the Royal Statistical Society A* 152: 341.

Dozel, M., Krischer, W., Pottier, P. and Simon, R. (1984) *Leaching of Low and Medium Level Waste Packages under Disposal Conditions*, London: Graham and Trotman for the Commission of the European Communities.

Draper, E.L. and Stanford, R.E.L. (1982) 'US utilities concerned about waste storage', *Nuclear Engineering International*, 27 (331): 15–19.

Dunster, H.J. (1959) 'The disposal of radioactive wastes into coastal waters', *Nuclear Energy Engineer* 13 (138): 540–3; 13 (139): 588–92.

Dunster, H.J. (1970) 'The avoidance of pollution in the disposal of radioactive wastes', *Atom* 162: 65–70.

Dunster, H.J. (1980) 'Some reactions to the accident at Three Mile Island', *Nuclear Energy* 19 (3): 139–46.

Dunster, H.J. (1981) 'The assessment of the risks of energy; an iconoclastic view', *Energy World* No. 88: 2–8.

Dunster, H.J. (1985) 'The disposal of radioactive waste from a nuclear power programme', *Nuclear Engineering International* 19 (220): 744–6.

Eaker, L.H. (1983) 'International legal and political considerations concerning the seabed disposal of nuclear waste', *Nuclear Law Bulletin* 31: 40–63.

Economist (1959) 'Can nuclear power compete?', 9 May: 543–5.

Edinger, V. (1964) *The Hanford Story*, Washington Public Power Supply System (mimeograph).

Edison Electric Institute (1967) *Statistical Yearbook*, New York: Edison Electric Institute.

Edmonds, J. and Reilly, J. (1983) 'Global energy production and use to the year 2050', *Energy* 8 (6): 419–32.

Eggington, J. (1970) 'Nuclear dump sitting on an earthquake', *Observer*, 15 March.

Eggleston, W. (1965) *Canada's Nuclear Story*, Toronto: Clarke, Irwin.

Eichholz, G.G. (1982) 'Small populations at risk and the linear hypothesis', *Health Physics* 42 (6): 857–8.

Eisenbud, M. (1981) 'The status of radioactive waste management: needs for reassessment', *Health Physics* 40: 429–37.

Electricité de France (1982) *Annual Report*, Paris: Electricité de France.

Electricity Consumers Council (1982) *Nuclear Power and the Economic Interests of Consumers*, London: Consumers Council.

Emmings, A. (1989) 'The next step for radioactive waste management', *Atom* 391: 6–8.

Energy Commission (1978) *Coal and Nuclear Power Station Costs*, London: Energy Commission, Paper No. 6.

Energy Policy Staff, Office of Science and Technology (1968) *Considerations Affecting Steam Power Plant Site Selection*, Washington DC: US Government Printing Office.

Environmental Risk Assessment Unit, School of Environmental Sciences, University of East Anglia (1988) *Responses to the Way Forward*, Harwell: NIREX.

Ergen, W.K. (1969) 'German practices with respect to reactor siting', *Nuclear Safety* 10 (5): 377–9.

Ergen, W.K. (1971) 'Consistency in siting practices regarding nuclear plants', in P.G. Voilleque (ed.) *Health Physics Aspects of Nuclear Facility Siting*, vol. i, Idaho Falls, ID: Burton R. Baldwin Publications, pp. 1–9.

Ester, P., Mindell, C., Van Der Linden, J. and Van Der Pligt, J. (1983) 'The influence of living near a nuclear power plant on beliefs about nuclear energy', *Zeitschrift fur Umweltpolitik* 6: 349–62.

Etemad, S. (1979) 'Human errors and accident procedures', *Guardian*, 8 November.

Evans, N. and Hope, C. (1984) *Nuclear Power: Futures, Costs and Benefits*, Cambridge: Cambridge University Press.

Evison, E.F. (1984) *Earthquake Prediction, Proceedings of the International Symposium on Earthquake Prediction*, 2–6 April, Paris: Terrapublications.

Fabrikant, J.I. (1981) 'Health effects of the nuclear accident at Three Mile Island', *Health Physics* 40: 151–61.

Falls, O.B. (1973) 'A survey of the market for nuclear power in developing countries', *Energy Policy* 1 (3): 225–42; 'A survey of nuclear power in developing countries', *IAEA Bulletin* 15 (5): 26–38.

Farmer, F.R. (1962) 'The evaluation of power reactor sites', Report DPR/INF 266, UKAEA.

Farmer, F.R. (1965) 'Reactor safety analysis as related to reactor siting', *Proceedings of the 3rd International Conference on the Peaceful Uses of Atomic Energy, Geneva, 1964*, vol. 13, Geneva: United Nations Organization, pp. 405–9.

Farmer, F.R. (1967a) 'Reactor safety and siting – a proposed risk criterion', *Nuclear Safety* 8 (6): 539–48.

Farmer, F.R. (1967b) 'Siting criteria – a new approach', *Symposium on the Containment and Siting of Nuclear Power Plants*, Vienna: IAEA, pp. 303–18.

Farmer, F.R. (1972) 'The environment and atomic energy in the United Kingdom', *Atom* 186: 60–70.

Fath, H.E.S. and Hashem, H.H. (1988) 'The nuclear energy alternative in Arab countries', *Energy* 13 (12): 871–82.

Faux, E. and Stone, G.N. (1963) 'Experience in planning and siting nuclear power stations for the Central Electricity Generating Board', *IAEA Symposium on Criteria for Guidance in the Selection of Sites for the Construction of Reactors and Nuclear Research Centres*, SM 39/36, Bombay: IAEA.

Feldman, S.L., Bernstein, M.A. and Noland, R.B. (1988) 'The prospects of completing unfinished US nuclear power plants', *Energy Policy* 16 (3): 270–9.

Fells, I. (1983) 'The options until 2030', *Energy World* no. 89: 15–21.

Fells, I. (1989) 'Energy and the environment', *Energy World* no. 168: 7–10.

Fimarwsr, O. (1982) 'Sweden's road to a firm nuclear programme', *Nuclear Europe*

1: 9–12.
Firebaugh, M.W. (1981) 'Public attitudes and information on the nuclear option', *Nuclear Safety* 22 (2): 147–55.
Fischer, D.W. (1981) 'Planning for large-scale accidents: learning from the Three Mile Island accident', *Energy* 6: 93–108.
Fishlock, D. (1982) 'Where nuclear power stations break down', *Financial Times*, 29 September: 11.
Fishlock, D. (1986) 'Counting the real cost of nuclear energy', *Financial Times*, 8 September.
Flood, M. and Grove-White, R. (1976) *Nuclear Prospects – A Comment on the Individual, the State and Nuclear Power*, London: Friends of the Earth in association with the Council for the Protection of Rural England and the National Council for Civil Liberties.
Flowers Commission (1976) *Royal Commission on Environmental Pollution, Sixth Report, Nuclear Power and the Environment*, Chairman, Sir Brian Flowers, Cmnd 6618, London: HMSO.
Ford, S.B. (1981) 'Uranium exploration and production, a review of the US situation', *Transactions of the American Nuclear Society* 36: 21–8.
Fortune Magazine (1967) 'Nuclear power goes critical', March: 117–24.
Fothergill, S., Gudgin, G. and Mason, N. (1983) 'The economic consequences of the Sizewell B nuclear power station', Department of Applied Economics, University of Cambridge.
Fox, Mr Justice R.W., Presiding Commissioner (1976, 1977) *Ranger Uranium Environmental Inquiry, First and Second Reports*, Canberra.
Fremlin, J.H. (1985) *Power Production: What are the Risks?*, Bristol: Adam Hilger.
Fremlin, J.H. (1986) 'The risks of electricity production', Proceedings of a Conference on Radioactive Waste Management: UK Policy Examined, London, 24–25 April 1986, London: IBC Technical Services, pp. 194–203.
Fremlin, J.H. and Wilson, C.K. (1982) 'Radiation doses to the public from the nuclear industry', *Progress in Nuclear Energy* 10 (2): 221–41.
Froment, R. and Lerat, S. (1989) *La France a L'Aube des Annees 90*, vol. 1, Montreuil: Breal, pp. 225–50.
Fryer, D.R.H. (1969) *Siting Policy and Safety Evaluation in the United Kingdom, British Nuclear Energy Society Symposium on Safety and Siting, 28 March 1969*, London: Institute of Civil Engineers, pp. 23–35.
Gagarinskii, A. Yu (1989) 'Chernobyl today, state of research', *Nuclear Safety* 30 (1): 18–22.
Gammon, K.M. (1979) 'CEGB experience in selecting and developing nuclear power station sites', *Newsletter* III, CEGB.
Gammon, K.M. and Pedgrift, G.F. (1983) 'Changes in the investigation and selection of sites for nuclear power stations', *Nuclear Engineering* 22 (1): 41–5.
Gardner, M.J., Snee, M.P., Hall, A.J., Powell, C.A., Downes, S. and Terrell, J.D. (1990a) 'Results of case-control study of leukaemia and lymphoma among young people near Sellafield nuclear plant in West Cumbria', *British Medical Journal* 300 (17 February): 423–28.
Gardner, M.J., Snee, M.P., Hall, A.J., Powell, C.A., Downes, S. and Terrell, J.D. (1990b) 'Methods and basic data of case-control study of leukaemia and lymphoma among young people near Sellafield nuclear plant in West Cumbria', *British Medical Journal* 300 (17 February): 429–33.
Gast, P.F. (1973) 'Divergent public attitudes toward nuclear and hydroelectric plant safety', *Transactions of the American Nuclear Society* 16: 40–1.
Gauvenet, A. (1986) 'Radiation and chemicals: a risk comparison', *Nuclear Europe* 6 (12): 6–9.
George, K.D. (1959) 'The economics of nuclear power', M.A. Thesis, Department

of Economics, University College of Wales, Aberystwyth.

Gerende, L.J., Taylor, F.H., Sumpter, E.H., Schwarzback, R.H. and Rogers, K.D. (1974) 'Infant and neonatal mortality rates during pre- and post-reactor periods for geographic areas adjacent to Shippingport, Pennsylvania', *Health Physics* 26: 431–8.

Gifford, F.A. (1972) 'Atmospheric transport and dispersion over cities', *Nuclear Safety* 13 (5): 391–401.

Gillette, R. (1974) 'Nuclear safety: calculating the odds of disaster', *Science* 185 (4154): 838.

Gillette, R. (1978) 'Russia's quest for nuclear power', *Guardian*, 17 October.

Ginier, J. (1965) 'L'énergie nucléaire en France,' *Information Géographique* 29 (1): 9–20.

Gittus, J.H. (1987) 'The Chernobyl accident and its consequences', *Atom* 368: 29–9.

Gittus, J.H., Hicks, D., Bonell, P.G., Clough, P.N., Dunbar, I.H., Egan, M.J., Hall, A.N., Nixon, W., Bulloch, R.S., Luckhurst, D.P. and Maccabee, A.R. (1987) 'The Chernobyl accident and its consequences', Report NOR 4200, UKAEA.

Glasstone, S. (1967) *Sourcebook on Atomic Energy*, New York: Van Nostrand Reinhold, 3rd edn.

Gofman, J.W. and Tamplin, A.R. (1970) 'The radiation effects controversy', *Bulletin of Atomic Scientists* 26 (2): 8–12.

Gofman, J.W. and Tamplin, A.R. (1971) *Poisoned Power: The Case Against Nuclear Power Plants*, Emmaus, PA: Rodale Press.

Golding, D. and Kasperson, R.E. (1988) 'Emergency planning and nuclear power, *Land Use Policy* 5 (1): 19–36.

González, A.J. (1983) 'The basic safety standards for radiation protection', *International Atomic Energy Authority Bulletin* 25 (3): 19–25.

González, A.J. and Webb, G.A.M. (1988) 'Bridging the gap between radiation protection and safety: the control of probabilistic exposures', *International Atomic Energy Bulletin* 30 (3): 35–41.

Goodman, G.T. and Rowe, W.D. (1980) *Energy Risk Management*, London: Academic Press.

Gowing, M. (1964) *Britain and Atomic Energy 1939–1945* (vol. 1 of the official history of British atomic energy), London: Macmillan.

Gowing, M. (1974) *Independence and Deterrence: Britain and Atomic Energy 1945–52* (vol. 2 of the official history of British atomic energy), London: Macmillan.

Gowing, M. (1978) *Reflections on Atomic Energy History*, The Rede Lecture, Cambridge: Cambridge University Press.

Graeub, R. (1972) *The Gentle Killers: Nuclear Power Stations* (transl. Peter Bostock), London: Abelard-Schuman.

Green, H.P. (1974) 'Internalizing the costs associated with catastrophic accidents in energy systems', draft report to the Ford Foundation, Energy Policy Project (mimeograph).

Green, P. and Daly, P. (1988) *Fallout over Chernobyl: A Review of the Official Radiation Monitoring Programme in the UK*, London: Friends of the Earth.

Greenwood, T., Rathjens, G.W. and Ruina, J. (1976) *Nuclear Power and Weapons Proliferation*, Adelphi Papers no. 130, London: International Institute for Strategic Studies.

Gregory, A.R. (1984) 'Plans for nuclear power station decommissioning', in *Decommissioning of Radioactive Facilities*, papers presented at a seminar organized by the Nuclear Energy Committee of the Power Industries Division of the Institution of Mechanical Engineers, 7 November 1984, London: Institution of Mechanical Engineers.

Gregory, G. (1983) 'Energy for Japan's new industrial frontier', *Energy* 8 (6): 481–90.

Griffiths, T. and Gausden, R. (1965) 'Development of licensing and inspection of nuclear installations in the UK', *Proceedings of the 3rd International Conference on the Peaceful Uses of Atomic Energy, Geneva 1964*, vol. 13, Geneva: United Nations Organization, pp. 334–40.

Grimes, B.K., Ramos, S.L. and Weiss, B.H. (1982) 'Emergency planning and preparedness since Three Mile Island', *Progress in Nuclear Energy* 10: 363–86.

Groueff, S. (1967) *Manhattan Project*, Boston, MA: Little, Brown.

Guerou, J., Lane, J.A., Maxwell, I.R. and Menke, J.R. (1986) *The Economics of Nuclear Power*, vol. 1, Oxford: Pergamon.

Häfele, W. (1977) 'Energy options open to mankind beyond the turn of the century', *Proceedings of the International Conference on Nuclear Power and its Fuel Cycle*, vol. 1, Vienna: IAEA, pp. 58–81.

Häfele, W. (1981) *Energy in a Finite World: a Global Systems Analysis*, 2 vols, Cambridge, MA: Ballinger.

Haire, T.P. and Shaw, J. (1979) 'Nuclear power plant licensing procedures in the United Kingdom', *Progress in Nuclear Energy* 4: 161–82.

Hake, G. (1969) 'Comparison of Canadian and US siting policies', *Nuclear Safety* 10: 365–72.

Hake, G. and Palmer, J.F. (1970) 'Power reactor siting in various countries', Report AECL 3740, Atomic Energy of Canada Ltd.

Hall, Tony (1986) *Nuclear Politics: the History of Nuclear Power in Britain*, Harmondsworth: Penguin.

Harris, D.J. and Davies, B.C.L. (1980) 'European energy policy and planning: the role of the institutions', *National Westminster Bank Quarterly Review* November: 23–33.

Hart, D. (1983) *Nuclear Power in India: A Comparative Analysis*, London: Allen & Unwin.

Hart, J.C. (1973) Consequences of effluent release, *Nuclear Safety* 14: (5): 482–506.

Hart, J.C. (1973) Management of radioactive aqueous wastes from AEC fuel reprocessing operations, *Nuclear Safety* 14 (5): 482–506.

Harvey, H. and Newland, E.V. (1969) 'Energy patterns to the year 2000', in *Technological Forecasting*, Edinburgh: Edinburgh University Press.

Hasson, J. (1965) *The Economics of Nuclear Power*, London: Longman.

Hawkes, N., Lean, G., Leigh, D., McKie, R., Pringle, P. and Wilson, A. (1986) *The Worst Accident in the World*, London: Pan and Heinemann.

Hayes, D. (1976) *Nuclear Power: the Fifth Horseman*, Washington, DC: Worldwatch Institute, Worldwatch Paper 6.

Hayes, D. (1978) 'Energy in the developing world', *Dialogue* 11 (3): 26–34.

Haywood, L.R. (1967) 'Evolution of nuclear power in Canada', Address to the Congress of Canadian Engineers, Montreal, 29 May 1967, Report AECL 2886, Atomic Energy of Canada Ltd.

Heafield, W. and Barlow, P. (1988) 'Management of wastes from the nuclear fuel cycle', *Nuclear Energy* 27 (6): 367–76.

Health and Safety Executive (1981 onwards) *Quarterly Statement on Nuclear Incidents*, London: Health and Safety Executive.

Health and Safety Executive (1982) *Emergency Plans for Civil Nuclear Installations*, London: Directorate of Information and Advisory Services, Health and Safety Executive.

Health and Safety Executive (1986) *The Leakage of Radioactive Liquor into the Ground, British Nuclear Fuels Ltd. Windscale, 15 March 1978*, London: Health and Safety Executive.

Healy, J.W. (1982) 'The ICRP dose limitation system – solution or problem?', *Health Physics* 42 (4): 407–13.

Hebel, W., Bruggman, A., Donato, A., Furrer, J. and White, I.F. (1986) 'Management of nuclear airborne wastes and tritium retention', in R. Simon (ed.), *Radioactive Waste Management and Disposal, Proceedings of the Second European Community Conference, Luxembourg, 22–26 April 1985*, Cambridge: Cambridge University Press for the Commission of the European Communities.

Hedley, D. (1986) *World Energy: The Facts and the Future*, London: Euromonitor Press, 2nd edn.

Henderson, P.D. (1975) *India: The Energy Sector*, Oxford: Oxford University Press for the World Bank.

Hill, Sir John (1974a) 'The nuclear fuel industry', *Atom* 207: 2–35.

Hill, Sir John (1974b) 'Future trends in nuclear power generation', *Transactions of the Royal Philosophical Society, Series A* 276: 181–93.

Hill, M.D. (1987) 'Radiological safety assessments', ch. 10 in N.A. Chapman and I.G. McKinley, *The Geological Disposal of Nuclear Waste*, Chichester: Wiley, pp. 192–9.

Hill, M.D. and White, I.F. (1982) 'Radiological impact of disposal of UK solid radioactive wastes, present and future', *Nuclear Energy* 21 (4): 225–33.

Hill, M.D., Wrixon, A.D. and Webb, G.A.M. (1988) 'Protection of the public and workers in the event of accidental releases of radioactive materials into the environment', UK National Radiological Protection Board (mimeograph).

Hinton, C., Brown, F.H.S. and Rotherham, L. (1960) 'The economics of nuclear power in Great Britain', in *World Power Conference Sectional Meeting, Madrid, 5–9 June 1960*, Madrid: Sucs. Rivadeneyra, pp. 1–24.

Hockley, G.J. (1983) 'Smokescreen hanging over CEGB coal figures', *Guardian*, 3 May.

Hoegberg, L. (1988) 'The Swedish nuclear safety programme', *Nuclear Safety* 29 (4): 421–35.

Hogerton, J.F. (1965) *Background Information on Atomic Power Safety*, New York: Atomic Industrial Forum.

Hogerton, J.F. (1968) 'The arrival of nuclear power', *Scientific American* 218 (2): 21–31.

Hohenemser, C., Kasperson, R. and Kates, R. (1977) 'The distrust of nuclear power', *Science*, 196: 25–34.

Holcombe, W.F. (1982) 'A history of ocean disposal of packaged low-level radioactive waste', *Nuclear Safety* 23 (2): 183–97.

Holmes, A. (1988) *Electricity in Europe: Opening the Market*, London: Financial Times.

House of Commons (1967) Select Committee on Science and Technology *United Kingdom Nuclear Reactor Programme*, London: HMSO.

House of Commons (1977) *Nuclear Power and the Environment: the Government's response to the Sixth Report of the Royal Commission on Environmental Pollution*, House of Commons Sessional Papers, May, London: HMSO.

House of Commons (1980) *Report of the House of Commons Select Committee on Science and Technology, Hazardous Waste Disposal*, London: HMSO.

House of Commons (1981) *First Report from the Select Committee on Energy*, vol. 1, London: HMSO.

House of Commons (1986) Environment Committee First Report, Session 1985–6, *Radioactive Waste*, vol. 1, London: HMSO.

House of Commons (1989) Energy Committee Sixth Report, *Energy Policy Implications of the Greenhouse Effect*, London: HMSO.

House of Lords (1986) Select Committee on the European Communities, Eighteenth Report 1985–6, *Nuclear Power in Europe*, HL 227–1, London: HMSO.

House of Lords (1988) Select Committee on the European Communities, Nineteenth Report Session 1987–8, *Radioactive Waste Management*, HL 99, London: HMSO.

Howarth, J.M. and Sandalls, F.J. (1987) 'Decontamination and reclamation of agricultural land following a nuclear accident', Report AERE 126666, UKAEA.

Hubert, P. (1983) 'The low dose controversy and radiological risk assessment', *Health Physics* 45 (1): 144–9.

Hueper, W.C. (1971) 'Public health hazards', *Health Physics* 21 (5): 689–707.

Huggard, A.J. (1987) 'The Japanese nuclear power programme', *Atom* 372: 2–6.

Huggett, H.L. (1982) 'Elements of nuclear plant capital costs: variables – trends – industry initiatives', *Transactions of the American Nuclear Society* 4, Suppl. 1: 16–17.

Hull, A.P. (1974) 'Comparing effluent releases from nuclear and fossil-fuelled power plants', *Nuclear News* 17 (5): 51–5.

Hunt, G.J., Hewett, C.J. and Shepherd, J.G. (1982) 'The identification of critical groups and its application to fish and shellfish consumers in the coastal areas of the north-east Irish Sea', *Health Physics* 43 (6): 875–89.

Iansiti, E. and Sennis, C. (1967) 'Nuclear plant siting from the point of view of engineering safety evaluation', *Containment and Siting of Nuclear Power Plants*, Vienna: IAEA, pp. 345–63.

IAEA (International Atomic Energy Agency) (1959 onwards) *Directory of Nuclear Reactors: Power and Research Reactors in Member States*, Vienna: IAEA, Cumulative.

IAEA (1961) *Regulations for the Safe Transport of Radioactive Materials*, Vienna: IAEA Safety Series No. 6 (Revised 1964, 1967, 1973, 1983).

IAEA (1963) *Radiological Health and Safety in Mining and Milling of Nuclear Materials, 26–31 August 1963*, 2 vols, Vienna: IAEA Proceedings Series.

IAEA (1965) *Radioactive Waste Disposal into the Ground*, Vienna: IAEA Safety Series 15.

IAEA (1967) *Containment and Siting of Nuclear Power Plants, Proceedings of a Symposium, Vienna, 3–7 April 1967*, Vienna: IAEA.

IAEA (1970a) *Management of Low and Intermediate Level Radioactive Wastes, Aix-en-Provence Symposium, 7–11 September 1970*, Vienna: IAEA Proceedings Series STI/PUB/264.

IAEA (1970b) *Economic Integration of Nuclear Power Stations in Electric Power System, Vienna Symposium 5–9 October 1970*, Vienna: IAEA Proceedings Series STI/PUB/266.

IAEA (1970c) *Nuclear Energy Costs and Economic Development, Proceedings of Istanbul Symposium, October 1969*, Unpublished.

IAEA (1971a), *Inhalation Risks from Radioactive Contaminants, Report of a Panel held in Vienna 30 November–4 December 1970*, Vienna: IAEA Technical Reports Series 142/STI/DOC 10/142.

IAEA (1971b) *The Recovery of Uranium, Sao Paulo Symposium, 17–21 August 1970*, Vienna: IAEA Proceedings Series STI/PUB/262.

IAEA (1972a) *Assessment of Radioactive Contamination in Man*, Vienna: IAEA, Proceedings Series STI/PUB/290.

IAEA (1972b) *Disposal of Radioactive Wastes into Rivers, Lakes and Estuaries*, Vienna: IAEA Safety Series No. 36.

IAEA (1972c) *Earthquake Guidelines for Reactor Siting*, Vienna: IAEA. Technical Reports Series No. 139.

IAEA (1973a) *Market Survey for Nuclear Power in Developing Countries, General Report*, Vienna: IAEA.

IAEA (1973b) *Radioactive Contamination of the Marine Environment, Seattle Symposium, 10–14 July 1973*, Vienna: IAEA Proceedings Series STI/PUB/313.

IAEA (1973c) *Principles and Standards of Reactor Safety, Proceedings of a Symposium Convened by the IAEA in Jülich, 5–9 February 1973*, Vienna: IAEA Proceedings Series STI/PUB/342.

IAEA (1975) *Radon in Uranium Mining: Proceedings of a Panel*, Vienna: IAEA.

IAEA (1977) *Regional Nuclear Fuel Cycle Centres, Report of the IAEA Study Project*, 2 vols, Vienna: IAEA.

IAEA (1978) *Safety in Nuclear Power Plant Siting*, Vienna: IAEA 50/C/S.

IAEA (1980) *Site Selection and Evaluation for Nuclear Power Plant with respect to Population Distribution*, Vienna: IAEA 50/SG/S4.

IAEA (1981) *Underground Disposal of Radioactive Wastes, Basic Guidance*, Vienna: IAEA Safety Series No. 54.

IAEA (1983a) *Proceedings of the International Conference on Nuclear Power Experience organized by the International Atomic Energy Agency, Vienna, 13–17 September 1982*, 6 vols, Vienna: IAEA.

IAEA (1983b) *Bulletin* 25: December.

IAEA (1984) *Management of Tritium at Nuclear Facilities*, Vienna: IAEA.

IBC Technical Services (1986) *Proceedings of a Conference on Radioactive Waste Management: UK Policy Examined, London, 24–25 April 1986*, London: IBC.

IBC Technical Services (1987) *Nuclear Risks: Re-assessing the Principles and Practice after Chernobyl, Papers of an International Conference on Nuclear Risks, London, 1–2 December 1986*, London: IBC.

ICRP (International Commission on Radiological Protection) (1960) *Permissible Dose for Internal Radiation*, Oxford: Pergamon, ICRP Publication 2.

ICRP (1966) *The Evaluation of Risks from Radiation*, Oxford: Pergamon, ICRP Publication 8.

ICRP (1966) *Recommendations of the ICRP*, Oxford: Pergamon, ICRP Publication 9.

ICRP (1968) *Evaluation of Radiation Doses to Body Tissues from Internal Contamination due to Occupational Exposure, Report of Committee 4*, Oxford: Pergamon, ICRP Publication 10.

ICRP (1977) *Recommendations of the International Commission*, Oxford: Pergamon, ICRP Publication 26.

ICRP (1984) *Principles Limiting the Exposure of the Public to Natural Radiation* Oxford: Pergamon.

IEA/OECD (International Energy Agency/Nuclear Energy Agency, Organization for Economic Co-operation and Development) (1982) *Nuclear Energy Prospects to 2000*, Paris: OECD.

IEA/OECD (1985) *Electricity in IEA Countries*, Paris: OECD.

Iida, K. and Iwasaki, T. (1983) *Tsunamis: Their Science and Engineering, Proceedings of the 1981 International Tsunami Symposium, Japan*, Amsterdam: Reidel, Kluwer.

Ilbery, B.W. (1981a) 'Nuclear power in Western Europe', *Tijdschrift voor Economische en Sociale Geografie* 72 (4): 242–51.

Ilbery, B.W. (1981b) 'The diffusion of nuclear power in Europe', *Geography* 66 (4): 297–9.

Imahori, A. (1981) 'Occupational radiation exposure at nuclear power plants in Japan', *Health Physics* 40: 317–22.

Ince, M. (1984) *Sizewell Report: What Happened at the Inquiry?*, London: Pluto Press.

Inhaber, H. (1978) *Risk of Energy Production*, Ottawa: Atomic Energy Control Board, 119 REV-1, 2nd ed.

Inouye, Goro (1973) 'Development and utilisation of atomic energy in Japan', *Nuclear Engineering International* 18 (206): 541.

Institution of Mechanical Engineers (1984) *Seismic Qualification of Safety Related Nuclear Plant and Equipment,* Papers presented at a seminar organized by the Nuclear Energy Committee of the Power Industries Division, 3 April 1984, London: Institution of Mechanical Engineers.

Ion, S.E., Watson, R.H. and Loch, E.P. (1989) 'Fabrication of nuclear fuel', *Nuclear Energy* 28 (1): 21–8.

Ipponmatsu, T. (1973) 'Long-term electricity programme and the role of nuclear power in Japan', *Nuclear Engineering International* 18 (206): 542–3.

Irish, E.R. and Cooley, C.R. (1980) 'Status of technologies related to the isolation of radioactive wastes in geologic repositories', *Radioactive Waste Management*: 1 (2): 121–46.

Jeffrey, J.W. (1982) 'The real cost of nuclear electricity in the UK', *Energy Policy* 10 (2): 76–100.

Jeffery, J.W. (1988) 'The collapse of nuclear economics', *Ecologist* 18 (1): 9–13.

Jersey Power and Light Company (1964) 'Report on economic analysis for Oyster Creek nuclear generating station', Jersey Power and Light Company, 17 February.

Johansson, T.B. and Steen, P. (1981) *Radioactive Waste from Nuclear Power Plants,* Berkeley, CA: University of California Press.

Johns, R. (1982) 'Middle East's uneven nuclear progress', *Financial Times,* 16 June.

Jones, M.J. (1976) *Geology, Mining and Extraction Processing of Uranium,* London, Institution of Mining and Metallurgy.

Jones, P.D. and Warrick, R.A. (1988) 'Greenhouse effect and the climate', *Atom* 381: 13–15.

Jones, P.M.S. (1984) 'Discounting and nuclear power' *Atom* 338: 8–11.

Jones, P.M.S. (1987a) 'An international comparison of electricity generation costs', *Atom* 364: 6–7.

Jones, P.M.S. (1987b) *Discounting and Nuclear Power: Policy and Prospects,* Chichester: Wiley.

Jones, P.M.S. (1989) 'The economics of the nuclear fuel cycle', *Nuclear Energy* 28 (1): 51–5.

Jones, S.B. (1951) 'The economic geography of atomic energy,' *Economic Geography* 27: 268–74.

Jordan, G. (1988) 'Fast reactors: Dounreay and the future', *Nuclear Europe* 8 (5): 18–20.

Judd, A.M. (1981) *Fast Breeder Reactors,* Oxford: Pergamon.

Jungk, R. (1979) *The Nuclear State,* London: John Calder.

Karam, R. and Morgan, K.Z. (1984) *Environmental Impacts of Nuclear Power Plants,* Oxford: Pergamon.

Kasperson, R.E., Burton, I., Whyte, A., Hohenemser, C., Kates, R.W. and Guy, K. (1977) 'Risk and rare events: a three-nation study of disaster prevention in nuclear energy programmes 1974–77', Ford Foundation Grant 740–0593, Final Report.

Kates, R.W. (1978) *Risk Assessment of Environment Hazard,* Chichester: Wiley.

Kathren, R.L. (1974) 'Nuclear power and public opinion', *Health Physics* 26 (6): 483–8.

Katz, J.E. and Marwah, O.S. (1982) *Nuclear Power in Developing Countries,* Lexington, MA: D.C. Heath.

Kavanagh, M.T. and Gualtieri, G. (1966) 'Transport of irradiated fuel from Latina to Windscale', *Atom* 113: 68–76.

Keeney, S.M. (1977) *Nuclear Power Issues and Choices,* Cambridge, MA: Ballinger.

Keepin, B. and Kats, G. (1988) 'Greenhouse warming: comparative analysis of nuclear and efficiency abatement strategies', *Energy Policy* 16 (6): 561–83.

Keepin, B. and Wynne, B. (1984) 'Technical analysis of the IIASA energy scenarios' (editorial), *Nature, London* 313 (5996): 691–5.

Keepin, B., Wynne, B. and Thompson, M. (1984) 'A technical appraisal of the IIASA energy scenarios', *Policy Sciences* 17 (3): 199–339.

Kemeny, J.G. (Chairman) (1980) *The Accident at Three Mile Island, The Need for Change: The Legacy of TMI, Report of the President's Commission*, New York: Pergamon.

Kemp, R. (1986) *The Legitimation of Nuclear Power in Britain*, Cambridge: Polity Press.

Kemp, R. and O'Riordan, T. (1988) 'Planning for radioactive waste disposal: some central considerations', *Land Use Policy* 5 (1): 37–44.

Kenny, A.W. (1970) 'Disposal of radioactive waste in Great Britain', *Proceedings of the Institution of Civil Engineers, Supplement 11*: 267–88.

Ketchum, B.H., Kester, Dana R. and Park, P.K. (1981) *Ocean Dumping of Industrial Wastes*, London: Plenum.

King, M. and Yang, C.T. (1981) 'Future economy of electric power generated by nuclear and coal-fired power plants', *Energy* 6: 263–75.

Kingshott, A.L. (1960) 'Nuclear energy and the European Common Market', *Nuclear Energy* 124 (142): 104–8.

Kirschen, W. *et al.* (1984) *Testing, Evaluation and Shallow Land Burial of Low and Medium Radioactive Waste Forms*, New York: Harwood Academic.

Klement, A.W. (1982) *CRC Handbook of Environmental Radiation*, Boca Raton, FL: CRC Press.

Knowles, A.N. (1985) 'Decommissioning wastes' *Proceedings of the Conference on Radioactive Waste Management: Technical Hazards and Public Acceptance, London, 5–6 March 1985*, London: Oyez.

Komanoff, C. (1981a) 'Sources of nuclear regulatory requirements', *Nuclear Safety* 22 (4): 435–48.

Komanoff, C. (1981b) *Power Plant Cost Escalation: Nuclear and Coal Capital Costs, Regulation and Economics*, New York: Van Nostrand Reinhold.

Konstantionov, L.V. and González, A.J. (1989) 'The radiological consequences of the Chernobyl accident', *Nuclear Safety* 30 (1): 53–69.

Krischer, W. and Simon, R.A. (eds) (1984) *Testing, Evaluation and Shallow Land Burial of Low and Medium Radioactive Waste Forms, Proceedings of a Seminar held in Geel, Belgium, September 1983*, New York: Harwood Academic.

Kushner, M.P. (1974) 'Nuclear fuel fabrication for commercial electric power generation, *IEEE Transactions on Power Apparatus and Systems* 93 (1): 44–7.

Lambert, R. (1983) 'Now – a question of survival', *Financial Times*, 25 August.

Langham, W.H. (1972) 'The biological implications of the transuranium elements for man', *Health Physics* 22: 943–52.

Laughton, A.S., Roberts, L.E.J., Wilkinson, Sir Denys and Gray, D.A. (1986) *The Disposal of Long-Lived and Highly Radioactive Wastes, Proceedings of a Royal Society Discussion Meeting, 30–31 May 1985, London, Philosophical Transactions of the Royal Society of London, Series A* 319: 1–189.

Lave, L.B. and Freeburg, L.C. (1973) 'Health effects of electricity generation from coal, oil and nuclear fuel', *Nuclear Safety* 14 (5): 409–28.

Layfield, Sir Frank (1987) *Sizewell B Public Inquiry Report, Summary of Conclusions and Recommendations*, London: HMSO.

Lean, G. and Leigh, D. (1989) 'Sword over Sellafield', *Observer*, 28 May.

Lee, A.J. and Ramster, J.W. (eds) (1981) *Atlas of the Seas Around the British Isles*, London: Ministry of Agriculture, Fisheries and Food.

Lee, T.R., Brown, J. and Henderson, J. (1984) 'The public's attitudes towards nuclear power in the South West', *Atom* 336: 8–11.

Libby, L.M. (1979) *The Uranium People*, New York: Crane Russak.

Lindell, B. (1985) *Concepts of Collective Dose in Radiological Protection* Paris: OECD.

Logan, S.E. (1974) 'Deep self-burial of radioactive wastes by rock-melting capsules', *Nuclear Technology* 21: 111–25.

Loprieno, N. (1986) 'Radiation knows no frontiers', *European Environment Review* 1 (1): 2–9.

Lowry, D. (1988) 'Corrupt to the core: nuclear waste – the hidden agenda', *Environment Now* 7: 27–9.

Luykz, F. and Fraser, G. (1982) 'Radioactive effluents from nuclear power stations in the European Community, 1974–78', *Nuclear Safety* 23 (5): 578–86.

MacAvoy, P.W. (1969) *Economic Strategy for Developing Nuclear Breeder Reactors*, Cambridge, MA: MIT Press.

McCarthy, G.J. (1979) *Scientific Basis for Nuclear Waste Management*, vol. 1, London: Plenum.

McCormick, M.J. (1982) 'Changes in the nuclear power industry after TMI', *Nuclear Energy* 10: 245–8.

McCracken, S. (1982) *The War Against the Atom*, New York: Basic Books.

MacGill, S.M. (1987) *The Politics of Anxiety: Sellafield's Cancer Link Controversy*, London: Pion.

McIntyre, H.C. (1975) 'National-uranium heavy-water reactors (CANDU)', *Scientific American* 233 (4): 17–27.

McKay, W.A., Johnson, C.E. and Branson, J.R. (1988) 'Environmental radioactivity in Caithness and Sutherland, Part 3: Initial measurements and modelling in inshore waters', *Nuclear Energy* 27 (5): 321–35.

MacLean, A. (1978) *Goodbye California*, London: Collins.

MacQueen, J.F. and Howells, G. (1978) 'Waste-heat disposal – a cool look at warm water', *CEGB Research* 7: 33–44.

McVeigh, J.C. (1984) *Energy around the World*, Oxford: Pergamon.

MAFF/Welsh Office (Ministry of Agriculture, Fisheries and Food and Welsh Office) (1987) *Radionuclide Levels in Food, Animals and Agricultural Products: Post-Chernobyl Monitoring in England and Wales*, London: HMSO.

Maine, F. (Chairman) (1976) *The Nuclear Option for Canada: the Renewable Energy Resources*, Papers presented in the House of Commons at two forums of the Parliamentary and Scientific Committee of the Association of the Scientific, Engineering and Technological Community of Canada, 16 November, ER–77–2, Ottawa.

Malay, F. and Chave, C.T. (1964) *Containment for Siting in Densely Populated Areas* New York: American Society of Mechanical Engineers, 64–WA/NE–1.

Manning, D.T. (1982) 'Post-TMI perceived risk from nuclear power in three communities', *Nuclear Safety* 23 (4): 379–84.

Marcus, F.R. and Aysee, J. (1965) 'Transport of irradiated nuclear fuel on the European continent', *Proceedings of the 3rd International Conference on the Peaceful Uses of Atomic Energy, Geneva 1964* vol. 13, Geneva: United Nations Organization, pp. 282–9.

Marei, A.N. (1972) 'Effects of natural factors on Cesium-137', *Health Physics* 22: 9–15.

Margerison, T.A. (1988) 'Public attitudes to nuclear power', *Nuclear Europe* 8 (5): 26–7.

Marley, W.G. and Fry, T.M. (1955), 'Radiological hazards from an escape of fission products and the implications in power reactor location', *Proceedings of the 1st International Conference on the Peaceful Uses of Atomic Energy, New York, 1954*, Geneva: United Nations Organization, pp. 102–5.

Marples, D.R. (1987) *Chernobyl and Nuclear Power in the USSR*, London: Macmillan in conjunction with the Canadian Institute of Ukrainian Studies, University of Alberta.

Marples, D.R. (1988) *The Social Impact of the Chernobyl Disaster*, London: Macmillan.

Marsh, G.P. (1982) 'Materials for high-level waste containment', *Nuclear Energy* 21 (4): 253–65.

Marshall, W. (1979) 'The use of plutonium', Fifth Chancellor's Lecture, 29 October 1979, University of Salford.

Masters, R. (1987) 'Finding a catalyst for co-operation', *Nuclear Engineering International* 32 (398): 39–45.

Mather, J.D., Chapman, N.A., Black, J.H. and Lintern, B.C. (1982) 'The geological disposal of high-level radioactive waste – a review of the Institute of Geological Science's research programme', *Nuclear Energy* 21 (3): 167–73.

Mathieson, R.S. (1980) 'Nuclear power in the Soviet bloc', *Annals of the Association of American Geographers* 70 (2): 271–9.

Maull, H. (1980) *Europe and World Energy*, London: Butterworths.

Mays, C.W. (1973) 'Cancer induction in man from internal radioactivity', *Health Physics* 25: 585–92.

Medvedev, Z. (1976) 'Two decades of dissidence: nuclear disaster in the Soviet Union', *New Scientist* 72: 264.

Melber, B.D. (1982) 'The impact of TMI upon the public acceptance of nuclear power', *Nuclear Energy* 10: 387–98.

Ministry of Supply and Services (1982) *Nuclear Industry Review: Problems and Prospects 1981–2000*, Ottawa: Department of Energy, Mines and Resources.

Mitchell, R.C. (1980) 'Public opinion and nuclear power before and after Three Mile Island', *Resources for the Future* 64: 5–7.

Mole, A.H. (1976) 'The Flowers Report, opportunities missed', *Nature, London* 264: 494–6.

Monckton, N. (1989) 'Ensuring the safety of deep repositories', *Nuclear Engineering International* 34 (414): 30–3.

Moore, R. (1966) *Niels Bohr: The Man, His Science and the World They Changed*, New York: Knopf.

Mooz, W.E. (1981) 'Cost analysis for LWR power plants', *Energy* 6: 197–225.

Morell, D. (1983) *Siting Hazardous Waste Facilities*, London: Harper & Row.

Moss, N. (1981) *The Politics of Uranium*, London: Andre Deutsch.

Mounfield, P.R. (1961) 'The location of nuclear power stations in the United Kingdom', *Geography* 46 (2): 139–56.

Mounfield, P.R. (1967) 'Nuclear power in the United Kingdom: a new phase', *Geography* 52 (3): 310–17.

Mounfield, P.R. (1981) 'Another look at nuclear power: the uranium market', *Geography* 66 (1): 15–28.

Mounfield, P.R. (1985a) 'Nuclear power in Western Europe', *Geography* 70 (4): 315–27.

Mounfield, P.R. (1985b) 'Developing Britain's energy strategy', *Standard Chartered Bank Review* November: 1–8.

Mounfield, P.R. (1986) 'Energy for the future?' *Standard Chartered Bank Review* June: 2–8.

Mounfield, P.R. and Humphrys, G. (1985) 'National energy policies and Welsh mining communities', *Cambria* 12 (2): 61–88.

Mukherjee, S.K. (1981) 'Energy policy and planning in India', *Energy* 6 (8): 823–51.

Mullenbach, P. (1963) *Civilian Nuclear Power*, New York: Twentieth Century Fund.

Munn, R.E. (1959) The application of air pollution climatology to town planning, *International Journal of Air and Water Pollution* 1: 276–87.

Munn, R.E. and Cole, A.F.W. (1967) Some strong wind downwash diffusion measurements at Douglas Point, Ontario, Canada, *Atmospheric Environment* 1: 601–4.

Murdock, S.H., Leistritz, F.L. and Hamm, R.R. (1983) *Nuclear Waste: Socio-Economic Dimensions of Long-Term Storage*, Boulder CO: Westview.

Murray, C.N., Barbreau, A. and Burdett, J.R. (1986) 'The feasibility of heat generating waste disposal into deep ocean sedimentary formations', in R. Simon (ed.), *Radioactive Waste Management and Disposal*, Cambridge: Cambridge University Press.

Nader, R. and Abbotts, J. (1977) *The Menace of Atomic Energy*, New York: Norton.

Nair, S., Bell, J.N.B. and Minski, M.J. (1986) 'Nuclear power and the terrestrial environment: the transport of radioactivity through foodchains to man', *CEGB Research* July: 3–16.

Narita, T. (1973) 'Policy for development and utilisation of atomic energy in Japan', *Nuclear Engineering International* 18 (206): 551–2.

National Academy of Sciences (1980) *Energy in Transition 1985–2010*, Final Report of the Committee on Nuclear and Alternative Energy Systems, National Research Council, San Francisco, CA: W.H. Freeman.

National Academy of Sciences (1984) *Social and Economic Aspects of Radioactive Waste Disposal: Considerations for Institutional Management*, study by the Panel on Social and Economic Aspects of Radioactive Waste Management, Washington DC: National Research Council, National Academy Press.

National Committee for Radiological Protection (1987) *Ionizing Radiation Exposure of the Population of the United States*, Washington, DC: NCRP Report 93.

National Radiological Protection Board (1987) *Report on the Effect of Chernobyl*, Brussels Commissariat à l'Energie Atomique.

Nealy, S.M., Melber, B.D. and Rankin, W.L. (1983) *Public Opinion and Nuclear Energy*, Lexington, MA: D.C. Heath.

Nehrt, L.C. (1966) *International Marketing of Nuclear Power Plants*, Bloomington, IN: Indiana University Press.

Neil, B.C.J. (1974) 'Sources and nature of radioactivity in nuclear reactor systems', *Ontario Hydro Research Quarterly* 26 (3): 3–7.

Nelkin, D. and Pollark, M. (1981) *The Atom Besieged: Extra-Parliamentary Dissent in France and Germany*, Cambridge MA: MIT Press.

Nero, A.V. (1979) *A Guide Book to Nuclear Reactors*, Berkeley CA: University of California Press.

Newsletter (NA) (1982a) *Nuclear Energy* 21 (2): 90.

Newsletter (NA) (1982b) 'No Urals disaster', *Nuclear Energy* 21 (3): 151.

Newmark, N.M. and Rosenblueth, E. (1971) *Fundamentals of Earthquake Engineering*, London: Prentice-Hall.

NIREX (Nuclear Industry Radioactive Waste Executive) (1983a) *First Report to the Secretary of State for the Environment*, Harwell: NIREX.

NIREX (1983b) *Disposal of Low-Level Waste in the Atlantic: Some Questions Answered*, Harwell: NIREX.

NIREX (1987) *The Disposal of Radioactive Waste: A Consultative Document*, Harwell: NIREX.

NIREX (1989) *Going Forward: The Development of a National Disposal Centre for Low and Intermediate Level Radioactive Waste*, Harwell: NIREX.

Odell, P.R. (1985) 'Natural gas in Western Europe: major expansion in prospect' EURICES Paper 85–4, Erasmus University, Rotterdam.

OECD (Organization for Economic Co-operation and Development) (1972) *Nuclear*

Legislation: Analytical Study (Regulations Governing Nuclear Installations and Radiation Protection), Paris: OECD.

OECD (1974) *Energy Prospects to 1985*, 2 vols, Paris: OECD.

OECD (1985) *Metrology and Monitoring of Radon, Thoron and Their Daughter Products*, Paris: OECD.

OECD/INEA (Organization for Economic Co-operation and Development/International Nuclear Energy Agency (1985) *Electricity in INEA Countries: Issues and Outlook*, Paris: OECD.

OECD/NEA (Organization for Economic Co-operation and Development/Nuclear Energy Agency) (1972) *Radioactive Waste Management Practices in Western Europe*, Paris: OECD.

OECD/NEA (1977) *Objectives, Concepts and Strategies for the Management of Radioactive Waste arising from Nuclear Power Programmes*, Paris: OECD.

OECD/NEA (1983a) *The Costs of Generating Electricity in Nuclear and Coal-fired Power Stations, Report by an Expert Group*, Paris: OECD.

OECD/NEA (1983b) *Geological Disposal of Radioactive Waste: In Situ Experiments in Granite, Proceedings of the NEA Workshop, Stockholm, 25–27 October 1983 (Stripa Project)*, Paris: OECD.

OECD/NEA (1983c) *Uranium Extraction Technology: Current Practice and New Developments in Ore Processing*, Paris: OECD.

OECD/NEA (1984a), *Geological Disposal of Radioactive Waste: an Overview of the Current Status of Understanding and Development*, Paris: OECD.

OECD/NEA (1984b), *Seabed Disposal of High Level Radioactive Waste: A Status Report on the NEA Co-ordinated Research Programme*, Paris: OECD.

OECD/NEA (1984c) *Nuclear Power and Public Opinion*, Paris: OECD.

OECD/NEA (1984d) *Long Term Radiological Aspects of the Management of Wastes from Uranium Mining and Milling, Report by an Expert Group*, Paris: OECD.

OECD/NEA (1984e) *Radiation Protection: The NEA's Contribution*, Paris: OECD.

OECD/NEA (1985a) *Summary of Nuclear Power and Fuel Cycle Data in OECD Member Countries*, Paris, OECD.

OECD/NEA (1985b) *Review of the Continued Suitability of the Dumping Site for Radioactive Waste in the North-east Atlantic*, Paris: OECD.

OECD/NEA (1985c) *The Economics of the Nuclear Fuel Cycle, Report by an Expert Group*, Paris: OECD.

OECD/NEA (1985d) *Radioactive Waste Disposal: In Situ Experiments in Granite, Proceedings of the Stockholm Symposium, 4–6 June 1985*, Paris: OECD.

OECD/NEA (1985e) *Storage with Surveillance versus Immediate Decommissioning for Nuclear Reactors, Proceedings of an NEA Workshop, 22–24 October 1984*, Paris: OECD.

OECD/NEA (1985f) *Management of High-Level Radioactive Waste: A Survey of Demonstration Activities*, Paris: OECD.

OECD/NEA (1986a) *Severe Accidents in Nuclear Power Plants*, Paris: OECD.

OECD/NEA (1986b) *Licensing Systems and Inspection of Nuclear Installations*, Paris: OECD.

OECD/NEA (1986c) *Co-ordinated Research and Environmental Surveillance Programme related to Sea Disposal of Radioactive Waste, CRESP Activity Report 1981–1985*, Paris: OECD.

OECD/NEA (1986d) *Projected Costs of Generating Electricity from Nuclear and Coal-Fired Power Stations for Commissioning in 1995, Report by an Expert Group*, Paris: OECD.

OECD/NEA (1987) *Nuclear Energy and its Fuel Cycle: Prospects to 2025*, Paris: OECD.

OECD/NEA/IAEA (Organization for Economic Co-operation and Development/ Nuclear Energy Agency/International Atomic Energy Agency) (1965 onwards) *Uranium Resources, Production and Demand* (the 'Red Book'), Paris: OECD, initially annual, currently biennial.

OECD/NEA/IAEA (1982a) *Uranium Exploration Methods*, Paris: OECD.

OECD/NEA/IAEA (1982b) *Proceedings of the Symposium on Uranium Exploration, Review of the NEA/IAEA Research and Development Programme*, Paris: OECD.

OECD/NEA/IEA (Organization for Economic Co-operation and Development/ Nuclear Energy Agency/International Energy Agency) (1989) *Projected Costs of Generating Electricity from Power Stations for Commissioning in the Period 1995–2000*. Paris: OECD.

Okrent, D. (1981) *Nuclear Reactor Safety: On the History of the Regulatory Process*, Madison, WI: University of Wisconsin Press.

Olds, F.C. (1974) 'Power plant capital costs going out of sight', *Power Engineering* 78 (8): 36–43.

Ontario Hydro (1959) 'Douglas Point nuclear generating station, preliminary report on siting and environmental data', Ontario Hydro, Toronto.

Ontario Hydro (1966) Pickering nuclear generation station. preliminary report on siting and environmental data', Ontario Hydro, Toronto.

Openshaw, S. (1982) 'The geography of reactor siting policies in the U.K.', *Transactions of the Institute of British Geographers (New Series)* 7 (27): 150–62.

Openshaw, S. (1986) *Nuclear Power: Siting and Safety*, London: Routledge & Kegan Paul.

Openshaw, S. (1988a) 'Planning Britain's long-term nuclear power expansion', *Land Use Policy* 5 (1): 7–18.

Openshaw, S. (1988b) 'Making nuclear power more publicly acceptable', *Nuclear Energy* 27 (2): 131–6.

Oppenheimer, R. (1957) 'The environs of atomic power;, in *Atoms for Power: United States Policy in Atomic Energy Development*, background papers prepared for the use of participants, 12th American Assembly, Columbia University, New York, 17–20 October 1957.

O'Riordan, T. (1984) 'The Sizewell B inquiry and a national energy strategy', *Geographical Journal* 150 (2): 171–82.

Orlowski, S.M. (1986) 'Radioactive waste management and disposal strategies in the European Community', in *Radioactive Waste Management: UK Policy Examined, Proceedings of a Conference, London, 24–25 April 1986*, London: IBC.

Ott, K.O. and Spinrad, B.I. (eds) (1985) *Nuclear Energy: A Sensible Alternative*, New York: Plenum.

Owen, K. (1974) 'Japan's nuclear plan holds lesson for UK', *The Times*, 5 April.

Palmer, J. and Garfield, A. (1988) 'Illegal dump evidence at atomic plants', *Guardian*, 23 February.

Pasciak, W., Branagan, E.F., Congel, F.J. and Fairobent, J.E. (1981) 'A method for calculating doses to the population from Xe-133 releases during the Three Mile Island Accident', *Health Physics* 40: 457–65.

Pasqualetti, M.J. (1988) 'Decommissioning at ground level', *Land Use Policy* 5 (1): 45–61.

Pasqualetti, M.J. and Pijawka, D. (eds) (1984) *Nuclear Power: Assessing and Managing Hazardous Technologies*, Boulder, CO: Westview.

Passant, F.H. (1985) 'Radioactive waste from nuclear power stations', in *Radioactive Waste Management: Technical Hazards and Public Acceptance, Proceedings of a Conference, London*, 3–6 March 1985, London: Oyez.

Pattenden, N.J., Cambray, R.S. and Playford, K. (1989) 'Environmental activity in

Caithness and Sutherland, Part 4: Radionuclide deposits in coastal soils 1978–83', *Nuclear Energy* 28 (2): 111–27.

Patterson, W.C. (1973) *Nuclear Reactors*, London: Earth Island.

Patterson, W.C. (1983) *Nuclear Power*, Harmondsworth: Penguin.

Pearce, F. (1988) 'Penney's Windscale thoughts', *New Scientist*, 7 January: 34–5.

Pentreath, R.J. (1980) *Nuclear Power, Man and the Environment*, London: Taylor & Francis.

Pentreath, R.J. (1982) 'Principles, practice and problems in the monitoring of radioactive wastes disposed of into the marine environment', *Nuclear Energy* 21 (4): 235–44.

Perla, H.F. (1973) 'Power plant siting concepts for California', *Nuclear News* 16 (13): 47–51.

Petruk, W. and Sibbald, T.I.T. (1985) *Geology of Uranium Deposits*, Montreal: CIM Publications.

Phung, D.L. (1980) 'Cost comparison of energy projects: discounted cash flow, and revenue requirement methods', *Energy* 5: 1053–72.

Pick, H. (1979) 'Nuclear power expansion on agenda in Prague', *The Guardian* 24 May.

Pigford, T.H. (1974) 'Environmental aspects of nuclear energy production', *Annual Review of Nuclear Science* 24 (December): 515–59.

Pijawka, K.D. (1982) 'Public response to the Diabolo Canyon nuclear generating station', *Energy* 7 (8): 667–80.

Pijawka, K.D. and Chalmers, J. (1983) 'Impacts of nuclear generating plants on local areas', *Economic Geography* 59 (1): 66–80.

Piper, H.B. (1972) *Indexed Bibliography on Nuclear Facility Siting*, Oak Ridge, TN: Nuclear Safety Information Centre, Oak Ridge National Laboratory.

Pochin, Sir Edward (1983) *Nuclear Radiation: Risks and Benefits*, Oxford: Clarendon Press, Monographs on Science, Technology and Society no. 2.

Pochin, Sir Edward (1985) 'Biological considerations involved in waste management', in *Radioactive Waste Management: Technical Hazards and Public Acceptance, Proceedings of a Conference, London, 5–6 March 1985, London*: Oyez.

Pocock, R.F. (1977) *Nuclear Power: Its Development in the United Kingdom*, Woking: Gresham Press for the Institution of Nuclear Engineers.

Polach, J.G. (1970) 'The development of energy in East Europe', *Economic Developments in Countries of Eastern Europe*, Sub-Committee on Foreign Economic Policy of the Joint Economic Committee, 91 US Congress, 2 Sess. Reprinted as Resources for the Future Reprint No.85.

Poneman, D. (1982) *Nuclear Power in the Developing World*, London: Allen & Unwin.

Potemans, M. (1982) 'Nuclear energy in Belgium', *Nuclear Europe* 3, 10–12.

Prescott-Clarke, P. and Hedges, A. (1987) *Radioactive Waste Disposal, The Public's View*, London: SCPR.

Probert, D. and Tarrant, C. (1989) 'Environmental risks of power generation from fossil fuels and nuclear facilities', *Applied Energy*, 32 (3): 171–206.

Pryde, P.R. (1978) 'Nuclear energy development in the Soviet Union', *Soviet Geography: Review and Translation* 19 (2): 75–83.

Puiseux, L. (1981) *Le Babel Nucléaire: Energie et développement*, Paris: Galilee.

Quinton, A. (1961) 'Siting of power reactors in the UK', *Nuclear Energy* September: 374–7.

Qin Tun-Luo (1981) 'Some aspects of energy policy in China', *Energy* 6 (8): 745–7.

Ramberg, B. (1980) *Nuclear Power Plants as Weapons for the Enemy: an Unrecognized Military Peril*, Berkeley, CA: University of California Press.

Ramsay, W.C. (1976) 'Radon from uranium mill tailings: a radiation hazard to the general population', *Environmental Management* 1 (2): 139–45.

Rawstron, E.M. (1951) 'The distribution and location of steam-driven power stations in Great Britain', *Geography* 36: 249–63.

Rawstron, E.M. (1955) 'Changes in the geography of electricity production in Great Britain', *Geography* 40: 92–7.

Reisch, F. (1987) 'The Chernobyl accident – its impact on Sweden', *Nuclear Safety* 28 (1): 29–36.

Richardson, J.A. (1973) Radioactive waste quantities produced by light water reactors and methods of storage, transportation and disposal, *Journal of the British Nuclear Energy Society* 12 (2): 199–211.

Roberts, A. and Medvedev, Z. (1977) *Hazards of Nuclear Power*, Nottingham: Spokesman Books.

Roberts, J.T. (1973) 'Uranium enrichment: supply, demand and costs', *Bulletin of the International Atomic Energy Authority* 15 (5): 14–24.

Roberts, L.E.J. (1987) 'The risk factor: acceptance and acceptability', *Nuclear Energy* 26 (6): 349–59.

Robinson, J.B. (1982) 'Bottom-up methods and low-down results: changes in the estimate of future energy demands', *Energy* 7 (7): 627–35.

Rolph, E.S. (1979) *Nuclear Power and the Public Safety – A Study in Regulation*, Lexington, MA: D.C. Heath.

Rose, K.S.B. (1982) 'Review of health studies at Kerala', *Nuclear Energy* 21 (6): 399–408.

Rotblat, J. (ed.) (1977) *Nuclear Reactors: To Breed or not to Breed*, London: Taylor & Francis.

Rothman, S. and Lichter, S.R. (1982) 'The nuclear energy debate: scientists, the media and the public', *Public Opinion* August–September: 47–52.

Roxburgh, I.S. (1988) *Geology of High-Level Nuclear Waste Disposal*, London: Chapman and Hall.

Roy, R. (1982) *Radioactive Waste Disposal, The Waste Package*, Oxford: Pergamon.

Royal Society (1981) 'Assessment and perception of risk', *Proceedings of the Royal Society of London, Series A*, 376 1764: 3–206.

Rummery, T.E. and McLean, D.R. (1981) 'Perspective on nuclear fuel waste disposal from a Canadian point of view', *Transactions of the American Nuclear Society* 36: 47–57.

Sabato, J.A. (1982) 'Major issues in the transfer of technology to developing countries', *Transactions of the American Nuclear Society* 42: 9–11.

Sagan, L.A. and Fry, S.A. (1980) 'Radiation accidents: a conference review', *Nuclear Safety* 21 (5): 562–7.

Sagers, M.J. and Green, M.B. (1982) 'Spatial efficiency in Soviet electrical transmission', *Geographical Review* 72 (3): 291–303.

Saunders, P.A.H. (1988) 'Safety Studies Nirex Radioactive Waste Disposal Research and Safety Assessment', Report NSS/G100, UKAEA.

Schumacher, D. and McVeigh, J.C. (1983) *Energy Options: Choosing for the Future*, London: Macmillan.

Schurr, S.H., Darmstadter, J., Perry, H., Ramsay, W. and Russell, M. (1981) *Energy in America's Future: The Choices Before Us*, Baltimore, MD: Johns Hopkins University Press for Resources for the Future.

Scorer, R.S. and Barrett, C.F. (1962) 'Gaseous pollution from chimneys' *International Journal of Air and Water Pollution* 5: 49–64.

Scottish Consumer Campaign (1981) *Cheap Electrickery: The Real Cost of Nuclear Power in Scotland*, Edinburgh: Scottish Consumer Campaign.

Scuricini, G.B. (1972) 'Uranium enrichment in Europe: past experience and prospects', *Journal of the British Nuclear Energy Society* 11 (4): 350–9.

Searby, P.J. (1971) 'Present worth evaluations as an aid to nuclear power decisions', *Atom* 178: 185–97; *Journal of the British Nuclear Energy Society* 10 (3): 177–89.

Semenov, B.A. (1983) 'Nuclear power in the Soviet Union', *International Atomic Energy Agency Bulletin* 25 (2): 47–59.

Shaw, J. (1972) 'The nuclear siting controversy in the United States of America', *Journal of the Institute of Nuclear Engineers* May–June: 85–90.

Shaw, J. and Palabrica, R.J. (1974) 'A critical review and comparison of the nuclear power plant siting policies in the United Kingdom and the USA', *Annals of Nuclear Science Engineering* 1: 241–54.

Siddall, E. (1957) *Reactor Safety Standards and Their Attainment* Report 498, Atomic Energy of Canada Ltd.

Sills, D.L., Wolf, C.P. and Shelanski, V.B. (eds) (1982) *Accident at Three Mile Island: The Human Dimension,* Boulder, CO: Westview.

Simon, R. (ed.) (1986) *Radioactive Waste Management and Disposal, Proceedings of the 2nd European Community Conference, Luxembourg, 22–26 April 1985,* Cambridge: Cambridge University Press for the Commission of the European Communities.

Simpson, P.R. Brown, G.C., Plant, J. and Ostle, D. (1979) 'Uranium mineralization and granite magmatism in the British Isles', *Philosophical Transactions of the Royal Society of London, Series A* 291: 385–412.

Simpson, R.H. and Riehl, H. (1981) *The Hurricane and its Impact,* San Francisco CA: W.H. Freeman.

Skeets, T.T.H. (1958) 'Uranium from South Africa', *Nuclear Energy Engineer* 12 (125): 328–31.

Skvortsov, S. and Siderenko, V. (1958) 'Atomic power plants', in I.T. Aladeev (ed.) *Atoms for Peace,* Moscow: Soviet Academy of Sciences and Asia Publishing House.

Slade, D.H. (ed.) (1968) 'Meteorology and atomic energy', USAEC Report TID-24190, Springfield, VA: Clearinghouse for Federal Scientific and Technical Information.

Slovic, P.B., Fischoff, B. and Lichtenstein, S. (1980) 'Images of disaster: perception and acceptance of risks from nuclear power', in G.T. Goodman and W.D. Rowe (eds), *Energy Risk Management,* London: Academic Press.

Slovic, P., Fischoff, B. and Lichtenstein, S. (1981) 'Informing the public about the risks from ionising radiation', *Health Physics* 41: 589–98.

Smith, C.B. (1973) 'Power plant safety and earthquakes', *Nuclear Safety* 14 (6): 586–96.

Smith, F.B. (1986) 'The passage of the Chernobyl nuclear cloud over Europe and the UK' Report Met O.14, Meteorological Office, Bracknell (mimeograph).

Smith, F.B. (1989) 'Debris from the Chernobyl nuclear disaster: how it came to the UK, and its consequences to agriculture', *Journal of the Institute of Energy* 62 (450): 3–13.

Smith, F.B. and Clark, A.J. (1986) 'Deposition of radionuclides from the Chernobyl cloud', Meteorological Office, Bracknell and NRPB, Didcot (mimeograph).

Smith, H. (1988) 'The International Commission on Radiological Protection: historical overview', *International Atomic Energy Agency Bulletin* 30 (30): 42–4.

Smith, H.A. (1965) 'Present and future developments for bulk supply of electricity in Ontario', presented at the Annual Meeting of the Engineering Institute of Canada, Toronto, 28 May 1965 (mimeograph).

Smith, P.G. and Douglas, A.J. (1986) 'Mortality of workers at the Sellafield plant of British Nuclear Fuels', *British Medical Journal* 293: 845–54.

Starr, C. and Greenfield, M.A. (1973) 'Public health risks of thermal power plants', *Nuclear Safety* 14 (4): 267–74.

Stello, V. (1978) 'US Nuclear Regulatory Commission's view of decontamination and decommissioning', in M.M. Osterhout (ed.), *Decontamination and Decommissioning of Nuclear Facilities,* New York: Plenum.

Sternglass, E.J. (1973) *Low Level Radiation,* London: Earth Island.

Stinson, R.C. (1987) 'Ushering in the age of global technology', *Nuclear Engineering International* 32 (398): 46–7.

Stobbs, J.J. and Taormina, A.M. (1982) 'The current status and long-term perspectives of fissile material availability in the world', *Transactions of the American Nuclear Society* 40: 1–7.

Surrey, J. (1974) 'Japan's uncertain energy prospects', *Energy Policy* September: 204–30.

Surrey, J. (ed.) (1984) *The Urban Transportation of Irradiated Fuel,* London: Macmillan.

Surrey, J. (1988) 'Nuclear power: an option for the Third World', *Energy Policy* 16 (5): 461–79.

Sweet, C. (1980) *The Fast Breeder Reactor: Needs, Costs and Risks,* London: Macmillan.

Sweet, C. (1983) *The Price of Nuclear Power,* London: Heinemann. Syndicat CFDT de l'Energie Atomique (1975) *L'Electronucléaire en France,* Paris: du Seuil.

Takashima, Y. (1981) 'Uranium enrichment', *Transactions of the American Nuclear Society* 36: 29–39.

Tamplin, A.R. (1969) 'Infant mortality and the environment', *Bulletin of Atomic Scientists* 25: 23–9.

Taylor, L.S. (1980) 'Some nonscientific influences on radiation protection standards and practice: the 1990 Sievert Lecture', *Health Physics* 39: 851–74.

Taylor, L.S. (1982) 'Planning for radiation emergencies', *Health Physics* 43 (3): 435.

Taylor, R. (1987) 'Health risks of low level radiation', paper given at the Physics Centre, University of Leicester, 10 March 1987.

Teitelbaum, P.D. (1963) *Energy Cost Comparisons* Washington, DC: Resources for the Future Inc. (first published in vol. 1 of *Papers Prepared for the United Nations Conference on the Application of Science and Technology for the benefit of the Less Development Areas,* Washington, DC: US Government Printing Office, 1963). This remains the best balanced short introduction to the factors that have to be considered in estimating electricity generating costs.

Teller, E. (1970) *Perils of the Peaceful Atom: The Myth of Safe Nuclear Power Plants,* London: Gollancz.

Teller, E. (1979) *Energy from Heaven and Earth,* San Francisco CA: W.H. Freeman.

Timm, M. (1987) 'Significance of nuclear for Germany's energy supply', *Nuclear Europe* 7 (1–2): 13–15.

Touraine, A., Hegedus, Z., Dubet, F. and Wieviorka, M. (1983) *Anti-Nuclear Protest: The Opposition to Nuclear Energy in France,* Cambridge: Cambridge University Press.

Tucker, A. (1984) 'How a nuclear energy forecast ran out of credibility', *Guardian,* 29 November.

UKAEA (United Kingdom Atomic Energy Authority), Economics and Programming Branch (1968) 'Prospects for nuclear power in overseas countries 1970–1985, L.O. Report 4, Economics and Programming Branch, UKAEA.

University of Surrey (for DoE) (1985) *Public Perceptions of Aspects of Radioactive Waste Management,* Guildford: University of Surrey.

UNO (United Nations Organization) (1963) *Methods of Calculating the Cost of Electric Power Produced by Thermal Power Stations,* Geneva: UNO, Publication ST/ECE/EP/15.

UNO (1971) *World Energy Requirements and Resources in the Year 2000, 4th International Conference on the Peaceful Uses of Atomic Energy,* Geneva: Resources and Transport Division, Department of Economic and Social Affairs, UNO, Publication A/CONF.49/P/420..

UNO (1986) *Yearbook of World Energy Statistics 1984,* Geneva: UNO.

UNSCEAR (United Nations Scientific Committee on the Effects of Atomic Radiation) (1982) *Ionizing Radiation: Sources and Biological Effects, Report to the General Assembly,* New York: UNSCEAR.

Uranium Institute (1977) *Uranium Supply and Demand,* Proceedings of the 2nd International Symposium on Uranium Supply and Demand, 22–24 June 1977, London: Mining Journal Books.

Uranium Institute (1978) *Government Influence on International Trade in Uranium,* London: Uranium Institute.

Uranium Institute (1981) *Uranium and Nuclear Energy: 1980, Proceedings of the 5th International Symposium,* London, 2–4 September 1980, Guildford: Westbury House.

USAEC (US Atomic Energy Commission) (1957) *Theoretical Possibilities and Consequences of Major Accidents in Large Power Plants* Washington, DC: USAEC, WASH 740, 58–544 C47N3.

USAEC (1961) *Reactor Site Criteria,* Title 10, Part 100, Washington, DC: USAEC, 26/FR/1224.

USAEC (1963) *Calculation of Distance Factors for Power and Test Reactor Sites,* Washington, DC: USAEC, TID-14844.

USAEC (1967) *Licensing of Production and Utilization Facilities, General Design Criteria for Nuclear Power Plants, Construction Permits,* Title 10, Part 50, Washington, DC: USAEC, 32/FR/10213–10218.

USAEC (1979) *Comparative Risk-Cost Benefit Study of Alternative Sources of Electrical Energy,* Washington, DC: USAEC, WASH 1224.

USAEC/ORNL (US Atomic Energy Commission/Oak Ridge National Laboratory) (1970) *Siting of Fuel Reprocessing Plants and Waste Management Facilities,* Oak Ridge, TN: Oak Ridge National Laboratory, ORNL–4451.

US Congress (1972) *Selected Materials on the Calvert Cliffs Decision: Its Origin and Aftermath,* Joint Committee on Atomic Energy.

US National Academy of Sciences, Biological Effects of Ionizing Radiations (BEIR) Committee (1980) *The Effects on Populations of Exposure to Low Levels of Ionizing Radiation* (BEIR III), Washington, DC: National Academy Press.

USNRC (US Nuclear Regulatory Commission) (1975) *The Reactor Safety Study: An Evaluation of the Risk of Accidents in Commercial Central Electric Nuclear Power Stations of the USA* (the Rasmussen Report), Washington, DC: USNRC and USAEC WASH 1400, NUREG 75/014.

USNRC (US Nuclear Regulatory Commission) (1979) *Population Dose and Health Impact of the Accident at the Three Mile Island Nuclear Station,* Washington, DC: Interagency Dose Assessment Group, USNRC.

US Public Health Service (1957) *Control of Radon Daughters in Uranium Mines and Calculations on Biologic Effects,* Washington, DC: US Public Health Service, Publication 494.

USSR State Committee on the Utilization of Atomic Energy (1986) *The Accident at the Chernobyl Nuclear Power Plant and its Consequences,* Part 1, *General Material,* Report presented at the IAEA Experts' Meeting, Vienna, 25–29 August 1986, Vienna: IAEA.

Van der Pligt, J. (1985) 'Public attitudes to nuclear energy: salience and anxiety', *Journal of Environmental Psychology* 5: 87–97.

Van der Pligt, J., Van Der Linden, J. and Ester, P. (1982) 'Attitudes to nuclear energy: beliefs, values and false consensus', *Journal of Environmental*

Psychology 2: 221–31.
Vann, H.E., Whitman, M.J. and Bowers, H.I. (1971) 'Factors affecting historical and projected capital costs of nuclear power plants in the USA', *Proceedings of the 4th International Conference on the Peaceful Uses of Atomic Energy*, Geneva: United Nations Organization.
Vasiliev, V.A. (1982) 'Nuclear power in the USSR', *Nuclear Energy* 21: 113–18.
Vennart, J. (1981) 'Limits for intakes of radionuclides by workers: ICRP publication 30', *Health Physics* 40 (4): 477–84.
Verbeet, L. and Gregory, R.W. (1986) 'Power generation from coal – what does it cost?' Report ICEAS/E10, International Energy Agency (IEA), London.
Von Bonka, H. (1974) 'Die ortsabhängige mittlere natürliche Strahlenbelastung der Bevölkerung in der Bundesrepublik Deutschland', *Atomkernenergie* 23 (2): 137–50.
Voskoboinik, D. (1959) *Nuclear Power*, Moscow: Foreign Languages Publishing House.
Wagner, H., Ziegler, E. and Closs, K-D. (1982) *Risikoaspekte der nuklearen Entsorgun*, Baden-Baden: Nomos-Verlagsgesellschaft.
Wakstein, C. (1987) *Possible Effects of a Magnox Flask Accident in North London*, London: ALARM.
Weaver, F. (1979) 'The promise and peril of nuclear energy', *National Geographic Magazine* April: 459–93.
Webb, G.A.M. (1974) 'Radiation exposure to the public – the current levels in the UK', Report NRPB R24, National Radiological Protection Board.
Webb, G.A.M. and Dunster, H.J. (1985) 'An integrated approach to the limitation of radiation risk from nuclear power', in *Interface Questions in Nuclear Health and Safety, Proceedings of NEA Seminar, 16–18 April 1985*, Paris: OECD, pp. 306–13.
Webber, D.J. (1982) 'Is nuclear power just another environmental issue? An analysis of California voters', *Environment and Behaviour* 14 (1): 72–83.
Weinberg, A. (1972) 'Social institutions and nuclear energy', *Science* 177: 27–34.
Whicker, F.W. and Schultz, V. (eds) (1982) *Radioecology, Nuclear Energy and the Environment*, 2 vols, Boca Raton, FL: CRC Press.
White, M.G. and Miller, S.R. (1983) 'Perspectives and issues in DoE remedial action programmes', *Transactions of the American Nuclear Society* 45: 45–6.
Whyte, A. (1977) 'Public perception of nuclear risk in Canada, UK and USA', *Conference on Socio-Psychological Aspects of Nuclear Energy, Paris, 13–15 January 1977.*
Wilkinson, M. (1987) 'Coal resources lead to a shift in strategy', *Financial Times* 18 December.
Williams, G.P.L. and Page, R.D. (1951) *Nuclear Fuel in Canada*, Ottawa: AECL.
Williams, R. (1986) *The Nuclear Power Decisions*, London: Croom Helm.
Wilson, D. (1986) *Soviet Energy to 2000*, London: Economist Intelligence Unit.
Wilson, J. (1989) *The Electricity Industry in England and Wales*, London: UBS Phillips and Drew.
Windeyer, Sir Brian (1974) 'Radiological protection and nuclear power', *Journal of the Institution of Nuclear Engineers* 15 (2): 35–46.
Wing, S.P. (1932) 'Wind bracing in steel buildings', *Communications of the American Society of Civil Engineers*, August.
Woo, T. and Castore, C. (1980) 'Expectancy-value and selective exposure as determinants of attitude toward a nuclear power plant', *Journal of Applied Social Psychology* 10: 224–34.
Yadigaroglu, G., Reinking, A.G. and Schrock, V.E. (1972) 'Spent fuel transportation risks', *Nuclear News* 15 (11): 71–5.
Yeats, R.S., Clark, M.N., Kellor, E.A. and Rockwell, T.K. (1981) 'Active fault

hazard in southern California: ground rupture versus seismic shaking', *Bulletin of the Geological Society of America* 1 (92): 189–96.

Yoshioka, T. (1973) 'Nuclear power in Japan', *Journal of the British Nuclear Energy Society* 12: 135–7.

Young, K. (1987) 'British nuclear reactions', in R. Jowell, S. Witherspoon and L. Brook (eds) *British Social Attitudes: The 1987 Report*, London: Gower for Social and Community Planning Research, ch. 4, pp. 71–8.

Yulish, C. (1973) 'Low level radiation: a summary of responses to ten years of allegations by Dr Ernest Sternglass', presented at 5th International Conference on Science and Society, Herceg-Novi, Yugoslavia, July 1973.

Yusuf, M. (1965) 'Introduction of Roappur nuclear power plant in EPWADA grid', *Proceedings of the Third International Conference on the Peaceful Uses of Atomic Energy, 1964*, Geneva: United Nations Organization.

Zeigler, D.J., Brunn, S.D. and Johnson, J.H. (1981) 'Evacuation from a nuclear technology disaster', *Geographical Review* 72 (1): 1–16.

Zimmermann, H.A. (1982) 'Nuclear versus fossil: one company's view', *Transactions of the American Nuclear Society* 41 (suppl. 1): 18–19.

Zinn, W.H., Pittman, F.K. and Hogerton., J.F. (1964) *Nuclear Power*, USA, New York: McGraw-Hill.

Zito, T. (1976) 'Nuclear material "lost" in US plants,' *Guardian* 29 July.

Index

Printed and bound by CPI Group (UK) Ltd, Croydon, CR0 4YY

21/10/2024

01777088-0018